MONOPOLE ELEMENTS ON CIRCULAR GROUND PLANES

The Artech House Microwave Library

Introduction to Microwaves by Fred E. Gardiol
Lossy Transmission Lines by Fred E. Gardiol
Microwaves Made Simple: Principles and Applications by W. Stephen Cheung, Frederic H. Levien, et al.
Microwaves Made Simple — the Workbook by Salvatore Algeri, W. Stephen Cheung, and Lawrence A. Stark
Microwave Tubes by A. S. Gilmour, Jr.
Electric Filters by Martin Hasler and Jacques Neirynck
Nonlinear Circuits by Martin Hasler and Jacques Neirynck
Microwave Technology by Erich Pehl
Receiving Systems Design by Stephen J. Erst
Microwave Mixers by Stephen A. Maas
Feedback Maximization by B.J. Lurie
High-Power Microwave Sources, Victor L. Granatstein and Igor Alexeff, eds.
Applications of GaAs MESFETs by Robert Soares, et al.
GaAs Processing Techniques by Ralph E. Williams
GaAs FET Principles and Technology by James V. DiLorenzo and Deen D. Khandelwal
E-Plane Integrated Circuits by P. Pramanick and P. Bhartia
Evanescent Mode Microwave Components by George F. Craven and Richard F. Skedd
Filters with Helical and Folded Helical Resonators by Peter Vizmuller
Design Guide for Monopulse Elements on Circular Ground Planes by Melvin M. Weiner et al.
Dielectric Resonators, Darko Kajfez and Pierre Guillon, eds.
Modern Spectrum Analyzer Theory and Applications by Morris Engelson
Microwave Materials and Fabrication Techniques by Thomas S. Laverghetta
Handbook of Microwave Testing by Thomas S. Laverghetta
Microwave Measurements and Techniques by Thomas S. Laverghetta
Linear Active Circuits: Design and Analysis by William Rynone, Jr.
The Design of Impedance-Matching Networks for Radio-Frequency and Microwave Amplifiers by Pieter L.D. Abrie
Microwave Filters, Impedance Matching Networks, and Coupling Structures by G.L. Matthaei, Leo Young, and E.M.T. Jones
Analysis, Design, and Applications of Fin Lines by Bharathi Bhat and Shiban Koul
Microwave Engineer's Handbook, 2 vol., Theodore Saad, ed.
Handbook of Microwave Integrated Circuits by R.K. Hoffmann
Microwave Integrated Circuits, Jeffrey Frey and Kul Bhasin, eds.
Computer-Aided Design of Microwave Circuits by K.C. Gupta, Ramesh Garg, and Rakesh Chadha
Microstrip Lines and Slotlines by K.C. Gupta, R. Garg, and I.J. Bahl
Advanced Mathematics for Practicing Engineers by Kurt Arbenz and Alfred Wohlhauser
Microwave Circuit Design Using Programmable Calculators by J. Lamar Allen and Max Medley, Jr.
Stripline Circuit Design by Harlan Howe, Jr.
Microwave Transmission Line Filters by J.A.G. Malherbe
Microwave Diode Control Devices by Robert V. Garver
Tables for Active Filter Design by Mario Biey and Amedeo Premoli
Design Tables for Discrete Time Normalized Lowpass Filters by Arild Lacroix and Karl-Heinz Witte
Active Filter Design by Arthur B. Williams
Laser Applications by W.V. Smith
Microwave Remote Sensing, 3 vol., by F.T. Ulaby, R.K. Moore, and A.K. Fung

MONOPOLE ELEMENTS ON CIRCULAR GROUND PLANES

Melvin M. Weiner
Stephen P. Cruze
Cho-Chou Li
Warren J. Wilson

Artech House

Library of Congress Cataloging-in-Publication Data

Design guide for monopole elements on circular ground planes.

Bibliography: p.
Includes index.
1. Radio—Antennas. 2. Radio frequency modulation.
I. Weiner, M. M.
TK6565.A6D47 1987 621.38′028′3 87-1030
ISBN 0-89006-252-8

ARTECH HOUSE, INC.
685 Canton Street
Norwood, MA 02062

International Standard Book Number: 0-89006-252-8
Library of Congress Catalog Card Number: 87-1030

To Sandra, Steven, and Robert
—M.M.W.

Contents

Preface

The US Air Force SINCGARS VHF-FM radio is a frequency-hopping antijamming radio that utilizes an electrically short antenna to minimize aerodynamic drag on airborne platforms. The development of optimally efficient, electronically tuneable antennas for this radio is of interest. Although the antenna ground plane is platform-dependent, it is usually small compared to an RF wavelength. A circular ground plane provides a standardized ground-plane geometry with which to model and evaluate candidate antennas. Accordingly, a VHF antenna range with an eight-foot diameter circular ground plane has been constructed at the MITRE Corp. to evaluate candidate antennas.

The electrical properties of a monopole element at the center of a circular ground plane of finite radius are of interest for *qualifying the antenna range, establishing antenna standards* with which to measure test antennas, and *modeling candidate antennas*. Our survey of the literature revealed that, although this antenna has the simplest geometry of any monopole antenna, its properties are neither well understood nor standardized, particularly for ground-plane radii that are small or comparable to a wavelength. This book attempts to address this deficiency.

M. Weiner wrote Chapters 1 to 5. S. Cruze contributed to Sec. 3.5 and wrote program MONOPL in App. B.3. C. Li contributed to Sec. 4.2, edited program RICHMD1 in App. B.2 and program AWADALLA in App. B.5, and wrote program BARDEEN in App. B.1, program RICHMD2 in App. B.2, and program MONOSTOR in App. B.4. W. Wilson contributed to Sec. 2.3 and Sec. 3.3. J. E. Kriegel derived the correct form of the continued fractions given in equations (3.5.4) and (3.5.5), and contributed to the evaluation of the limits in (3.3.22) to

(3.3.24). W. C. Corrieri skillfully performed the measurements discussed in Chapter 5. K. Pamidi contributed to the development of equation (3.3.16).

The authors are grateful to Professor Alfred Leitner of Rensselaer Polytechnic Institute for helpful conversations regarding his method of oblate spheroidal wave functions, Professor Jack Richmond of Ohio State University for helpful conversations and a magnetic a tape of his method of moments program, and Professor Kamal Awadalla of Menoufia University (Egypt) for helpful correspondence, including a listing of his program for the method of moments combined with the geometric theory of diffraction.

After the authors had obtained results by using the integral equation method and the method of oblate spheroidal functions, it was possible to confirm the correctness of Professor Richmond's method-of-moment results, which were subsequently published in the open literature [2].

This book is based on the report ESD-TR-86-241 prepared by the MITRE Corporation under Project No. 6480, Contract F19628-84-C-0001. The contract was sponsored by the Electronic Systems Division, Air Force Systems Command, Hanscom Air Force Base, MA.

Part I
Theory and Results

Chapter 1
Introduction

Monopole antennas are commonly employed in airborne and ground-based communication systems at a wide range of frequencies. The electrical properties of such antennas are dependent upon the geometry of both the monopole element and the ground plane. Typically, the monopole element may be electrically short (length is much less than a quarter-wavelength) or near-resonant (length approximately a quarter-wavelength), and it may be thin (length-to-radius ratio is much greater than 10^4) or relatively thick (length-to-radius ratio of 10^1 to 10^4). In addition, the ground-plane dimensions may vary from a fraction of a wavelength to many wavelengths. Therefore, it is desirable to know how the input impedance and radiation pattern of the antenna change as the dimensions of the monopole element and the ground plane vary. The directional gain on or near the radio horizon (the ground plane is assumed to be horizontal) is of particular interest because the maximum operational range of a communication system often depends on the directivity on the radio horizon.

This study is restricted to a monopole geometry consisting of a vertical cylindrical element at the center of a perfectly conducting, infinitely thin, circular ground plane in free space. This geometry is of interest because its radiation pattern is uniform in the azimuthal direction and because its electrical characteristics are primarily a function of only three parameters, namely, the *element length,* the *element radius,* and the *ground-plane radius,* when each is normalized to the excitation wavelength. Therefore, this geometry is conducive to analysis, experimental verification, and standardization.

A typical feed for the monopole antenna is a coaxial line with its inner conductor connected through a hole in the ground plane to the vertical monopole element and its outer conductor connected by means of a flange to the ground plane. Typically, the inner conductor's diameter is equal to the monopole element's diameter and the outer conductor's diameter is equal to the ground-plane hole diameter. Unless stated otherwise, such a feed will be assumed in this study. The ratio of the coaxial line's outer-to-inner conductor diameters affects the antenna's input impedance, but only significantly for a relatively thick monopole element on a very small ground plane.

For the idealized case of a ground plane of infinite extent and infinite conductivity, the monopole antenna may be modeled by the method of images as a dipole with one-half the input impedance and double the peak directivity of the dipole. The infinite ground plane prevents monopole radiation into the hemisphere below the ground plane, but allows a radiation pattern identical to that of the dipole in the upper hemisphere. However, for a monopole element mounted on a ground plane of finite extent, the outer edge of the ground plane diffracts incident radiation in all directions, and consequently modifies the currents on the ground plane and the vertical element from those of an infinite ground plane. At the outer edge of the ground plane, the currents on its top and bottom faces are equal in magnitude but opposite in direction because the net current must be zero at the edge. Outer-edge diffraction becomes increasingly significant with decreasing size of the ground plane because of the increasing magnitude of the currents on the ground-plane faces at the outer edge. Edge diffraction can alter the input impedance by more than 100% and directive gain in the plane of the ground plane by more than 6 dB from the values for a ground plane of infinite extent.

Theoretical models exist for predicting the effects of diffraction by the outer edge of the ground plane. The existing models may be classified into two categories, distinguished by whether the current distribution on the monopole element is initially known or unknown.

When the monopole element is very thin and not too long, its current distribution is approximately sinusoidal and independent of the radius of the ground plane. Consequently, the element's current distribution can be initially specified and we need only determine the ground plane's current distribution. For this category of monopoles, the theoretical models reported in the literature essentially consist of Bardeen's integral equation method for a ground-plane radius that is small compared to a wavelength [1], Richmond's method of moments (ground plane only) for a ground-plane radius that is not too large compared to a wavelength [2], Leitner and Spence's method of oblate spheroidal wave functions for a groundplane radius that is comparable to a wavelength [3–5], Tang's scalar theory of diffraction [6] and the geometric theory of diffraction (GTD) for a ground-plane radius that is large compared to a wavelength, and Storer's variational method for a ground-plane radius that is very large compared to a wavelength [7, 8].

When the monopole element is relatively thick, its current distribution is no longer sinusoidal, and the current distribution on both the monopole element and the ground plane consequently need to be determined as a function of the ground-plane radius. For this category of monopoles, the theoretical models reported in the literature essentially consist of Richmond's method of moments for ground-plane radius that is not too large compared to a wavelength [2] and Awadalla-Maclean's method of moments (monopole element only) combined with the geometric theory of diffraction for ground-plane radius that is large or comparable to a wavelength [9, 10]. Thiele and Newhouse have also reported a model that combines the method of moments with the geometric theory of diffraction, but their computer program is unavailable [11].

Each of the existing models is valid for different and sometimes overlapping sets of limited values of ground-plane radii. Some of the models require extensive numerical computation. Therefore, the collection of models taken as a whole has several deficiencies. In the open literature, results for input impedance and directive gain have never been assembled as a continuous function of ground-plane radius for the entire range of values from zero to infinite wavenumbers. In regions where models overlap, it is sometimes unclear which models are more accurate. In some models, numerical results have been reported for only a few values of ground-plane radius. In one model, Bardeen's integral equation, the base of the monopole element has not been allowed to be in the same plane as the ground plane, which is our present case of interest. Computer programs are not available for some of the older models because they were developed before the advent of computers. One of the most versatile of the models, Richmond's method of moments, gives only the input impedance and not the radiation pattern. In another model, Leitner and Spence's oblate spheroidal wave functions, one of the published algorithms for computing the eigenvalues is incorrect. Finally, extensive numerical results for small ground planes and resonant monopoles with finite ground planes are not to be found in the open literature.

This book is our attempt to correct these deficiencies. Computer programs and numerical results are presented for all of the models. The induced

EMF method is utilized to determine the input impedance of a thin, idealized monopole element in the absence of a ground plane. In Bardeen's integral equation method, the excitation function for the ground-plane currents is extended to include the singularity that occurs when the base of the monopole element is in the same plane as the ground plane. Richmond's method of moments is extended to give the far-field radiation pattern. In Leitner and Spence's method of oblate spheroidal wave functions, we have corrected the published continued-fraction algorithm for computing the eigenvalues. Numerical results for input impedance and directive gain are presented as a continuous function of ground-plane radius for arbitrary radius. Numerical results of various models are compared, and we identify the suspected best available results. Extensive numerical results are given for small ground planes and resonant monopoles on finite ground planes. New experimental data is presented and compared with numerical results.

Circuit representations of the monopole antenna are developed in Chapter 2. Theoretical models and numerical results are presented in Chapter 3 for the case in which the current distribution on the monopole element is initially known. In Chapter 4, theoretical models and numerical results are presented for the case in which the current distribution on both the monopole element and ground plane are initially unknown. The theoretical models are compared with experimental data in Chapter 5. Computer printouts of directive gain and the far-field elevation pattern are given in Appendix A. Computer programs of the theoretical models are given in Appendix B.

Chapter 2
Circuit Representation

2.1 GEOMETRY AND COORDINATE SYSTEMS

Consider a monopole element of length h and radius b, which is located in free space at the center of an infinitely thin, circular ground plane of radius a and infinite conductivity (see Fig. 1). The ground-plane radius, when expressed in wavenumbers, is given by

$$\epsilon \equiv ka \tag{2.1.1}$$

where

$k = 2\pi/\lambda =$ wavenumber (m^{-1});
$\lambda =$ excitation wavelength (m).

The monopole element and ground plane have current distributions in real time given by

$$I_z(z, t) = \mathrm{Re}[I(z)e^{j\omega t}], \quad I_\rho(\rho, t) = \mathrm{Re}[I(\rho)e^{j\omega t}] \tag{2.1.2}$$

where

$\omega =$ radian frequency of the excitation $= 2\pi c/\lambda$ (rad/s);
$c =$ wave velocity in free space $= 2.9979 \times 10^8$ m/s;
$I(z), I(\rho) =$ element and ground-plane current amplitude distributions, respectively (A). The subscripts z and ρ are suppressed in our notation.

A field point $P(r, \theta, \phi)$, expressed in spherical coordinates, is shown in Fig. 1. The field is uniform in the azimuthal direction ϕ. The relationships between spherical, cylindrical, and oblate spheroidal coordinate systems are shown in Table 1. In the far-field, the elevation angle θ is related to the oblate spheroidal angular coordinate η by $\eta \rightarrow \cos\theta$ as the spheroidal radial coordinate $\xi \rightarrow \infty$.

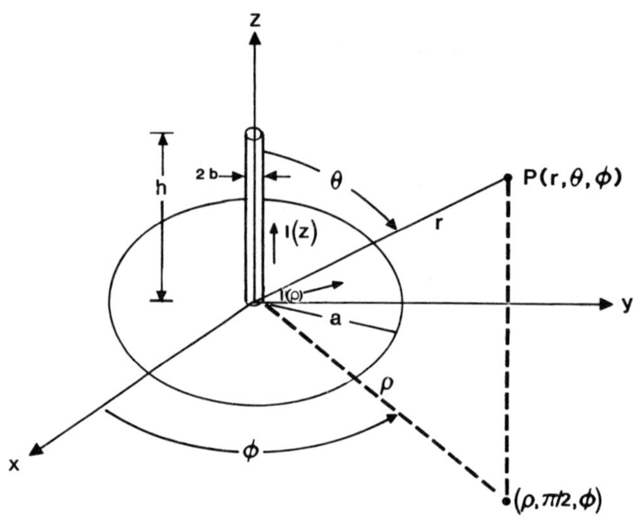

Fig. 1 Monopole Element at the Center of a Circular Ground Plane

2.2 DIRECTIVE GAIN AND INPUT IMPEDANCE

At a far-field observation point $P(r, \theta, \phi)$, the numeric directive gain $d(\theta, \phi)$ of the antenna is defined as the ratio of its radiated power density $s(\theta, \phi)$ to its radiated power density averaged over all directions. The radiated power density averaged over all directions is equivalent to the power density radiated by a hypothetical "isotropic" (ideal) antenna. Accordingly, the directive gain $d(\theta, \phi)$, expressed in spherical coordinates with the origin at the antenna, is given by

$$d(\theta, \phi) = \frac{s(\theta, \phi)}{(1/4\pi) \int_0^{2\pi} \int_0^{\pi} s(\theta, \phi) \sin\theta \, d\theta \, d\phi} \qquad (2.2.1)$$

Table 1 Spherical, Cylindrical, and Oblate Spheroidal Coordinates

COORDINATES	*Spherical* (r, θ, ϕ)	*Cylindrical* (ρ, z, ϕ)	*Oblate Spheroidal* (ξ, η, ϕ)
Radius	r	$\rho = r \sin\theta$	$\rho = a[(1 - \eta^2)(1 + \xi^2)]^{1/2}$
Elevation	θ	$z = r \cos\theta$	$z = a\eta\xi$
Azimuth	ϕ	ϕ	ϕ

Note 1: In Table 1, the notation ξ, η is that of Leitner and Spence $(L + S)$, *Franklin Institute Journal,* Vol. 249, No. 4, pp. 299–321, April 1950. This notation is related to that of Abramowitz and Stegun $(A + S)$, *Handbook of Mathematical Functions,* National Bureau of Standards, Applied Mathematical Series, No. 55, p. 752, June 1964, by

$$\xi_{L+S} = (\xi_{A+S}^2 - 1)^{1/2}$$
$$\eta_{L+S} = (1 - \eta_{A+S}^2)^{1/2}$$

Note 2: The cylindrical coordinates (ρ, z, ϕ) are related to the rectangular coordinates (x, y, z) by

$$x = \rho \cos\phi$$
$$y = \rho \sin\phi$$
$$z = z$$

For antenna patterns that are uniform in the azimuthal direction, such as for the antenna geometry of Fig. 1, (2.2.1) reduces to

$$d(\theta) = \frac{2s(\theta)}{\int_0^{\pi} s(\theta) \sin\theta \, d\theta} = \frac{2s'(\eta)}{\int_{-1}^{1} s'(\eta) \, d\eta} = d'(\eta) \tag{2.2.2}$$

where

$s(\theta) = (1/2)\sqrt{\mu_0/\epsilon_0}\,|H_\phi|^2 = (1/2)\sqrt{\epsilon_0/\mu_0}\,|E_\theta|^2$

H_ϕ, E_θ = far-field magnetic and electric field intensities, respectively;

$s'(\eta)$, $d'(\eta)$ = radiation power density and directive gain, respectively, in oblate spheroidal coordinates;

$\sqrt{\mu_0/\epsilon_0}$ = wave impedance in free space.

The numeric directive gain $d(\theta, \phi)$ is related to the directive gain $D(\theta, \phi)$, expressed in dBi, by

$$D(\theta, \phi) = 10 \log_{10} d(\theta, \phi) \quad \text{(dBi)} \tag{2.2.3}$$

The total time-averaged radiated power P_{total} of the antenna is given by

$$\begin{aligned} P_{total} &= \int_0^{2\pi} \int_0^{\pi} s(\theta, \phi)\, r^2 \sin\theta \, d\theta \, d\phi \\ &= 2\pi r^2 \int_0^{\pi} s(\theta) \sin\theta \, d\theta = 2\pi r^2 \int_{-1}^{1} s'(\eta) \, d\eta \end{aligned} \tag{2.2.4}$$

The antenna radiation resistance R, referred to the complex amplitude $I(z = 0)$ of the antenna base current, is defined by

$$R \equiv 2P_{total}/|I(z = 0)|^2 \tag{2.2.5}$$

By substituting (2.2.6) and (2.2.7) into (2.2.4), we have

$$\begin{aligned} d(\theta) &= s(\theta)\, 8\pi r^2/[R\, |I(z = 0)|^2] \\ &= s'(\eta)\, 8\pi r^2/[R\, |I(z = 0)|^2] = d'(\eta) \end{aligned} \tag{2.2.6}$$

The antenna input impedance, Z_{in}, is given by

$$Z_{in} = V(z = 0)/I(z = 0) = R_{in} + jX_{in} \tag{2.2.7}$$

where

$V(z = 0)$ = complex amplitude of the excitation voltage across the aperture of the coaxial line feed to the antenna (V);
R_{in} = input resistance (Ω);
X_{in} = input reactance (Ω).

The input resistance R_{in} is related to the radiation resistance R by

$$R_{in} = R + R_{ohmic} \tag{2.2.8}$$

where R_{ohmic} is the ohmic loss resistance of the antenna for finite conductivity of either the monopole element or the ground plane. Here, $R_{ohmic} = 0$ because the monopole element and ground plane are assumed to be of infinite conductivity. Accordingly,

$$R_{in} = R, \quad R_{ohmic} = 0 \tag{2.2.9}$$

Equation (2.2.9) is a statement that the antenna is assumed to have an efficiency of unity.

2.3 RELATIONSHIP BETWEEN RADIATION RESISTANCE AND DIRECTIVE GAIN ON THE HORIZON

For a vertical monopole element with a finite ground plane, the far-field radiated power density on the radio horizon, $s(\theta = \pi/2)$, is determined only by the current distribution on the monopole element (and not the ground-plane current) because only the vertical monopole element has a component of electron acceleration normal to the radio horizon. (This statement is not true for a ground plane of infinite extent because a far-field point on the radio horizon is then on the ground plane.) Identical monopole elements with identical current distributions, but mounted on ground planes of different finite radii, will consequently have identical far-field radiated power densities on the radio horizon. Accordingly,

$$s_\epsilon(\theta = \pi/2) = s_0(\theta = \pi/2) \tag{2.3.1}$$

for identical monopole element, identical element current distribution, and ground plane of finite extent, where the subscript ϵ denotes the radius in wavenumbers of the ground plane of arbitrary but finite extent ($\epsilon < \infty$), and the subscript 0 denotes a ground plane of zero extent ($\epsilon = 0$).

If we substitute (2.2.6) into (2.3.1) and compute the quantity $s(\theta = \pi/2)|I(z = 0)|^2/8\pi r^2$, we obtain the following relationship between radiation resistance R and numeric directive gain on the horizon $d(\theta = \pi/2)$:

$$d_\epsilon(\theta = \pi/2)\, R_\epsilon = d_0(\theta = \pi/2)\, R_0 = \text{constant} \tag{2.3.2}$$

for identical monopole elements, identical element current distributions, and ground plane of finite extent.

In (2.3.1) and (2.3.2), the condition of identical element current distributions for ground planes of different radii is generally not satisfied by the use of monopole antennas. The element current distribution $I(z)$ is generally dependent on the ground-plane current $I(\rho)$, which, in turn, is a function of the ground-plane radius. However, for monopole elements that are sufficiently thin electrically and not too long, the element current distribution $I(z)$ is approximately sinusoidal and independent of the ground-plane current $I(\rho)$ (see Sec. 3.1). Expressions for $d_0(\theta = \pi/2)$ and R_0, for elements with a sinusoidal current distribution, are determined in Sec. 3.2. Substitution of those expressions into (2.3.2), for the case of an infinitely thin monopole element ($b = 0$), yields

$$d(\theta = \pi/2)\, R = (\eta/4\pi)\, [1 - \cos(kh)]^2/\sin^2(kh); \quad b = 0 \qquad (2.3.3)$$

for a sinusoidal element current distribution and ground planes of finite extent, where

η = wave impedance of free space = 376.73 Ω;
h = length of the monopole element.

The condition $b = 0$ may be removed from (2.3.3) without substantially altering the result because the radiation pattern and radiation resistance of electrically thin elements that are not too long are weakly dependent upon the element radius (see Sec. 3.2).

2.4 CHARACTERIZATION OF CURRENTS

The characterization of the currents on the monopole element, ground plane, and coaxial line feed is illustrated in Fig. 2. The physical realization, circuit representation, and two circuit idealizations of the currents are shown in Figs. 2(a), 2(b), 2(c), and 2(d), respectively. In Fig. 2(b), the coaxial line excitation of Fig. 2(a) is replaced by equivalent electric and magnetic currents on a conductor completely enclosing the coaxial line. In Fig. 2(c), the normalized ferrite attenuation distance $h_1/\lambda \ll 1$ is idealized to be zero. In Fig. 2(d), the magnetic frill M_ϕ is assumed to be negligible.

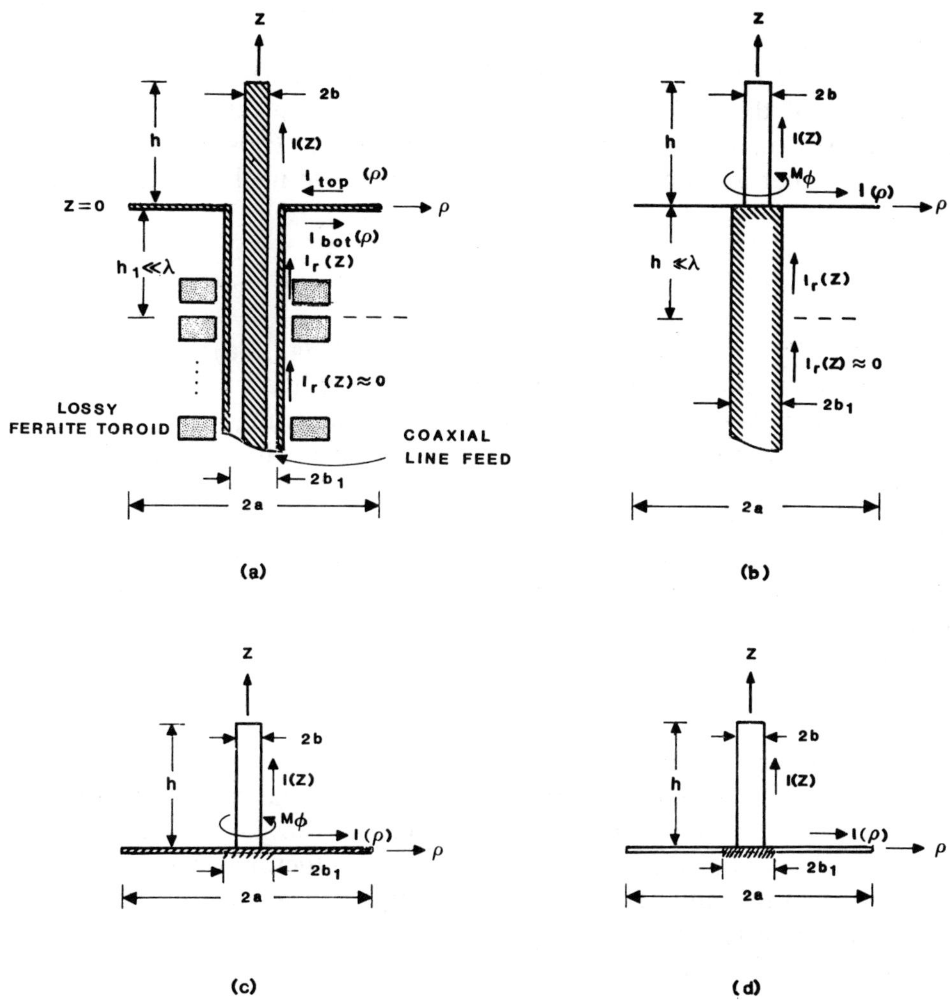

Fig. 2 Characterization of Currents on Monopole Element, Ground Plane, and Coaxial Line Feed
(a) Physical Realization
(b) Circuit Representation
(c) Idealization of Circuit ($h_1/\lambda = 0$)
(d) Idealization of Circuit ($h_1/\lambda = 0$, $kb_1 \ll 1$, where $I(z)$ or $I(\rho)$ initially known)

The currents of interest are the element current $I(z)$ (positive in the $+z$ direction), the return current $I_r(z)$ on the outside of the coaxial line outer conductor (positive in the $+z$ direction), the current $I_{\text{bot}}(\rho)$ on the bottom face of the ground plane (positive in the $+\rho$ direction, and the current $I_{\text{top}}(\rho)$ on the top face of the ground plane (positive in the $-\rho$ direction). A net ground-plane current $I(\rho)$ (positive in the $-\rho$ direction) is defined as

$$I(\rho) = I_{\text{bot}}(\rho) - I_{\text{top}}(\rho) \tag{2.4.1}$$

In the physical realization of the currents (Fig. 2(a)), lossy ferrite toroids are mounted along the outside of the coaxial line's outer conductor. Such a procedure is commonly utilized on antenna ranges to reduce the radiation or pickup of unwanted signals from currents induced on the outside of the transmitter or receiver coaxial cables. For a sufficient number of ferrite toroids near the terminal end of the coaxial line, the return current $I_r(z)$ is approximately zero at distances greater than the $1/e$ attenuation distance h_1 from the termination of the line. Here, we assume the ferrite toroids to be sufficiently lossy, so that

$$h_1 \ll \lambda \tag{2.4.2}$$

where λ is the excitation wavelength.

The radii of the outer and inner conductors of the coaxial line are b_1 and b, respectively, where b is also the radius of the monopole element. The wall thickness of the coaxial line's outer conductor is assumed to be much less than its diameter. Consequently, the return current $I_r(z)$ occurs at the radial coordinate $\rho \approx b_1$.

The constraints on the various currents are given as follows below.

Element:

$$I(z = h) = 0 \tag{2.4.3}$$

Ground Plane:

$$I(\rho = a) = 0 \tag{2.4.4}$$

Coaxial Line:

$$I_r(z) \approx 0, \quad -\infty < z < -h_1 \tag{2.4.5}$$

Element Ground-Plane:

$$I(z = 0) = -I(\rho = b) \tag{2.4.6}$$

Ground-Plane Coaxial Line:

$$I_{bot}(\rho = b_1) = I_r(z = 0) \tag{2.4.7}$$

The element and ground-plane constraints are a consequence of an open circuit at the end of the element and the ground plane. The coaxial line constraint is a consequence of the lossy ferrites. The element ground-plane constraint is a consequence of conservation of charge (Kirchhoff's current law) at a node. The ground-plane coaxial line constraint is a consequence of conservation of charge along a conductor.

By the use of the equivalence principle [12], the coaxial line feed excitation may be replaced by equivalent tangential field excitations defined along a surface completely enclosing the coaxial line. At field points external to this surface, the equivalent field excitations will give the same field as the original source excitation. In the circuit representation of the monopole antenna currents (Fig. 2(b)), the coaxial aperture excitation is replaced by an equivalent surface magnetic current density (magnetic frill) M_ϕ, which sits on top of a thick ground plane of radius b_1. The magnetic frill M_ϕ is defined to be positive in the positive azimuthal direction, and is given by

$$M_\phi = \begin{cases} -V(0)/[\rho \ln(b_1/b)], & b \leq \rho \leq b_1 \\ 0, \quad \rho < b, \ \rho > b_1 \end{cases} \tag{2.4.8}$$

Equation (2.4.8) is derived as follows. The radial field E of the coaxial line aperture, assuming a TEM-mode excitation, is given by

$$E_\rho = \begin{cases} V(0)/\rho \ln(b_1/b), & b \leq \rho \leq b_1 \\ 0, \quad \rho < b \end{cases} \tag{2.4.9}$$

where $V(0)$ is the positive voltage at $z = 0$ across the aperture with the coaxial outer conductor at zero potential. By way of the equivalence principle [12], an aperture field may be replaced by a magnetic frill **M** that sits on top of a ground plane congruent with the aperture surface, and which is defined as

$$\mathbf{M} = \mathbf{E}_{\text{tangential}} \times \mathbf{n} \tag{2.4.10}$$

where **n** is the outward normal to the aperture surface and $\mathbf{E}_{\text{tangential}}$ is the tangential field at the aperture surface. By substituting (2.4.9) into (2.4.10), we have

$$\mathbf{M} = (\hat{u}_\rho \, E_\rho) \times \hat{u}_z = -E_\rho \hat{u}_\phi = \hat{u}_\phi M_\phi \tag{2.4.11}$$

where M_ϕ is given by (2.4.8). Equation (2.4.8) agrees with the result obtained by Richmond [2].

In the circuit representation of Fig. 2(b), the net ground-plane current $I(\rho)$ is the same as defined by (2.4.1), with the additional current constraint:

$$I_{\text{bot}}(\rho) = 0, \quad 0 \leq \rho < b_1 \tag{2.4.12}$$

If the circuit representation of the monopole antenna in Fig. 2 is now idealized by setting the ferrite $1/e$ attenuation distance $h_1/\lambda = 0$, then,

$$I_r(z) \approx 0; \; z < 0, \quad h_1/\lambda = 0 \tag{2.4.13}$$

Consequently, the coaxial line's outer conductor may be removed from the circuit as shown in Fig. 2(c). The ground-plane coaxial-line current constraint of (2.3.7) is not disturbed by such an idealization. Because it has already been assumed that $h_1 \ll \lambda$, the idealization $h_1 = 0$ does not appreciably alter either the radiation pattern or the input impedance of the monopole antenna, provided that the monopole length $h \gg h_1$. Finite currents on an aperture that is small compared to the excitation wavelength contribute little to the far-field and input impedance. (Compare with the results for an electrically small dipole.) Experimental radiation patterns and measurements of input impedance (see Chapter 5) confirm that the use of lossy ferrite toroids along the coaxial line's outer conductor yields results which are in close agreement with theoretical results for the circuit idealization condition of (2.4.13)—even for electrically small ground planes.

In the idealization of the monopole antenna circuit, the magnetic frill M_ϕ may be removed (Fig. 2(d)) without appreciably altering the radiation pattern or input impedance, provided that

$$kb_1 \ll 1 \tag{2.4.14}$$

and $I(z)$ or $I(\rho)$ is initially known, which constitute the conditions for neglecting the magnetic frill.

The condition $kb_1 \ll 1$ corresponds to the condition for TEM-mode excitation of the coaxial line and negligible power radiated from the coaxial line aperture. If either $I(z)$ or $I(\rho)$ is initially known, then the one which is not known may be determined from the other, without requiring knowledge of the original coaxial line excitation or its equivalent, provided that the field radiated by the known current distribution is the predominant field incident on the conductor of unknown current distribution. When neither $I(z)$ nor $I(\rho)$ is known, then the original source excitation or its equivalent (in this case, the magnetic frill M_ϕ) must be specified to determine the unknown radiated fields.

Chapter 3

Models in which the Current Distribution on the Monopole Element is Initially Known

3.1 BOUNDARY CONDITIONS

The current amplitudes, $I(z)$ and $I(\rho)$, on the monopole element and ground plane, respectively, are generally complex and initially unknown quantities (see Fig. 1). Consider now the case where the current distribution on the monopole element is assumed to be sinusoidal. For such a case and for the waveform dependence given in (2.1.2), we have

$$I(z) = \frac{I(0)}{\sin(kh)} \sin[k(h - z)], \quad 0 \leqslant z \leqslant h \tag{3.1.1}$$

where

h = monopole length (m);
$k = 2\pi/\lambda$ = wavenumber (m^{-1});
$I(0)$ = current amplitude of the monopole base current at $z = 0$ (A).

From (3.1.1)

$$\arg I(z) = \text{constant}, \quad 0 \leqslant z \leqslant h \tag{3.1.2}$$

Although a sinusoidal distribution of current is not possible, even for an infinitely thin antenna, (3.1.1) is most likely a fair approximation to the current if the monopole element is sufficiently thin electrically and not too long [13]. For a center-fed dipole of radius b and total length $2h$, Elliot [14] gives examples where the current distribution is approximately sinusoidal and of approximately constant argument for $b/\lambda = 1.0 \times 10^{-4}$ and $h/\lambda = 0.125, 0.25$. However, for $b/\lambda = 1.0 \times 10^{-4}$ and $h/\lambda = 0.375, 0.5$, Elliot demonstrates that the current distribution is no longer sinusoidal near the center of the dipole, nor is arg z approximately constant. Balanis [15] shows that for $b/\lambda = 2.0 \times 10^{-4}$ and $h/\lambda = 0.25, 0.5$, the current distribution is not sinusoidal near the center of the dipole. Elliot and Balanis also demonstrate that for $b/\lambda = 1.0 \times 10^{-2}$ and $h/\lambda = 0.125, 0.25, 0.375, 0.50$, the current distribution is neither sinusoidal, nor of constant phase, and that the deviations from (3.1.1) and (3.1.2) increase with increasing values of h/λ and b/λ. On the basis of the above results, it appears that (3.1.1) and (3.1.2) are approximately valid for the conditions:

$$b/\lambda < 10^{-4}, \quad h/\lambda \leqslant 0.25 \tag{3.1.3}$$

In addition to the constraint on $I(z)$ given by (3.1.1), assume that the return current $I_r(z)$ on the outside of the coaxial line's outer conductor (see Fig. 2) is given by

$$I_r(z) \approx 0, \quad z < 0 \tag{3.1.4}$$

The constraint of (3.1.4) corresponds to the idealized condition that the ferrite toroids have a $1/e$ current attenuation distance h_1, given by

$$h_1/\lambda = 0 \tag{3.1.5}$$

It should be noted that (3.1.4) and (3.1.5) do not alter the constraint $I_r(z = 0) = I_{\text{bot}}(\rho = b_1)$ given by (2.3.7), nor do they impose any constraints on $I_{\text{bot}}(\rho = b_1)$, where $I_{\text{bot}}(\rho = b_1)$ is the current on the bottom of the ground plane at a radius equal to that of the outer conductor.

Combining the current constraints given by (3.1.1) and (3.1.4) with those given by (2.4.3) to (2.4.7) and (2.4.12), the current constraints on the monopole antenna are given by

Element:

$$I(z) = [I(0)/\sin(kh)] \sin[k(h - z)], \quad 0 \leqslant z \leqslant h \tag{3.1.6}$$

Ground Plane:

$$I(\rho = a) = 0 \tag{3.1.7}$$

$$I_{\text{bot}}(\rho) = 0, 0 \leqslant \rho \leqslant b_1 \tag{3.1.8}$$

Coaxial Line:

$$I_r(z) \approx 0, z < 0 \tag{3.1.9}$$

Element Ground Plane:

$$I(z = 0) = -I(\rho = b) \tag{3.1.10}$$

Ground-Plane Coaxial Line:

$$I_{\text{bot}}(\rho) = I_r(z = 0) \tag{3.1.11}$$

In Chapter 3, we will assume that all models satisfy the current constraints given by (3.1.6) to (3.1.11). The results are expected to be approximately correct if the monopole element is sufficiently thin electrically and not too long (conditions (3.1.3)) and if the ferrite toroids are sufficiently lossy ($h_1/\lambda \ll 1$, idealized by condition (3.1.5)). For these conditions, the circuit representation of the monopole antenna is shown in Fig. 2(d).

For the current constraints of (3.1.6) to (3.1.11), the total magnetic and electric field intensities $\mathbf{H}^{(\text{total})}$, $\mathbf{E}^{(\text{total})}$ at an arbitrary field point $P(x, y, z)$ external to the element-excitation-source points are simply the vector sum of the fields resulting from the element current and the current induced on the ground plane by the fields incident by the element.

Accordingly,

$$\begin{aligned} \mathbf{H}^{(\text{total})} &= \mathbf{H}^{(e)} + \mathbf{H}^{(g)}\,[\mathbf{E}^{(e)}, \mathbf{H}^{(e)}] \\ \mathbf{E}^{(\text{total})} &= \mathbf{E}^{(e)} + \mathbf{E}^{(g)}\,[\mathbf{E}^{(e)}, \mathbf{H}^{(e)}] \end{aligned} \tag{3.1.12}$$

for the current constraints of (3.1.6) to (3.1.11), where

$\mathbf{H}^{(e)}, \mathbf{E}^{(e)}$ = magnetic and electric field intensities, respectively, generated by the element current;

$\mathbf{H}^{(g)}, \mathbf{E}^{(g)}$ = magnetic and electric field intensities, respectively, generated by the ground-plane current induced by the element incident fields.

The element fields $\mathbf{H}^{(e)}, \mathbf{E}^{(e)}$ are determined in Sec. 3.2. The ground-plane fields $\mathbf{H}^{(g)}, \mathbf{E}^{(g)}$ are determined in Sec. 3.3 through Sec. 3.8 for ground-plane radii of various extents.

3.2 INDUCED EMF METHOD, GROUND PLANE OF ZERO EXTENT

3.2.1 Concept of a Ground Plane of Zero Extent

Consider a monopole antenna excited by a coaxial line with an outer conductor of radius b_1 that is terminated by free space, rather than by a ground plane (Fig. 3(a)). The ground plane for such an antenna is denoted as being of *zero extent.* As was shown in Sec. 2.4, the coaxial line excitation may be replaced by an equivalent magnetic current (frill) M_ϕ sitting on top of a thick ground plane of radius $\rho = b_1$ (Fig. 3(b)). For sufficiently lossy ferrite toroids along the outside of the coaxial line, the current on the exterior of the coaxial line's outer conductor may be neglected (Fig. 3(c)). The magnetic frill may be removed from the circuit without appreciably affecting the results because $kb_1 \ll 1$ for the assumed sinusoidal current distribution on the monopole element (see (3.1.3) and the discussion concerning the circuit idealization of Fig. 2(d)). Finally, the ground plane of radius b_1 may be removed from the circuit without appreciably affecting the results because a finite current on an electrically small conductor does not radiate an appreciable amount of power compared with that radiated by a monopole element of length $h \gg b_1$. The circuit idealization of a monopole antenna with a ground plane of zero extent is therefore an electrically thin monopole element with no ground plane (Fig. 3(d)).

The near fields, far fields, and input impedance of an electrically thin monopole element are derived and summarized in the remainder of Sec. 3.2. The input impedance is derived by way of the induced EMF method.

3.2.2 Near Fields

Consider a monopole element of length h with a sinusoidal current distribution $I(z') = [I(0)/\sin(kh)] \sin[k(h - z')]$, $0 \le z' \le h$, at points

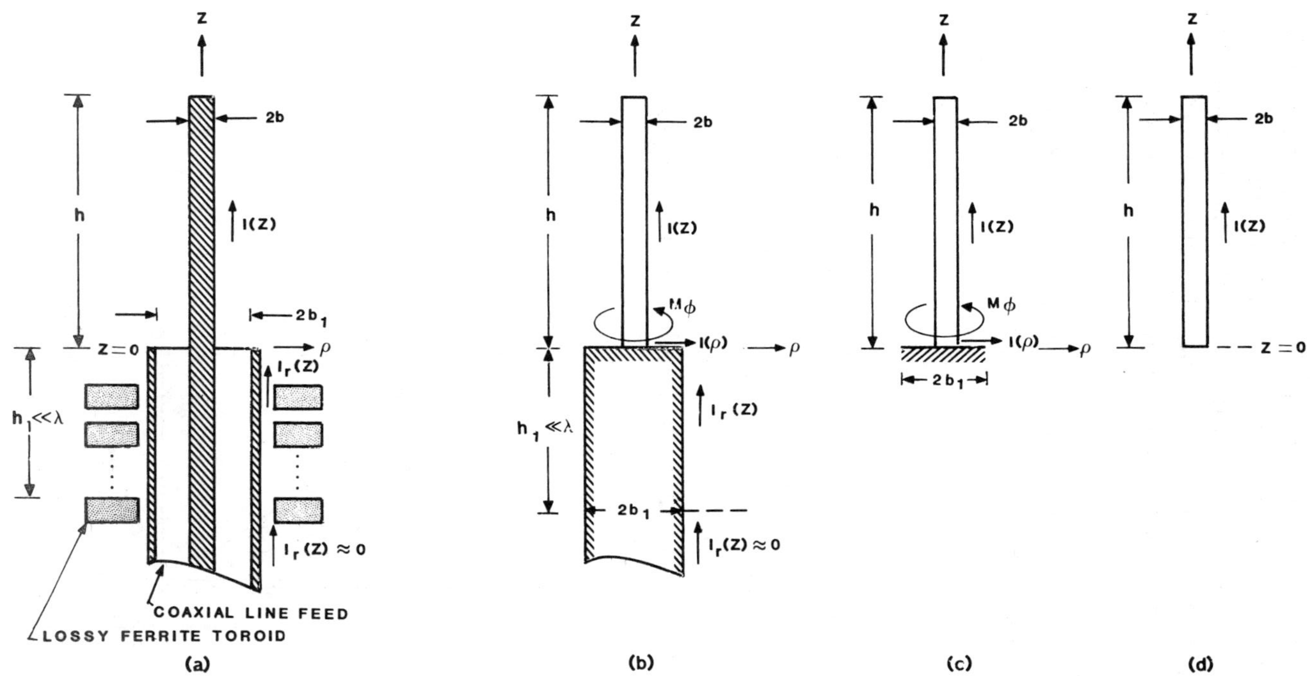

Fig. 3 Characterization of Currents for Monopole Antenna with Ground Plane of Zero Extent

(a) Physical Realization

(b) Circuit Representation

(c) Idealization of Circuit ($h_1/\lambda = 0$)

(d) Idealization of Circuit ($h_1/\lambda = 0$, $kb_1 \ll 1$, where $I(z)$ or $I(\rho)$ initially known)

$Q(x', y', z')$ on the surface of the element (Fig. 4). For an electrically thin element with a known current distribution, the fields at arbitrary points $P(x, y, z)$ external to the element may be determined almost exactly by approximating the source points to lie on the element axis, that is, $Q(x', y', z') \approx Q(0, 0, z')$. For the current waveform of (2.1.2), the magnetic vector potential **A** is given by [16]:

$$\mathbf{A}(x, y, z) = \hat{u}_z A_z = \hat{u}_z \frac{\mu_0 I(0)}{4\pi \sin(kh)} \int_0^h \frac{\sin[k(h - z')] \exp(-jk\overline{PQ})}{\overline{PQ}} dz' \tag{3.2.1}$$

where

$$\begin{aligned} \mu_0 &= \text{permeability of free space} = 4\pi \times 10^{-7} \text{ H/m;} \\ \overline{PQ} &= [(x - x')^2 + (y - y')^2 + (z - z')^2]^{1/2} \\ &\approx [x^2 + y^2 + (z - z')^2]^{1/2} \\ &= [\rho^2 + (z - z')^2]^{1/2}; \end{aligned}$$

$\hat{u}_\rho, \hat{u}_\phi, \hat{u}_z$ = unit vectors in the radial, azimuthal, and axial cylindrical directions, respectively.

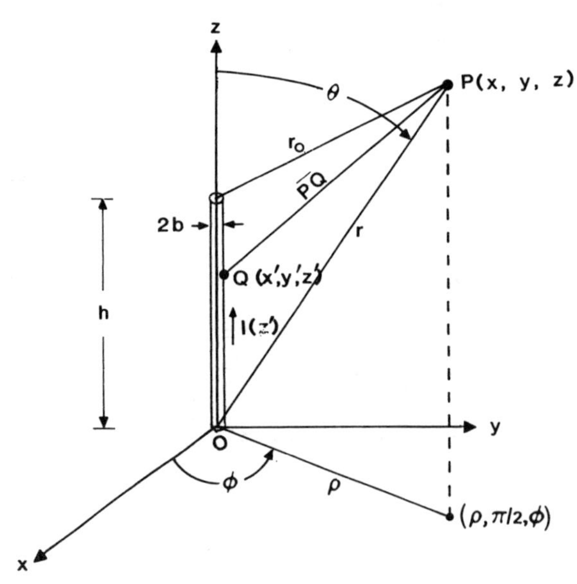

Fig. 4 Monopole Element Geometry

The magnetic and electric field intensities, **H** and **E**, respectively, are given by

$$\mathbf{H} = \frac{1}{\mu_0}(\nabla \times \mathbf{A}) = -\hat{u}_\phi \frac{1}{\mu_0}\frac{\partial A_z}{\partial \rho} \tag{3.2.2}$$

$$\mathbf{E} = \frac{1}{j\omega\epsilon_0}(\nabla \times \mathbf{H}) = -\hat{u}_\rho \frac{1}{j\omega\epsilon_0}\frac{\partial H_\phi}{\partial z} + \hat{u}_z \frac{1}{j\omega\epsilon_0}\frac{1}{\rho}\frac{\partial}{\partial \rho}(\rho H_\phi) \tag{3.2.3}$$

where ϵ_0 = permittivity of free space = 8.854×10^{-12} F/m.

Exact closed-form expressions of **A, H, E,** and radiation resistance were first obtained for an infinitely thin element by Brillouin [17] and are summarized by Stratton [18]. The magnetic and electric fields are given by

$$H_\phi = \frac{jI(0)}{4\pi\rho \sin(kh)}\left[\exp(-jkr_0) - \cos(kh)\exp(-jkr)\frac{-jz}{r}\sin(kh) \times \exp(-jkr)\right] \tag{3.2.4}$$

$$\begin{aligned} E_\rho &= \frac{jI(0)\eta}{4\pi\rho \sin(kh)}\left\{\frac{(z-h)\exp(-jkr_0)}{r_0} - \frac{z}{r}\cos(kh)\exp(-jkr) + \sin(kh)\frac{\partial}{\partial z}\left[\frac{z}{kr}\exp(-jkr)\right]\right\} \\ &= \frac{jI(0)\eta}{4\pi\rho \sin(kh)}\left[\frac{(z-h)\exp(-jkr_0)}{r_0} - \frac{z}{r}\cos(kh)\exp(-jkr) + \sin(kh)\exp(-jkr)\left(\frac{1}{kr} - \frac{z^2}{kr^3} - \frac{jz^2}{r^2}\right)\right] \end{aligned} \tag{3.2.5}$$

$$E_z = \frac{-jI(0)\eta}{4\pi \sin(kh)}\left\{\frac{\exp(-jkr_0)}{r_0} - \cos(kh)\frac{\exp(-jkr)}{r} + \sin(kh)\frac{\partial}{\partial z}\left[\frac{\exp(-jkr)}{r}\right]\right\}$$

$$E_z = \frac{-jI(0)\eta}{4\pi \sin(kh)} \left[\frac{\exp(-jkr_0)}{r_0} - \cos(kh) \frac{\exp(-jkr)}{r} - z \sin(kh) \exp(-jkr)\left(\frac{j}{r^2} + \frac{1}{kr^3}\right) \right] \quad (3.2.6)$$

where

$$\eta = k/\omega\epsilon_0 = \text{wave impedance of free space} = 376.73\ \Omega;$$
$$r_0 = [\rho^2 + (z - h)^2]^{1/2};$$
$$r = (\rho^2 + z^2)^{1/2}.$$

Equations (3.2.4) to (3.2.6) are identical to the results given by Stratton [$\xi_2 = h$, $\xi_1 = 0$, $l = h$, $\alpha = kh$, $r_2 = r_0$, $r_1 = r$, $I_0 = -I(0)/[\sin(kh)]$ after the substitution of $-j$ for j to account for the $\exp(-j\omega t)$ waveform of Stratton, instead of the $\exp(j\omega t)$ waveform of (2.1.2). The fields given by (3.2.4) to (3.2.6) are exact for an infinitely thin element, and they are almost exact for an electrically thin element with the same sinusoidal current distribution.

3.2.3 Far Fields

Consider a field point P at a sufficiently large radial distance r, which satisfies both the far radiation zone and Fraunhofer diffraction conditions given by

$$h \ll r, \quad kr \gg 1 \quad (3.2.7)$$

for the far radiation zone conditions, and

$$kh^2/2r \ll 2\pi \quad (3.2.8)$$

for the Fraunhofer diffraction condition. For these conditions, we write

$$\begin{aligned}
\exp(-jkr_0) &\approx \exp[-kj(r - h\cos\theta)], \\
&\qquad h \ll r,\ (kh^2/2r) \ll 2\pi \\
1/r_0 &\approx 1/r, \quad h \ll r \\
z - h &\approx z = r\cos\theta, \quad h \ll r \\
(1/kr) - (z^2/kr^3) - (jz^2/r^2) &\approx -j\cos^2\theta, \quad kr \gg 1 \\
(j/r^2) + (1/kr^3) &\approx (j/r^2), \quad kr \gg 1
\end{aligned} \quad (3.2.9)$$

For the approximations of (3.2.9), the "exact" fields given by (3.2.4) to (3.2.6) reduce to the far fields given by

$$H_\phi = \frac{jI(0)\exp(-jkr)}{4\pi r \sin\theta \sin(kh)} [\exp(jkh\cos\theta) - \cos(kh) - j\cos\theta \sin(kh)] \quad (3.2.10)$$

$$E_\rho = \frac{jI(0)\eta \cos\theta \exp(-jkr)}{4\pi r \sin\theta \sin(kh)} [1 - \cos(kh) - j\cos\theta \sin(kh)] \quad (3.2.11)$$

$$E_z = \frac{-jI(0)\,\eta \exp(-jkr)}{4\pi r \sin(kh)} [1 - \cos(kh) - j\cos\theta \sin(kh)] \quad (3.2.12)$$

The resultant electric field $\mathbf{E} = \hat{u}_\rho E_\rho + \hat{u}_z E_z$ reduces in the far field to

$$\mathbf{E} = \hat{u}_\theta E_\theta \text{ (far field)} \quad (3.2.13)$$

where

$$E_\theta = \frac{jI(0)\,\eta \exp(-jkr)}{4\pi r \sin\theta \sin(kh)} [1 - \cos(kh) - j\cos\theta \sin(kh)]$$

The time-averaged Poynting vector $\mathbf{S} = (1/2)\,(\mathbf{E} \times \mathbf{H}) = (1/2)(\hat{u}_\theta E_\theta \times \hat{u}_\phi H_\phi) = \hat{u}_r s(\theta)$, with a time-averaged radiated power density $s(\theta)$ given by

$$s(\theta) = |E_\theta|^2/(2\eta) = (\eta/2)|H_\phi|^2 = \frac{\eta|I(0)|^2 f(\theta)}{32\pi^2 r^2 \sin^2(kh)} \quad (3.2.14)$$

where

$$f(\theta) \equiv \frac{[\cos(kh\cos\theta) - \cos(kh)]^2 + [\sin(kh\cos\theta) - \cos\theta \sin(kh)]^2}{\sin^2\theta}$$

The direction of maximum radiated power is $\theta = \pi/2$ rad.

By substituting (3.2.14) into (2.2.4) and (2.2.5), the radiation resistance R, referred to the base current, is given for an infinitely thin element by

$$R = \frac{\eta \int_0^{\pi} f(\theta)\, \sin\theta \, d\theta}{8\pi \sin^2(kh)} \ (\Omega)$$

$$= \frac{\eta}{4\pi \sin^2(kh)} [\mathrm{Cin}(2kh) - \sin^2(kh)], \quad b = 0 \tag{3.2.15}$$

where Cin(z) is the modified cosine integral [19]:

$$\int_0^z \frac{dt}{t} (1 - \cos t)$$

The result given for the definite integral of (3.2.15) is readily obtained by letting $t = \cos\theta$, and noting that

$$\frac{1}{1 - t^2} = \frac{1}{2}\left(\frac{1}{x} + \frac{1}{y}\right)$$

where $x = 1 + t$ and $y = 1 - t$.

By substituting (3.2.14) and (3.2.15) into (2.2.8), the numeric directive gain $d(\theta)$ is given by

$$d(\theta) = \frac{f(\theta)}{\mathrm{Cin}(2kh) - \sin^2(kh)} \tag{3.2.16}$$

where $f(\theta)$ is as defined in (3.2.14).

For electrically short monopole elements ($kh \ll 1$) and for $\eta = 376.73 \approx 120\pi$ (Ω), (3.2.14) to (3.2.16) reduce to

$$f(\theta) = \frac{1}{4} (kh)^4 \sin^2\theta, \quad kh \ll 1 \tag{3.2.17}$$

$$s(\theta) = \frac{15}{16} \frac{|I(0)|^2 (kh)^2 \sin^2\theta}{\pi r^2}, \quad kh \ll 1 \tag{3.2.18}$$

$$R = 5(kh)^2 \ (\Omega), \quad kh \ll 1, \ b = 0 \tag{3.2.19}$$

$$d(\theta) = \frac{3}{2} \sin^2\theta, \quad kh \ll 1 \tag{3.2.20}$$

The relative power-radiation pattern $s(\theta)/s_{max} = s(\theta)/s(\pi/2)$, radiation resistance R, and the directive gain (directivity) $d(\theta)$ are tabulated in App. A.1, Tables A.1-1, A.1-14, A.1-26, and A.1-38 for $h/\lambda = 1/4, 1/10, 1/40$, and 1/100, respectively. The numeric directive gain pattern $d(\theta)$ for a quarter-wave monopole element (see Fig. 8(a) in Sec. 3.9) is similar to that of a half-wave dipole, except that its peak directivity is less (1.88 dBi versus 1.76 dBi) and its 3 dB beamwidth is more (94° versus 78°).

3.2.4 Input Impedance

The input impedance may be determined by the induced EMF method introduced by Brillouin [17]. With reference to Fig. 4, the input impedance Z_{in} is given by [20]:

$$Z_{in} = -(1/I_m) \int_0^h I_z(\rho = b, z = z') \, E_z(\rho = b, z = z') \, dz', \quad b > 0 \tag{3.2.21}$$

where

$$I_m = I(0)/\sin(kh)$$

The induced EMF method is indeterminate for collinear current elements unless the elements are of radius $b > 0$. For sufficiently thin elements, the current I_z is given by (3.1.1) and the electric near field E_z is given almost exactly by (3.2.6). By substituting (3.1.1) and (3.2.6) into (3.2.21), we have

$$Z_{in} = \frac{j\eta}{4\pi \sin(kh)} \int_0^h dz' \sin[k(h - z')] \left\{ \frac{\exp(-jkr_0)}{r_0} - \cos(kh) \frac{\exp(-jkr)}{r} + \sin(kh) \frac{\partial}{\partial z'} \left[\frac{\exp(-jkr)}{r} \right] \right\} \tag{3.2.22}$$

where

$$r_0 = [b^2 + (z' - h)^2]^{1/2}$$
$$r = (b^2 + z'^2)^{1/2}$$

Each of the three terms of the integrand of (3.2.22) may be integrated using the methods summarized by Stratton [18]. However, the third term of the integrand should first be integrated by parts. Accordingly,

$$Z_{in} = R_{in} + jX_{in} = Z_1 + Z_2 + Z_3 \tag{3.2.23}$$

where

$$\begin{aligned} Z_1 &= \frac{j\eta}{4\pi \sin^2(kh)} \int_0^h dz' \left\{ \sin[k(h - z)] \frac{\exp(-jkr_0)}{r_0} \right\} \\ &= R_1 + jX_1 \end{aligned} \tag{3.2.24}$$

$$R_1 = \frac{\eta}{4\pi \sin^2(kh)} \left[\frac{1}{2} \operatorname{Cin}(x_1) + \frac{1}{2} \operatorname{Cin}(x_2) - \operatorname{Cin}(x_3) \right] \tag{3.2.25}$$

$$X_1 = \frac{\eta}{4\pi \sin^2(kh)} \left[\frac{1}{2} \operatorname{Si}(x_1) + \frac{1}{2} \operatorname{Si}(x_2) - \operatorname{Si}(x_3) \right] \tag{3.2.26}$$

$$\begin{aligned} Z_2 &= \frac{j\eta}{4\pi \sin^2(kh)} \int_0^h -dz' \left\{ \sin[k(h - z')] \cos(kh) \frac{\exp(-jkr)}{r} \right\} \\ &= R_2 + jX_2 \end{aligned} \tag{3.2.27}$$

$$\begin{aligned} R_2 = \frac{\eta}{4\pi \sin^2(kh)} &\left\{ \frac{-\sin(2kh)}{4} [\operatorname{Si}(x_1) - \operatorname{Si}(x_2)] \right. \\ &\left. + \cos^2(kh) \left[-\operatorname{Cin}(x_3) + \frac{1}{2} \operatorname{Cin}(x_2) + \frac{1}{2} \operatorname{Cin}(x_1) \right] \right\} \end{aligned} \tag{3.2.28}$$

$$\begin{aligned} X_2 = \frac{\eta}{4\pi \sin^2(kh)} &\left\{ \frac{-\sin(2kh)}{4} [\operatorname{Cin}(x_2) - \operatorname{Cin}(x_1) + \ln(x_1/x_2)] \right. \\ &\left. + \frac{\cos^2(kh)}{2} [\operatorname{Si}(x_1) + \operatorname{Si}(x_2) - 2 \operatorname{Si}(x_3)] \right\} \end{aligned} \tag{3.2.29}$$

$$Z_3 = \frac{j\eta}{4\pi \sin^2(kh)} \int_0^h dz' \left\{ \sin[k(h - z')] \sin(kh) \frac{\partial}{\partial z'} \left[\frac{\exp(-jkr)}{r} \right] \right\}$$
$$= R_3 + jX_3 \tag{3.2.30}$$

$$R_3 = \frac{\eta}{4\pi \sin^2(kh)} \left\{ \frac{\sin^2(kh)}{2} \left[\text{Cin}(x_1) + \text{Cin}(x_2) - 2\,\text{Cin}(x_3) + \frac{2(x_1 + x_2)(\sin x_1 + \sin x_2)}{x_1^2 + x_2^2 + 2(kb)^2} - \frac{4x_3 \sin x_3}{x_3^2 + (kb)^2} \right] - \frac{\sin(2kh)}{4} \times \left[\text{Si}(x_2) - \text{Si}(x_1) - \frac{2(x_1 + x_2)(\cos x_1 - \cos x_2)}{x_1^2 + x_2^2 + 2(kb)^2} \right] \right\} \tag{3.2.31}$$

$$X_3 = \frac{\eta}{4\pi \sin^2(kh)} \left\{ \frac{\sin^2(kh)}{2} \left[\text{Si}(x_1) + \text{Si}(x_2) - 2\,\text{Si}(x_3) + \frac{2(x_1 + x_2)(\cos x_1 + \cos x_2)}{x_1^2 + x_2^2 + 2(kb)^2} - \frac{4x_3 \cos x_3}{x_3^2 + (kb)} \right] \right\} + \frac{\sin(2kh)}{4} \left[-\text{Cin}(x_1) + \text{Cin}(x_2) + \ln(x_1/x_2) - \frac{2(x_1 + x_2)(\sin x_1 - \sin x_2)}{x_1^2 + x_2^2 + 2(kb)^2} \right] \right\} \tag{3.2.32}$$

where

$$x_1 = k[(b^2 + h^2)^{1/2} + h]$$
$$x_2 = k[(b^2 + h^2)^{1/2} - h]$$
$$x_3 = kb$$

As before, $\text{Cin}(x)$ is the modified cosine integral [19]:

$$\int_0^x \frac{dt}{t} (1 - \cos t)$$

and Si(x) is the sine integral [19]:

$$\int_0^x \frac{\sin t}{t}\,dt$$

Then, summing (3.2.23) to (3.2.32), we have

$$\begin{aligned} R_{in} = \frac{\eta}{4\pi \sin^2(kh)} \Bigg\{ & \mathrm{Cin}(x_1) + \mathrm{Cin}(x_2) - 2\,\mathrm{Cin}(x_3) \\ & + \frac{\sin(2kh)}{2} \left[\frac{(x_1 + x_2)(\cos x_1 - \cos x_2)}{x_1^2 + x_2^2 + 2x_3^2} \right] \\ & + \sin^2(kh) \left[\frac{(x_1 + x_2)(\sin x_1 + \sin x_2)}{x_1^2 + x_2^2 + 2x_3^2} - \frac{\sin x_3}{x_3} \right] \Bigg\} \end{aligned} \qquad (3.2.33)$$

$$\begin{aligned} X_{in} = \frac{\eta}{4\pi \sin^2(kh)} \Bigg\{ & \mathrm{Si}(x_1) + \mathrm{Si}(x_2) - 2\,\mathrm{Si}(x_3) \\ & - \frac{\sin(2kh)}{2} \left[\frac{(x_1 + x_2)(\sin x_1 - \sin x_2)}{x_1^2 + x_2^2 + 2x_3^2} \right] \\ & + \sin^2(kh) \left[\frac{(x_1 + x_2)(\cos x_1 + \cos x_2)}{x_1^2 + x_2^2 + 2x_3^2} - \frac{\cos x_3}{x_3} \right] \Bigg\} \end{aligned} \qquad (3.2.34)$$

where

$$x_1 = k[(b^2 + h^2)^{1/2} + h], \quad x_2 = k[(b^2 + h^2)^{1/2} - h], \quad x_3 = kb$$

For $b = 0$, (3.2.33) and (3.2.34) reduce to

$$R_{in} = \frac{\eta}{4\pi \sin^2(kh)} [\mathrm{Cin}(2kh) - \sin^2(kh)], \quad b = 0 \qquad (3.2.35)$$

$$X_{in} = \frac{\eta}{4\pi \sin^2(kh)} \left[\mathrm{Si}(2kh) - \frac{\sin^2(kh)}{kb} \right], \quad b \to 0 \qquad (3.2.36)$$

Equation (3.2.35) agrees with the result for radiation resistance given by (3.2.15). A comparison of (3.2.33) with (3.2.35) reveals that the input resistance is relatively insensitive to the monopole element radius b for $kb \ll 1$. The input resistance given by (3.2.35) is plotted in Fig. 5. For $kh = \pi/2$ rad, $R_{in} = 19.4\ \Omega$.

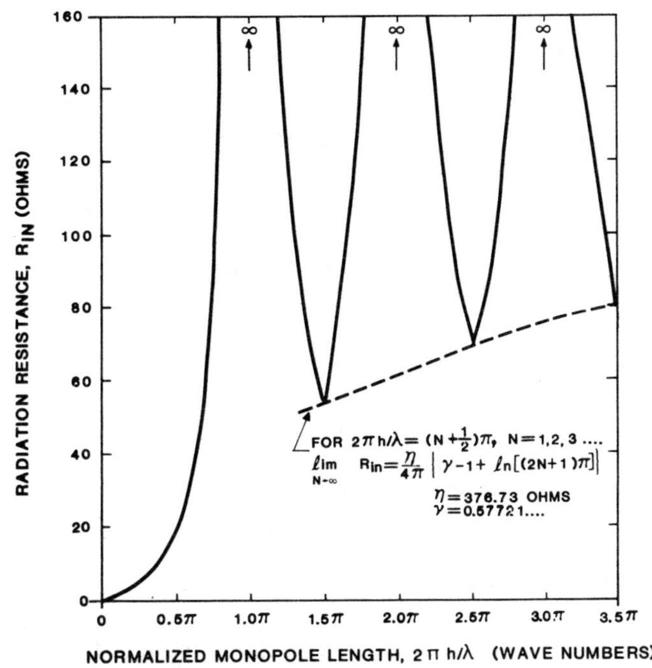

Fig. 5 Radiation Resistance of Electrically Thin ($kb \ll 1$) Monopole Elements

The input reactance is sensitive to the element radius b as can be seen in (3.2.36). For $b = 0$, the input reactance is given by

$$X_{in} = \begin{cases} -\infty;\ kh \neq n\pi,\ n = 1, 2, 3, \ldots;\ b = 0 \\ \infty;\ kh = n\pi,\ n = 1, 2, 3, \ldots;\ b = 0 \end{cases} \tag{3.2.37}$$

From (3.2.36), resonance ($X_{in} = 0$) occurs for

$$(kh)_{\text{resonance}} \approx n\pi \pm [kb\ \text{Si}(2n\pi)]^{1/2}, \quad n = 1, 2, 3, \ldots \tag{3.2.38}$$

The input resistance at resonance, for kh given by (3.2.38), is found from (3.2.35) to be

$$(R_{\text{in}})_{\text{resonance}} \approx \frac{\eta}{4\pi}\left[\frac{\text{Cin}(2n\pi)}{kb\ \text{Si}(2n\pi)} - 1\right] \tag{3.2.39}$$

Minima of input resistance occur for kh, given by

$$(kh)_{\text{minima}} \approx \left(N + \frac{1}{2}\right)\pi, \quad N = 1, 2, 3, \ldots \tag{3.2.40}$$

The input resistance at these minima element lengths is found from (3.2.35) to be

$$(R_{\text{in}})_{\text{minima}} \approx \frac{\eta}{4\pi}\,[\gamma - 1 + \ln[(2N + 1)\pi]] \tag{3.2.41}$$

where

$$\gamma = \text{Euler's constant} = 0.57721$$

The accuracy of (3.2.40) increases as $N \rightarrow \infty$.

3.2.5 Summary of Results

The input impedance and directive gain properties of quarter-wave and electrically short monopole elements with zero, large, and infinite extent ground planes are compared in Table 4 of Sec. 3.9. The peak directivity is approximately 3 dB *less* with ground planes of zero extent than with ground planes of large extent. However, the directive gain on the horizon is approximately 3 dB *more* with ground planes of zero extent than with ground planes of large, but finite, extent. The radiation resistance with ground planes of zero extent is approximately one-half that with ground planes of large extent.

Unlike dipole elements in free space where the first resonance occurs for dipole half-lengths that are approximately equal to a quarter-wavelength, monopole elements with ground planes of zero extent have a first resonance for an element length that is approximately equal to a half-wavelength.

3.3 INTEGRAL EQUATION: $0 \leq ka \leq 2.75$

The fields generated by the monopole element were determined in Sec. 3.2. These fields impinge on the ground plane and induce a ground-plane current. For sufficiently small ground planes, the fields generated by the ground-plane current may be determined by Bardeen's integral equation method [1].

In Bardeen's method, a cylindrically symmetrical electromagnetic wave (generated by the element) is incident on the ground-plane disk. The fields generated by the induced ground-plane current are required and the solution depends on solving an integral equation of the first kind. For arbitrary ground-plane radii, the integral equation cannot be readily solved because it contains two integrals. However, for sufficiently small ground-plane radii, Bardeen neglects one of the integrals so that the equation may be solved explicitly.

Although Bardeen gives a general formulation of the solution for the resulting single-integrand integral equation, his only explicit results are for the case in which the incident wave is generated by an infinitely thin dipole element having a base that is at a nonzero height above the center of the ground plane. Bardeen restricted his solution to elements at a nonzero height above the ground plane in order to avoid having a source point (the base of the element) located at a near-field point of interest, which, for the integral equation method, includes the center of the ground plane. The total field is then given by the vector sum of the incident and induced fields (see (3.1.12)).

The present case of interest is that of a monopole element having a base that is in the plane of the ground plane (i.e., at a vertical height $v = 0$ above the center of the ground plane (see Fig. 6). For this case, the total magnetic field may be determined by first evaluating the field with the element at an arbitrary height $v > 0$, and then by evaluating the resulting expression in the lim $v \rightarrow 0$. By using such a procedure, we avoid an indeterminate expression for the field generated by the induced ground-plane current.

Accordingly, the total magnetic field intensity $H_\phi^{(\text{total})}$ (see (3.1.12)), in the lim $v \rightarrow 0$, is given by

$$\lim_{v \rightarrow 0} H_\phi^{(\text{total})} = \lim_{v \rightarrow 0} H_\phi^{(e)} + \lim_{v \rightarrow 0} H_\phi^{(g)} \tag{3.3.1}$$

where $H_\phi^{(e)}$ and $H_\phi^{(g)}$ are the magnetic field intensities generated by the element and ground-plane currents, respectively. In the following evaluation of the two terms of (3.3.1), the fields are assumed to have an $e^{j\omega t}$ time dependence (see (2.1.2)), unlike the $e^{-j\omega t}$ time dependence assumed by Bardeen.

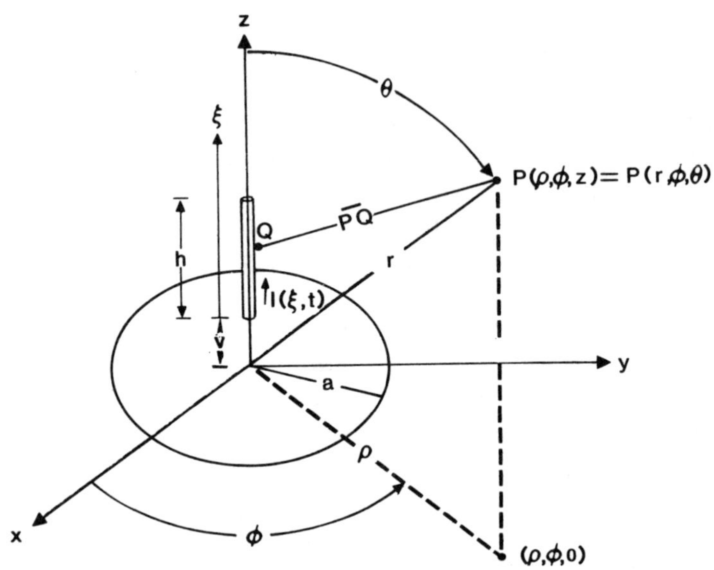

Fig. 6 Thin Monopole Element, the Base of which is above the Center of a Circular Ground Plane

The first term of (3.3.1) is given exactly, for an infinitely thin element, by (see (3.2.4)):

$$\lim_{\nu\to 0} H_\phi^{(e)} = \frac{jI(0)}{4\pi\rho\,\sin(kh)}\left[\exp\left(-jkr_0\right) - \cos(kh)\exp\left(-jkr\right) - \frac{jz}{r}\sin(kh)\exp\left(-jkr\right)\right] \tag{3.3.2}$$

In the far field, (3.3.2) reduces to (see (3.2.10)):

$$\lim_{\nu \to 0} H_\phi^{(e)} = A_0[\mathrm{j} \exp(\mathrm{j}kh \cos\theta) + \mathrm{j} \cos(kh) + \cos\theta \sin(kh)], \quad \text{(far field)} \tag{3.3.3}$$

where

$$A_0 = \frac{-I(0)\, \mathrm{e}^{-\mathrm{j}kr}}{4\pi r \sin\theta \sin(kh)}$$

The second term of (3.3.1) may be evaluated by utilizing Bardeen's equation [1, Eq. (31)] for $H_\phi^{(g)}$, which he obtained as a solution to his single-integrand integral equation. For an infinitely thin element and sufficiently small ground planes ($ka \lessapprox 1$), $H_\phi^{(g)}$ is given by Bardeen's equation [1, Eq. (31)] as

$$H_\phi^{(g)} = \mathrm{sgn}(z) \frac{-\mathrm{j}\omega\epsilon_0}{k\rho} \int_0^a F(s)K(s)\mathrm{d}s, \quad ka \lessapprox 1 \tag{3.3.4}$$

where

$$K(s) = \frac{\mathrm{e}^{-\mathrm{j}kA_1} \sinh(kB_1) \sinh(ks) - \mathrm{e}^{-\mathrm{j}kA_2} \sinh(kB_2) \sinh(ka)}{\sinh(ka)}$$

$$\mathrm{sgn}(z) = \begin{cases} +1, & z \geqslant 0 \\ -1, & z < 0 \end{cases}$$

ϵ_0 = permittivity of free space (F/m);

$$A_1 - \mathrm{j}B_1 = \{\rho^2 + [-\mathrm{j}a + z\, \mathrm{sgn}(z)]^2\}^{1/2} = [r^2 - \mathrm{sgn}(z)\, 2\mathrm{j}ra \cos\theta - a^2]^{1/2}$$

$$A_2 - \mathrm{j}B_2 = \{\rho^2 + [-\mathrm{j}s + z\, \mathrm{sgn}(z)]^2\}^{1/2} = [r^2 - \mathrm{sgn}(z)\, 2\mathrm{j}rs \cos\theta - s^2]^{1/2}$$

s = dummy variable having the dimension of length;

$F(s)$ = excitation function related to the radial electric near-field intensity $E_r^{(e)}|_{z=0}$, which is incident on the ground plane.

The second term of (3.3.1) is therefore given by

$$\lim_{\nu \to 0} H_\phi^{(g)} = \mathrm{sgn}(z) \frac{-\mathrm{j}\omega\epsilon_0}{k\rho} \lim_{\nu \to 0} \int_0^a F(s)K(s)\, \mathrm{d}s, \quad ka \lessapprox 1$$

$$\lim_{\nu\to 0} H_{\phi}^{(g)} = \text{sgn}(z)\,\frac{-\mathrm{j}\omega\epsilon_0}{k\rho}\int_0^a \left[\lim_{\nu\to 0} F(s)\right] K(s)\,\mathrm{d}s, \quad ka \lessdot 1 \tag{3.3.5}$$

The excitation function $F(s)$ is given by Bardeen [1, Eq. (35)] as

$$F(s) = -\frac{2}{\pi}\int_0^\infty ph(p)\,\frac{\sin[s(p^2 - k^2)^{1/2}]}{(p^2 - k^2)^{1/2}}\,\mathrm{d}p \tag{3.3.6}$$

where

p = dummy variable having the dimensions of $(\text{length})^{-1}$;
$h(p)$ = function of the radial electric near-field intensity $E_\rho^{(e)}|_{z=0}$, which is incident on the ground plane.

The function $h(p)$ is given by Bardeen [1, Eq. (33)] as

$$-E_\rho^{(e)}|_{z=0} = \int_0^\infty h(p)\,J_1(\rho p)\,\mathrm{d}p \tag{3.3.7}$$

where $J_1(x)$ = Bessel function of the first kind.

The electric field intensity $E_\rho^{(e)}|_{z=0}$, incident on the ground plane, is given by

$$E_\rho^{(e)}\Big|_{z=0} = \hat{u}_\rho \cdot \frac{[\nabla x(\nabla x\mathbf{A}^{(e)})]}{\mathrm{j}\omega\,\mu_0\epsilon_0}\Bigg|_{z=0} \tag{3.3.8}$$

where

$\mathbf{A}^{(e)}$ = magnetic vector potential generated by the monopole source points;
μ_0 = permeability of free space (H/m);
$\hat{u}_\rho$ = unit vector in the cylindrical radial direction.

The magnetic vector potential $\mathbf{A}^{(e)}$ for the element base at an arbitrary height v is given by [16]:

$$\mathbf{A}^{(e)} = \hat{u}_z\,(\mu_0/4\pi)\int_0^h I(\xi)(1/\overline{PQ})\exp(-\mathrm{j}k\overline{PQ})\,\mathrm{d}\xi \tag{3.3.9}$$

where

$$\begin{aligned}
\hat{u}_z &= \text{unit vector along the } z \text{ axis;}\\
\overline{PQ} &= \text{distance from source point } Q \text{ to field point } P;\\
&\approx [\rho^2 + (z - \xi - v)^2]^{1/2},\ Q(x', y', \xi) \approx Q(0, 0, \xi);\\
I(\xi) &= \text{monopole current distribution} = [I(0)/\sin(kh)]\sin[k(h - \xi)].
\end{aligned}$$

In order to facilitate the evaluation of $h(p)$ in (3.3.7), it is convenient to express the factor $(1/\overline{PQ})\exp(-jk\overline{PQ})$ in terms of Bessel functions. By using Sommerfeld's formula [21] and the dummy variable p introduced in (3.3.6), we write

$$\frac{e^{-jk\overline{PQ}}}{\overline{PQ}} = \int_0^\infty \frac{J_0(p\rho)\exp[-|z - \xi - v|\sqrt{p^2 - k^2}]\,p\,dp}{(p^2 - k^2)^{1/2}};$$
$$-(\pi/2) < \arg\sqrt{p^2 - k^2} \leqslant (\pi/2), \quad -\pi < \arg k \leqslant 0 \tag{3.3.10}$$

By substituting (3.3.10) into (3.3.9), we have

$$\begin{aligned}
\mathbf{A}^{(e)} &= \frac{\mu_0 I(0)}{4\pi\sin(kh)}\hat{u}_z \int_0^h d\xi \sin[k(\xi - h)]\\
&\times \int_0^\infty \frac{J_0(p\rho)\exp[-|z - \xi - v|\sqrt{p^2 - k^2}]}{(p^2 - k^2)^{1/2}}\,p\,dp;\\
&-(\pi/2) < \arg\sqrt{p^2 - k^2} \leqslant (\pi/2),\\
&-\pi < \arg k \leqslant 0
\end{aligned} \tag{3.3.11}$$

Because we are interested in evaluating $E_\rho^{(e)}|_{z=0}$, consider the case of $z < \xi + v$. By substituting (3.3.11) for such a case into (3.3.8), and noting that

$$\mathbf{A}^{(e)}\hat{u}_z = A_z \text{ and } \frac{\partial J_0(p\rho)}{\partial\rho} = -pJ_1(p\rho)$$

$$E_\rho^{(e)}\Big|_{z=0} = \frac{1}{-j\omega\mu_0\epsilon_0}\frac{\partial^2 A_z}{\partial z\partial\rho}\Big|_{z=0}$$

$$E_\rho^{(e)}\bigg|_{z=0} = \frac{I(0)}{-\mathrm{j}4\pi\omega\epsilon_0 \sin(kh)} \int_0^\infty p^2 J_1(p\rho) \times \exp\left[(z - v)\sqrt{p^2 - k^2}\right] B(p)\, \mathrm{d}p\bigg|_{z=0} \tag{3.3.12}$$

where

$$B(p) = \int_0^h \sin[k(h - \xi)] \exp\left[-\xi \sqrt{p^2 - k^2}\right] \mathrm{d}\xi = \frac{k \exp\left[-h\sqrt{p^2 - k^2}\right] + \sqrt{p^2 - k^2}\sin(kh) - k\cos(kh)}{p^2} \tag{3.3.13}$$

By substituting (3.3.13) into (3.3.12), and letting $z = 0$, we have

$$E_\rho^{(e)}|_{z=0} = \frac{I(0)}{-\mathrm{j}4\pi\omega\epsilon_0 \sin(kh)} \int_0^\infty \mathrm{d}p\{J_1(p\rho)\, \mathrm{e}^{-v\sqrt{p^2 - k^2}} \times [\mathrm{e}^{-h\sqrt{p^2-k^2}} + \sqrt{p^2 - k^2}\sin(kh) - k\cos(kh)]\} \tag{3.3.14}$$

Comparison of (3.3.14) with (3.3.7) yields

$$h(p) = \frac{I(0)}{\mathrm{j}4\pi\omega\epsilon_0 \sin(kh)} [k\mathrm{e}^{-(h+v)\sqrt{p^2-k^2}} + \sqrt{p^2 - k^2}\sin(kh)\, \mathrm{e}^{-v\sqrt{p^2-k^2}} - k\cos(kh)\, \mathrm{e}^{-v\sqrt{p^2-k^2}}] \tag{3.3.15}$$

By substituting (3.3.15) into (3.3.6), we have

$$F(s) = \frac{-2Ck}{\pi}\left[I_1 + \frac{\sin(kh)}{k} I_2 - \cos(kh)\, I_3\right] \tag{3.3.16}$$

where

$$C \equiv \frac{I(0)}{\mathrm{j}4\pi\omega\epsilon_0 \sin(kh)}$$

$$I_1 = \int_0^\infty \frac{p\ \mathrm{e}^{-(h+v)\sqrt{p^2-k^2}}\ \sin(s\sqrt{p^2 - k^2})\ \mathrm{d}p}{(p^2 - k^2)^{1/2}}$$

$$I_2 = \int_0^\infty p \sin(s\ \sqrt{p^2 - k^2})\ \mathrm{e}^{-v\sqrt{p^2-k^2}}\ \mathrm{d}p$$

$$I_3 = \int_0^\infty \frac{p \sin(s\sqrt{p^2 - k^2})\ \mathrm{e}^{-v\sqrt{p^2-k^2}}\ \mathrm{d}p}{(p^2 - k^2)^{1/2}}$$

We introduce in integrals I_1, I_2, and I_3 a change in variable p to y, given by

$$y = (p^2 - k^2)^{1/2}, \quad -(\pi/2) < \arg y \leqslant (\pi/2) \tag{3.3.17}$$

where the condition on y follows from the conditions of (3.3.10).

It follows (from Integral 2.663.1 [22]) that I_1 is given by

$$\begin{aligned} I_1 &= \int_{jk}^\infty \mathrm{e}^{-(h+v)y} \sin(sy)\ \mathrm{d}y \\ &= \frac{\mathrm{e}^{-(h+v)y}}{(h + v)^2 + s^2} [-(h + v) \sin(sy) - s \cos(sy)] \Big|_{jk}^{\infty} \\ &= \frac{-\mathrm{e}^{-k(h+v)}}{(v + h)^2 + s^2} [-\mathrm{j}(h + v) \sinh(ks) - s \cosh(ks)] \end{aligned} \tag{3.3.18}$$

It follows (from Integral 2.667.5 [22]) that I_2 is given by

$$\begin{aligned} I_2 &= \int_{jk}^\infty y \sin(sy)\ \mathrm{e}^{-vy}\ \mathrm{d}y = \frac{\mathrm{e}^{-vy}}{v^2 + s^2} \Bigg[\left(-vy + \frac{s^2 - v^2}{v^2 + s^2}\right) \sin(sy) \\ &\quad - \left(sy + \frac{2vs}{s^2 + v^2}\right) \cos(sy)\Bigg] \Bigg|_{jk}^{\infty} \\ &= \frac{-\mathrm{e}^{-\mathrm{j}kv}}{s^2 + v^2} \Bigg[kv \sinh(ks) - \mathrm{j}ks \cosh(ks) \\ &\quad - \frac{\mathrm{j}(v^2 - s^2) \sinh(ks) - 2\ vs \cosh(ks)}{s^2 + v^2}\Bigg] \end{aligned} \tag{3.2.19}$$

It follows (from Integral 2.663.1 [22]) that I_3 is given by

$$
\begin{aligned}
I_3 &= \int_{jk}^{\infty} \sin(sy)\, e^{-vy}\, dy \\
&= e^{-vy} \left[\frac{-v \sin(sy) - s \cos(sy)}{s^2 + v^2} \right] \Bigg|_{jk}^{\infty} \\
&= -e^{-jkv} \left[\frac{-jv \sinh(ks) - s \cosh(ks)}{s^2 + v^2} \right]
\end{aligned}
\tag{3.3.20}
$$

Consider now the limits of I_1, I_2, and I_3 as $v \to 0$. Thus,

$$
\lim_{v \to 0} I_1 = \frac{e^{-jkh}}{h^2 + s^2} [-jh \sinh(ks) - s \cosh(ks)] \tag{3.3.21}
$$

$$
\begin{aligned}
\lim_{v \to 0} I_2 = & -\pi k \delta(s) \sinh(ks) + \frac{jk \cosh(ks)}{s} \\
& + j\pi^2 \delta^2(s) \sinh(ks) - \frac{j \sinh(ks)}{s^2} \\
& - \pi \delta'(s) \cosh(ks)
\end{aligned}
\tag{3.3.22}
$$

$$
\lim_{v \to 0} I_3 = j\pi \delta(s) \sinh(ks) + \frac{\cosh(ks)}{s} \tag{3.3.23}
$$

where [23]

$$
\delta(x) = \text{Dirac delta function} = \lim_{v \to 0} \left[\frac{1}{\pi} \frac{v}{v^2 + x^2} \right]
$$

$$
\int_a^b \delta(x - x_0)\, dx = \begin{cases} 1, & a < x_0 < b \\ 1/2, & a = x_0 < b \text{ or } a < x = b \\ 0, & x_0 < a \text{ or } x_0 > b \end{cases}
$$

$$
\delta'(x) = \frac{d}{dx} [\delta(x)] = \lim_{v \to 0} \frac{\partial}{\partial x} \left[\frac{1}{\pi} \frac{v}{v^2 + x^2} \right] = \lim_{v \to 0} \left[-\frac{1}{\pi} \frac{2vx}{(v^2 + x^2)^2} \right]
$$

In (3.3.22) and (3.3.23), those terms containing the product $\delta(s)\sinh(ks)$ may be set equal to zero because (from Eq. (B.2-12) [23]) $x\delta(x) = 0$.

The lim $(v \rightarrow 0)$ $F(s)$ is found by substituting (3.3.21) to (3.3.23) into (3.3.16). Accordingly,

$$\lim_{v\to 0} F(s) = \frac{2Ck}{\pi}\frac{e^{-jkh}}{h^2+s^2}[-jh\sinh(ks) - s\cosh(ks)] + \frac{2C\sin(kh)}{\pi}\left[\frac{-jk\cosh(ks)}{s} - \frac{j\sinh(ks)}{s^2} + \pi\delta'(s)\cosh(ks)\right] + \frac{2Ck\cos(kh)}{\pi}\frac{\cosh(ks)}{s} \tag{3.3.24}$$

The magnetic field intensity generated by the ground-plane current, for the case $v \rightarrow 0$, is found by substituting (3.3.24) into (3.3.5). The total magnetic field is explicitly determined by numerical evaluation of (3.3.1) to (3.3.5).

In the far field, the factor $K(s)$ and the parameters A_1, A_2, B_1, and B_2 in (3.3.4) and (3.3.5) reduce to a simpler form.

In the far field, when $r \rightarrow \infty$, then $a \ll r$ and $s \ll r$. For these conditions, we write

$$A_1 - jB_1 \approx [r^2 - \operatorname{sgn}(z)\, 2jra\cos\theta - a^2\cos^2\theta]^{1/2} = r - \operatorname{sgn}(z)\, ja\cos\theta,\ a \ll r \tag{3.3.25}$$

$$A_2 - jB_2 \approx [r^2 - \operatorname{sgn}(z)\, 2jra\cos\theta - s^2\cos^2\theta]^{1/2} = r - \operatorname{sgn}(z)\, js\cos\theta,\ s \ll r \tag{3.3.26}$$

If we equate the real and imaginary parts in (3.3.25) and (3.3.26), then

$$A_1 = A_2 = r, \quad a \ll r,\ s \ll r \tag{3.3.27}$$

$$B_1 = \operatorname{sgn}(z)\, a\cos\theta, \quad a \ll \rho \tag{3.3.28}$$

$$B_2 = \operatorname{sgn}(z)\, s\cos\theta, \quad s \ll \rho \tag{3.3.29}$$

By substituting (3.3.27), (3.3.28), and (3.3.29) into (3.3.4) and (3.3.5), the magnetic far-field intensity due to the ground-plane current, for the case $v \to 0$, is given by

$$\lim_{v\to 0} H_\phi^{(g)} = \frac{-j\omega\epsilon_0 \exp(-jkr)}{kr \sin\theta} \int_0^a \left[\lim_{v\to 0} F(s)\right] g(s)\, ds$$

$$\text{far field, } ka \lesssim 1 \qquad (3.3.30)$$

where

$$g(s) = \frac{\sinh(ka \cos\theta)\sinh(ks) - \sinh(ks\cos\theta)\sinh(ka)}{\sinh(ka)}$$

and lim ($v \to 0$) $F(s)$ is given by (3.3.24).

Equation (3.3.30) can be further reduced by utilizing the properties of the Dirac delta function (from Eq. (B.2-7) [23]) and the relation:

$$\int_0^a \delta'(x)\, f(x)\, dx = -\frac{1}{2} f'(0^+)$$

It follows that (3.3.30) reduces to

$$\lim_{v\to 0} H_\phi^{(g)} = \frac{2A_0}{\pi} \int_0^a ds\, g(s) \left\{ \frac{e^{-jkh}}{h^2 + s^2} [-jh \sinh(ks) - s\cosh(ks)] + \sin(kh) \frac{-j\cosh(ks)}{s} + \frac{j}{ks^2}\sinh(ks) + \frac{\cos(kh)\cosh(ks)}{s} \right\}$$

$$- A_0 \sin(kh) \left[\frac{\sinh(ka\cos\theta) - \cos\theta \sinh(ka)}{\sinh(ka)} \right];$$

$$\text{far field, } ka \lesssim 1 \qquad (3.3.31)$$

where A_0 and $g(s)$ are defined in (3.3.3) and (3.3.30), respectively.

The total magnetic far-field intensity generated by the element and ground-plane currents, for the case $v \to 0$, is found by substituting (3.3.3) and (3.3.31) into (3.3.1). The resulting expression for lim ($v \to 0$) $H_\phi^{(\text{total})}$ is in a form suitable for computer evaluation.

The radiated power density $s(\theta)$, directive gain $d(\theta)$, and radiation resistance R are found from (2.2.2) and (2.2.5). These quantities are computed in program *BARDEEN* of App. B.1. Computer printouts of the relative power radiation pattern $s(\theta)/s_{max}$, radiation resistance R, and the directive gain $d(\theta)$ are given in Tables A1-1–A1-13, A1-14–A1-25, A1-26–A1-37, A1-38–A1-49 for normalized element lengths $h/\lambda = 0.25$, 0.1, 0.040, 0.01, respectively, and normalized ground-plane radii = 0, 0.25, 0.50, . . . , 3.0 wavenumbers.

The radiation patterns have no appreciable change in shape for ground-plane radii $0 \leq ka \leq 1.75$, and resemble that of a dipole in free space, with peak gains approximately in the direction of the horizon and directivities less than that of a dipole for which the total element length is twice that of the monopole element. For a quarter-wave monopole element, the directive gain on the horizon deceases from 1.88 dBi for $ka = 0$ to 1.23 dBi for $ka = 1.75$.

In Table 5 of Sec. 3.9, the radiation resistances obtained by the integral equation method are compared with those from the method of moments for a quarter-wave element and ground-plane radii $0 \leq ka \leq 3.0$. The values obtained with the integral equation method differ from those obtained using the method of moments by less than 1% for $0 \leq ka \leq 1.75$ and by less than 10% for $0 \leq ka \leq 2.75$. These results suggest that the integral equation method is accurate for ground-plane radii $0 \leq ka \leq 1.75$, and also that it is useful for $0 \leq ka \leq 2.75$. We suspect that the best available results are obtained by using the integral equation method for $0 \leq ka \leq 1.0$.

3.4 METHOD OF MOMENTS: $0 < ka \leq 14$

We saw in Sec. 3.3 that the integral equation method, which is used for determining the fields generated by the ground-plane current when the current distribution on the monopole element is initially specified, is accurate only for ground-plane radii of less than approximately 1.75 wavenumbers. We will show in Sec. 3.5 that the algorithms utilized in the oblate spheroidal wave-function method are accurate only for ground-plane radii which are no smaller than 3.0 wavenumbers. A question thus arises. What method is accurate over a range of ground-plane radii that includes the region $1.75 < ka < 3.0$?

We have found that Richmond's method of moments [2] is the only method in the currently available literature to be accurate over a range of ground-plane radii that includes the region $1.75 < ka < 3.0$. Although this method is primarily intended to be used when the current distribution on the monopole element is initially unknown, it is also applicable when the element is specified to have a sinusoidal current distribution.

Richmond's method of moments is discussed in Sec. 4.2. In this method, the element is subdivided into N equal segments and the ground plane is subdivided into M concentric annular zones of equal width. The unknown current distributions on the element and ground plane are expanded as a series of $N+M$ overlapping sinusoidal dipole modes (sinusoidal-Galerkin method), each with an unknown current distribution. The $N+M$ currents are determined by the Crout substitution method of solution of a $(N + M) \times (N + M)$ matrix. The numbers of subdivisions, N and M, are limited by the cost of computation time and the precision of the computer. The accuracy of the solution can be appreciably decreased if either N or M is too small or too large. The method of moments converges to a solution when an increase or decrease of unity in the value of N or M does not appreciably alter the solution for input impedance.

A particularly useful property of the sinusoidal-Galerkin method is the sinusoidal current distribution that is imposed on the element by setting $N = 1$.

The input impedance and radiation pattern of thin quarter-wave elements ($b/\lambda = 10^{-6}$, $h/\lambda = 0.25$), for ground-plane radii $0 < ka \leqslant 14$ wavenumbers, were determined by Richmond's method of moments utilizing MITRE's programs *RICHMD1* and *RICHMD2*, which are discussed in Sec. 4.2 and listed in App. B.5. With $N = 1$, convergent solutions were obtained for values of M given by

$M = 3$, $ka = 0.25$
$M = 7$, $ka = 0.50$
$M = 16$, $ka = 0.75, 1.0, \ldots, 8.5$
$M = 3\,ka$, $ka = 8.75, 9.0, \ldots, 14.0$

Computer printouts of the input admittance, ground-plane current distribution, radiation resistance, directive gain, and radiation patterns are given in App. A.5.

The input resistance and reactance, as determined by the method of moments, or other methods which give similar results, are plotted for thin quarter-wave elements on ground planes of radii $0 \leqslant ka \leqslant 14$ in Figs. 9

and 10 of Sec. 3.9. The directive gain on the horizon, peak directivity, and elevation angle of the peak directivity, given by the computer printouts of App. A.5, are plotted in Figs. 11–13 of Sec. 3.9.

The input impedance of thin elements with a sinusoidal current distribution were also determined for element lengths $h/\lambda = 0.1$ and 0.025 and ground-plane radii $0 < ka \leqslant 8.0$ wavenumbers by utilizing program RICHMD1 with $N = 1$. The radiation resistance of these elements is compared in Fig. 14 of Sec. 3.9 with those of a quarter-wave element. In Fig. 14, the radiation resistance is normalized to the value of radiation resistance of each element for $ka = 0$.

Values of the radiation resistance for quarter-wave elements are determined in App. A.5 by computing the input impedance (program RICHMD1) and also by the far-field radiation pattern (program RICHMD2). The values determined from the two methods differ by less than 1% for small ground-plane radii and less than 3% for the larger ground-plane radii.

The values of radiation resistance obtained from program RICHMD2 are compared in Table 5 of Sec. 3.9 with the values obtained by the integral equation method and the oblate spheroidal wave-function method. Richmond's method of moments gives useful results over the entire range of $0 < ka \leqslant 14$, and it provides good agreement with the integral equation method for $0 < ka \leqslant 1.75$ and the oblate spheroidal wave-function method for $3.0 \leqslant ka \leqslant 6.5$. Whereas the method of moments gives useful results in the regions $1.75 < ka < 3.0$ and $6.5 < ka \leqslant 14$, the other methods fail in these regions. For $ka > 14$, Richmond's method of moments is not as useful because of increased computation time and decreased accuracy. We suspect that for thin monopole elements the best available results are obtained by the method of moments for $1.25 \leqslant ka \leqslant 2.75$ and $6.75 \leqslant ka \leqslant 14$. For relatively thick monopole elements, Richmond's method of moments gives the best available results for ground-plane radii $0 < ka \leqslant 14$, as discussed in Sec. 4.2.

3.5 OBLATE SPHEROIDAL WAVE FUNCTIONS: $3.0 \leqslant ka \leqslant 6.5$

Oblate spheroidal coordinates (see Table 1 in Sec. 2.1) are particularly convenient for handling the boundary conditions of the magnetic field intensity on the ground plane. The requirement for constant tangential magnetic field intensity across the upper and lower hemispheres at the groundplane interface may be specified at all points of the ground-plane disk by a boundary condition at only the oblate *radial* coordinate $\xi = 0$.

For ground-plane radii of the same order of magnitude as the excitation wavelength, Leitner and Spence [3–5] utilized oblate spheroidal wave functions to determine the ground-plane current induced by a thin quarter-wave element with a sinusoidal current distribution. Leitner and Spence give numerical values of the ground-plane current distribution, radiation resistance, and far-field power density (at constant element base current) for ground-plane radii $ka = 3, 4, 5$, and $\sqrt{42}$ wavenumbers. The complex current distributions are reported on both the top and bottom faces of the ground plane.

In this section, we report the results of a computer program *MONOPL,* based on the theory of Leitner and Spence, which calculates the directivity pattern and radiation resistance for ground-plane radii including the cases of $ka = 3, 4, 5$, and $\sqrt{42}$ wavenumbers. Our results are consistent with, but not identical to, the results reported by Leitner and Spence. The form of the solution, corrections, regions of validity for the calculations, accuracy, and numerical results of program MONOPL are discussed in the remainder of this section.

3.5.1 Form of Solution

For a quarter-wave monopole ($h = \lambda/4$), the far-field power density $s'(\eta)$ and radiation resistance R, expressed in oblate spheroidal wave functions, are given with reference to Fig. 1 of Sec. 2.1, by [5]:

$$s'(\eta) = [I^2(0)/8\pi^2 r^2]\,(\mu_0/\epsilon_0)^{1/2}\,\epsilon^4 \left| \sum_{l=1}^{\infty} B_l u_{l1}(\eta) \right|^2 \qquad (3.5.1)$$

$$R = (1/2\pi)\,(\mu_0/\epsilon_0)^{1/2}\,\epsilon^4 \sum_{l=1}^{\infty} N_{l1}\,|B_l|^2 \qquad (3.5.2)$$

where

$(\mu_0/\epsilon_0)^{1/2}$ = wave impedance in free space (Ω);
$\epsilon \equiv ka = 2\pi a/\lambda$ = ground-plane radius (wavenumbers);
$u_{l1}(\eta)$ = angular oblate spheroidal wave function of order 1 and degree l;

$$N_{l1} = \int_{-1}^{1} u_{l1}^2(\eta)\, d\eta = \text{norm of } u_{l1}(\eta);$$

$$_2w_{l1}(\xi) = (1 + \xi^2)^{1/2}\, {}_2v_{l1}(\xi)$$

$$B_l = (-1)^{l-1}\, {}_2w_{l1}(\pi/2\epsilon)/\alpha_{l1} N_{l1} q_{l1}\, {}^{(3)}v'_{l1}(0)$$

The reader is referred to Leitner and Spence [3] for a definition of the oblate spheroidal wave functions α_{l1}, q_{l1}, ${}^{(3)}v'_{l1}(\xi)$, and ${}_2v_{l1}(\xi)$.

Substitution of (3.5.1) and (3.5.2) into (2.2.4) yields the directivity $d'(\eta)$, given by

$$d'(\eta) = 2 \left| \sum_{l=1}^{\infty} B_l u_{l1}(\eta) \right|^2 \Big/ \sum_{l=1}^{\infty} N_{l1} \left| B_l \right|^2 \tag{3.5.3}$$

3.5.2 Corrections

Problems were encountered in generating the eigenvalues of the oblate spheroidal wave functions. These problems were narrowed down to the continued fractions that generate the eigenvalues. Said continued fractions were derived from the recurrence relations, which, in turn, come from the differential equation. A discrepancy was discovered between our derived continued fractions and the published continued fractions [3]. The corrected continued fractions appear below.

For $l - m$ even:

$$\gamma_{lm} = \cfrac{-(l+m)(l-m)\ \epsilon^2}{2(2l-1)+\gamma_{lm} + \cfrac{(l+m-2)(l-m-2)\ \epsilon^2}{4(2l-3)+\gamma_{lm} + \cfrac{(l+m-4)(l-m-4)\ \epsilon^2}{6(2l-5)+\gamma_{lm} + \cdots}}}$$

$$+ \cfrac{(l+m+2)(l-m+2)\ \epsilon^2}{2(2l+3)-\gamma_{lm} + \cfrac{(l+m+4)(l-m-4)\ \epsilon^2}{4(2l+5)-\gamma_{lm} + \cfrac{(l+m+6)(l-m+6)\ \epsilon^2}{6(2l+7)-\gamma_{lm} + \cdots}}} \tag{3.5.4}$$

For l − m odd:

$$\gamma_{lm} = \cfrac{-(l+m-1)(l-m-1)\,\epsilon^2}{2(2l-1)+\gamma_{lm} + \cfrac{(l+m-3)(l-m-3)\,\epsilon^2}{4(2l-3)+\gamma_{lm} + \cfrac{(l+m-5)(l-m-5)\,\epsilon^2}{6(2l-5)+\gamma_{lm} + }}}$$

$$+ \cfrac{(l+m+1)(l-m+1)\,\epsilon^2}{2(2l+3)-\gamma_{lm} + \cfrac{(l+m+3)(l-m+3)\,\epsilon^2}{4(2l+5)-\gamma_{lm} + \cfrac{(l+m+5)(l-m+5)\,\epsilon^2}{6(2l+7)-\gamma_{lm} + }}} \tag{3.5.5}$$

The eigenvalues can also be expressed in terms of a series expansion, which has the form:

$$\gamma_{lm} = \sum_{k=1}^{\infty} f_{2k}^{lm}\, \epsilon^{2k} \tag{3.5.6}$$

The first two terms in this expansion were checked against the continued fractions and found to agree. This is important because the continued fraction method by which values of the eigenvalues are obtained depends on the accuracy of the roots γ_{lm} in (3.5.6).

3.5.3 Regions of Calculation Validity

A lower bound on the value of $\epsilon = 2\pi a/\lambda$ for which the calculations are valid depends on the following equations (see Eqs. (34) and (48), [3]):

$$^{(1)}v_{l1}(\xi) = (1 + \xi^2)^{1/2} \sum_{n=0,1}^{\infty} a_n^{l1}\, \xi^n, \quad \text{for } \xi = \pi/2\epsilon \tag{3.5.7}$$

and

$$g_{l1}(\xi) = (1 + \xi^2)^{-1/2} \sum_{n=0,1}^{\infty}{}' \, b_n^{l1} \, \xi^n, \quad \text{for } \xi = \pi/2\epsilon \tag{3.5.8}$$

where the prime indicates summation over alternate n, starting with $n = 0$ if $(l - 1)$ is even, and $n = 1$ if $(l - 1)$ is odd.

Equation (3.5.7) is the expression for the first-order radial spheroidal wave function of the first kind. Equation (3.5.8) is used in another expression to obtain the first-order radial spheroidal wave function of the second kind. As ϵ becomes small, ξ becomes large, so more terms are needed in (3.5.7) and (3.5.8). In theory, (3.5.7) and (3.5.8) converge for all real values of $\xi = \pi/2\epsilon$. Computationally, however, because of the finite accuracy of the computer, (3.5.7) and (3.5.8) will not converge for all real values of $\xi = \pi/2\epsilon$. To be cautious, we can restrict ξ so that it is less than unity. We obtain from this assumption the computational constraint $\xi = \pi/2\epsilon < 1$, which implies a lower bound on ϵ, given by $\epsilon > \pi/2 \approx 1.57$.

An even tighter lower bound on ϵ is obtained by observing what happens in the algorithm that is used to derive the eigenvalues. However, this algorithm is not well behaved for values of ϵ less than $\epsilon = 2.5$. Therefore, a lower bound for ϵ is given by $\epsilon \geqslant 2.5$. For accurate values of radiation resistance, a lower bound for ϵ would be $\epsilon \geqslant 3.0$ (see Table 5 of Sec. 3.9).

An upper bound on ϵ is obtained by observing what happens in the continued fraction algorithm that is used to derive the eigenvalues. For values of ϵ greater than 6.5, the series expansion for the eigenvalues, given by (3.5.6), does not give a sufficiently accurate answer by using only the first four terms. The resulting eigenvalues are so far from the correct eigenvalues as to cause the continued fraction algorithm of (3.5.4) and (3.5.5) to converge to a root that is not the correct value. As a consequence, erroneous values of radiation resistance can be obtained for $\epsilon > 6.5$ (see Table 5 of Sec. 3.9).

Consequently, the useful range of ϵ for Leitner and Spence's method of oblate spheroidal wave functions is $2.5 \leqslant \epsilon \leqslant 6.5$. We suspect that the best available results are obtained by the oblate spheroidal wave-function method for $3.0 \leqslant \epsilon \leqslant 6.5$.

Suggestions for extending the method of oblate spheroidal wave functions to the entire range $0 \leqslant \epsilon < \infty$ are as follows.

$0 \leqslant \epsilon < 3.0$

In determining the eigenvalues γ_{lm}, bypass the subroutine IMPROV. In determining the radial eigenfunctions ${}_2w_{11}(\xi)$, substitute (65) and (66) for (34) and (48) of [3]. Program MONOPL now utilizes (34) and (48) of [3].

$3 \leqslant \epsilon \leqslant 6.5$

Use Program MONOPL in its present form.

$6.5 < \epsilon < \infty$

In determining the eigenvalue γ_{lm}, use the asymptotic expansion given by (21.8.2) of [19] to obtain a raw value and then use subroutine IMPROV to obtain a refined value. In determining the eigenfunctions, use the present routines in Program MONOPL.

3.5.4 Accuracy

The calculation of directivity and radiation resistance is comprised of many equations. Some of these equations involve series expansions. When we varied the number of terms in these series, the radiation resistance was found to vary only in the fifth or sixth significant term.

Another problem mentioned previously was the accuracy by which (3.5.6) computes the eigenvalues. Because of computational reasons, we used only the first four terms in (3.5.6). As a result, a raw eigenvalue is computed by using (3.5.6), which is then used as an initial guess in the continued fractions. The continued fractions have many roots in γ_{lm}. The number of roots is dependent on the number of fractional terms used in the continued fraction. However, only one of these roots can be the eigenvalue. If the raw eigenvalue is sufficiently far from the correct eigenvalue, then the continued fraction will converge to a root that is not an eigenvalue. This will produce incorrect results.

The computed eigenvalues were checked against published values [24]. The computed values were found to be within the percentage error of the published values.

The computed eigenvalues of directivity on the horizon and radiation resistance, for different values of ground-plane extent, were found to agree with the relationship given by (2.3.3) to at least five decimal places with the free-space wave impedance $\eta = 376.73037\ \Omega$.

3.5.5 Numerical Results

The far-field power density $s'(\eta)$, radiation resistance R, and directive gain $d'(\eta)$ of quarter-wave elements given by (3.5.1), (3.5.2), and (3.5.3), respectively, were numerically evaluated by program MONOPL written in FORTRAN 77 language for use on a DEC PDP-11/70 computer. The program listing is given in App. B.3. Numerical values were obtained for the cases of $2.5 \leq ka \leq \sqrt{42}$.

The computed eigenvalues are given in Table 2. Computer printouts of the directivity patterns are given in App. A.3 for $ka = 3, 4, 5$, and $\sqrt{42}$. The patterns are plotted in Fig. 8 of Sec. 3.9 as polar graphs on the same linear scale. The total radiated power is held constant in these plots.

The directional gain on the horizon, radiation resistance, and peak directivity are summarized in Table 3.

Table 2 Eigenvalues γ_{l1} for $2\pi a/\lambda = 3, 4, 5, \sqrt{42}$

	γ_{l1}			
l	$2\pi a/\lambda = \epsilon = 3$	$\epsilon = 4$	$\epsilon = 5$	$\epsilon = \sqrt{42}$
1	6.7305791912	11.0907992567	15.056611686	21.687439619
2	4.8215414322	8.1301387077	11.872162474	17.738230003
3	5.0070741733	9.4059031099	15.750367134	28.127723151
4	4.7772522576	8.6287184908	13.694959245	23.237303490
5	4.6896340053	8.4462229270	13.439315734	23.585994650
6	4.6368101196	8.3210439564	13.161309568	22.624607030
7	4.6032991570	8.2427831207	12.999910517	22.234563635
8	4.5807102835	8.1899480807	12.890952874	21.960565094
9	4.5647761329	8.1526212369	12.814155253	21.770549914
10	4.5531234272	8.1252813999	12.757940027	21.632114595

Table 3 Directive Gain on Horizon and Radiation Resistance of Quarter-Wave Element, $2.5 \leq 2\pi a/\lambda \leq \sqrt{42}$

Ground-Plane Radius, $\epsilon = 2\pi a/\lambda$ (Wave-numbers)	*Directive Gain on Horizon*		*Radiation Resistance, R* (Ω)		*Peak Directivity*	
	$d(\pi/2)$ (numeric)	$D(\pi/2) = 10 \log_{10} d(\pi/2)$ (dBi)	Program MONOPL	Leitner*	D_{peak} (dBi)	θ_{peak} (degrees)
2.5	1.09745	0.40383	27.31731			
3.0	0.79995	−0.96938	37.47646	39.32	2.52250	46
3.4	0.66233	−1.78923	45.26299			
3.5	0.65086	−1.86515	46.06128			
3.6	0.64812	−1.88346	46.25589			
3.7	0.65316	−1.84980	45.89866			
4.0	0.70254	−1.53326	42.67237	42.10	3.89943	40
4.5	0.82743	−0.82271	36.23191			
5.0	0.92059	−0.35934	32.56526	31.87	3.37175	36
5.25	0.93879	−0.27430	31.93379			
5.5	0.93294	−0.30146	32.13410			
6.0	0.85088	−0.70134	35.23337			
$\sqrt{42}$	0.74637	−1.27045	40.16661	39.88	2.55213	56

*A. Leitner, Research Report No. EM-19, New York University, New York, April 1950.

3.6 SCALAR DIFFRACTION THEORY AND GEOMETRIC THEORY OF DIFFRACTION: $6.5 \leqslant ka < \infty$

For an element on a ground plane of sufficiently large radius, Tang [6] utilized a scalar theory of diffraction to calculate the far-field elevation pattern. For elevation angles near the horizon ($\theta \approx \pi/2$), the magnetic far-field intensity is determined by linear extrapolation to the result for the element itself given by (3.2.10) with $\theta = \pi/2$ rad. Tang's method for the radiation pattern is more accurate than that obtained by the variational method of Sec. 3.7 because it includes an additional term in the expansion for the total magnetic field intensity. Because the variational method is useful for ground-plane radii as small as $ka = 30$ wavenumbers, Tang's method should be useful for even smaller ground-plane radii, provided that $ka \gg 1$.

The geometric theory of diffraction (GTD) is another method applicable for sufficiently large ground-plane radii. In GTD, the fields are determined by ray optics (an incident ray plus a reflected ray) and diffraction by the edge. However, the effect of edge diffraction is only approximated in GTD because the edge is treated point by point as though it were a straight knife-edge of infinite extent. Therefore, GTD may be applied to an element at the center of a circular ground plane only when said ground plane is of sufficiently large radius. The GTD method is reviewed by Balanis [27], who also gives a computer program for calculating the diffraction coefficient.

In Sec. 4.3, the method of moments combined with GTD yields results for input impedance that are useful for ground planes of radii $ka \geqslant 6$ wavenumbers and accurate for $ka > 8$. Therefore, when the element current distribution is constrained to be sinusoidal, the GTD method is expected to give useful results for ground-plane radii $ka \geqslant 6.5$ wavenumbers, over which range the oblate spheroidal wave-function method does not give useful results.

3.7 VARIATIONAL METHOD: $30 \leqslant ka < \infty$

For an element on a ground plane of very large radius ($ka \geqslant 30$ wavenumbers), differences in the input impedance and radiation pattern from the values for a ground plane of infinite extent may be determined by utilizing a variational method of Storer [7, 8].

With reference to Fig. 1 of Sec. 2.1, for a sinusoidal current distribution on the element as given by (3.1.1), the input impedance difference is (from Eq. (20) [7]):

$$Z_{in} - Z_\infty = \frac{-j\eta \exp(j2ka)}{4\pi ka}\left[\frac{1 - \cos(kh)}{\sin(kh)}\right]^2 \times \left\{1 + \frac{\exp[j(2ka + 3\pi/4)]}{(4\pi ka)^{1/2}}\right\}^{-1}, \quad ka \geqslant 30 \tag{3.7.1}$$

where

Z_{in} = input impedance for an element on a ground plane of radius ka wavenumbers (Ω);
Z_∞ = input impedance for the same element on a ground plane of radius $ka = \infty$ wavenumbers (Ω);
η = wave impedance in free space = 376.73037 Ω.

Because $(4\pi ka)^{1/2} \gg 1$ for $ka \geqslant 30$, the difference of input resistance $R_{in} - R_\infty$ and input reactance $X_{in} - X_\infty$ are given approximately by

$$R_{in} - R_\infty \approx \left[\frac{1 - \cos(kh)}{\sin(kh)}\right]^2 \frac{\sin(2ka)}{4\pi ka}, \quad ka \geqslant 30 \tag{3.7.2}$$

$$X_{in} - X_\infty \approx \left[\frac{1 - \cos(kh)}{\sin(kh)}\right]^2 \frac{\cos(2ka)}{4\pi ka}, \quad ka \geqslant 30 \tag{3.7.3}$$

The maxima, nulls, and minima of $R_{in} - R$ occur for values of ka given approximately by

$$ka = \begin{cases} \left(2N + \frac{1}{2}\right)(\pi/2) \text{ (maxima)}, & N = 0, 1, 2, \ldots \\ N\pi/2 \text{ (nulls)}, & N = 0, 1, 2, \ldots \\ \left(2N + \frac{3}{2}\right)(\pi/2) \text{ (minima)}, & N = 0, 1, 2, \ldots \end{cases} \tag{3.7.4}$$

The input impedance $Z_{in} - Z_\infty$ given by (3.7.1) is calculated in computer program *MONOSTOR,* the listing of which is given in App. B.4.

For very thin quarter-wave elements, $R_\infty = 36.54\ \Omega$ and $X_\infty = 21.26\ \Omega$ (see Sec. 3.8). The numeric directivity on the horizon, $d(\pi/2)$, is related to the radiation resistance $R = R_{in}$ by (2.3.2). Computer printouts of the maxima and minima of $R_{in} - R_\infty$, as well as for $X_{in} - X_\infty$, are given in App. A.4. For very thin quarter-wave elements, App. A.5 also gives computer printouts of R_{in}, X_{in}, $d(\pi/2)$, and $D(\pi/2) = 10 \log_{10} d(\pi/2)$.

Differences in the far-field radiation pattern from that for a ground plane of infinite extent are given by Storer [8]. For the waveform of (2.1.2) and a sinusoidal current distribution $I(z)$ for the elements (see Fig. 1 of Sec. 2.1), the difference in the far-field magnetic field intensity is (from Eq. (6) [8]):

$$H_\phi - H_\phi|_{ka=\infty} = \mathrm{sgn}(\theta) \left\{ \frac{\mathrm{j}I(0) \exp(-\mathrm{j}kr)\,[1 - \cos(kh)]}{[\sin(kh)\, 2\pi r]\,[2\pi\sqrt{2}]} \right\} \times \left\{ \int_0^{2\pi} \frac{\exp[-\mathrm{j}ka(1 - \sin\theta \cos\phi)] \cos\phi}{(1 - \sin\theta \cos\phi)^{1/2}}\, d\phi \right\}, \quad ka \gtrsim 30 \tag{3.7.5}$$

where

H_ϕ = magnetic far-field intensity for an element on a ground plane of radius ka wavenumbers (A/m);

$H_\phi|_{ka=\infty}$ = magnetic far-field intensity for an element on a ground plane of infinite extent (A/m);

and

$$\mathrm{sgn}\,\theta = \begin{cases} +1, & 0 \le \theta < \pi/2 \\ 0 & \\ -1, & \pi/2 < \theta \le \pi \end{cases}$$

The magnetic field intensity $H_\phi|_{ka=\infty}$ is given by (see Sec. 3.8):

$$H_\phi|_{ka=\infty} = \begin{cases} \dfrac{\mathrm{j}I(0) \exp(-\mathrm{j}kr)}{\sin(kh)\, 2\pi r} \left[\dfrac{\cos(kh \cos\theta) - \cos(kh)}{\sin\theta} \right], & 0 \le \theta \le \pi/2 \\ 0, \ \pi/2 < \theta \le \pi & \end{cases} \tag{3.7.6}$$

The electric far-field intensity $E_\theta = \eta H_\phi$, where H_ϕ is given by (3.7.5) and η is the free-space wave impedance.

One of the distinguishing features of the far-field radiation pattern for ground plane of large radii is the occurrence of a fine-structured lobing pattern, $H_\phi - H_\phi|_{ka=\infty}$, superimposed on the pattern for a ground plane of infinite extent. The lobing pattern, given by (3.7.5), is symmetrical about the horizon (except for a phase shift of π rad) with the most prominent lobes near the zenith and nadir directions. The nth maximum of the lobing pattern in each quadrant decreases with increasing values of n, where $n = 1$ corresponds to the lobe nearest to either the zenith or nadir direction.

The lobing structure is characterized [8] by the elevation angle θ_n of the nth maxima, the angular separation $\Delta\theta$ between lobes, the angle θ_{max} within which all prominent maxima occur, and the number N of prominent lobes. The elevation angle θ_n in radians is given approximately by

$$\theta_n \approx \begin{cases} 0.59\ (\pi/ka);\ n = 1, \quad ka \gtrsim 30 \\ (n + 0.75)(\pi/ka);\ n = 2, 3, \ldots, N; \quad ka \gtrsim 30 \end{cases} \tag{3.7.7}$$

The angular separation $\Delta\theta$ in radians is given approximately by

$$\Delta\theta \approx \begin{cases} 1.16\ (\pi/ka);\ n = 1, \quad ka \gtrsim 30 \\ \pi/ka;\ n = 2, 3, \ldots, N; \quad ka \gtrsim 30 \end{cases} \tag{3.7.8}$$

Prominent lobes are defined [8] as lobes having maxima that are less than one-fourth the amplitude for the case of a ground plane of infinite extent. The angle θ_{max} in radians is given approximately by

$$\theta_{max} \approx 1.87\ (\pi/ka)^{1/3} \left[\frac{1 - \cos(kh)}{kh \sin(kh)}\right]^{2/3}, \quad ka \gtrsim 30 \tag{3.7.9}$$

The number N of prominent lobes is determined by

$$\theta_{max} = \theta_1 + (\Delta\theta)_{n=1} + (N - 2)(\Delta\theta)_{n\neq 1}, \quad ka \gtrsim 30 \tag{3.7.10}$$

where θ_1, $\Delta\theta$, and θ_{max} are given by (3.7.7) to (3.7.9), respectively.

Solving for N, we have

$$N = [\theta_{max}/(\pi/ka)] - 0.25 = 1.87(\pi/ka)^{-2/3} - 0.25, \quad ka \gtrsim 30 \tag{3.7.11}$$

3.8 METHOD OF IMAGES: $ka = \infty$

For the idealized case of a monopole element mounted on a ground plane of infinite extent and infinite conductivity, the monopole antenna may be modeled, by using the method of images, as a dipole with one-half of the input impedance and double the peak numeric directivity of the dipole [25]. The infinite ground plane prevents monopole radiation into the hemisphere below the ground plane, but it allows a radiation pattern identical to that of the dipole in the upper hemisphere.

In this section, we assume that the current has a waveform as given by (2.1.2), and a current distribution $I(z)$ on the element and its image (see Fig. 7) of the form:

$$I(z) = \begin{cases} [I(0)/\sin(kh)] \sin[k(h - z)], & 0 \leq z \leq h \text{ (element)} \\ [I(0)/\sin(kh)] \sin[k(h + z)], & -h \leq z \leq 0 \text{ (image)} \end{cases} \quad (3.8.1)$$

The near fields, far fields, and input impedance of an electrically thin element on a ground plane of infinite extent are summarized in the remainder of this section.

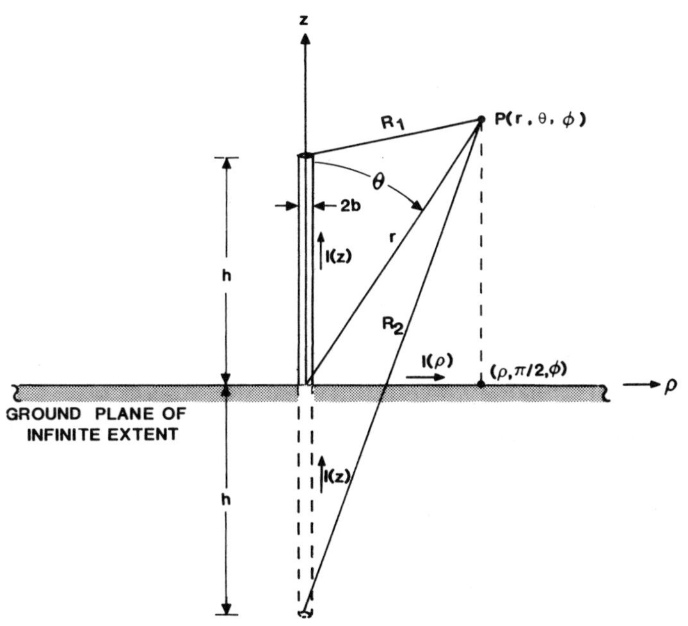

Fig. 7 Method of Images for a Monopole Element at the Center of a Circular Ground Plane of Infinite Extent

3.8.1 Near Fields

The exact magnetic field intensity $\mathbf{H} = \hat{u}_\phi H_\phi$ and electric field intensity $\mathbf{E} = \hat{u}_\rho E_\rho + \hat{u}_z E_z$ for an infinitely thin element are given (from Eqs. (7-15) and (7-17) [25]) as

$$H_\phi = \frac{jI(0)}{4\pi\rho \sin(kh)} [\exp(-jkR_1) + \exp(-jkR_2) - 2\cos(kh)\exp(-jkr)], \quad z \geq 0 \tag{3.8.2a}$$

$$= 0, \quad z < 0 \tag{3.8.2b}$$

$$E_\rho = \frac{j\eta I(0)}{4\pi\rho \sin(kh)} \left[\frac{(z-h)\exp(-jkR_1)}{R_1} + \frac{(z+h)\exp(-jkR_2)}{R_2} - 2z\cos(kh)\frac{\exp(-jkr)}{r}\right], \quad z \geq 0 \tag{3.8.3a}$$

$$= 0, \quad z < 0 \tag{3.8.3b}$$

$$E_z = \frac{-j\eta I(0)}{4\pi \sin(kh)} \left[\frac{\exp(-jkR_1)}{R_1} + \frac{\exp(-jkR_2)}{R_2} - 2\cos(kh)\frac{\exp(-jkr)}{r}\right], \quad z \geq 0 \tag{3.8.4a}$$

$$= 0, \quad z < 0 \tag{3.8.4b}$$

where

$$R_1 = [\rho^2 + (z-h)^2]^{1/2}$$
$$R_2 = [\rho^2 + (z+h)^2]^{1/2}$$

The magnetic field intensities at the top and bottom surfaces of the ground plane are respectively given by (3.8.2) as

$$H_\phi|_{z=0+} = \frac{jI(0)}{2\pi\rho \sin(kh)} [\exp(-jkR_3) - \cos(kh)\exp(-jk\rho)] \quad \text{(top of ground plane)} \tag{3.8.5}$$

$$H_\phi|_{z=0-} = 0 \text{ (bottom of ground plane)} \tag{3.8.6}$$

where

$$R_3 = (\rho^2 + h^2)^{1/2}$$

The incremental ground-plane current $\Delta I(\rho)$ contained within a differential azimuthal angle $d\phi$ may be determined from (3.8.5), (3.8.6), and Ampere's circuit law applied to a closed path along the top and bottom surfaces of the ground plane for arc lengths within $d\phi$. Because the ground-plane current $I(\rho)$ is defined to be positive in the $+\rho$ direction (see Fig. 1 of Sec. 2.1), the path is taken in the clockwise direction. Accordingly,

$$\begin{aligned}\Delta I(\rho) &= (-\hat{u}_\phi \cdot \mathbf{H}_{\text{top}} + \hat{u}_\phi \cdot \mathbf{H}_{\text{bottom}})\ \rho\ d\phi \\ &= [-H_\phi|_{z=0+} + H_\phi|_{z=0-}]\ \rho\ d\phi \end{aligned} \tag{3.8.7}$$

The total ground-plane current $I(\rho)$ is given by

$$\begin{aligned} I(\rho) &= \int_0^{2\pi} \Delta I(\rho)\ d\phi = 2\pi\rho[-H_\phi|_{z=0+} + H_\phi|_{z=0-}] \\ &= \frac{jI(0)}{\sin(kh)} [\cos(kh) \exp(-jk\rho) - \exp(-jkR_3)] \end{aligned} \tag{3.8.8}$$

Equation (3.8.8) agrees with that given by Richmond (Eq. (33) [2]).

3.8.2 Far Fields

For the far-field conditions given by (3.2.7) and (3.2.8), the relations (3.8.2) to (3.8.4) reduce to the far fields $\mathbf{H} = \hat{u}_\phi H_\phi$ and $\mathbf{E} = \hat{u}_\theta E_\theta = \hat{u}_\theta \eta H_\theta$, where (from Eq. (4-62) [25]), H_ϕ is given by

$$H_\phi = \begin{cases} \dfrac{jI(0) \exp(-jkr)}{\sin(kh)\ 2\pi r} \left[\dfrac{\cos(kh \cos\theta) - \cos(kh)}{\sin\theta}\right], & 0 \leq \theta \leq \pi/2 \\ 0, \quad \pi/2 < \theta \leq \pi \end{cases} \tag{3.8.9}$$

The time-averaged radiated power density $s(\theta) = (\eta/2)|H_\phi|^2$ is given by

$$s(\theta) = \begin{cases} \dfrac{\eta |I(0)|^2 \, f(\theta)}{8\pi^2 r^2 \sin^2(kh)}, & 0 \leq \theta \leq \pi/2 \\ 0, & \pi/2 < \theta \leq \pi \end{cases} \tag{3.8.10}$$

where

$$f(\theta) = \left[\frac{\cos(kh \cos\theta) - \cos(kh)}{\sin\theta} \right]^2$$

The direction of maximum radiated power is $\theta = \pi/2$ rad.

The input impedance of a monopole element of length h on a ground plane of infinite extent is one-half that of a dipole of total length $2h$ in free space. Accordingly, the radiation resistance R of an infinitely thin element on a ground plane of infinite extent is given by (compare with Eqs. (4-70) and (4-79) [25]):

$$R = \frac{\eta}{4\pi \sin^2(kh)} \left\{ \text{Cin}\,(2kh) + \frac{1}{2} \sin(2kh)\,[\text{Si}\,(4kh) - 2\,\text{Si}\,(2kh)] + \frac{1}{2} \cos(2kh)\,[2\,\text{Cin}\,(2kh) - \text{Cin}\,(4kh)] \right\}, \quad b = 0 \tag{3.8.11}$$

where Si (x) and Cin (x) are the sine integral and modified cosine integrals, respectively, which are defined according to (3.2.32). Equation (3.8.9) is also approximately valid for thin elements ($kb \ll 1$). (For example, compare (3.2.33) and (3.2.35) for a thin monopole element in the absence of a ground plane.)

By substituting (3.8.10) and (3.8.11) into (2.2.6), the numeric directive gain $d(\theta)$ is given by

$$d(\theta) = \frac{4f(\theta)}{B} \tag{3.8.12}$$

where

$$B \equiv \text{Cin}\,(2kh) + \frac{1}{2} \sin(2kh)[\text{Si}\,(4kh) - 2\,\text{Si}\,(2kh)] + \frac{1}{2} \cos(2kh)[2\,\text{Cin}\,(2kh) - \text{Cin}\,(4kh)]$$

For a quarter-wave monopole element ($kh = \pi/2$), (3.8.11) and (3.8.12) reduce to

$$R = \frac{\eta}{4\pi}\frac{\text{Cin}(2\pi)}{2} = 36.5395\ \Omega; \quad kh = \pi/2,\ b = 0 \tag{3.8.13}$$

$$f(\theta) = \frac{\cos^2[(\pi/2)\cos\theta]}{\sin^2\theta}, \quad kh = \pi/2 \tag{3.8.14}$$

$$d(\theta) = \frac{4\cos^2[(\pi/2)\cos\theta]}{\sin^2\theta}\,\frac{2}{\text{Cin}(2\pi)}, \quad kh = \pi/2 \tag{3.8.15}$$

The peak directivity $d(\pi/2) = 3.2818$ (numeric) corresponding to $D(\pi/2) = 10\log_{10} d(\pi/2) = 5.1612$ dBi. The directional gain is plotted in Fig. 8(f) of Sec. 3.9.

For an electrically short monopole element ($kh \ll 1$) and $\eta \approx 120\pi$ (Ω), (3.8.11) and (3.8.12) reduce to

$$R = 10\,(kh)^2, \quad (kh) \ll 1 \tag{3.8.16}$$

$$f(\theta) = \frac{(kh)^4}{4}\sin^2\theta, \quad (kh) \ll 1 \tag{3.8.17}$$

$$d(\theta) = 3\sin\theta, \quad (kh) \ll 1 \tag{3.8.18}$$

The peak directivity $d(\pi/2) = 3.0$ (numeric) corresponding to $D(\pi/2) = 10\log_{10} d(\pi/2) = 4.7712$ dBi.

3.8.3 Input Impedance

The input resistance R_{in} is given by (3.8.11), which is exact for an infinitely thin element and approximately correct for thin elements, provided that the current distribution of the elements is sinusoidal.

The input reactance X_{in} for thin elements ($kb \ll 1$) is given by (compare with Eqs. (7-33) and (7-30) [25]):

$$X_{in} = \frac{\eta}{4\pi \sin^2(kh)} \left\{ \text{Si}\,(2kh) + \cos(2kh)[\text{Si}\,(2kh) - \frac{1}{2}\text{Si}\,(4kh)] \right.$$
$$- \sin(2kh)[\ln(h/b) - \text{Cin}\,(2kh) + \frac{1}{2}\text{Cin}\,(4kh)$$
$$\left. + \frac{1}{2}\text{Cin}\,(kb^2/h)] \right\} \tag{3.8.19}$$

For a quarter-wave element, in the limit $b \to 0$, (3.8.19) gives an input reactance $X_{in} = 21.2576\ \Omega$. In (3.8.19), $X_{in} = 0$ for element lengths that are slightly less than $kh = (2n - 1)(\pi/2)$, $n = 1, 2, 3, \ldots$, which are approximately one-half of the resonant lengths for a monopole element in the absence of a ground plane (see (3.2.38)).

3.9 SUMMARY OF RESULTS

In Chapter 3, a sinusoidal current distribution is assumed on the monopole element. Although such a current distribution is never exactly realized, even for an infinitely thin element (see Sec. 3.1), it is a useful approximation for sufficiently thin elements. As an example, for a quarter-wave element of radius $b = 10^{-6}\lambda$, comparison of the input impedances computed by determining the actual current distribution of the elements with that computed by assuming a sinusoidal current distribution yields a difference of no more than 5% for ground-plane radii ka = 6–14 (compare Table 7 of Sec. 4.5 with Tables A2-24–A2-42 of Appendix A). The assumption of a sinusoidal current distribution of the elements allows computationally simpler models to determine input impedance and radiation patterns than those models which follow in Chapter 4. The results of these simpler models are summarized in this section.

The electrical properties of electrically short and quarter-wave monopole elements on zero, large, and infinite extent ground planes are compared in Table 4 with the corresponding properties of electrically short and half-wave dipoles. The peak directivity of a quarter-wave monopole is 1.88 dBi and 5.16 dBi for zero and infinite extent ground planes, respectively. The directivity on the horizon of a quarter-wave monopole is 1.88 dBi and −0.86 dBi for ground planes of zero extent and very large but finite extent, respectively. Slightly smaller directivities are obtained for electrically short elements than is the case for quarter-wave elements.

Table 4 Electrical Properties of Very Thin Monopole Elements on Ground Planes of Zero, Large, and Infinite Extent

Antenna Structure	*Ground-Plane Radius (wave-numbers),* $2\pi a/\lambda$	*Element Length (Wave-lengths),* h/λ	*Peak Directivity**		*Directivity on Horizon*		*Input Impedance*		*Reference Equations*
			$d(\theta_p)$ (numeric)	$D(\theta_p)$ (dBi)	$d(\theta=\pi/2)$ (numeric)	$D(\theta=\pi/2)$ (dBi)	*Radiation Resistance* (Ω)	*Reactance‡* (Ω)	
Monopole	0	$\ll 1$	1.500	1.761	1.500	1.761	$20\pi^2(h/\lambda)^2$	$-\infty$	(3.2.19), (3.2.20)
	0	0.25	1.543	1.882	1.543	1.882	19.43	$-\infty$	(3.2.15), (3.2.16)
	$\gg 1$, FINITE	$\ll 1$	3.000	4.771	0.750	−1.249	$40\pi^2(h/\lambda)^2$	$-\infty$	(3.7.1), (3.8.18), (2.3.2)
	$\gg 1$, FINITE	0.25	3.282	5.161	0.820	−0.859	36.54	21.26	(3.7.1), (3.8.15), (2.3.2)
	∞	$\ll 1$	3.000	4.771	3.000	4.771	$40\pi^2(h/\lambda)^2$	$-\infty$	(3.8.16), (3.8.18), (3.8.19)
	∞	0.25	3.282	5.161	3.282	5.161	36.54	21.26	(3.8.13), (3.8.15), (3.8.19)
Dipole†	0	$\ll 1$	1.500	1.761	1.500	1.761	$80\pi^2(h/\lambda)^2$	$-\infty$	Balanis [15], pp. 108, 112, 131, 132
	0	0.25	1.641	2.151	1.641	2.151	73.08	42.52	
Idealized isotropic radiator			1.000	0	1.000	0	—	—	—

*Direction of peak directivity is $\theta_p = \pi/2$ rad, except for monopole elements on ground planes of finite but nonzero extent, in which case $\theta_p < \pi/2$ rad.
†Very thin dipole of half-length h.
‡Reactance is in the limit of an infinitely thin element.

The radiation resistances obtained by different methods are compared in Table 5 for a thin quarter-wave element on a ground plane of radius $0 \leq ka \leq 8.5$ wavenumbers. The suspected best available results are obtained by the integral equation method for $0 \leq ka \leq 1.0$, by the method of moments ($N = 1$) for $1 < ka < 3.0$, by the oblate spheroidal wave-function method for $3.00 \leq ka \leq 6.5$, and by the method of moments ($N = 1$) for $6.5 < ka \leq 14$. The result obtained by the method of moments is in good agreement with the suspected best available results obtained by other methods.

Table 5 Radiation Resistance of a Thin Quarter-Wave Element at the Center of a Circular Ground Plane of Radius $0 \leq ka \leq 8.5$ Wavenumbers (Sinusoidal Current Distribution Assumed on Element)

Normalized Ground-Plane Radius, $2\pi a/\lambda$	*Radiation Resistance (Ω)*		
	Integral Equation Method	*Oblate Spheroidal Wave-Function Method*	*Method of Moments*
0	*19.43		
0.25	*19.48		19.49
0.50	*19.62		19.62
0.75	*19.86		19.86
1.00	*20.23		20.21
1.25	20.76		*20.71
1.50	21.51		*21.25
1.75	22.59		*22.44
2.00	24.15		*23.89
2.25	26.46		*25.99
2.50	29.95	27.32	*29.02
2.75	35.44	31.92	*33.24
3.00	44.60	*37.48	38.62
3.25		*43.01	44.12
3.50		*46.06	47.57
3.75		*45.55	47.35
4.00		*42.67	44.43
4.25		*39.23	40.58
4.50		*36.23	37.13
4.75		*34.00	34.46

Table 5 (cont'd.)

Normalized Ground-Plane Radius, $2\pi a/\lambda$	*Radiation Resistance* (Ω)		
	Integral Equation Method	*Oblate Spheroidal Wave-Function Method*	*Method of Moments*
5.00		*32.57	32.68
5.25		*31.93	31.70
5.50		*32.13	31.53
5.75		*33.23	32.26
6.00		*35.23	34.04
6.25		*37.85	36.94
6.50		*40.33	40.56
6.75		30.12	*43.53
7.00		30.09	*44.20
7.25			*42.30
7.50			*39.10
7.75			*35.96
8.00			*33.50
8.25			*31.88
8.50			*31.16

*Suspected best available result.

The numeric directive gain patterns of a quarter-wave element on ground planes of radii $ka = 0, 3, 4, 5, \sqrt{42}$, and ∞ wavenumbers are plotted in Fig. 8. These polar graphs of directive gain should not be confused with the polar graphs of radiated power density plotted by Leitner and Spence [5] for constant base current. It should be noted that the peak directivity and its direction are not monotonic functions of the ground-plane radius.

The input impedance of a thin quarter-wave element is plotted in Figs. 9 and 10 for ground-plane radii $0 \leqslant ka \leqslant 14$ wavenumbers. The input resistance varies between 19.4 and 46.1 Ω, and asymptotically approaches 36.5 Ω for increasingly large ground-plane radii. The input reactance varies between $-\infty$ and 32.5 Ω, and asymptotically approaches 21.3 Ω for increasingly large ground-plane radii.

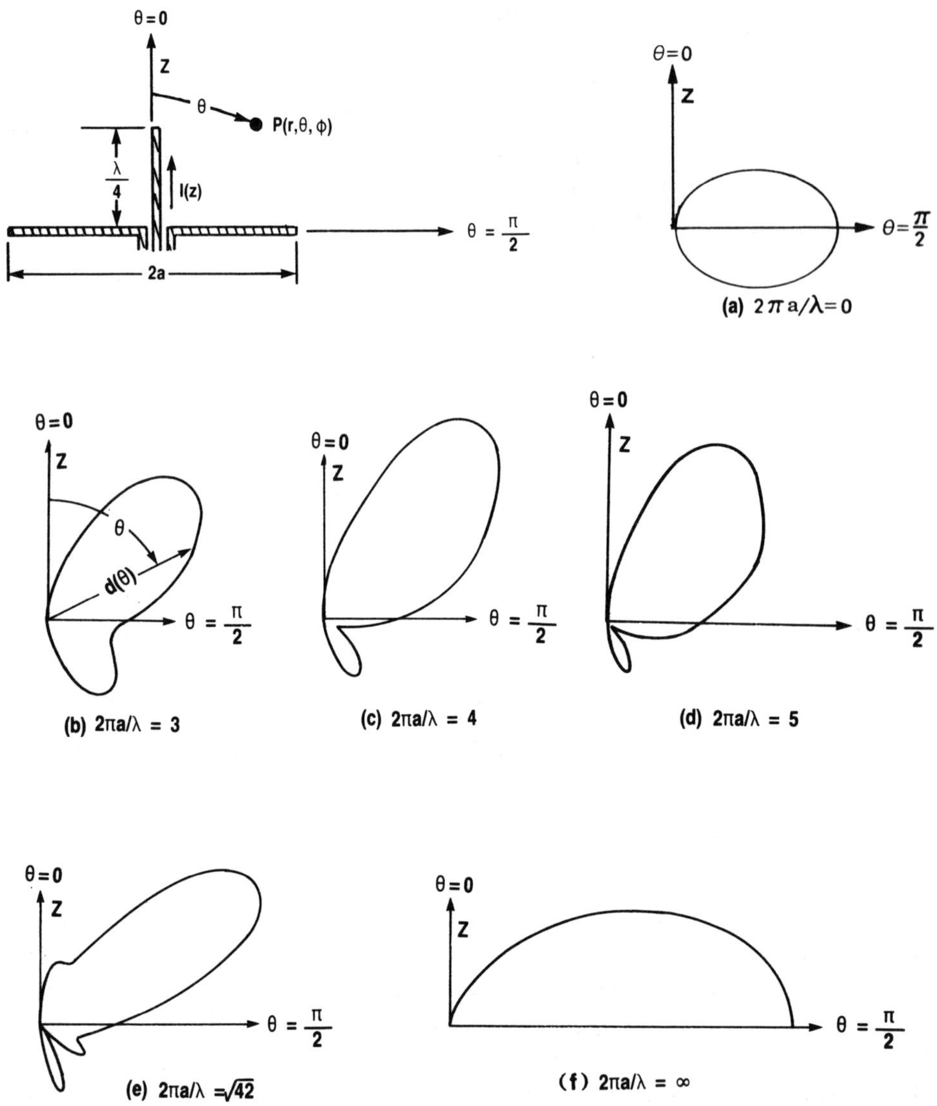

Fig. 8 Elevation Directive Gain Patterns, for any Azimuthal Direction, of a Quarter-Wave Element Mounted on a Ground Plane of Radius a (The Patterns are Polar Graphs on the Same Linear Scale)

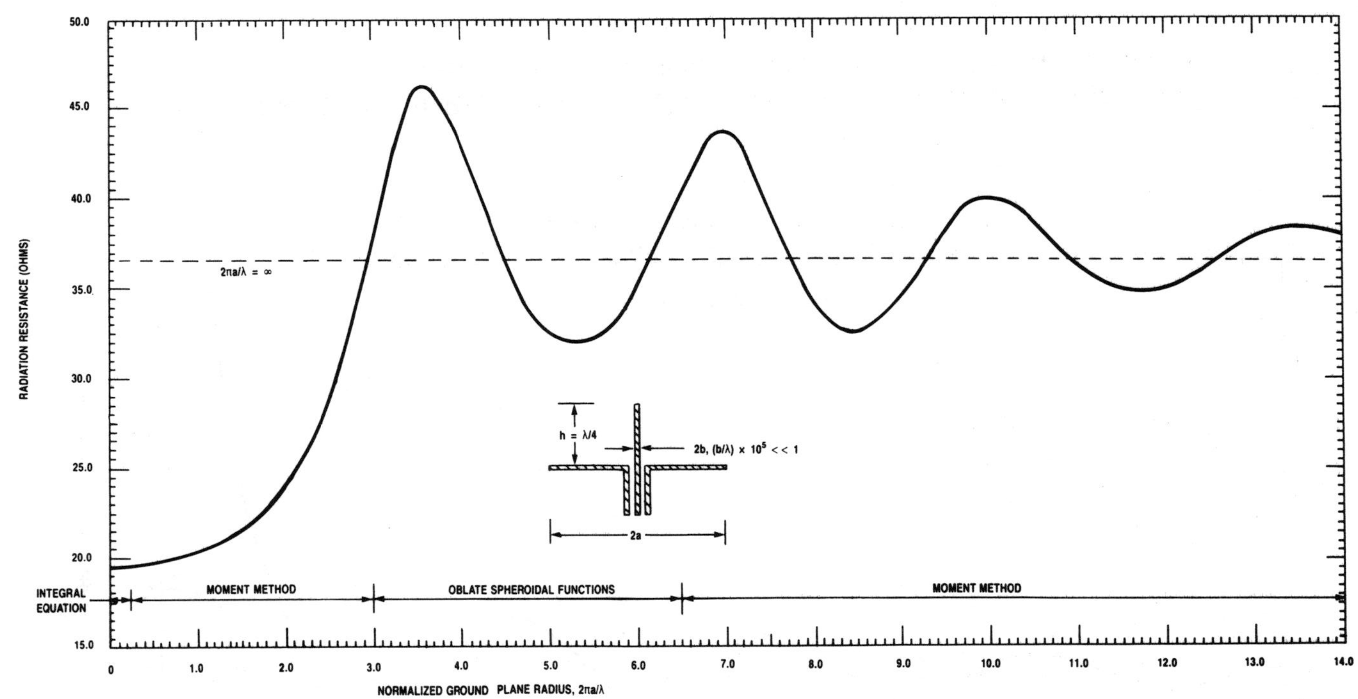

Fig. 9 Radiation Resistance of a Thin Quarter-Wave Element at the Center of a Circular Ground Plane of Radius $0 \leqslant ka \leqslant 14$ Wavenumbers

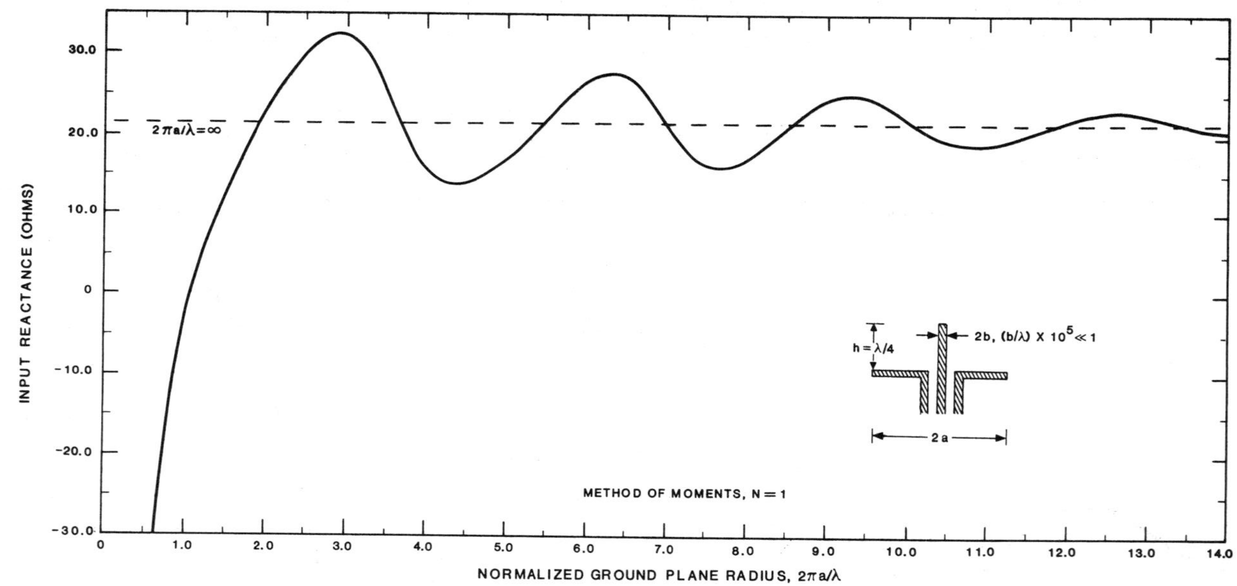

Fig. 10 Input Reactance of a Thin Quarter-Wave Element at the Center of a Circular Ground Plane of Radius $0 \leq ka \leq 14$ Wavenumbers

The directive gain on the horizon, peak directivity, and elevation angle of the peak directivity for a quarter-wave element are plotted in Figs. 11–13 with ground-plane radii of $0 \leq ka \leq 14$ wavenumbers. The directive gain on the horizon varies between 1.9 and −1.9 dBi, and asymptotically approaches −0.86 dBi for increasingly large but finite ground-plane radii. The peak directivity varies between 1.0 and 5.3 dBi, and asymptotically approaches 5.2 dBi for increasingly large ground-plane radii. The elevation angle of the peak directivity varies between 33° and 90°, and asymptotically approaches 90° for increasingly large ground-plane radii.

The radiation resistances for thin elements of length $h/\lambda = 0.25, 0.1$, and 0.025 with ground-plane radii of $0 \leq ka \leq 8$ wavenumbers are compared in Fig. 14. The radiation resistance of each element is normalized to that for a ground plane of zero extent. The normalized radiation resistance, as a function of the ground-plane radius, is approximately independent of the element length. For an electrically short, thin element ($h/\lambda \leq 0.1$) with length that is small or comparable to the ground-plane radius, the input reactance (not shown) is approximately independent of the ground-plane radius.

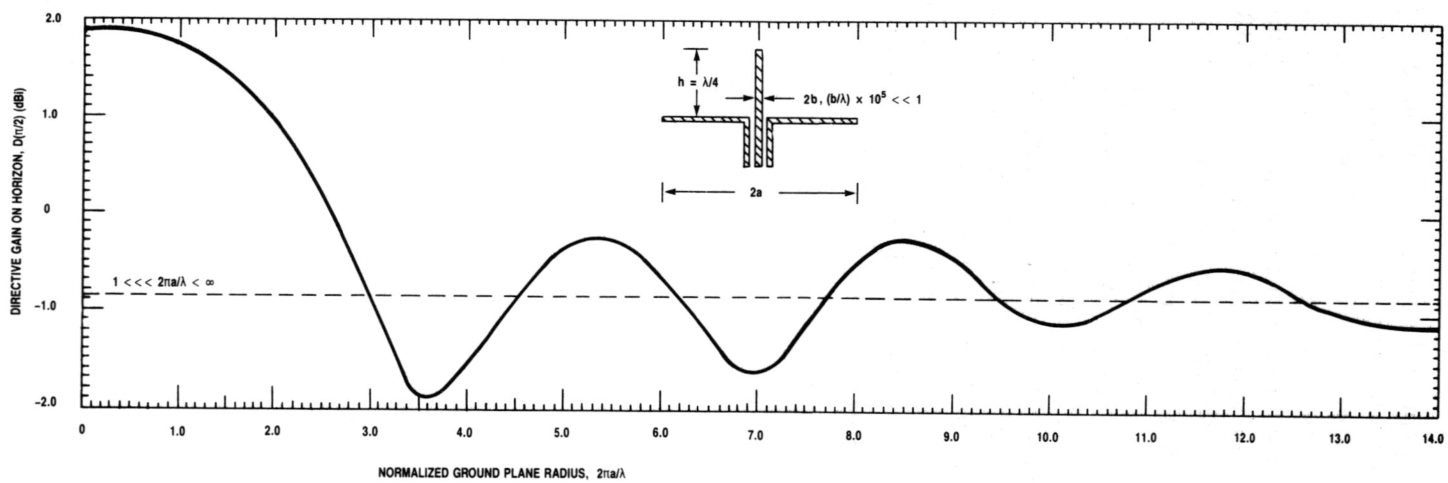

Fig. 11 Directive Gain on the Horizon of a Thin Quarter-Wave Element at the Center of a Circular Ground Plane of Radius $0 \leqslant ka \leqslant 14$ Wavenumbers

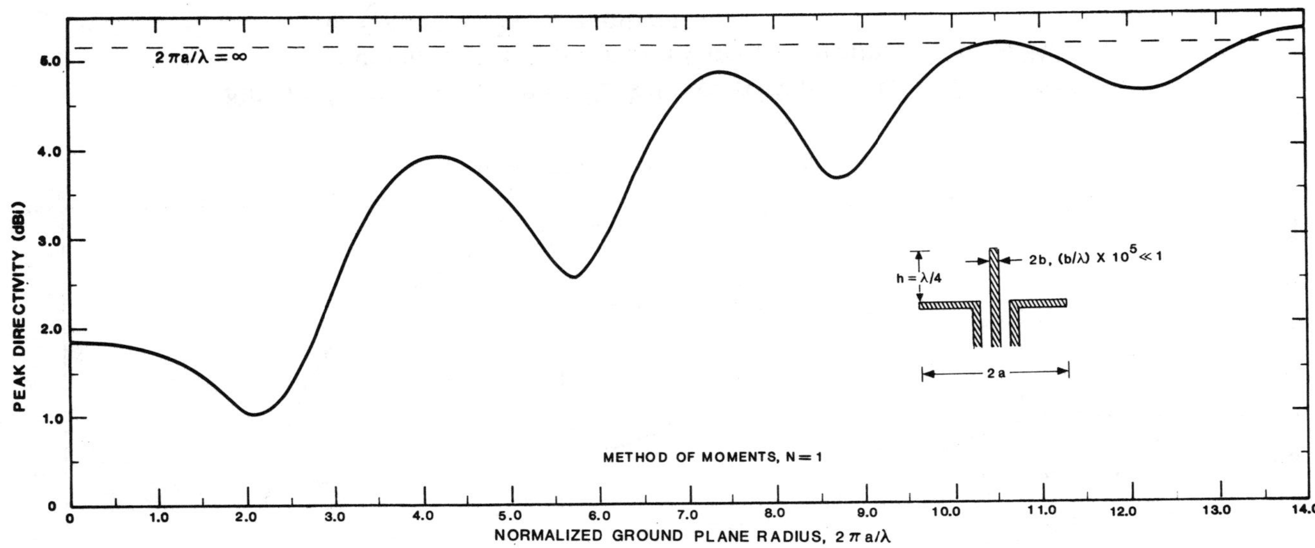

Fig. 12 Peak Directivity of a Thin Quarter-Wave Element at the Center of a Circular Ground Plane of Radius $0 \leqslant ka \leqslant 14$ Wavenumbers

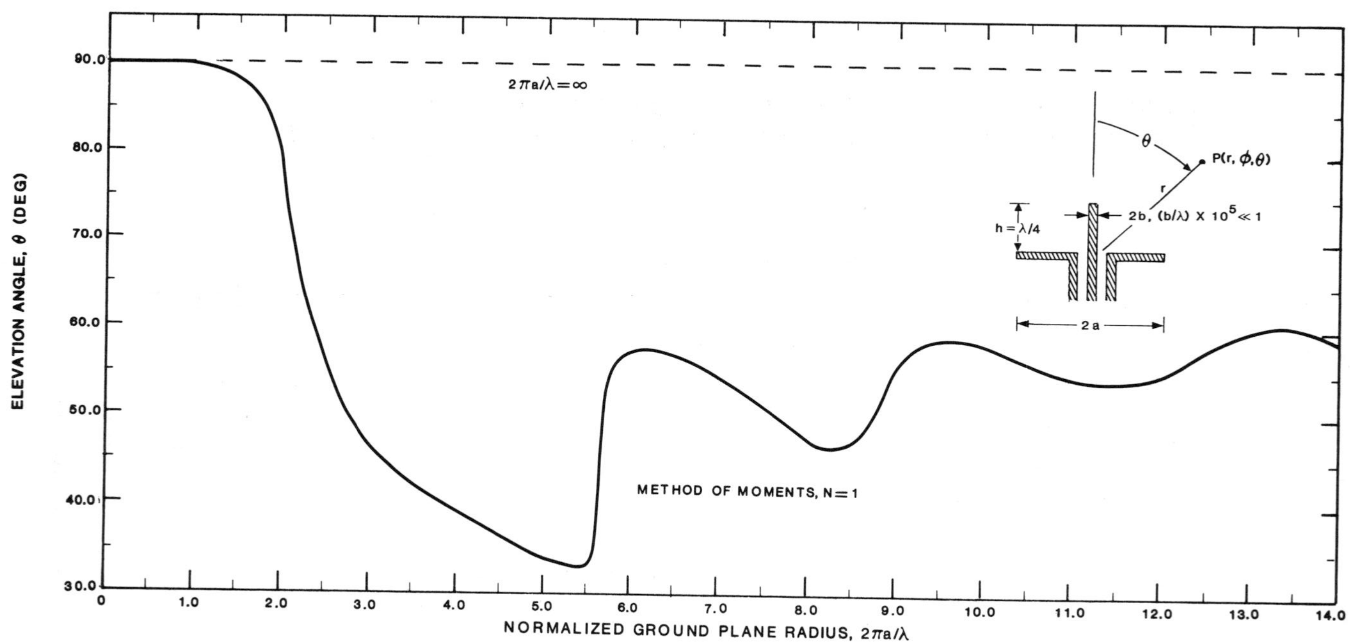

Fig. 13 Elevation Angle of Peak Directivity for a Thin Quarter-Wave Element at the Center of a Circular Ground Plane of Radius $0 \leqslant ka \leqslant 14$ Wavenumbers

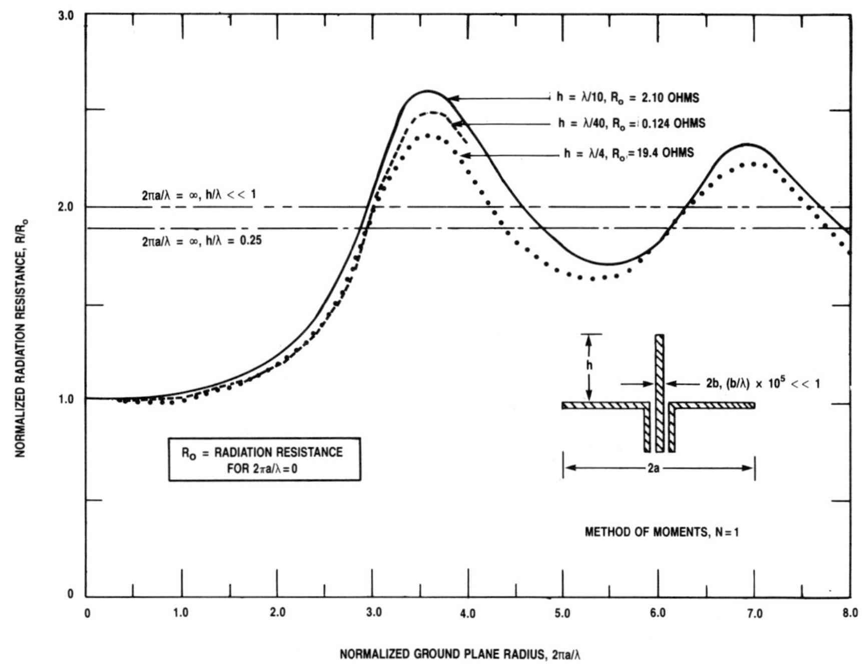

Fig. 14 Radiation Resistance for Thin Elements of Various Lengths at the Center of a Circular Ground Plane in Free Space

Chapter 4
Models in which the Current Distributions on the Monopole Element and Ground Plane are Both Initially Unknown

4.1 BOUNDARY CONDITIONS

In Chapter 3, the total field at an arbitrary field point could be simply expressed as the vector sum of an incident field and an induced field (see (3.1.12)) because the incident field was specified from the initially known current distribution on the monopole element. In this chapter, such a procedure is not possible because the current distributions on the element and ground plane are both initially unknown. Instead, the total field is determined here by representing the unknown current distribution on either the element, ground plane, or both by a finite series of overlapping modes with unknown current amplitudes. The current amplitudes are determined by matrix algebra, subject to boundary conditions that comprise current constraints and the excitation voltage across the coaxial line feed.

The antenna geometry is shown in Fig. 1 of Sec. 2.1. The current waveform is given by (2.1.2). The models which follow in this chapter are based on the current characterization and circuit idealization of Fig. 2(c) from Sec. 2.4.

The coaxial line feed and the excitation voltage, which is not explicitly shown in Fig. 1, are characterized in Fig. 2(c) by a surface magnetic current density (magnetic frill) M_ϕ, which sits on top of a thick ground plane of radius b_1, where b_1 is the radius of the outer conductor of the coaxial line feed. The magnetic frill M_ϕ is defined over the ground plane by (2.4.8).

Constraints on the various circuit currents of Fig. 2(c) are given by (2.4.3) to (2.4.7). These constraints, together with the magnetic frill, constitute the boundary conditions on the current amplitudes.

4.2 METHOD OF MOMENTS: $0 < ka \leqslant 14$

When the element and ground-plane current distribution are initially unknown, the current distributions may be determined by a sinusoidal-Galerkin method of moments as employed by Richmond [2]. The antenna geometry is shown in Fig. 1 of Chapter 2.

In the method of moments, the element is divided into N equal segments (see Fig. 15) of length d' given by

$$d' = h/N \tag{4.2.1}$$

where N is a positive integer and h is the element length. The ground plane is divided into M concentric annular zones (see Fig. 16) of width d, given by

$$d = (a - b)/M > b_1 - b \tag{4.2.2}$$

where M is a positive integer, a is the ground-plane radius, b is the element radius, and b_1 is the radius of the outer conductor of the coaxial line feed. The ground plane extends from the coaxial line's inner conductor because of the equivalent circuit representation of the coaxial line excitation in Fig. 2(b). The element current distribution $I(z)$ and ground-plane current distribution $I(\rho)$ are the sum of the current distributions on each element segment and ground-plane annular zone, respectively, which are given by

$$I(z) = \sum_{n=1}^{N} I_n(z) \tag{4.2.3}$$

$$I(\rho) = \sum_{m=1}^{M} I'_m(\rho) \tag{4.2.4}$$

where $I_n(z)$ and $I'_m(\rho)$ are the current distributions on the nth segment and mth annular zone, respectively.

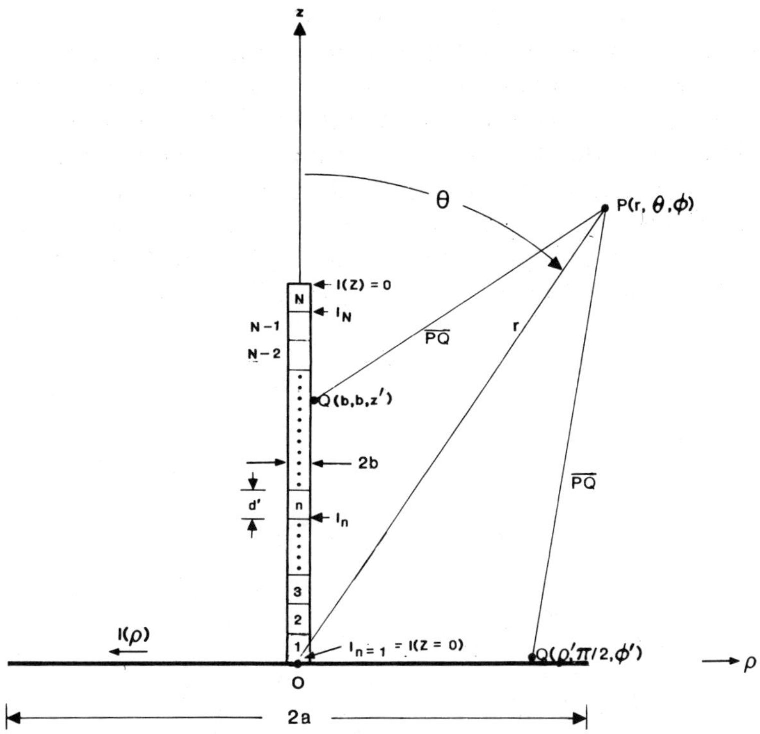

Fig. 15 Element with N Equal Segments of Length d'

In the sinusoidal-Galerkin method, the current distribution $I_n(z)$ on the nth element segment is approximated by

$$I_n(z) \approx \frac{I_n \sin[k(nd' - z)] + I_{n+1} \sin\{k[z - (n - 1)]d'\}}{\sin(kd')};$$

$$(n - 1)d' \leqslant z \leqslant nd'; \; n = 1, 2, 3, \ldots, N \qquad (4.2.5)$$

where $I_{n=1} = I(z = 0)$ and $I_{N+1} = 0$.

The current distribution $I'_m(\rho)$ on the mth annular zone is approximated by

$$I'_m(\rho) \approx \frac{I'_m \sin\{k[md + b - \rho]\} + I'_{m-1} \sin\{k[\rho - (m - 1)]\, d - b\}}{\sin(kd)};$$

$$b + (m - 1)d \leqslant \rho \leqslant b + md; \; m = 1, 2, 3, \ldots, M \qquad (4.2.6)$$

where $I'_{m=1} = -I_{n=1} = -I(z = 0)$ and $I'_{M+1} = 0$.

The current distribution on each segment and annular zone is therefore the sum of two overlapping dipole modes, except for the segment and annular zone adjacent to the base of the element. The *base* mode may be thought of as a dipole mode comprising an element monopole mode ($n = 1$) and a ground-plane monopole mode ($m = 1$) with terminals along the circumference of the element's base. The element and ground-plane dipole modes are functionally tied to each other by means of the base-mode amplitude constraint $I_{n=1} = -I'_{m=1} = I(z = 0)$.

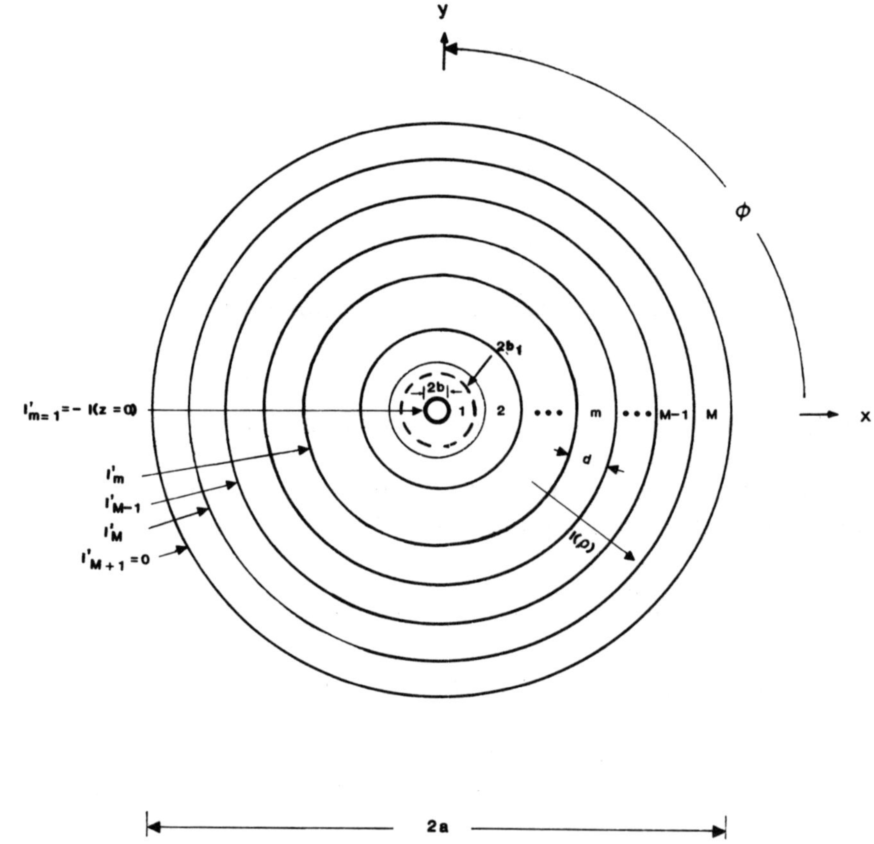

Fig. 16 Ground Plane with M Annular Zones of Width d

The element and ground-plane current distributions are represented in (4.2.3) to (4.2.6) as a series of $N + M$ overlapping sinusoidal dipole modes with $N + M$ unknown complex current amplitudes. The $N + M$ amplitudes are determined in Richmond's method of moments by the Crout substitution method of solution of a $(N + M) \times (N + M)$ matrix, subject to the boundary conditions discussed in Sec. 4.1. The constraint $I_{n=1} = -I'_{m=1}$ reduces the matrix size to $(N + M - 1) \times (N + M - 1)$.

The numbers of subdivisions, N and M, are limited by the cost of computation time and the precision of the computer. The accuracy of the solution can be appreciably decreased if either N or M is too small or too large. The method of moments converges to a solution when an increase or decrease of unity in the value of N or M does not appreciably alter the solution for input impedance.

A method-of-moments computer program for a monopole element at the center of a circular ground plane in free space was obtained from Professor J. H. Richmond of Ohio State University. The program computes the input impedance and the $N + M$ complex current amplitudes on the element and ground plane for a voltage input $V(0) = 1$ V. The program, written in FORTRAN IV language and in single precision, was edited and converted by the MITRE Corp. to double precision for use on a DEC VAX 11/780 computer. A listing of the MITRE version, designated *RICHMD1,* is given in App. B.2.

The authors' experience with this program suggests the following constraints on its use. At least double precision is required to give convergent results. Meaningful results were obtained for element radii of $b/\lambda \geqslant 10^{-10}$ and ground-plane radii of $ka > 0.25$. As a rule of thumb, $N = 2\text{–}3\ kh$ and $M = 2\text{–}3\ ka$ give reasonably accurate and convergent results. The amount of central processing unit (CPU) time on the VAX 11/780 is approximately two minutes for $N + M = 20$, six minutes for $N + M = 30$, 34 minutes for $N + M = 60$, and 50 minutes for $N + M = 75$.

The versatility of a sinusoidal-Galerkin method of moments is illustrated by the case of $N = 1$. A sinusoidal current distribution is imposed on the element by setting $N = 1$ (see (4.2.5)).

The element and ground-plane current distributions obtained from program RICHMD1 can be utilized to obtain the far-field radiation pattern in the following manner.

The magnetic field intensity $\mathbf{H}(P)$ at a far-field point $P(r, \theta, \phi)$ and for an $e^{j\omega t}$ waveform is related to the magnetic vector potential $\mathbf{A}(P)$ by [26]:

$$\mathbf{H}(P) = -j(1/\eta)\omega\hat{u}_r \times \mathbf{A}(P) \tag{4.2.7}$$

where $\hat{u}_r$ is a unit vector in the radial direction and η is the free-space wave impedance. The magnetic vector potential $\mathbf{A}(P)$ at the far-field point P resulting from a current source point $Q(x', y', z') = Q(b, b, z')$ on the element or a source point $Q(\rho', \pi/2, \phi')$ on the ground plane (see Fig. 4) is given by [26]:

$$\begin{aligned}\mathbf{A}(P) &= (\mu_0/4\pi)\int_v J(Q)\exp(-jk\overline{PQ})(1/\overline{PQ})\,dv \\ &\approx [\mu_0 \exp(-jkr)/4\pi r]\int_v J(Q)\exp(jk\hat{u}_r \cdot \mathbf{OQ})\,dv\end{aligned} \tag{4.2.8}$$

where

$\overline{PQ} = [r^2 - 2r\hat{u}_r \cdot \mathbf{OQ} + x'^2 + y'^2 + z'^2]^{1/2} \approx \mathbf{r} - \hat{u}_r \cdot \mathbf{OQ}$, $r \gg (x'^2 + y'^2 + z'^2)^{1/2}$;

$\hat{u}_r \cdot \mathbf{OQ} = x' \sin\theta \cos\phi + y' \sin\theta \sin\phi + z' \cos\theta$;

μ_0 = permeability of free space $= 4\pi \times 10^{-7}$ (H/m);

$J(Q)$ = source surface current density at an arbitrary source point Q on either the element or ground plane (A/m);

$r = \overline{OP}$ = radial distance between the origin $O(0, 0, 0)$ and the far-field point $P(r, \theta, \phi)$;

dv = differential surface area containing the current source points Q(m^2).

By substituting (4.2.3) into (4.2.8), the magnetic far-field intensity $H_\phi^{(e)}$, resulting from the monopole current distribution, is given by

$$H_\phi^{(e)} = \frac{j\exp(-jkr)\sin\theta}{4\pi r}\int_0^h \sum_{n=1}^{N} I_n(z')\exp(jkz'\cos\theta)\,dz'$$

$$H_\phi^{(e)} = \frac{\mathrm{j} \exp(-\mathrm{j}kr)}{4\pi r \sin\theta \sin(kd')} \left\{ \left[\sum_{n=1}^{N} I_n \exp(\mathrm{j}nkd' \cos\theta) \right] \right.$$

$$\times [1 - \cos(kd') \cos(kd' \cos\theta) - \cos\theta \sin(kd') \sin(kd' \cos\theta)$$

$$+ \mathrm{j} \cos(kd') \sin(kd' \cos\theta) - \mathrm{j} \cos\theta \sin(kd') \cos(kd' \cos\theta)]$$

$$- \left\{ \sum_{n=2}^{N} I_n \exp[\mathrm{j}(n - 2) kd' \cos\theta] \right\} [\cos(kd') \cos(kd' \cos\theta)$$

$$+ \cos\theta \sin(kd') \sin(kd' \cos\theta) - 1 + \mathrm{j} \cos(kd') \sin(kd' \cos\theta)$$

$$\left. - \mathrm{j} \cos\theta \sin(kd') \cos(kd' \cos\theta)] \right\} \qquad (4.2.9)$$

By substituting (4.2.4) into (4.2.8), the magnetic far-field intensity $H_\phi^{(g)}$, resulting from the ground-plane current distribution, is given by

$$H_\phi^{(g)} = \frac{-\mathrm{j} \exp(-\mathrm{j}kr)}{4\pi r} \int_0^a \int_0^{2\pi} \frac{1}{2\pi} \sum_{m=1}^{M} I'_m(\rho')$$

$$\times \exp[\mathrm{j}k\rho' \cos(\phi - \phi)] \cos\theta \cos(\phi' - \phi)\, d\phi'\, d\rho' \qquad (4.2.10)$$

Equation (4.2.10) is not readily evaluated when (4.2.6) is substituted into it. An approximate simplified solution can be found when each annular zone of the ground plane is subdivided into X smaller annular zones of width $\Delta\rho = d/X$ so that the current distribution in each subdivided zone is almost constant and approximately equal to its average current. The total number J of subdivided annular zones on the entire ground plane is given by

$$J = XM \qquad (4.2.11)$$

The average current I_u in the uth subdivided zone is given by

$$I_u = \frac{1}{\Delta\rho} \int_{(u-1)\Delta\rho+b}^{u\Delta\rho+b}$$

$$\times \frac{I'_m \sin[k(md + b - \rho] + I'_{m+1} \sin\{k[\rho - (m-1)d - b]\}}{\sin(kd)} d\rho,$$

$$[b + (m - 1)d] < \rho < (b + md) \qquad (4.2.12)$$

The current distribution $I(\rho)$ on the ground plane is therefore given approximately by

$$I(\rho) \approx \sum_{u=1}^{J} I_u \qquad (4.2.13)$$

By substituting $I_u(\rho')$ for $I_m(\rho')$ in (4.2.10), we have

$$H_\phi^{(g)} \approx \frac{-\mathrm{j} \exp(-\mathrm{j}kr)}{4\pi r} \int_0^a \int_0^{2\pi}$$

$$\times \frac{1}{2\pi} \left\{ \sum_{u=1}^{J} I_u(\rho') \exp[\mathrm{j}k\rho' \cos(\phi' - \phi)] \right\}$$

$$\times \cos\theta \cos(\phi' - \phi)\, d\phi'\, d\rho'$$

$$= \frac{\exp(-\mathrm{j}kr)\cos\theta}{4\pi r \sin\theta} \sum_{u=1}^{J} I_u(\rho')$$

$$\times \{J_0[k \sin\theta\, ((u - 1)\Delta\rho + b)]$$

$$- J_0[k \sin\theta\, (u\Delta\rho + b)]\} \qquad (4.2.14)$$

where J_0 is the zero-order Bessel function of the first kind.

The total magnetic far-field intensity $H_\phi^{(\text{total})}$ is given by

$$H_\phi^{(\text{total})} = H_\phi^{(e)} + H_\phi^{(g)} \qquad (4.2.15)$$

where $H_\phi^{(e)}$ and $H_\phi^{(g)}$ are given by (4.2.9) and (4.2.14), respectively.

The time-averaged radiated power density $s(\theta)$, directive gain $d(\theta)$, and radiation resistance R are found from (2.2.2) and (2.2.5), where $H_\phi^{(\text{total})}$ is given by (4.2.15) and $I(z = 0) = I_{n=1}$. Numerical evaluation of

$s(\theta)$, $d(\theta)$, and R is implemented by means of MITRE's computer program *RICHMD2,* written in FORTRAN language for use on a DEC VAX 11/780 computer. A listing of program RICHMD2 is given in App. B.2.

A discussion of the results obtained by the method of moments for a sinusoidal current distribution on the element ($N = 1$) is given in Sec. 3.4.

Computer printouts of the directive gain and far-field radiation pattern of the experimental monopole antennas discussed in Chapter 5 were obtained by the method of moments and are given in App. A.2.

The input impedance determined by the method of moments ($N = 4$, $M = 3ka$) for a quarter-wave element of radius $b = 10^{-6}\lambda$ and ground-plane radii of $6 \leq ka \leq 30$ wavenumbers is compared in Table 7 of Sec. 4.5 with that obtained by the method of moments combined with the geometric theory of diffraction. For $ka > 14$, Richmond's method of moments is inaccurate. In addition to the large amount of CPU time required (greater than 30 minutes for $ka > 14$), convergent results are difficult to obtain for $ka > 14$. It is most likely that the usefulness of Richmond's method of moments could be extended to significantly larger ground-plane radii if a more efficient method of matrix inversion were utilized, rather than the Crout method employed by Richmond. When the element's current distribution is not assumed, Richmond's method of moments gives the best currently available results for ground-plane radii of $ka \leq 14$ wavenumbers.

The lower limit of ka has not been firmly established for accurate results with Richmond's method of moments. For an element segment number $N > 1$, we have obtained useful results for ka as small as 0.25 wavenumbers, but not for ground planes of zero extent. For $N = 1$, accurate results were obtained, subject to the constraint of a sinusoidal current distribution of the elements, for ground planes of zero extent.

The radiation resistance for resonant (zero input reactance) elements of various diameters is compared in Fig. 17 of Sec. 4.5 with the radiation resistance of an infinitely thin quarter-wave element for ground-plane radii of $0 \leq ka \leq 8.5$ wavenumbers. For $ka \rightarrow 0$, the radiation resistance of the resonant elements is appreciably different from that of the infinitely thin quarter-wave element because an infinitely thin element on a ground plane of zero extent is resonant for an element length that is equal to 0.5λ, not 0.25λ (see (3.2.38)).

The lengths and radiation resistances for resonant elements of various diameters, for ground planes of radii $0.25 \leq ka \leq 7$ wavenumbers, are plotted in Figs. 18 and 19 of Sec. 4.5.

4.3 METHOD OF MOMENTS COMBINED WITH GEOMETRIC THEORY OF DIFFRACTION: $8.5 \leqslant ka < \infty$

The method of moments, when used to determine the current distributions on both the element and the ground plane, can require considerable computation time for large ground-plane radii. For example, Richmond's method of moments requires approximately one-half hour of CPU time on the VAX 11/780 for a ground-plane radius $ka = 20$ with $N = 4$ element segments and $M = 60$ ground-plane annular zones (see Sec. 4.2). Although it may be possible to reduce the computation time by use of a more efficient program than the one employed by Richmond, it is of interest to find an alternative method for large ground planes.

Awadalla and Maclean [9, 10] have reduced the computation time for large ground planes by combining the method of moments with the geometric theory of diffraction. The element's current distribution is determined by the method of moments and the effect on input impedance due to ground-plane edge diffraction is determined by GTD.

The antenna geometry is shown in Fig. 1 of Sec. 2.1. The method of Awadalla and Maclean is described [9] for determining the input impedance and the radiation pattern [10].

A fictitious magnetic edge current is defined and expressed in terms of a GTD coefficient to account for diffraction by the edge of the ground plane. The GTD method is valid only for sufficiently large ground-plane radii. Unfortunately, this method, as applied by Awadalla and Maclean, does not determine the ground plane's current distribution.

The ground plane's current distribution is idealized by Awadalla and Maclean in determining the radiation pattern. The current distribution on the top of the ground plane is assumed to be that for an infinite ground plane, but it is defined only over the finite extent of the ground plane. The net current at the edge of the ground plane is set equal to zero. This constraint satisfies the boundary condition given by (2.4.4), and requires, at the edge of the ground plane, that the current on the bottom be equal, but in the opposite direction, to that on the top. The current distribution on the bottom of the ground plane is then assumed to decrease linearly from the edge of the ground plane to zero at the base of the element.

The method of moments employed by Awadalla and Maclean is not a sinusoidal-Galerkin method. Consequently, we cannot impose a sinusoidal current distribution on the element by setting the number of element segments to be $N = 1$ as a test case for purposes of comparison with other models.

A program listing was obtained from Professor K. H. Awadalla and edited at the MITRE Corp. A listing of the edited program *AWADALLA* is given in App. A.5. Program AWADALLA is written in FORTRAN language for use on a DEC VAX 11/780 computer. The CPU time for $N = 30$ element segments is less than 10 seconds.

Program AWADALLA was utilized to obtain the input impedance, directive gain, and far-field elevation pattern of a quarter-wave element of radius $b = 10^{-6}\lambda$ on ground plane of radii $8 \leq ka \leq 50$ wavenumbers. The results for input impedance are given in Table 6. These results are compared in Table 7 of Sec. 4.5 with those obtained by the method of moments ($N = 4$). As concerns the input impedance, the method of moments combined with GTD gives useful results for $ka \geq 6$, accurate results for $ka > 8$, and the suspected best available results for $14 < ka < \infty$.

Table 6 Input Impedance of a Quarter-Wave Element of Radius $b = 10^{-6}\lambda$ at the Center of a Circular Ground Plane of Radius $8 \leq ka \leq 50$ Wavenumbers
(Method of Moments Combined with Geometric Theory of Diffraction)

ka	*Input Resistance* (Ω)	*Input Reactance* (Ω)
8	35.96	18.28
9	36.55	24.45
10	41.45	21.82
11	37.54	18.67
12	36.30	23.08
13	40.49	22.51
14	38.44	19.17
15	36.36	22.18
16	39.72	22.82
17	38.99	19.70
18	36.58	21.51
19	39.09	22.89
20	36.87	21.05
22	38.58	22.84
23	39.45	20.63
24	37.20	20.73
25	38.17	22.67
26	39.48	21.06

Table 6 (cont'd)

ka	*Input Resistance* (Ω)	*Input Reactance* (Ω)
27	37.55	20.56
28	37.87	22.46
29	39.41	21.40
30	37.88	20.47
31	37.67	22.19
32	39.27	21.68
33	38.19	20.49
34	37.55	21.93
35	39.08	21.90
36	38.45	20.56
37	37.51	21.65
38	38.87	22.06
39	38.66	20.68
40	37.54	21.42
41	38.64	22.13
42	38.82	20.86
43	37.62	21.21
44	38.43	22.14
45	38.92	21.05
46	37.75	21.04
47	38.23	22.10
48	38.96	21.26
49	37.91	20.93
50	38.06	21.99

A sample computer printout of directive gain and the far-field elevation pattern obtained by program AWADALLA for $ka = 20$ is given in App. A.5. Unfortunately, the idealizations made by Awadalla and Maclean in characterizing the ground-plane current distribution yield unrealistic peak directivities and elevation patterns. For example, a peak directivity of 7.5 dBi was obtained for $ka = 49$. This result for a thin quarter-wave element seems unlikely because such an element with a sinusoidal current distribution that is mounted on a ground plane of infinite extent has a peak directivity of only 5.2 dBi.

The method of moments combined with GTD by using the definition of a fictitious magnetic edge current has also been reported by Thiele and Newhouse [11] for computing input impedance and by Stutzman and Thiele [28] for computing the far-field radiation pattern. In the method of Stutzman and Thiele [28], the far-field pattern is determined without idealization of the ground-plane current distribution.

4.4 METHOD OF IMAGES: $ka = \infty$

For the idealized case of a monopole element mounted on a ground plane of infinite extent and infinite conductivity, a monopole antenna of length h may be modeled by the method of images as a dipole of total length $2h$, but with one-half of the input impedance and double the peak numeric directivity of the dipole (see Sec. 3.8). The infinite ground plane prevents radiation into the hemisphere below it, but the ground plane generates fields in the upper hemisphere that are identical to those of a dipole.

A detailed treatment of the fields and input impedance of a dipole is given by King [13]. An excellent summary of the current state of dipole theory, including plots of input impedance as a function of dipole length and radius, is given by Elliot [14].

4.5 SUMMARY OF RESULTS

In Chapter 4, the current distribution on the element is determined by the method of moments, rather than being assumed, as was the case in Chapter 3. The determination of the element's current distribution is particularly important for element radii of $b > 10^{-4}\lambda$.

The essential difference between the two models utilized in Chapter 4 is the treatment of the ground plane's current distribution. In Richmond's model, the ground plane's current distribution is determined by the method of moments. Useful results are obtained for $0 < ka \leq 14$ wavenumbers. In the model of Awadalla and Maclean for large but finite ground planes, the input impedance is accurately computed by the introduction of a fictitious magnetic edge current, which is determined by the GTD. However, in that model, the ground-plane current distribution is not determined, but rather idealized, thus causing unrealistic peak directivities and far-field radiation patterns.

The input impedances, determined by the method of moments and the method of moments combined with GTD, are compared in Table 7 for a quarter-wave element of radius $b = 10^{-6}\lambda$ and ground-plane radii of $6 \leq ka \leq 30$ wavenumbers. The suspected best available results for input impedance are obtained by the method of moments for $0 < ka \leq 8.5$ and by the method of moments combined with GTD for $8.5 \leq ka < \infty$. The method of moments combined with GTD is inaccurate for $ka \leq 8$ (approximately 6% error in input reactance for $ka = 8$ and approximately 7% error in input resistance for $ka = 6$). Richmond's method of moments is inaccurate for $ka > 14$ (approximately 11% error in input reactance and approximately 5% error in input resistance, both for $ka = 15$).

The radiation resistance for resonant (zero input reactance) elements of various diameters is compared in Fig. 17 with the radiation resistance of an infinitely thin quarter-wave element for ground plane of radii $0 \leq ka \leq 8.5$ wavenumbers. The radiation resistance of the resonant elements is not appreciably different from that of the quarter-wave element for $ka > 1$ wavenumber. However, as $ka \to 0$, the radiation resistances of the resonant elements become increasingly different from that of the infinitely thin element because such an element on a ground plane of zero extent is resonant for an element length that is equal to 0.5λ, not 0.25λ (see (3.2.38)).

The lengths and radiation resistance of resonant elements with various diameters for ground plane of radii $0.25 \leq ka \leq 7$ wavenumbers are plotted in Figs. 18 and 19. For these ground-plane radii, the element's resonant length h_{res}/λ varies from approximately 0.22λ to 0.34λ for element radii of $10^{-7} \leq b/\lambda < 10^{-2}\lambda$. The resonant radiation resistance, for these ground-plane and element radii, varies from approximately 21Ω to 65Ω.

Table 7 Input Impedance of a Quarter-Wave Element at the Center of a Circular Ground Plane of Radius $6 \leq ka \leq 30$ Wavenumbers ($h/\lambda = 0.25$, $b/\lambda = 10^{-6}$)

Normalized Ground-Plane Radius, ka	*Input Resistance* (Ω)			*Input Reactance* (Ω)		
	(1) *Method of Moments*	(2) *Method of Moments with GTD*	$\frac{\|(1)-(2)\|}{(1)} \times 100$	(1) *Method of Moments*	(2) *Method of Moments with GTD*	$\frac{\|(1)-(2)\|}{(1)} \times 100$
6	*35.2988	37.7258	6.88	*26.7927	26.6729	0.45
7	*45.7499	42.6747	6.72	*20.5664	20.2537	1.52
8	*35.7335	35.9640	0.65	*17.1839	18.2823	6.39
9	36.2886	*36.5496	0.72	24.7424	*24.4496	1.16
10	42.0987	*41.4538	1.53	21.5758	*21.8225	1.14
11	38.1401	*37.5375	1.58	18.8073	*18.6728	0.72
12	36.7590	*36.2962	1.26	22.0824	*23.0841	4.54
13	39.7689	*40.4928	1.82	22.6158	*22.5061	0.49
14	39.6986	*38.4400	2.60	20.3449	*19.1676	5.79
15	38.3512	*36.3630	5.18	20.0026	*22.1845	10.91
20	38.4250	*39.3019	2.28	22.2706	*20.1706	9.43
25	39.0594	*38.1746	2.26	21.5934	*22.6704	4.99
30	38.8703	*37.8798	2.55	21.0052	*20.4663	2.57

*Suspected best available result.
(1) Richmond's method of moments ($N = 4$, $M = 3\ ka$).
(2) Awadalla and Maclean, method of moments combined with geometric theory of diffraction ($N = 30$).

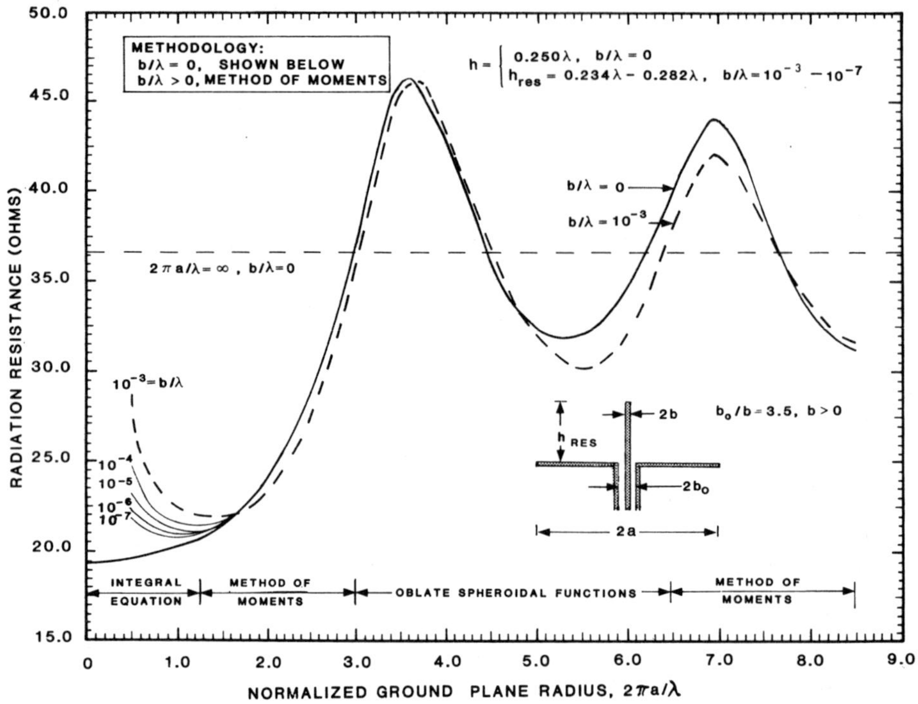

Fig. 17 Radiation Resistances of Resonant Elements of Various Diameters and an Infinitely Thin Quarter-Wave Element at the Center of a Circular Ground Plane of Radius $0 \leqslant ka \leqslant 8.5$ Wavenumbers

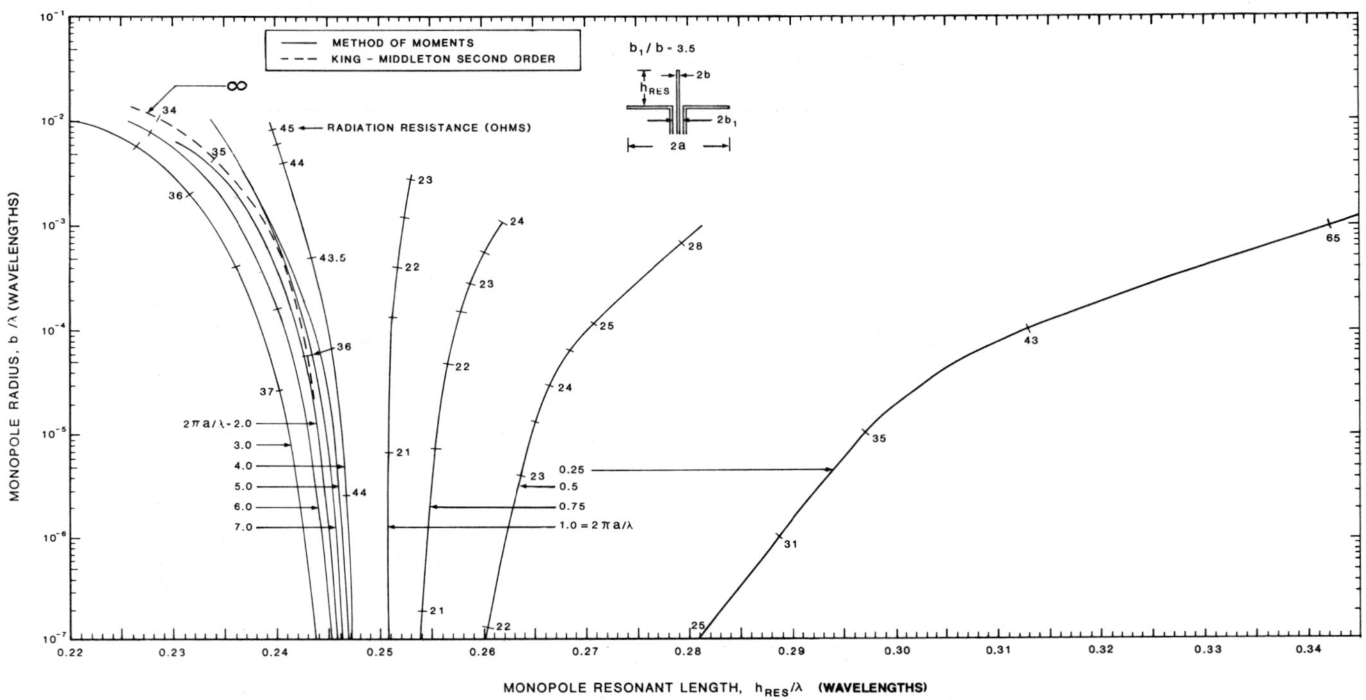

Fig. 18 Resonant Length of Elements of Various Diameters at the Center of a Circular Ground Plane of Radius $0.25 \leqslant ka \leqslant 7$ Wavenumbers

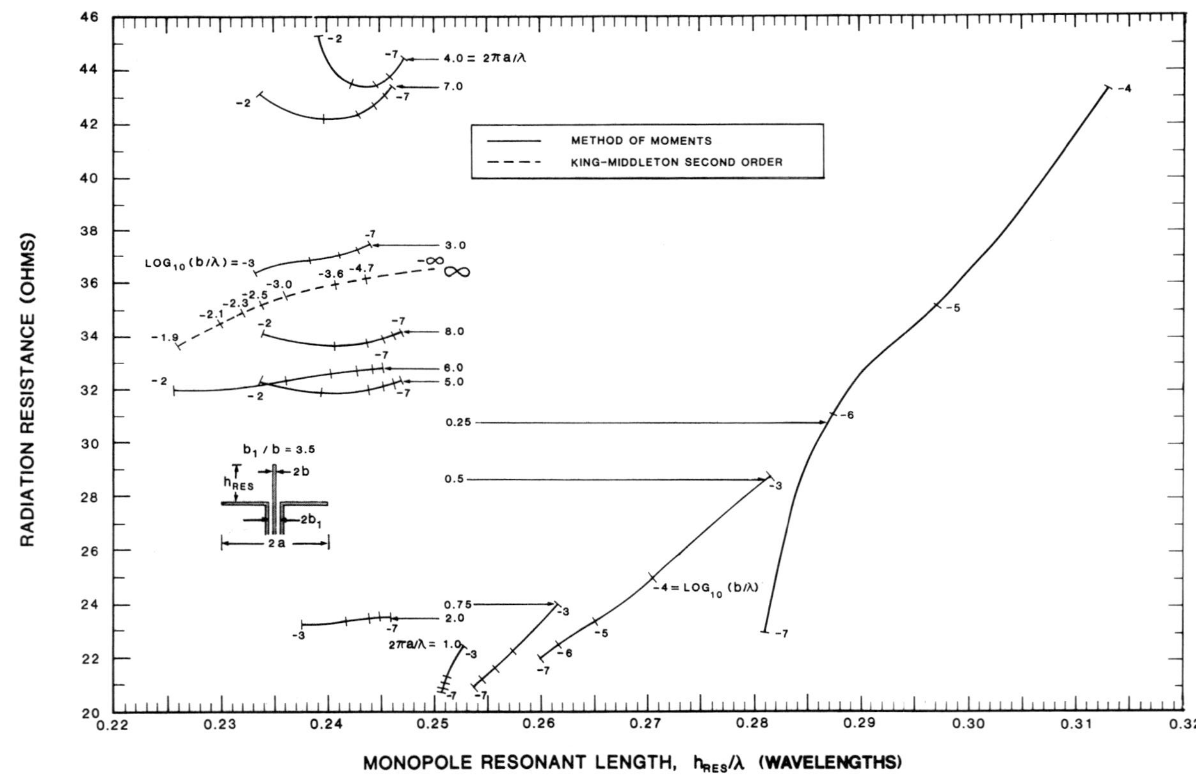

Fig. 19 Radiation Resistances and Lengths of Resonant Elements of Various Diameters at the Center of a Circular Ground Plane of Radius $0.25 \leqslant ka \leqslant 7$ Wavenumbers

Chapter 5
Comparison with Experimental Results

The input impedance and elevation patterns of several monopole antennas were measured, each at a different frequency within the 30–254 MHz band, on the MITRE Corp. VHF antenna range. Each antenna consisted of a tubular element of radius $b = 0.25$ inch mounted at the center of a ground plane of radius $a = 4$ feet, and fed by a 50-ohm RG-214 coaxial cable with a type N panel connector and a 50-ohm tapered adapter to the element. The length h of each element was approximately a quarter-wavelength. The exact length of each element was chosen to be resonant (input reactance equals zero ohms) for a ground plane of infinite extent. On a ground plane of finite extent, the input reactance is expected to approach zero asymptotically as $ka \rightarrow \infty$.

The VHF antenna range is located on a roof. The transmitting and receiving antennas are at a height of 27 feet above the roof and separated by approximately 40 feet. A conducting fence, 16 feet high and 48 feet wide, is located on the roof midway between the transmitting and receiving antennas to minimize multipath reflections from the roof. Lossy ferrite toroids (Ferronics, type-B material, 0.540-inch inner diameter × 0.870-inch outer diameter × 0.25 inch) are spaced along the receiving and transmitting cables to minimize currents on the outside of the cables. Outside RF interference is reduced to at least 40 dB below the desired signal as a result of the use of narrowband RF filters.

The monopole test antenna was operated in the receiving mode to obtain elevation patterns on the same platform used for input impedance measurements. The monopole ground plane was supported by a 10-foot wooden vertical mast, mounted on an antenna pedestal.

The input impedance was measured with a Hewlett-Packard HP-8754A network analyzer with a computerized printout. The measurement test set-up is shown in Fig. 20. The predicted input impedance of each experimental monopole antenna was determined by Richmond's method of moments, utilizing program RICHMD1. The predicted and measured input impedances are compared in Table 8. The measured input resistance differs from predicted values by approximately 1% to 10% over a range of normalized ground-plane radii = 0.77–6.5 wavenumbers. The measured input reactance differs by approximately 2Ω to 12Ω from the predicted values. Because some of the predicted values are near resonance, we do not make a percentage comparison for input reactance.

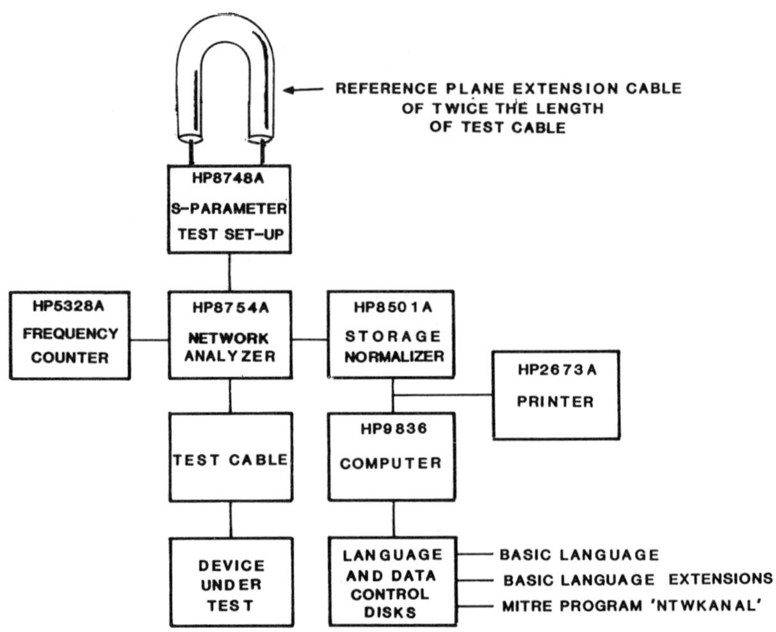

Fig. 20 Input Impedance Test Set-Up

Measured elevation patterns for most of the test monopole antennas specified in Table 8 are compared in Figs. 21–29 with theoretical patterns predicted by the method of moments (see App. A.2 for the computer printouts). Allowing for some multipath distortion by the VHF test range, the received patterns are in good agreement with the predicted pattern. We should note that the measured pattern for $ka = 0.766$ is not appreciably different from the predicted pattern for the monopole element itself. The effect of the ground plane on the shape of the pattern is not readily apparent until $ka \gtrsim 2.0$.

Table 8 Input Impedance of 0.50-Inch-Diameter Monopole Elements on an 8-Foot-Diameter Ground Plane (Comparison of method of moments predictions with MITRE measurements)

Frequency (MHz) [Wavelength, λ (in.)]	Element Length*, *h* (in.)	Normalized Ground-Plane Radius, *ka*	Normalized Element Radius, $(b/\lambda)\times 10^4$	Normalized Element Length, h/λ	Input Resistance, $R_{in}(\Omega)$: Predicted by Method of Moments (1)	Input Resistance, $R_{in}(\Omega)$: Measured by MITRE** (2)	Input Resistance, $R_{in}(\Omega)$: Percentage Difference $\frac{(1)-(2)}{(1)}\times 100$	Input Reactance*, $X_{in}(\Omega)$: Predicted by Method of Moments (3)	Input Reactance*, $X_{in}(\Omega)$: Measured by MITRE† (4)	Input Reactance*, $X_{in}(\Omega)$: Difference* $(3)-(4)(\Omega)$
30.0 [393.43]	94.26	0.766	6.35	0.2396	17.76	17.62	0.8	−35.97	−30.92	− 5.1
36.0 [327.86]	78.55	0.919	7.63	0.2396	18.35	18.57	−1.2	−25.48	−13.59	−11.9
43.0 [274.48]	65.46	1.097	9.11	0.2385	18.77	19.05	−1.5	−19.33	−16.38	− 3.0
54.0 [218.57]	52.07	1.379	11.43	0.2382	19.93	20.15	−1.1	−12.11	− 5.92	− 6.2
62.4 [189.15]	45.00	1.593	13.21	0.2379	20.80	22.82	−9.7	− 4.73	− 0.48	− 4.3
75.0 [157.37]	37.36	1.915	15.89	0.2374	22.95	23.23	−1.2	− 0.43	1.60	− 2.3
86.0 [137.24]	32.48	2.197	18.22	0.2366	25.35	27.63	−9.0	3.34	7.39	− 4.1
89.7 [131.58]	31.13	2.300	19.00	0.2366	26.59	28.16	−5.9	4.99	− 1.05	6.0
97.5 [121.05]	28.60	2.500	20.65	0.2363	29.41	31.22	−6.2	7.29	11.05	− 3.8
117.0 [100.88]	23.76	3.000	24.78	0.2355	39.27	40.50	−3.1	8.27	15.21	− 6.9
136.5 [86.47]	20.34	3.500	28.91	0.2352	45.76	46.23	−1.0	0.84	7.18	− 6.3
156.0 [75.66]	17.75	4.000	33.04	0.2346	40.39	38.59	4.7	− 8.16	− 1.09	− 7.1
175.5 [67.25]	15.77	4.500	37.17	0.2344	34.00	30.94	9.0	− 7.45	− 1.91	− 5.5
195.0 [60.53]	14.14	5.000	41.30	0.2335	30.54	28.58	6.4	− 4.36	0.51	− 4.9
214.5 [55.02]	12.82	5.500	45.44	0.2330	30.33	28.28	6.8	− 0.18	4.87	− 5.1

Table 8 (cont'd)

Frequency (MHz) [Wavelength, λ (in.)]	*Element Length*, h* (in.)	*Normalized Ground-Plane Radius, ka*	*Normalized Element Radius,* $(b/\lambda)\times 10^4$	*Normalized Element Length, h/λ*	*Input Resistance,* $R_{in}(\Omega)$			*Input Reactance*,* $X_{in}(\Omega)$		
					Predicted by Method of Moments (1)	*Measured by MITRE*** (2)		*Predicted by Method of Moments* (3)	*Measured by MITRE†* (4)	*Difference** (3) − (4)(Ω)
234.0 [50.44]	11.74	6.000	49.56	0.2328	33.69	31.44	6.7	3.64	5.56	− 1.9
253.5 [46.56]	10.87	6.500	53.69	0.2335	40.25	41.13	2.2	3.91	6.57	− 2.7

*The element lengths were selected to be resonant when mounted on a ground plane of infinite extent [see R.W.P. King, *Theory of Linear Antennas* (Cambridge, MA: Harvard University Press, 1956) Fig. 30.7a]. On ground planes of finite extent, the input reactance is expected to approach zero asymptotically as $ka \to \infty$.

†MITRE measurements were made on Jan. 7 and Jan. 10, 1985 utilizing computerized printout of network analyzer.

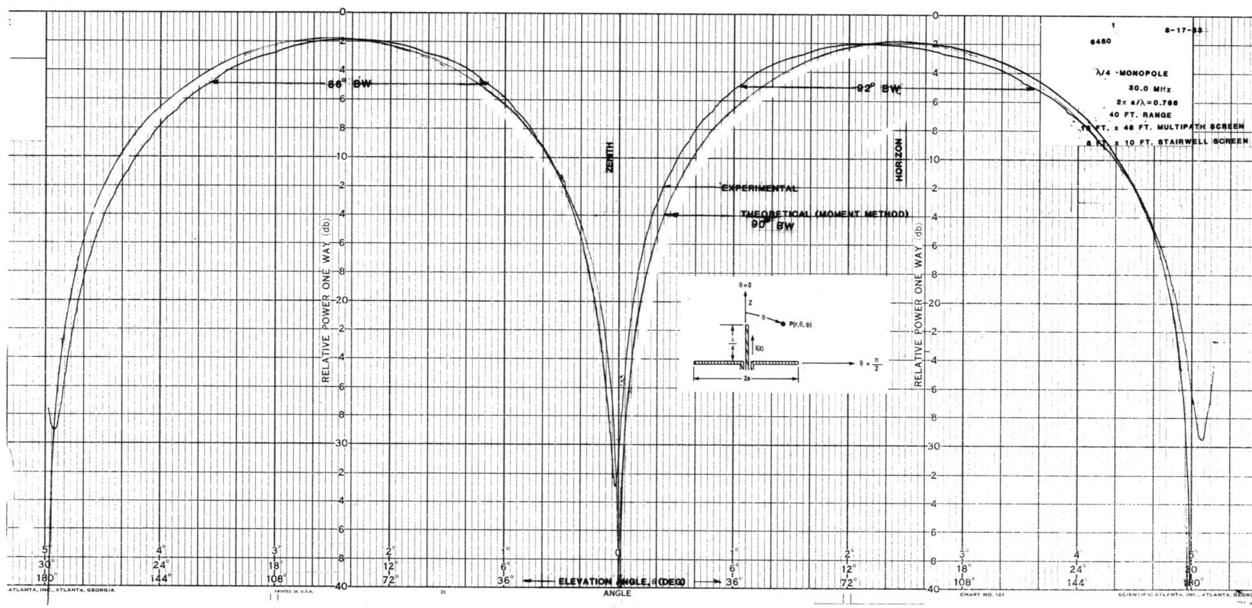

Fig. 21 Measured and Theoretical Elevation Patterns, 30 MHz
($h/\lambda = 0.2396$, $b/\lambda = 6.35 \times 10^{-4}$, $ka = 0.766$)

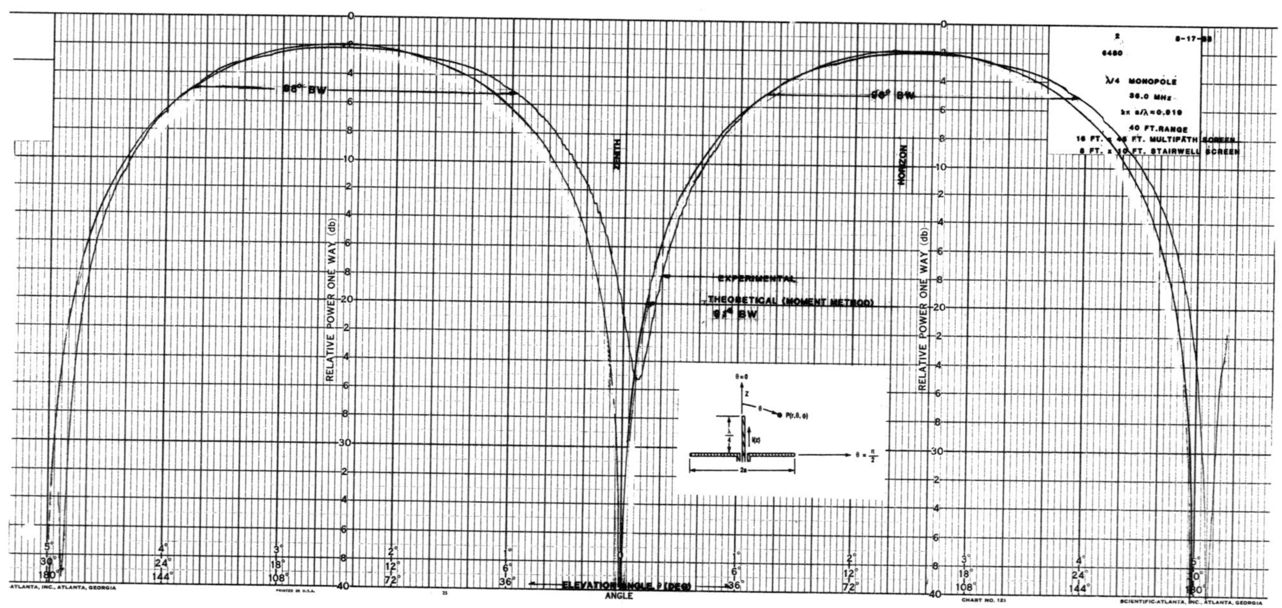

Fig. 22 Measured and Theoretical Elevation Patterns, 36 MHz
($h/\lambda = 0.2396$, $b/\lambda = 7.63 \times 10^{-4}$, $ka = 0.919$)

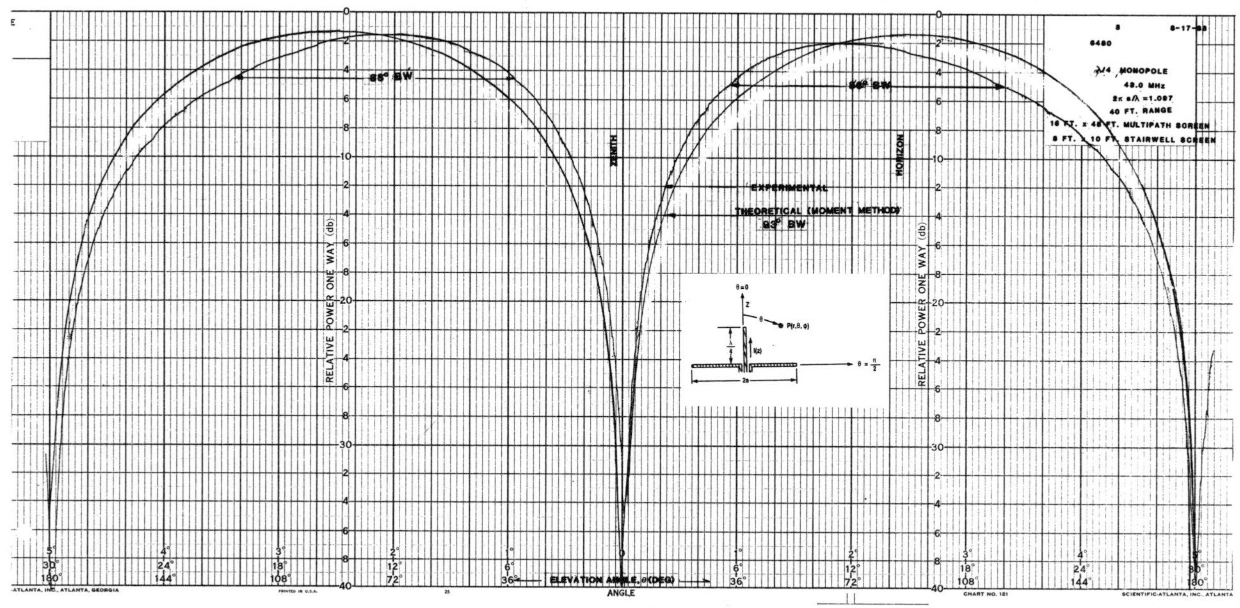

Fig. 23 Measured and Theoretical Elevation Patterns, 43 MHz
($h/\lambda = 0.2385$, $b/\lambda = 9.11 \times 10^{-4}$, $ka = 1.097$)

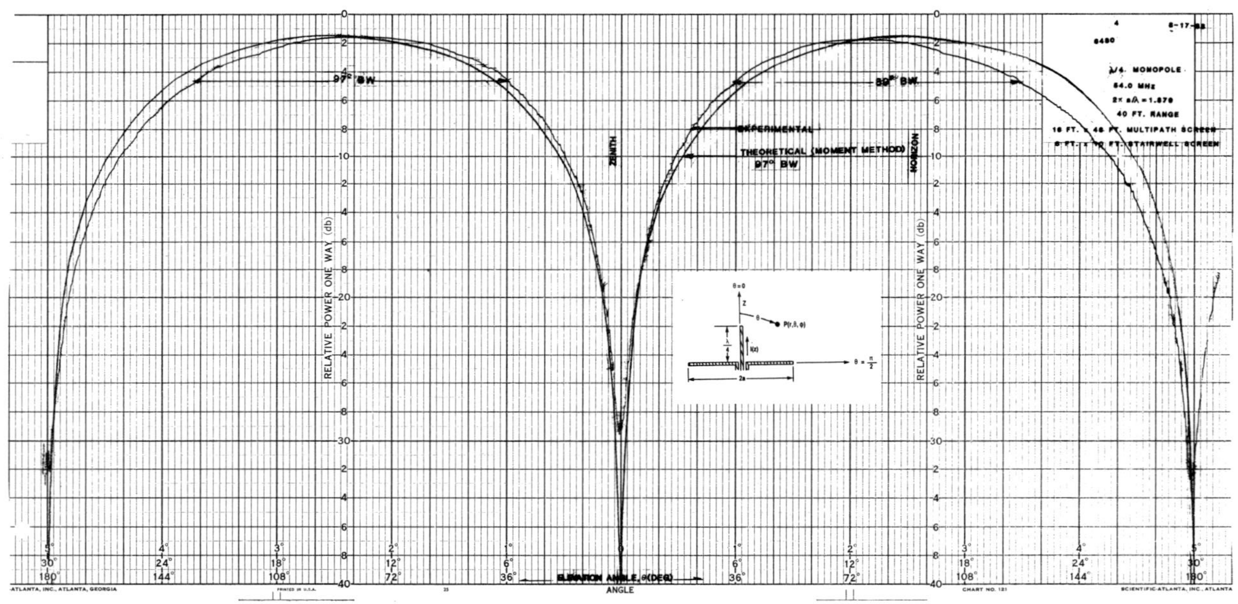

Fig. 24 Measured and Theoretical Elevation Patterns, 54 MHz
($h/\lambda = 0.2382$, $b/\lambda = 11.43 \times 10^{-4}$, $ka = 1.379$)

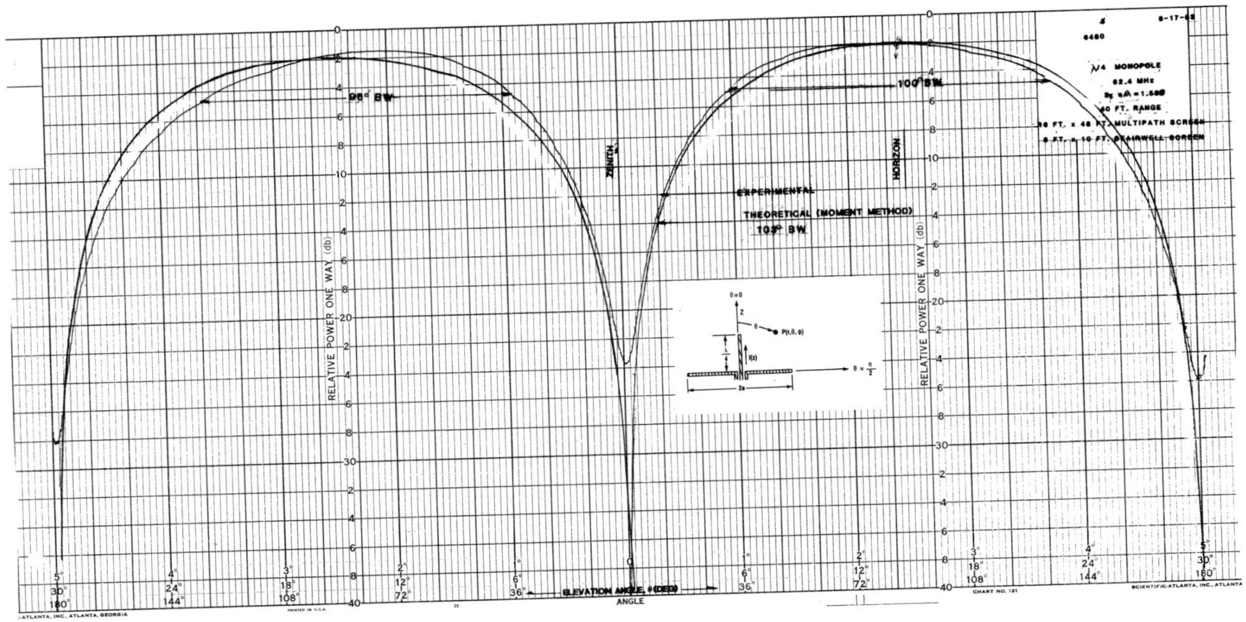

Fig. 25 Measured and Theoretical Elevation Patterns, 62.4 MHz ($h/\lambda = 0.2379$, $b/\lambda = 13.21 \times 10^{-4}$, $ka = 1.593$)

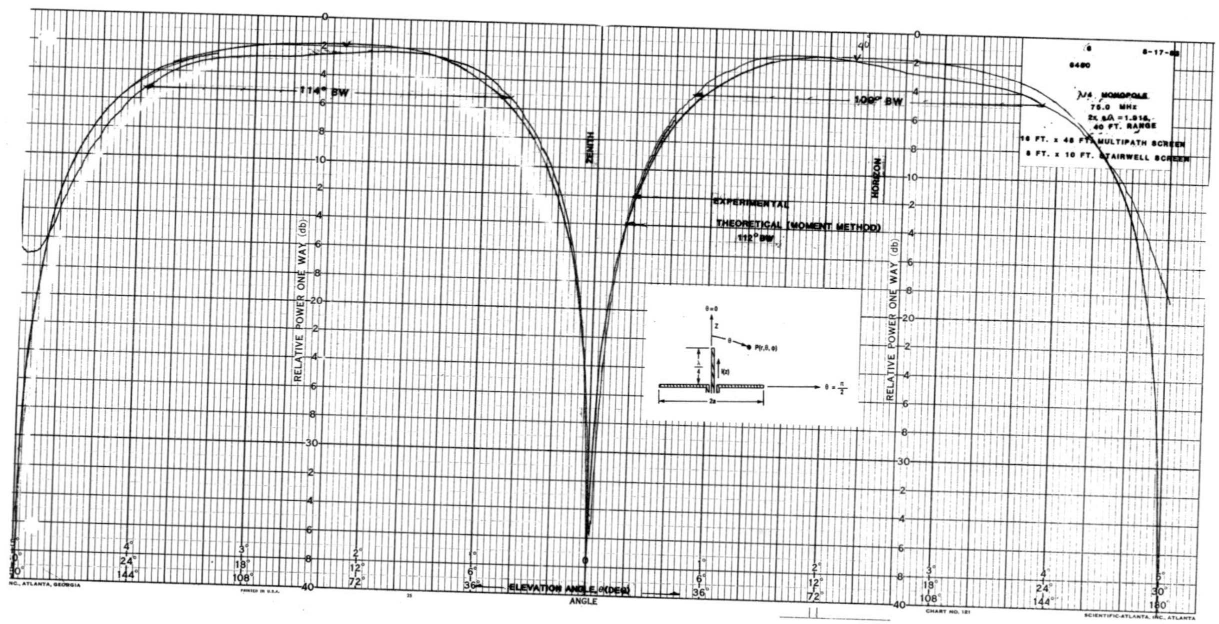

Fig. 26 Measured and Theoretical Elevation Patterns, 75 MHz ($h/\lambda = 0.2374$, $b/\lambda = 15.89 \times 10^{-4}$, $ka = 1.915$)

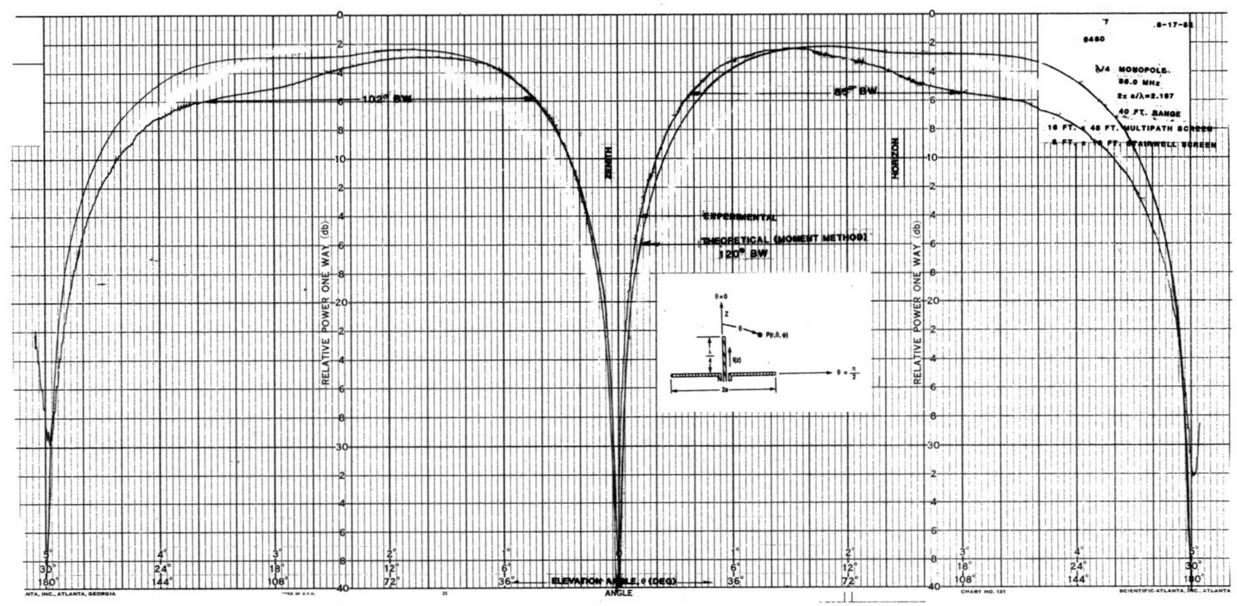

Fig. 27 Measured and Theoretical Elevation Patterns, 86 MHz ($h/\lambda = 0.2366$, $b/\lambda = 18.22 \times 10^{-4}$, $ka = 2.197$)

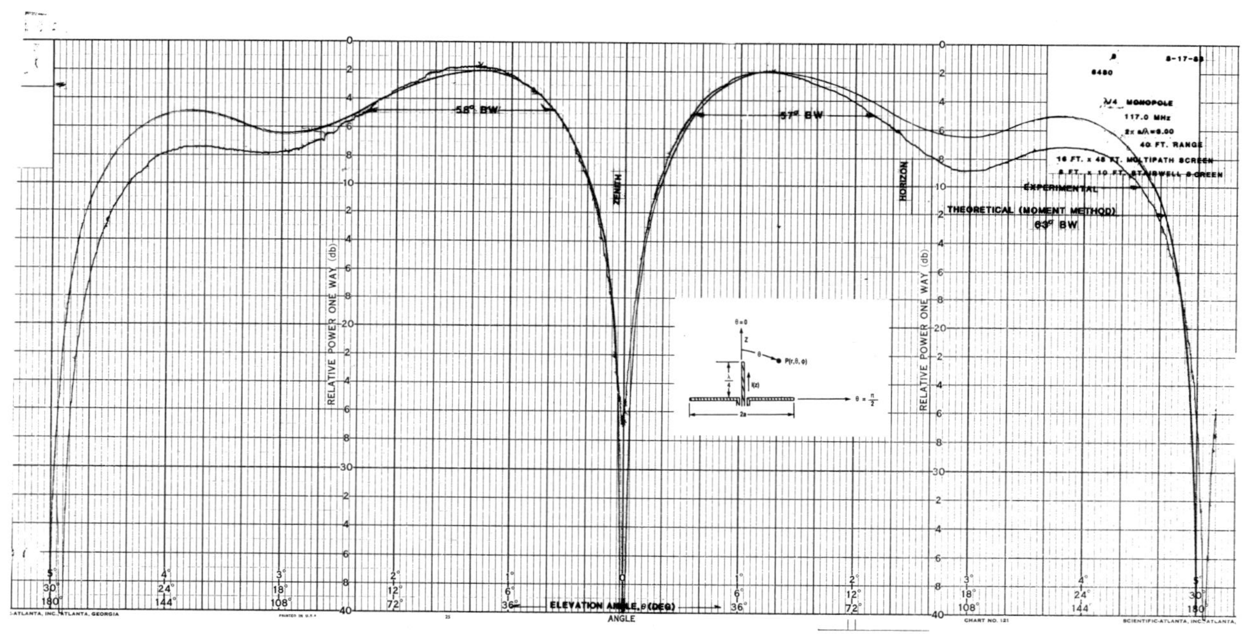

Fig. 28 Measured and Theoretical Elevation Patterns, 117 MHz ($h/\lambda = 0.2355$, $b/\lambda = 24.78 \times 10^{-4}$, $ka = 3.000$)

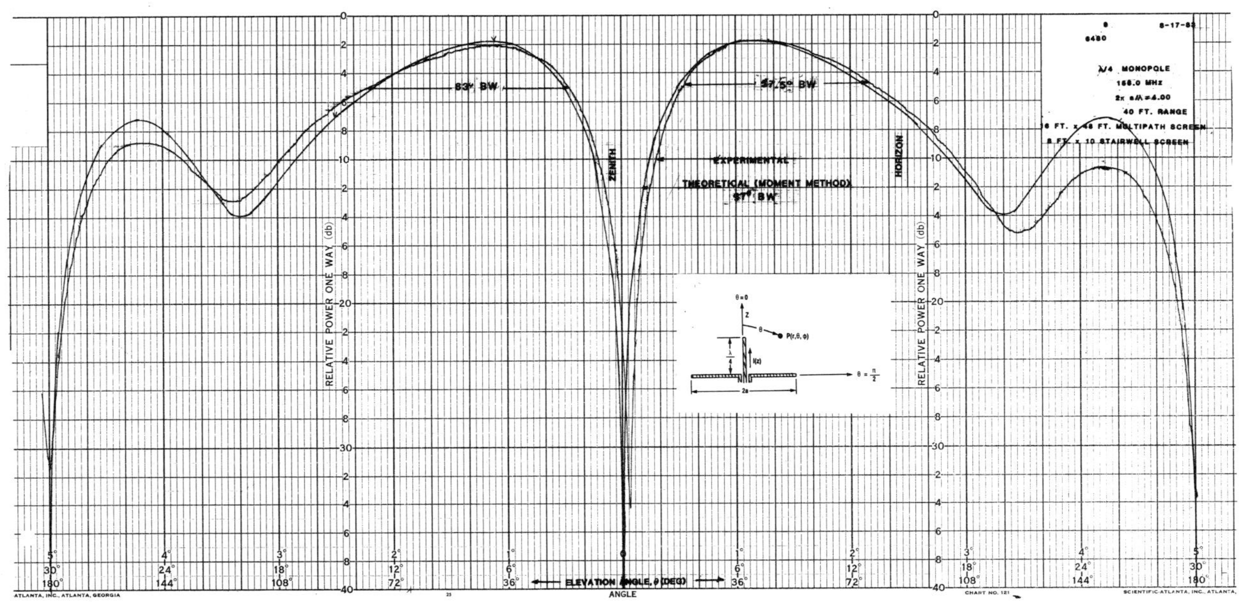

Fig. 29 Measured and Theoretical Elevation Patterns, 156 MHz ($h/\lambda = 0.2346$, $b/\lambda = 33.04 \times 10^{-4}$, $ka = 4.000$)

REFERENCES

1. J. Bardeen, "The Diffraction of a Circularly Symmetrical Electromagnetic Wave by a Co-axial Circular Disc of Infinite Conductivity," *Physical Review,* Vol. 36, Nov. 1930, pp. 1482–1488.
2. J.H. Richmond, "Monopole Antenna on Circular Disk," Technical Report 711639-1, Electro Science Laboratory, Ohio State University, July 1979. See also *IEEE Trans. Antennas and Propagation,* Vol. AP-32, No. 12, Dec. 1984, pp. 1282–1287.
3. A. Leitner and R.D. Spence, "The Oblate Spheroidal Wave Functions," *Franklin Institute J.,* Vol. 249, No. 4, April 1950, pp. 299–321.
4. A. Leitner, "Effect of a Circular Groundplane on Antenna Radiation," Mathematics Research Group, Research Report No. EM-19, New York University, April 1950.
5. A. Leitner and R.D. Spence, "Effect of a Circular Groundplane on Antenna Radiation," *J. Applied Physics,* Vol. 21, Oct. 1950, pp. 1001–1006.
6. C.L. Tang, "On the Radiation Pattern of a Base-Driven Antenna Over a Circular Conducting Screen," Technical Report No. 301, Cruft Laboratory, Harvard University, May 1959.
7. J.E. Storer, "The Impedance of an Antenna over a Large Circular Screen," *J. Applied Physics,* Vol. 22, Aug. 1951, pp. 1058–1066. See also Technical Report No. 119, Cruft Laboratory, Harvard University, Nov. 1950.
8. J.E. Storer, "The Radiation Pattern of an Antenna over a Circular Ground Screen," Technical Report No. 126, Cruft Laboratory, Harvard University, June 1951.
9. K.H. Awadalla and T.S.M. Maclean, "Input Impedance of a Monopole Antenna at the Center of a Finite Groundplane," *IEEE Trans. Antennas and Propagation,* Vol. AP-26, No. 2, March 1978, pp. 244–248.

10. K.H. Awadalla and T.S.M. Maclean, "Monopole Antenna at the Center of a Circular Groundplane: Input Impedance and Radiation Pattern," *IEEE Trans. Antennas and Propagation,* Vol. AP-27, No. 2, March 1979, pp. 151–153.
11. G.A. Thiele and T.H. Newhouse, "A Hybrid Technique for Combining Moment Methods with the Geometrical Theory of Diffraction," *IEEE Trans. Antennas and Propagation,* Vol. AP-23, No. 1, Jan. 1975, pp. 62–69.
12. R.F. Harrington, *Time-Harmonic Electromagnetic Fields,* New York: McGraw-Hill, 1961, pp. 106–112. Please note that the signs of the radial field E_ρ on p. 111 and the magnetic frill M_ϕ on p. 112 differ from those in (2.4.9) and (2.4.8), respectively, because in [12] the voltage at the outer conductor is positive with respect to that at the inner conductor.
13. R.W.P. King, *The Theory of Linear Antennas,* Cambridge, MA: Harvard University Press, 1956, p. 87.
14. R.S. Elliot, *Antenna Theory and Design,* Englewood Cliffs, NJ: Prentice-Hall, 1981, pp. 290–291.
15. C.A. Balanis, *Antenna Theory: Analysis and Design,* New York: Harper and Row, 1982, p. 315.
16. *Ibid.,* p. 101, Eq. (4-2).
17. L. Brillouin, "Origin of Radiation Resistance," *Radioélectricité,* Vol. 3, 1922, pp. 147–152.
18. J.A. Stratton, *Electromagnetic Theory,* New York: McGraw-Hill, 1941, pp. 454–457.
19. M. Abramowitz and I.A. Stegun, *Handbook of Mathematical Functions,* New York: Dover Publications, 1972, pp. 231, 244.
20. Balanis, *op. cit.,* p. 292.
21. W. Magnus and F. Oberhettigner, *Formulas and Theorems for the Functions of Mathematical Physics,* New York: Chelsea Publishing, 1954, p. 34.
22. I.S. Gradshteyn and I.M. Ryzhik, *Table of Integrals, Series and Products,* New York: Academic Press, 1965.
23. J.B. Thomas, *An Introduction to Statistical Communication Theory,* New York: John Wiley and Sons, 1969, Appendix B: The Dirac Delta Function.
24. Abramowitz and Stegun, *op. cit.,* p. 763.
25. Balanis, *op. cit.,* pp. 118–126, 133–135, 285–295.
26. Elliot, *op. cit.,* p. 29.

27. Balanis, *op. cit.,* pp. 502–522, 529–531.
28. W.J. Stutzman and G.A. Thiele, *Antenna Theory and Design,* New York: John Wiley and Sons, 1981, pp. 489–495, 500–507.

Part II
Appendices

Appendix A
Computer Printouts of Directive Gain and the Far-Field Elevation Pattern

A.1 INTEGRAL EQUATION

Tables A1-1–A1-13; $h/\lambda = 0.25,\ 0 \leqslant ka \leqslant 3.0$

TABLE A1-1

INTEGRAL EQUATION METHOD
EPSILON= 0.0 (NORMALIZED GROUNDPLANE RADIUS)
MONOPOLE LENGTH= 0.250000(WAVELENGTH)
INFINITELY THIN MONOPOLE RADIUS
RADIATION RESISTANCE = 19.434878 (OHMS)

ELEVATION ANGLE θ (DEG)	DIRECTIVITY (NUMERIC)	DIRECTIVITY (DBI)	RELATIVE POWER (DB)	ELEVATION ANGLE (DEG)	DIRECTIVITY (NUMERIC)	DIRECTIVITY (DBI)	RELATIVE POWER (DB)
0.	0.00000	-INFINITY	-INFINITY				
2.0	0.00163	-27.88102	-29.76341	92.0	1.54040	1.87635	-0.00604
4.0	0.00651	-21.86343	-23.74582	94.0	1.53399	1.85821	-0.02418
6.0	0.01463	-18.34664	-20.22903	96.0	1.52334	1.82796	-0.05443
8.0	0.02597	-15.85494	-17.73733	98.0	1.50853	1.78553	-0.09686
10.0	0.04050	-13.92589	-15.80828	100.0	1.48966	1.73087	-0.15152
12.0	0.05816	-12.35353	-14.23592	102.0	1.46686	1.66389	-0.21850
14.0	0.07892	-11.02803	-12.91042	104.0	1.44028	1.58446	-0.29793
16.0	0.10271	-9.88386	-11.76625	106.0	1.41009	1.49247	-0.38992
18.0	0.12946	-8.87877	-10.76116	108.0	1.37649	1.38774	-0.49464
20.0	0.15908	-7.98396	-9.86635	110.0	1.33971	1.27010	-0.61229
22.0	0.19147	-7.17889	-9.06128	112.0	1.29996	1.13930	-0.74309
24.0	0.22654	-6.44846	-8.33085	114.0	1.25751	0.99511	-0.88728
26.0	0.26417	-5.78120	-7.66359	116.0	1.21261	0.83722	-1.04517
28.0	0.30421	-5.16823	-7.05062	118.0	1.16554	0.66528	-1.21711
30.0	0.34653	-4.60256	-6.48495	120.0	1.11659	0.47892	-1.40347
32.0	0.39097	-4.07857	-5.96096	122.0	1.06603	0.27768	-1.60471
34.0	0.43735	-3.59167	-5.47406	124.0	1.01415	0.06104	-1.82135
36.0	0.48550	-3.13812	-5.02050	126.0	0.96126	-0.17158	-2.05397
38.0	0.53521	-2.71478	-4.59717	128.0	0.90764	-0.42085	-2.30324
40.0	0.58627	-2.31905	-4.20143	130.0	0.85358	-0.68756	-2.56995
42.0	0.63845	-1.94870	-3.83109	132.0	0.79936	-0.97259	-2.85498
44.0	0.69153	-1.60188	-3.48427	134.0	0.74525	-1.27697	-3.15936
46.0	0.74525	-1.27697	-3.15936	136.0	0.69153	-1.60188	-3.48427
48.0	0.79936	-0.97259	-2.85498	138.0	0.63845	-1.94870	-3.83109
50.0	0.85358	-0.68756	-2.56995	140.0	0.58627	-2.31905	-4.20143
52.0	0.90764	-0.42085	-2.30324	142.0	0.53521	-2.71478	-4.59717
54.0	0.96126	-0.17158	-2.05397	144.0	0.48550	-3.13812	-5.02050
56.0	1.01415	0.06104	-1.82135	146.0	0.43735	-3.59167	-5.47406
58.0	1.06603	0.27768	-1.60471	148.0	0.39097	-4.07857	-5.96096
60.0	1.11659	0.47892	-1.40347	150.0	0.34653	-4.60256	-6.48495
62.0	1.16554	0.66528	-1.21711	152.0	0.30421	-5.16823	-7.05062
64.0	1.21261	0.83722	-1.04517	154.0	0.26417	-5.78120	-7.66359
66.0	1.25751	0.99511	-0.88728	156.0	0.22654	-6.44846	-8.33085
68.0	1.29996	1.13930	-0.74309	158.0	0.19147	-7.17889	-9.06128
70.0	1.33971	1.27010	-0.61229	160.0	0.15908	-7.98396	-9.86635
72.0	1.37649	1.38774	-0.49464	162.0	0.12946	-8.87877	-10.76116
74.0	1.41009	1.49247	-0.38992	164.0	0.10271	-9.88386	-11.76625
76.0	1.44028	1.58446	-0.29793	166.0	0.07892	-11.02803	-12.91042
78.0	1.46686	1.66389	-0.21850	168.0	0.05816	-12.35353	-14.23592
80.0	1.48966	1.73087	-0.15152	170.0	0.04050	-13.92589	-15.80828
82.0	1.50853	1.78553	-0.09686	172.0	0.02597	-15.85494	-17.73733
84.0	1.52334	1.82796	-0.05443	174.0	0.01463	-18.34664	-20.22903
86.0	1.53399	1.85821	-0.02418	176.0	0.00651	-21.86343	-23.74582
88.0	1.54040	1.87635	-0.00604	178.0	0.00163	-27.88102	-29.76341
90.0	1.54255	1.88239	0.0	180.	0.00000	-INFINITY	INFINITY

TABLE A1-2

INTEGRAL EQUATION METHOD
EPSILON= 0.2500(NORMALIZED GROUNDPLANE RADIUS)
MONOPOLE LENGTH= 0.250000(WAVELENGTH)
INFINITELY THIN MONOPOLE RADIUS
RADIATION RESISTANCE = 19.480016 (OHMS)

ELEVATION ANGLE (DEG)	DIRECTIVITY (NUMERIC)	DIRECTIVITY (DBI)	RELATIVE POWER (DB)	ELEVATION ANGLE (DEG)	DIRECTIVITY (NUMERIC)	DIRECTIVITY (DBI)	RELATIVE POWER (DB)
0.	0.00000	-INFINITY	-INFINITY				
2.0	0.00164	-27.83892	-29.71123	92.0	1.53686	1.86634	-0.00598
4.0	0.00657	-21.82152	-23.69383	94.0	1.53052	1.84839	-0.02393
6.0	0.01477	-18.30504	-20.17735	96.0	1.52000	1.81845	-0.05387
8.0	0.02622	-15.81378	-17.68610	98.0	1.50538	1.77646	-0.09586
10.0	0.04088	-13.88530	-15.75762	100.0	1.48674	1.72236	-0.14996
12.0	0.05870	-12.31364	-14.18595	102.0	1.46421	1.65605	-0.21627
14.0	0.07964	-10.98895	-12.86126	104.0	1.43794	1.57742	-0.29490
16.0	0.10362	-9.84569	-11.71801	106.0	1.40810	1.48633	-0.38599
18.0	0.13057	-8.84163	-10.71395	108.0	1.37487	1.38262	-0.48970
20.0	0.16040	-7.94795	-9.82026	110.0	1.33847	1.26608	-0.60623
22.0	0.19301	-7.14411	-9.01642	112.0	1.29912	1.13650	-0.73581
24.0	0.22830	-6.41499	-8.28731	114.0	1.25707	0.99361	-0.87871
26.0	0.26613	-5.74914	-7.62145	116.0	1.21258	0.83710	-1.03521
28.0	0.30636	-5.13766	-7.00997	118.0	1.16591	0.66664	-1.20568
30.0	0.34886	-4.57354	-6.44586	120.0	1.11733	0.48181	-1.39050
32.0	0.39344	-4.05116	-5.92348	122.0	1.06713	0.28217	-1.59014
34.0	0.43995	-3.56594	-5.43825	124.0	1.01559	0.06720	-1.80512
36.0	0.48819	-3.11410	-4.98642	126.0	0.96300	-0.16372	-2.03603
38.0	0.53796	-2.69252	-4.56484	128.0	0.90965	-0.41124	-2.28356
40.0	0.58904	-2.29858	-4.17089	130.0	0.85582	-0.67617	-2.54849
42.0	0.64120	-1.93004	-3.80236	132.0	0.80179	-0.95940	-2.83171
44.0	0.69422	-1.58504	-3.45735	134.0	0.74783	-1.26196	-3.13427
46.0	0.74783	-1.26195	-3.13426	136.0	0.69422	-1.58505	-3.45736
48.0	0.80179	-0.95939	-2.83170	138.0	0.64120	-1.93005	-3.80237
50.0	0.85582	-0.67616	-2.54847	140.0	0.58904	-2.29859	-4.17090
52.0	0.90965	-0.41123	-2.28355	142.0	0.53796	-2.69254	-4.56485
54.0	0.96301	-0.16371	-2.03602	144.0	0.48819	-3.11411	-4.98643
56.0	1.01559	0.06720	-1.80511	146.0	0.43995	-3.56595	-5.43827
58.0	1.06713	0.28218	-1.59013	148.0	0.39344	-4.05118	-5.92349
60.0	1.11733	0.48182	-1.39049	150.0	0.34885	-4.57356	-6.44587
62.0	1.16591	0.66665	-1.20567	152.0	0.30636	-5.13767	-7.00999
64.0	1.21258	0.83711	-1.03520	154.0	0.26612	-5.74915	-7.62147
66.0	1.25708	0.99362	-0.87870	156.0	0.22830	-6.41501	-8.28732
68.0	1.29912	1.13651	-0.73581	158.0	0.19301	-7.14412	-9.01644
70.0	1.33847	1.26609	-0.60623	160.0	0.16040	-7.94796	-9.82028
72.0	1.37487	1.38262	-0.48969	162.0	0.13057	-8.84165	-10.71396
74.0	1.40810	1.48633	-0.38598	164.0	0.10362	-9.84571	-11.71802
76.0	1.43794	1.57742	-0.29489	166.0	0.07963	-10.98896	-12.86128
78.0	1.46422	1.65605	-0.21626	168.0	0.05870	-12.31365	-14.18597
80.0	1.48674	1.72236	-0.14996	170.0	0.04088	-13.88532	-15.75763
82.0	1.50538	1.77646	-0.09585	172.0	0.02622	-15.81380	-17.68611
84.0	1.52000	1.81845	-0.05387	174.0	0.01477	-18.30506	-20.17737
86.0	1.53052	1.84839	-0.02392	176.0	0.00657	-21.82153	-23.69385
88.0	1.53686	1.86634	-0.00598	178.0	0.00164	-27.83893	-29.71125
90.0	1.53897	1.87231	0.0	180.	0.00000	-INFINITY	-INFINITY

TABLE A1-3

INTEGRAL EQUATION METHOD
EPSILON= 0.5000(NORMALIZED GROUNDPLANE RADIUS)
MONOPOLE LENGTH= 0.250000(WAVELENGTH)
INFINITELY THIN MONOPOLE RADIUS
RADIATION RESISTANCE = 19.618316 (OHMS)

ELEVATION ANGLE (DEG)	DIRECTIVITY (NUMERIC)	DIRECTIVITY (DBI)	RELATIVE POWER (DB)	ELEVATION ANGLE (DEG)	DIRECTIVITY (NUMERIC)	DIRECTIVITY (DBI)	RELATIVE POWER (DB)
0.	0.00000	-INFINITY	-INFINITY				
2.0	0.00169	-27.71176	-29.55335	92.0	1.52609	1.83579	-0.00580
4.0	0.00677	-21.69493	-23.53652	94.0	1.51999	1.81841	-0.02318
6.0	0.01521	-18.17942	-20.02101	96.0	1.50988	1.78942	-0.05217
8.0	0.02698	-15.68950	-17.53109	98.0	1.49580	1.74875	-0.09284
10.0	0.04205	-13.76273	-15.60432	100.0	1.47786	1.69634	-0.14526
12.0	0.06035	-12.19313	-14.03472	102.0	1.45616	1.63208	-0.20951
14.0	0.08183	-10.87086	-12.71245	104.0	1.43082	1.55585	-0.28574
16.0	0.10641	-9.73037	-11.57196	106.0	1.40201	1.46750	-0.37409
18.0	0.13399	-8.72940	-10.57099	108.0	1.36989	1.36685	-0.47474
20.0	0.16447	-7.83912	-9.68071	110.0	1.33466	1.25369	-0.58790
22.0	0.19774	-7.03898	-8.88057	112.0	1.29652	1.12778	-0.71381
24.0	0.23368	-6.31384	-8.15543	114.0	1.25569	0.98884	-0.85275
26.0	0.27213	-5.65221	-7.49380	116.0	1.21242	0.83654	-1.00505
28.0	0.31295	-5.04520	-6.88679	118.0	1.16695	0.67052	-1.17108
30.0	0.35598	-4.48577	-6.32736	120.0	1.11953	0.49035	-1.35124
32.0	0.40103	-3.96826	-5.80985	122.0	1.07042	0.29556	-1.54603
34.0	0.44791	-3.48807	-5.32966	124.0	1.01991	0.08560	-1.75599
36.0	0.49643	-3.04142	-4.88301	126.0	0.96824	-0.14015	-1.98174
38.0	0.54637	-2.62515	-4.46674	128.0	0.91572	-0.38239	-2.22398
40.0	0.59750	-2.23659	-4.07818	130.0	0.86259	-0.64194	-2.48353
42.0	0.64960	-1.87352	-3.71511	132.0	0.80915	-0.91972	-2.76131
44.0	0.70242	-1.53402	-3.37561	134.0	0.75565	-1.21679	-3.05838
46.0	0.75571	-1.21644	-3.05803	136.0	0.70236	-1.53438	-3.37597
48.0	0.80921	-0.91938	-2.76097	138.0	0.64955	-1.87390	-3.71549
50.0	0.86266	-0.64161	-2.48320	140.0	0.59745	-2.23698	-4.07857
52.0	0.91578	-0.38207	-2.22366	142.0	0.54632	-2.62554	-4.46713
54.0	0.96831	-0.13984	-1.98143	144.0	0.49638	-3.04183	-4.88342
56.0	1.01997	0.08589	-1.75570	146.0	0.44787	-3.48849	-5.33008
58.0	1.07049	0.29584	-1.54576	148.0	0.40099	-3.96868	-5.81027
60.0	1.11960	0.49061	-1.35098	150.0	0.35594	-4.48619	-6.32778
62.0	1.16701	0.67076	-1.17083	152.0	0.31292	-5.04563	-6.88722
64.0	1.21249	0.83677	-1.00482	154.0	0.27210	-5.65265	-7.49424
66.0	1.25576	0.98906	-0.85254	156.0	0.23365	-6.31428	-8.15587
68.0	1.29658	1.12798	-0.71361	158.0	0.19772	-7.03943	-8.88102
70.0	1.33471	1.25388	-0.58771	160.0	0.16445	-7.83957	-9.68117
72.0	1.36994	1.36702	-0.47457	162.0	0.13397	-8.72986	-10.57145
74.0	1.40205	1.46765	-0.37394	164.0	0.10639	-9.73084	-11.57243
76.0	1.43086	1.55598	-0.28561	166.0	0.08182	-10.87133	-12.71292
78.0	1.45619	1.63219	-0.20940	168.0	0.06034	-12.19360	-14.03519
80.0	1.47789	1.69643	-0.14516	170.0	0.04204	-13.76320	-15.60479
82.0	1.49583	1.74882	-0.09277	172.0	0.02698	-15.68997	-17.53156
84.0	1.50990	1.78947	-0.05212	174.0	0.01521	-18.17989	-20.02148
86.0	1.52001	1.81845	-0.02314	176.0	0.00677	-21.69541	-23.53700
88.0	1.52609	1.83581	-0.00578	178.0	0.00169	-27.71223	-29.55382
90.0	1.52813	1.84159	0.0	180.	0.00000	-INFINITY	-INFINITY

TABLE A1-4

INTEGRAL EQUATION METHOD
EPSILON= 0.7500(NORMALIZED GROUNDPLANE RADIUS)
MONOPOLE LENGTH= 0.250000(WAVELENGTH)
INFINITELY THIN MONOPOLE RADIUS
RADIATION RESISTANCE = 19.860252 (OHMS)

ELEVATION ANGLE (DEG)	DIRECTIVITY (NUMERIC)	DIRECTIVITY (DBI)	RELATIVE POWER (DB)	ELEVATION ANGLE (DEG)	DIRECTIVITY (NUMERIC)	DIRECTIVITY (DBI)	RELATIVE POWER (DB)
0.	0.00000	-INFINITY	-INFINITY				
2.0	0.00178	-27.49547	-29.28383	92.0	1.50759	1.78283	-0.00553
4.0	0.00711	-21.47962	-23.26798	94.0	1.50189	1.76638	-0.02198
6.0	0.01597	-17.96572	-19.75408	96.0	1.49244	1.73897	-0.04939
8.0	0.02833	-15.47805	-17.26641	98.0	1.47929	1.70053	-0.08783
10.0	0.04411	-13.55415	-15.34251	100.0	1.46251	1.65098	-0.13738
12.0	0.06327	-11.98804	-13.77640	102.0	1.44218	1.59021	-0.19815
14.0	0.08571	-10.66985	-12.45821	104.0	1.41843	1.51807	-0.27029
16.0	0.11133	-9.53401	-11.32237	106.0	1.39136	1.43439	-0.35397
18.0	0.14002	-8.53824	-10.32660	108.0	1.36112	1.33897	-0.44939
20.0	0.17164	-7.65369	-9.44205	110.0	1.32788	1.23157	-0.55679
22.0	0.20607	-6.85979	-8.64815	112.0	1.29179	1.11193	-0.67643
24.0	0.24314	-6.14135	-7.92971	114.0	1.25306	0.97972	-0.80864
26.0	0.28269	-5.48687	-7.27523	116.0	1.21188	0.83459	-0.95377
28.0	0.32453	-4.88740	-6.67576	118.0	1.16846	0.67614	-1.11222
30.0	0.36848	-4.33589	-6.12425	120.0	1.12303	0.50391	-1.28445
32.0	0.41432	-3.82663	-5.61499	122.0	1.07582	0.31738	-1.47098
34.0	0.46185	-3.35498	-5.14334	124.0	1.02706	0.11597	-1.67239
36.0	0.51084	-2.91712	-4.70548	126.0	0.97702	-0.10099	-1.88935
38.0	0.56107	-2.50985	-4.29821	128.0	0.92592	-0.33424	-2.12260
40.0	0.61229	-2.13046	-3.91882	130.0	0.87405	-0.58465	-2.37301
42.0	0.66425	-1.77667	-3.56503	132.0	0.82164	-0.85319	-2.64155
44.0	0.71672	-1.44653	-3.23489	134.0	0.76896	-1.14095	-2.92931
46.0	0.76942	-1.13835	-2.92671	136.0	0.71627	-1.44921	-3.23757
48.0	0.82212	-0.85067	-2.63903	138.0	0.66383	-1.77942	-3.56778
50.0	0.87454	-0.58222	-2.37058	140.0	0.61189	-2.13327	-3.92163
52.0	0.92642	-0.33190	-2.12026	142.0	0.56070	-2.51273	-4.30109
54.0	0.97752	-0.09874	-1.88710	144.0	0.51050	-2.92006	-4.70842
56.0	1.02757	0.11813	-1.67023	146.0	0.46153	-3.35798	-5.14634
58.0	1.07633	0.31943	-1.46893	148.0	0.41403	-3.82967	-5.61803
60.0	1.12353	0.50586	-1.28250	150.0	0.36822	-4.33898	-6.12734
62.0	1.16895	0.67798	-1.11038	152.0	0.32430	-4.89054	-6.67890
64.0	1.21236	0.83631	-0.95205	154.0	0.28248	-5.49005	-7.27841
66.0	1.25352	0.98132	-0.80704	156.0	0.24296	-6.14457	-7.93293
68.0	1.29223	1.11341	-0.67495	158.0	0.20592	-6.86304	-8.65140
70.0	1.32829	1.23293	-0.55543	160.0	0.17152	-7.65698	-9.44534
72.0	1.36151	1.34020	-0.44816	162.0	0.13991	-8.54155	-10.32991
74.0	1.39171	1.43549	-0.35287	164.0	0.11124	-9.53734	-11.32570
76.0	1.41874	1.51904	-0.26932	166.0	0.08564	-10.67320	-12.46156
78.0	1.44246	1.59105	-0.19731	168.0	0.06322	-11.99141	-13.77977
80.0	1.46274	1.65168	-0.13668	170.0	0.04408	-13.55754	-15.34590
82.0	1.47948	1.70109	-0.08727	172.0	0.02830	-15.48145	-17.26981
84.0	1.49258	1.73939	-0.04897	174.0	0.01596	-17.96913	-19.75749
86.0	1.50199	1.76666	-0.02170	176.0	0.00711	-21.48304	-23.27140
88.0	1.50764	1.78297	-0.00539	178.0	0.00178	-27.49889	-29.28725
90.0	1.50951	1.78836	0.0	180.	0.00000	-INFINITY	-INFINITY

TABLE A1-5

INTEGRAL EQUATION METHOD
EPSILON= 1.0000(NORMALIZED GROUNDPLANE RADIUS)
MONOPOLE LENGTH= 0.250000(WAVELENGTH)
INFINITELY THIN MONOPOLE RADIUS
RADIATION RESISTANCE = 20.227316 (OHMS)

ELEVATION ANGLE (DEG)	DIRECTIVITY (NUMERIC)	DIRECTIVITY (DBI)	RELATIVE POWER (DB)	ELEVATION ANGLE (DEG)	DIRECTIVITY (NUMERIC)	DIRECTIVITY (DBI)	RELATIVE POWER (DB)
0.	0.00000	-INFINITY	-INFINITY				
2.0	0.00191	-27.18126	-28.89009	92.0	1.48032	1.70357	-0.00526
4.0	0.00764	-21.16680	-22.87562	94.0	1.47515	1.68837	-0.02046
6.0	0.01716	-17.65519	-19.36402	96.0	1.46662	1.66317	-0.04565
8.0	0.03040	-15.17073	-16.87955	98.0	1.45476	1.62791	-0.08091
10.0	0.04731	-13.25092	-14.95975	100.0	1.43962	1.58248	-0.12634
12.0	0.06777	-11.68977	-13.39860	102.0	1.42126	1.52675	-0.18208
14.0	0.09168	-10.37740	-12.08622	104.0	1.39976	1.46053	-0.24829
16.0	0.11890	-9.24820	-10.95703	106.0	1.37519	1.38364	-0.32518
18.0	0.14928	-8.25987	-9.96869	108.0	1.34767	1.29583	-0.41300
20.0	0.18266	-7.38351	-9.09234	110.0	1.31729	1.19682	-0.51200
22.0	0.21885	-6.59853	-8.30735	112.0	1.28419	1.08630	-0.62253
24.0	0.25765	-5.88969	-7.59852	114.0	1.24850	0.96390	-0.74492
26.0	0.29885	-5.24545	-6.95428	116.0	1.21038	0.82922	-0.87960
28.0	0.34223	-4.65683	-6.36565	118.0	1.16998	0.68180	-1.02703
30.0	0.38755	-4.11669	-5.82551	120.0	1.12749	0.52112	-1.18770
32.0	0.43458	-3.61930	-5.32813	122.0	1.08308	0.34661	-1.36222
34.0	0.48306	-3.15997	-4.86880	124.0	1.03696	0.15762	-1.55121
36.0	0.53275	-2.73480	-4.44363	126.0	0.98933	-0.04659	-1.75541
38.0	0.58337	-2.34054	-4.04936	128.0	0.94041	-0.26682	-1.97564
40.0	0.63468	-1.97442	-3.68325	130.0	0.89043	-0.50399	-2.21281
42.0	0.68642	-1.63411	-3.34294	132.0	0.83963	-0.75914	-2.46796
44.0	0.73832	-1.31757	-3.02640	134.0	0.78824	-1.03344	-2.74227
46.0	0.79012	-1.02307	-2.73189	136.0	0.73651	-1.32823	-3.03706
48.0	0.84158	-0.74907	-2.45789	138.0	0.68469	-1.64503	-3.35386
50.0	0.89243	-0.49424	-2.20306	140.0	0.63305	-1.98560	-3.69442
52.0	0.94245	-0.25740	-1.96623	142.0	0.58184	-2.35194	-4.06077
54.0	0.99139	-0.03753	-1.74636	144.0	0.53132	-2.74642	-4.45525
56.0	1.03903	0.16629	-1.54253	146.0	0.48175	-3.17179	-4.88061
58.0	1.08515	0.35489	-1.35393	148.0	0.43338	-3.63131	-5.34013
60.0	1.12953	0.52899	-1.17983	150.0	0.38647	-4.12886	-5.83768
62.0	1.17199	0.68924	-1.01959	152.0	0.34126	-4.66915	-6.37797
64.0	1.21233	0.83621	-0.87261	154.0	0.29799	-5.25791	-6.96674
66.0	1.25038	0.97042	-0.73840	156.0	0.25690	-5.90228	-7.61111
68.0	1.28598	1.09234	-0.61648	158.0	0.21821	-6.61123	-8.32006
70.0	1.31898	1.20237	-0.50646	160.0	0.18212	-7.39632	-9.10514
72.0	1.34923	1.30086	-0.40796	162.0	0.14884	-8.27277	-9.98159
74.0	1.37662	1.38815	-0.32067	164.0	0.11854	-9.26118	-10.97000
76.0	1.40104	1.46451	-0.24432	166.0	0.09140	-10.39045	-12.09927
78.0	1.42239	1.53017	-0.17865	168.0	0.06756	-11.70288	-13.41171
80.0	1.44057	1.58535	-0.12347	170.0	0.04716	-13.26408	-14.97290
82.0	1.45553	1.63022	-0.07861	172.0	0.03031	-15.18392	-16.89275
84.0	1.46720	1.66491	-0.04392	174.0	0.01711	-17.66842	-19.37725
86.0	1.47555	1.68953	-0.01930	176.0	0.00762	-21.18005	-22.88887
88.0	1.48052	1.70415	-0.00468	178.0	0.00191	-27.19453	-28.90335
90.0	1.48212	1.70883	0.0	180.	0.00000	-INFINITY	-INFINITY

TABLE A1-6

INTEGRAL EQUATION METHOD
EPSILON= 1.2500(NORMALIZED GROUNDPLANE RADIUS)
MONOPOLE LENGTH= 0.250000(WAVELENGTH)
INFINITELY THIN MONOPOLE RADIUS
RADIATION RESISTANCE = 20.756936 (OHMS)

ELEVATION ANGLE (DEG)	DIRECTIVITY (NUMERIC)	DIRECTIVITY (DBI)	RELATIVE POWER (DB)	ELEVATION ANGLE (DEG)	DIRECTIVITY (NUMERIC)	DIRECTIVITY (DBI)	RELATIVE POWER (DB)
0.	0.00000	-INFINITY	-INFINITY				
2.0	0.00211	-26.75595	-28.35253	92.0	1.44259	1.59143	-0.00514
4.0	0.00843	-20.74330	-22.33988	94.0	1.43804	1.57772	-0.01886
6.0	0.01890	-17.23472	-18.83129	96.0	1.43066	1.55537	-0.04121
8.0	0.03346	-14.75446	-16.35104	98.0	1.42046	1.52429	-0.07228
10.0	0.05200	-12.84004	-14.43662	100.0	1.40746	1.48436	-0.11221
12.0	0.07438	-11.28543	-12.88201	102.0	1.39168	1.43540	-0.16117
14.0	0.10044	-9.98072	-11.57730	104.0	1.37316	1.37721	-0.21936
16.0	0.13001	-8.86028	-10.45685	106.0	1.35193	1.30953	-0.28705
18.0	0.16286	-7.88176	-9.47834	108.0	1.32803	1.23207	-0.36451
20.0	0.19878	-7.01624	-8.61282	110.0	1.30151	1.14448	-0.45209
22.0	0.23752	-6.24306	-7.83963	112.0	1.27245	1.04639	-0.55018
24.0	0.27881	-5.54695	-7.14353	114.0	1.24090	0.93737	-0.65921
26.0	0.32238	-4.91631	-6.51289	116.0	1.20696	0.81692	-0.77966
28.0	0.36795	-4.34210	-5.93867	118.0	1.17072	0.68452	-0.91206
30.0	0.41523	-3.81712	-5.41370	120.0	1.13228	0.53955	-1.05702
32.0	0.46392	-3.33558	-4.93216	122.0	1.09179	0.38137	-1.21520
34.0	0.51372	-2.89272	-4.48929	124.0	1.04936	0.20923	-1.38734
36.0	0.56434	-2.48456	-4.08113	126.0	1.00515	0.02232	-1.57425
38.0	0.61549	-2.10777	-3.70435	128.0	0.95934	-0.18026	-1.77684
40.0	0.66688	-1.75953	-3.35611	130.0	0.91211	-0.39954	-1.99612
42.0	0.71822	-1.43740	-3.03397	132.0	0.86365	-0.63664	-2.23321
44.0	0.76926	-1.13927	-2.73585	134.0	0.81418	-0.89281	-2.48939
46.0	0.81973	-0.86331	-2.45989	136.0	0.76393	-1.16949	-2.76606
48.0	0.86938	-0.60792	-2.20449	138.0	0.71313	-1.46828	-3.06486
50.0	0.91798	-0.37166	-1.96823	140.0	0.66206	-1.79103	-3.38761
52.0	0.96532	-0.15328	-1.74985	142.0	0.61096	-2.13984	-3.73642
54.0	1.01120	0.04836	-1.54822	144.0	0.56013	-2.51714	-4.11372
56.0	1.05542	0.23425	-1.36232	146.0	0.50983	-2.92578	-4.52235
58.0	1.09782	0.40532	-1.19126	148.0	0.46035	-3.36907	-4.96565
60.0	1.13825	0.56236	-1.03421	150.0	0.41200	-3.85100	-5.44757
62.0	1.17656	0.70613	-0.89045	152.0	0.36506	-4.37632	-5.97290
64.0	1.21263	0.83728	-0.75930	154.0	0.31983	-4.95085	-6.54743
66.0	1.24635	0.95641	-0.64016	156.0	0.27658	-5.58177	-7.17835
68.0	1.27764	1.06407	-0.53250	158.0	0.23561	-6.27813	-7.87470
70.0	1.30639	1.16074	-0.43584	160.0	0.19717	-7.05153	-8.64810
72.0	1.33256	1.24685	-0.34972	162.0	0.16154	-7.91724	-9.51382
74.0	1.35606	1.32280	-0.27377	164.0	0.12895	-8.89593	-10.49250
76.0	1.37687	1.38892	-0.20765	166.0	0.09962	-10.01651	-11.61309
78.0	1.39493	1.44552	-0.15106	168.0	0.07377	-11.32134	-12.91792
80.0	1.41021	1.49284	-0.10373	170.0	0.05157	-12.87605	-14.47263
82.0	1.42269	1.53111	-0.06546	172.0	0.03319	-14.79056	-16.38713
84.0	1.43235	1.56050	-0.03607	174.0	0.01875	-17.27087	-18.86745
86.0	1.43918	1.58115	-0.01543	176.0	0.00836	-20.77950	-22.37607
88.0	1.44316	1.59315	-0.00342	178.0	0.00209	-26.79217	-28.38875
90.0	1.44430	1.59658	0.0	180.	0.00000	-INFINITY	-INFINITY

TABLE A1-7

INTEGRAL EQUATION METHOD
EPSILON= 1.5000(NORMALIZED GROUNDPLANE RADIUS)
MONOPOLE LENGTH= 0.250000(WAVELENGTH)
INFINITELY THIN MONOPOLE RADIUS
RADIATION RESISTANCE = 21.510639 (OHMS)

ELEVATION ANGLE (DEG)	DIRECTIVITY (NUMERIC)	DIRECTIVITY (DBI)	RELATIVE POWER (DB)	ELEVATION ANGLE (DEG)	DIRECTIVITY (NUMERIC)	DIRECTIVITY (DBI)	RELATIVE POWER (DB)
0.	0.00000	-INFINITY	-INFINITY				
2.0	0.00240	-26.20435	-27.64602	92.0	1.39196	1.43626	-0.00541
4.0	0.00956	-20.19397	-21.63564	94.0	1.38808	1.42414	-0.01753
6.0	0.02143	-16.68915	-18.13083	96.0	1.38206	1.40525	-0.03642
8.0	0.03790	-14.21416	-15.65583	98.0	1.37388	1.37947	-0.06220
10.0	0.05880	-12.30647	-13.74814	100.0	1.36353	1.34663	-0.09504
12.0	0.08395	-10.76003	-12.20171	102.0	1.35099	1.30652	-0.13516
14.0	0.11311	-9.46491	-10.90659	104.0	1.33624	1.25884	-0.18283
16.0	0.14603	-8.35545	-9.79712	106.0	1.31926	1.20330	-0.23838
18.0	0.18242	-7.38925	-8.83092	108.0	1.30002	1.13950	-0.30218
20.0	0.22196	-6.53734	-7.97901	110.0	1.27850	1.06702	-0.37466
22.0	0.26430	-5.77902	-7.22070	112.0	1.25469	0.98538	-0.45630
24.0	0.30910	-5.09897	-6.54064	114.0	1.22858	0.89404	-0.54764
26.0	0.35600	-4.48551	-5.92719	116.0	1.20016	0.79241	-0.64927
28.0	0.40462	-3.92955	-5.37123	118.0	1.16945	0.67983	-0.76184
30.0	0.45459	-3.42383	-4.86550	120.0	1.13647	0.55559	-0.88608
32.0	0.50554	-2.96246	-4.40413	122.0	1.10126	0.41890	-1.02277
34.0	0.55711	-2.54060	-3.98227	124.0	1.06388	0.26891	-1.17277
36.0	0.60895	-2.15421	-3.59588	126.0	1.02439	0.10467	-1.33701
38.0	0.66072	-1.79986	-3.24153	128.0	0.98291	-0.07485	-1.51652
40.0	0.71209	-1.47462	-2.91630	130.0	0.93956	-0.27077	-1.71245
42.0	0.76279	-1.17598	-2.61765	132.0	0.89447	-0.48436	-1.92603
44.0	0.81251	-0.90171	-2.34339	134.0	0.84782	-0.71698	-2.15866
46.0	0.86102	-0.64989	-2.09156	136.0	0.79980	-0.97019	-2.41187
48.0	0.90808	-0.41878	-1.86046	138.0	0.75063	-1.24571	-2.68739
50.0	0.95348	-0.20686	-1.64854	140.0	0.70057	-1.54549	-2.98717
52.0	0.99707	-0.01275	-1.45443	142.0	0.64987	-1.87174	-3.31342
54.0	1.03867	0.16480	-1.27688	144.0	0.59883	-2.22700	-3.66867
56.0	1.07818	0.32691	-1.11477	146.0	0.54775	-2.61419	-4.05587
58.0	1.11548	0.47461	-0.96706	148.0	0.49696	-3.03676	-4.47844
60.0	1.15049	0.60885	-0.83283	150.0	0.44681	-3.49876	-4.94044
62.0	1.18317	0.73047	-0.71121	152.0	0.39765	-4.00504	-5.44671
64.0	1.21346	0.84027	-0.60141	154.0	0.34983	-4.56148	-6.00315
66.0	1.24135	0.93895	-0.50272	156.0	0.30371	-5.17535	-6.61702
68.0	1.26683	1.02719	-0.41448	158.0	0.25967	-5.85575	-7.29743
70.0	1.28991	1.10558	-0.33609	160.0	0.21805	-6.61438	-8.05605
72.0	1.31059	1.17467	-0.26701	162.0	0.17920	-7.46654	-8.90822
74.0	1.32890	1.23493	-0.20674	164.0	0.14345	-8.43296	-9.87464
76.0	1.34488	1.28682	-0.15485	166.0	0.11111	-9.54261	-10.98428
78.0	1.35854	1.33072	-0.11095	168.0	0.08245	-10.83787	-12.27955
80.0	1.36992	1.36697	-0.07471	170.0	0.05775	-12.38443	-13.82610
82.0	1.37906	1.39585	-0.04583	172.0	0.03722	-14.29221	-15.73388
84.0	1.38599	1.41760	-0.02408	174.0	0.02105	-16.76728	-18.20895
86.0	1.39072	1.43240	-0.00927	176.0	0.00939	-20.27214	-21.71381
88.0	1.39329	1.44040	-0.00127	178.0	0.00235	-26.28255	-27.72422
90.0	1.39369	1.44167	0.0	180.	0.00000	-INFINITY	-INFINITY

TABLE A1-8

INTEGRAL EQUATION METHOD
EPSILON= 1.7500(NORMALIZED GROUNDPLANE RADIUS)
MONOPOLE LENGTH= 0.250000(WAVELENGTH)
INFINITELY THIN MONOPOLE RADIUS
RADIATION RESISTANCE = 22.587827 (OHMS)

ELEVATION ANGLE (DEG)	DIRECTIVITY (NUMERIC)	DIRECTIVITY (DBI)	RELATIVE POWER (DB)	ELEVATION ANGLE (DEG)	DIRECTIVITY (NUMERIC)	DIRECTIVITY (DBI)	RELATIVE POWER (DB)
0.	0.00000	-INFINITY	-INFINITY				
2.0	0.00281	-25.51382	-26.74553	92.0	1.32528	1.22307	-0.00863
4.0	0.01120	-19.50615	-20.73786	94.0	1.32207	1.21253	-0.01918
6.0	0.02509	-16.00584	-17.23755	96.0	1.31757	1.19775	-0.03396
8.0	0.04429	-13.53714	-14.76885	98.0	1.31177	1.17858	-0.05313
10.0	0.06859	-11.63751	-12.86921	100.0	1.30460	1.15478	-0.07692
12.0	0.09770	-10.10087	-11.33258	102.0	1.29601	1.12608	-0.10563
14.0	0.13130	-8.81726	-10.04897	104.0	1.28590	1.09209	-0.13962
16.0	0.16901	-7.72098	-8.95269	106.0	1.27420	1.05238	-0.17933
18.0	0.21040	-6.76961	-8.00132	108.0	1.26080	1.00646	-0.22525
20.0	0.25503	-5.93411	-7.16582	110.0	1.24559	0.95376	-0.27795
22.0	0.30243	-5.19374	-6.42545	112.0	1.22847	0.89365	-0.33806
24.0	0.35212	-4.53312	-5.76483	114.0	1.20933	0.82544	-0.40627
26.0	0.40360	-3.94052	-5.17223	116.0	1.18806	0.74837	-0.48334
28.0	0.45638	-3.40676	-4.63847	118.0	1.16456	0.66162	-0.57009
30.0	0.50997	-2.92451	-4.15622	120.0	1.13876	0.56431	-0.66740
32.0	0.56392	-2.48782	-3.71953	122.0	1.11057	0.45547	-0.77624
34.0	0.61777	-2.09175	-3.32346	124.0	1.07996	0.33407	-0.89763
36.0	0.67109	-1.73217	-2.96388	126.0	1.04689	0.19902	-1.03269
38.0	0.72351	-1.40556	-2.63727	128.0	1.01137	0.04911	-1.18260
40.0	0.77466	-1.10889	-2.34059	130.0	0.97343	-0.11696	-1.34866
42.0	0.82423	-0.83951	-2.07122	132.0	0.93313	-0.30058	-1.53229
44.0	0.87194	-0.59511	-1.82682	134.0	0.89057	-0.50331	-1.73502
46.0	0.91757	-0.37363	-1.60534	136.0	0.84590	-0.72682	-1.95853
48.0	0.96090	-0.17322	-1.40493	138.0	0.79928	-0.97299	-2.20470
50.0	1.00180	0.00779	-1.22391	140.0	0.75095	-1.24390	-2.47560
52.0	1.04014	0.17093	-1.06078	142.0	0.70115	-1.54188	-2.77359
54.0	1.07586	0.31756	-0.91414	144.0	0.65019	-1.86960	-3.10131
56.0	1.10892	0.44899	-0.78272	146.0	0.59840	-2.23011	-3.46182
58.0	1.13930	0.56639	-0.66532	148.0	0.54614	-2.62694	-3.85865
60.0	1.16704	0.67085	-0.56085	150.0	0.49383	-3.06426	-4.29597
62.0	1.19218	0.76341	-0.46830	152.0	0.44187	-3.54701	-4.77872
64.0	1.21479	0.84501	-0.38670	154.0	0.39074	-4.08116	-5.31287
66.0	1.23496	0.91653	-0.31517	156.0	0.34087	-4.67407	-5.90578
68.0	1.25280	0.97881	-0.25290	158.0	0.29276	-5.33491	-6.56662
70.0	1.26841	1.03261	-0.19910	160.0	0.24686	-6.07544	-7.30715
72.0	1.28192	1.07863	-0.15308	162.0	0.20366	-6.91104	-8.14275
74.0	1.29346	1.11751	-0.11419	164.0	0.16359	-7.86248	-9.09419
76.0	1.30313	1.14986	-0.08185	166.0	0.12709	-8.95880	-10.19051
78.0	1.31105	1.17619	-0.05551	168.0	0.09457	-10.24242	-11.47413
80.0	1.31734	1.19698	-0.03473	170.0	0.06639	-11.77905	-13.01076
82.0	1.32209	1.21262	-0.01909	172.0	0.04287	-13.67868	-14.91038
84.0	1.32539	1.22345	-0.00825	174.0	0.02428	-16.14737	-17.37907
86.0	1.32732	1.22975	-0.00196	176.0	0.01085	-19.64766	-20.87937
88.0	1.32792	1.23171	0.0	178.0	0.00272	-25.65532	-26.88703
90.0	1.32723	1.22946	-0.00224	180.	0.00000	-INFINITY	-INFINITY

TABLE A1-9

INTEGRAL EQUATION METHOD
EPSILON= 2.0000(NORMALIZED GROUNDPLANE RADIUS)
MONOPOLE LENGTH= 0.250000(WAVELENGTH)
INFINITELY THIN MONOPOLE RADIUS
RADIATION RESISTANCE = 24.149264 (OHMS)

ELEVATION ANGLE (DEG)	DIRECTIVITY (NUMERIC)	DIRECTIVITY (DBI)	RELATIVE POWER (DB)	ELEVATION ANGLE (DEG)	DIRECTIVITY (NUMERIC)	DIRECTIVITY (DBI)	RELATIVE POWER (DB)
0.	0.00000	-INFINITY	-INFINITY				
2.0	0.00340	-24.68049	-25.64850	92.0	1.23900	0.93071	-0.03729
4.0	0.01356	-18.67593	-19.64394	94.0	1.23640	0.92158	-0.04642
6.0	0.03033	-15.18081	-16.14882	96.0	1.23358	0.91168	-0.05632
8.0	0.05346	-12.71934	-13.68735	98.0	1.23050	0.90080	-0.06720
10.0	0.08262	-10.82898	-11.79699	100.0	1.22705	0.88863	-0.07938
12.0	0.11739	-9.30363	-10.27164	102.0	1.22314	0.87475	-0.09326
14.0	0.15728	-8.03330	-9.00130	104.0	1.21861	0.85864	-0.10937
16.0	0.20173	-6.95225	-7.92025	106.0	1.21331	0.83970	-0.12830
18.0	0.25015	-6.01801	-6.98602	108.0	1.20705	0.81723	-0.15077
20.0	0.30189	-5.20153	-6.16953	110.0	1.19963	0.79046	-0.17755
22.0	0.35629	-4.48200	-5.45000	112.0	1.19083	0.75852	-0.20949
24.0	0.41267	-3.84398	-4.81199	114.0	1.18045	0.72047	-0.24753
26.0	0.47036	-3.27570	-4.24371	116.0	1.16824	0.67533	-0.29267
28.0	0.52870	-2.76790	-3.73591	118.0	1.15399	0.62203	-0.34598
30.0	0.58706	-2.31320	-3.28121	120.0	1.13748	0.55942	-0.40859
32.0	0.64483	-1.90555	-2.87356	122.0	1.11849	0.48632	-0.48169
34.0	0.70146	-1.53995	-2.50795	124.0	1.09684	0.40145	-0.56656
36.0	0.75646	-1.21216	-2.18017	126.0	1.07238	0.30348	-0.66452
38.0	0.80936	-0.91858	-1.88658	128.0	1.04496	0.19100	-0.77701
40.0	0.85979	-0.65606	-1.62407	130.0	1.01449	0.06250	-0.90551
42.0	0.90743	-0.42187	-1.38987	132.0	0.98093	-0.08364	-1.05164
44.0	0.95202	-0.21355	-1.18156	134.0	0.94425	-0.24912	-1.21713
46.0	0.99336	-0.02893	-0.99694	136.0	0.90452	-0.43582	-1.40383
48.0	1.03133	0.13398	-0.83402	138.0	0.86183	-0.64578	-1.61378
50.0	1.06586	0.27701	-0.69100	140.0	0.81635	-0.88122	-1.84922
52.0	1.09694	0.40181	-0.56619	142.0	0.76831	-1.14463	-2.11264
54.0	1.12459	0.50996	-0.45805	144.0	0.71799	-1.43882	-2.40683
56.0	1.14892	0.60290	-0.36510	146.0	0.66574	-1.76695	-2.73496
58.0	1.17004	0.68202	-0.28599	148.0	0.61197	-2.13268	-3.10068
60.0	1.18812	0.74861	-0.21940	150.0	0.55715	-2.54026	-3.50827
62.0	1.20334	0.80389	-0.16412	152.0	0.50179	-2.99476	-3.96277
64.0	1.21592	0.84905	-0.11896	154.0	0.44645	-3.50223	-4.47024
66.0	1.22608	0.88519	-0.08282	156.0	0.39173	-4.07011	-5.03811
68.0	1.23406	0.91337	-0.05463	158.0	0.33825	-4.70765	-5.67566
70.0	1.24011	0.93461	-0.03340	160.0	0.28664	-5.42668	-6.39468
72.0	1.24447	0.94983	-0.01817	162.0	0.23754	-6.24264	-7.21065
74.0	1.24737	0.95994	-0.00807	164.0	0.19159	-7.17636	-8.14437
76.0	1.24903	0.96575	-0.00226	166.0	0.14939	-8.25692	-9.22493
78.0	1.24969	0.96801	0.0	168.0	0.11151	-9.52680	-10.49481
80.0	1.24951	0.96740	-0.00061	170.0	0.07849	-11.05174	-12.01975
82.0	1.24868	0.96453	-0.00348	172.0	0.05080	-12.94176	-13.90976
84.0	1.24735	0.95990	-0.00811	174.0	0.02882	-15.40295	-16.37095
86.0	1.24564	0.95393	-0.01408	176.0	0.01289	-18.89787	-19.86587
88.0	1.24364	0.94695	-0.02106	178.0	0.00323	-24.90230	-25.87031
90.0	1.24141	0.93917	-0.02884	180.	0.00000	-INFINITY	-INFINITY

TABLE A1-10

INTEGRAL EQUATION METHOD
EPSILON= 2.2500(NORMALIZED GROUNDPLANE RADIUS)
MONOPOLE LENGTH= 0.250000(WAVELENGTH)
INFINITELY THIN MONOPOLE RADIUS
RADIATION RESISTANCE = 26.457698 (OHMS)

ELEVATION ANGLE (DEG)	DIRECTIVITY (NUMERIC)	DIRECTIVITY (DBI)	RELATIVE POWER (DB)	ELEVATION ANGLE (DEG)	DIRECTIVITY (NUMERIC)	DIRECTIVITY (DBI)	RELATIVE POWER (DB)
0.	0.00000	-INFINITY	-INFINITY				
2.0	0.00425	-23.71543	-24.56194	92.0	1.12995	0.53060	-0.31591
4.0	0.01693	-17.71432	-18.56083	94.0	1.12788	0.52264	-0.32387
6.0	0.03780	-14.22493	-15.07145	96.0	1.12686	0.51869	-0.32782
8.0	0.06650	-11.77148	-12.61799	98.0	1.12680	0.51846	-0.32805
10.0	0.10253	-9.89140	-10.73791	100.0	1.12759	0.52150	-0.32501
12.0	0.14526	-8.37857	-9.22508	102.0	1.12907	0.52720	-0.31931
14.0	0.19396	-7.12297	-7.96949	104.0	1.13104	0.53478	-0.31173
16.0	0.24781	-6.05885	-6.90536	106.0	1.13328	0.54336	-0.30316
18.0	0.30594	-5.14370	-5.99021	108.0	1.13551	0.55190	-0.29461
20.0	0.36742	-4.34841	-5.19493	110.0	1.13744	0.55928	-0.28723
22.0	0.43130	-3.65216	-4.49867	112.0	1.13875	0.56427	-0.28224
24.0	0.49666	-3.03945	-3.88596	114.0	1.13909	0.56558	-0.28093
26.0	0.56254	-2.49845	-3.34496	116.0	1.13811	0.56183	-0.28469
28.0	0.62808	-2.01985	-2.86637	118.0	1.13542	0.55158	-0.29493
30.0	0.69243	-1.59622	-2.44273	120.0	1.13067	0.53336	-0.31316
32.0	0.75484	-1.22143	-2.06794	122.0	1.12347	0.50562	-0.34089
34.0	0.81463	-0.89041	-1.73693	124.0	1.11347	0.46679	-0.37973
36.0	0.87119	-0.59886	-1.44538	126.0	1.10033	0.41522	-0.43129
38.0	0.92404	-0.34309	-1.18960	128.0	1.08374	0.34923	-0.49728
40.0	0.97278	-0.11985	-0.96637	130.0	1.06342	0.26706	-0.57946
42.0	1.01711	0.07368	-0.77284	132.0	1.03917	0.16686	-0.67966
44.0	1.05683	0.24005	-0.60646	134.0	1.01081	0.04669	-0.79982
46.0	1.09184	0.38158	-0.46493	136.0	0.97825	-0.09550	-0.94201
48.0	1.12212	0.50037	-0.34614	138.0	0.94147	-0.26192	-1.10844
50.0	1.14773	0.59839	-0.24812	140.0	0.90054	-0.45499	-1.30150
52.0	1.16881	0.67746	-0.16906	142.0	0.85560	-0.67732	-1.52383
54.0	1.18558	0.73930	-0.10721	144.0	0.80689	-0.93184	-1.77835
56.0	1.19828	0.78558	-0.06093	146.0	0.75477	-1.22185	-2.06837
58.0	1.20722	0.81786	-0.02865	148.0	0.69966	-1.55112	-2.39764
60.0	1.21274	0.83768	-0.00883	150.0	0.64209	-1.92401	-2.77053
62.0	1.21521	0.84651	0.0	152.0	0.58269	-2.34566	-3.19217
64.0	1.21501	0.84580	-0.00071	154.0	0.52213	-2.82221	-3.66872
66.0	1.21254	0.83696	-0.00955	156.0	0.46119	-3.36117	-4.20768
68.0	1.20819	0.82136	-0.02516	158.0	0.40069	-3.97187	-4.81838
70.0	1.20235	0.80032	-0.04619	160.0	0.34149	-4.66619	-5.51270
72.0	1.19541	0.77515	-0.07136	162.0	0.28447	-5.45963	-6.30615
74.0	1.18771	0.74710	-0.09941	164.0	0.23051	-6.37308	-7.21959
76.0	1.17960	0.71734	-0.12917	166.0	0.18048	-7.43565	-8.28216
78.0	1.17138	0.68699	-0.15952	168.0	0.13521	-8.68986	-9.53637
80.0	1.16335	0.65709	-0.18942	170.0	0.09547	-10.20148	-11.04800
82.0	1.15573	0.62857	-0.21794	172.0	0.06194	-12.08057	-12.92708
84.0	1.14875	0.60225	-0.24426	174.0	0.03521	-14.53323	-15.37974
86.0	1.14257	0.57882	-0.26769	176.0	0.01577	-18.02205	-18.86856
88.0	1.13732	0.55884	-0.28768	178.0	0.00396	-24.02282	-24.86933
90.0	1.13310	0.54269	-0.30383	180.	0.00000	-INFINITY	-INFINITY

TABLE A1-11

INTEGRAL EQUATION METHOD
EPSILON= 2.5000(NORMALIZED GROUNDPLANE RADIUS)
MONOPOLE LENGTH= 0.250000(WAVELENGTH)
INFINITELY THIN MONOPOLE RADIUS
RADIATION RESISTANCE = 29.949409 (OHMS)

ELEVATION ANGLE (DEG)	DIRECTIVITY (NUMERIC)	DIRECTIVITY (DBI)	RELATIVE POWER (DB)	ELEVATION ANGLE (DEG)	DIRECTIVITY (NUMERIC)	DIRECTIVITY (DBI)	RELATIVE POWER (DB)
0.	0.00000	-INFINITY	-INFINITY				
2.0	0.00544	-22.64798	-23.64127	92.0	0.99690	-0.01348	-1.00677
4.0	0.02162	-16.65057	-17.64385	94.0	0.99529	-0.02050	-1.01378
6.0	0.04822	-13.16734	-14.16062	96.0	0.99613	-0.01683	-1.01012
8.0	0.08467	-10.72249	-11.71577	98.0	0.99933	-0.00290	-0.99619
10.0	0.13021	-8.85344	-9.84673	100.0	1.00475	0.02059	-0.97270
12.0	0.18390	-7.35407	-8.34735	102.0	1.01220	0.05267	-0.94061
14.0	0.24466	-6.11432	-7.10760	104.0	1.02145	0.09215	-0.90113
16.0	0.31129	-5.06840	-6.06168	106.0	1.03219	0.13761	-0.85567
18.0	0.38249	-4.17379	-5.16708	108.0	1.04411	0.18748	-0.80581
20.0	0.45694	-3.40136	-4.39465	110.0	1.05683	0.24003	-0.75325
22.0	0.53331	-2.73023	-3.72351	112.0	1.06991	0.29346	-0.69982
24.0	0.61026	-2.14487	-3.13815	114.0	1.08290	0.34588	-0.64740
26.0	0.68653	-1.63340	-2.62668	116.0	1.09531	0.39536	-0.59792
28.0	0.76094	-1.18648	-2.17977	118.0	1.10661	0.43993	-0.55335
30.0	0.83241	-0.79662	-1.78990	120.0	1.11625	0.47762	-0.51566
32.0	0.89999	-0.45764	-1.45093	122.0	1.12369	0.50645	-0.48684
34.0	0.96285	-0.16443	-1.15772	124.0	1.12835	0.52443	-0.46886
36.0	1.02032	0.08738	-0.90590	126.0	1.12968	0.52956	-0.46373
38.0	1.07190	0.30156	-0.69173	128.0	1.12715	0.51983	-0.47345
40.0	1.11722	0.48138	-0.51191	130.0	1.12027	0.49321	-0.50007
42.0	1.15605	0.62977	-0.36352	132.0	1.10856	0.44761	-0.54568
44.0	1.18832	0.74935	-0.24394	134.0	1.09166	0.38087	-0.61241
46.0	1.21409	0.84251	-0.15077	136.0	1.06924	0.29075	-0.70253
48.0	1.23352	0.91148	-0.08181	138.0	1.04109	0.17488	-0.81841
50.0	1.24690	0.95830	-0.03498	140.0	1.00710	0.03071	-0.96257
52.0	1.25457	0.98495	-0.00834	142.0	0.96727	-0.14451	-1.13779
54.0	1.25698	0.99328	0.0	144.0	0.92176	-0.35381	-1.34710
56.0	1.25462	0.98513	-0.00815	146.0	0.87084	-0.60060	-1.59389
58.0	1.24803	0.96226	-0.03102	148.0	0.81495	-0.88872	-1.88200
60.0	1.23778	0.92642	-0.06686	150.0	0.75464	-1.22260	-2.21589
62.0	1.22444	0.87936	-0.11392	152.0	0.69064	-1.60746	-2.60075
64.0	1.20860	0.82282	-0.17047	154.0	0.62380	-2.04951	-3.04280
66.0	1.19084	0.75853	-0.23475	156.0	0.55510	-2.55631	-3.54960
68.0	1.17173	0.68827	-0.30502	158.0	0.48560	-3.13724	-4.13053
70.0	1.15180	0.61378	-0.37950	160.0	0.41646	-3.80423	-4.79751
72.0	1.13158	0.53685	-0.45643	162.0	0.34891	-4.57282	-5.56610
74.0	1.11154	0.45923	-0.53405	164.0	0.28419	-5.46392	-6.45721
76.0	1.09211	0.38266	-0.61062	166.0	0.22353	-6.50670	-7.49999
78.0	1.07370	0.30882	-0.68446	168.0	0.16812	-7.74370	-8.73698
80.0	1.05665	0.23933	-0.75395	170.0	0.11910	-9.24072	-10.23400
82.0	1.04129	0.17570	-0.81759	172.0	0.07749	-11.10782	-12.10110
84.0	1.02785	0.11930	-0.87399	174.0	0.04415	-13.55115	-14.54443
86.0	1.01656	0.07132	-0.92196	176.0	0.01980	-17.03329	-18.02657
88.0	1.00757	0.03275	-0.96053	178.0	0.00498	-23.03004	-24.02333
90.0	1.00100	0.00433	-0.98896	180.	0.00000	-INFINITY	-INFINITY

TABLE A1-12

INTEGRAL EQUATION METHOD
EPSILON= 2.7500(NORMALIZED GROUNDPLANE RADIUS)
MONOPOLE LENGTH= 0.250000(WAVELENGTH)
INFINITELY THIN MONOPOLE RADIUS
RADIATION RESISTANCE = 35.435753 (OHMS)

ELEVATION ANGLE (DEG)	DIRECTIVITY (NUMERIC)	DIRECTIVITY (DBI)	RELATIVE POWER (DB)	ELEVATION ANGLE (DEG)	DIRECTIVITY (NUMERIC)	DIRECTIVITY (DBI)	RELATIVE POWER (DB)
0.	0.00000	-INFINITY	-INFINITY				
2.0	0.16697	-7.77357	-9.16490	92.0	0.83880	-0.76343	-2.15477
4.0	0.04084	-13.88912	-15.28045	94.0	0.83553	-0.78038	-2.17172
6.0	0.01774	-17.50997	-18.90130	96.0	0.83619	-0.77695	-2.16829
8.0	0.00974	-20.11249	-21.50383	98.0	0.84070	-0.75360	-2.14493
10.0	0.00608	-22.15993	-23.55127	100.0	0.84893	-0.71130	-2.10264
12.0	0.00411	-23.85916	-25.25049	102.0	0.86068	-0.65156	-2.04290
14.0	0.00294	-25.32105	-26.71238	104.0	0.87573	-0.57629	-1.96763
16.0	0.00218	-26.61237	-28.00370	106.0	0.89377	-0.48775	-1.87909
18.0	0.00167	-27.77678	-29.16812	108.0	0.91444	-0.38847	-1.77980
20.0	0.00130	-28.84480	-30.23614	110.0	0.93732	-0.28114	-1.67247
22.0	0.00104	-29.83895	-31.23028	112.0	0.96193	-0.16856	-1.55990
24.0	0.00084	-30.77677	-32.16810	114.0	0.98774	-0.05357	-1.44490
26.0	0.00068	-31.67266	-33.06399	116.0	1.01415	0.06103	-1.33031
28.0	0.00056	-32.53914	-33.93047	118.0	1.04051	0.17247	-1.21886
30.0	0.00046	-33.38779	-34.77913	120.0	1.06612	0.27807	-1.11326
32.0	0.00038	-34.23011	-35.62144	122.0	1.09024	0.37524	-1.01610
34.0	0.00031	-35.07836	-36.46970	124.0	1.11210	0.46145	-0.92989
36.0	0.00025	-35.94670	-37.33804	126.0	1.13090	0.53425	-0.85708
38.0	0.00021	-36.85281	-38.24414	128.0	1.14585	0.59127	-0.80007
40.0	0.00017	-37.82054	-39.21187	130.0	1.15615	0.63014	-0.76119
42.0	0.00013	-38.88502	-40.27635	132.0	1.16106	0.64853	-0.74280
44.0	0.00010	-40.10317	-41.49450	134.0	1.15986	0.64406	-0.74727
46.0	0.00007	-41.57970	-42.97103	136.0	1.15195	0.61432	-0.77701
48.0	0.00004	-43.54736	-44.93870	138.0	1.13678	0.55678	-0.83455
50.0	0.00002	-46.74541	-48.13675	140.0	1.11398	0.46878	-0.92255
52.0	1.37763	1.39133	0.0	142.0	1.08329	0.34747	-1.04387
54.0	1.36140	1.33985	-0.05148	144.0	1.04465	0.18971	-1.20163
56.0	1.33969	1.27006	-0.12128	146.0	0.99817	-0.00796	-1.39930
58.0	1.31335	1.18381	-0.20752	148.0	0.94418	-0.24946	-1.64079
60.0	1.28321	1.08298	-0.30836	150.0	0.88323	-0.53928	-1.93061
62.0	1.25010	0.96946	-0.42188	152.0	0.81608	-0.88268	-2.27401
64.0	1.21484	0.84520	-0.54614	154.0	0.74372	-1.28592	-2.67726
66.0	1.17821	0.71221	-0.67912	156.0	0.66733	-1.75662	-3.14795
68.0	1.14093	0.57260	-0.81873	158.0	0.58828	-2.30418	-3.69551
70.0	1.10371	0.42856	-0.96277	160.0	0.50809	-2.94057	-4.33191
72.0	1.06718	0.28239	-1.10895	162.0	0.42841	-3.68138	-5.07271
74.0	1.03192	0.13646	-1.25487	164.0	0.35095	-4.54754	-5.93887
76.0	0.99845	-0.00674	-1.39808	166.0	0.27744	-5.56824	-6.95957
78.0	0.96723	-0.14470	-1.53603	168.0	0.20960	-6.78605	-8.17739
80.0	0.93867	-0.27485	-1.66619	170.0	0.14905	-8.26680	-9.65814
82.0	0.91312	-0.39470	-1.78603	172.0	0.09726	-10.12057	-11.51190
84.0	0.89087	-0.50184	-1.89317	174.0	0.05555	-12.55350	-13.94484
86.0	0.87216	-0.59404	-1.98537	176.0	0.02496	-16.02821	-17.41954
88.0	0.85716	-0.66936	-2.06069	178.0	0.00628	-22.02051	-23.41184
90.0	0.84602	-0.72621	-2.11754	180.	0.00000	-INFINITY	-INFINITY

TABLE A1-13

INTEGRAL EQUATION METHOD
EPSILON= 3.0000(NORMALIZED GROUNDPLANE RADIUS)
MONOPOLE LENGTH= 0.250000(WAVELENGTH)
INFINITELY THIN MONOPOLE RADIUS
RADIATION RESISTANCE = 44.598673 (OHMS)

ELEVATION ANGLE (DEG)	DIRECTIVITY (NUMERIC)	DIRECTIVITY (DBI)	RELATIVE POWER (DB)	ELEVATION ANGLE (DEG)	DIRECTIVITY (NUMERIC)	DIRECTIVITY (DBI)	RELATIVE POWER (DB)
0.	0.00000	-INFINITY	-INFINITY				
2.0	0.26737	-5.72891	-6.58995	92.0	0.00007	-41.44714	-42.30817
4.0	0.06676	-11.75511	-12.61614	94.0	0.00006	-42.32503	-43.18607
6.0	0.02964	-15.28085	-16.14189	96.0	0.00004	-43.47261	-44.33365
8.0	0.01666	-17.78195	-18.64298	98.0	0.00003	-45.12107	-45.98211
10.0	0.01066	-19.72094	-20.58197	100.0	0.00002	-48.00829	-48.86933
12.0	0.00741	-21.30390	-22.16494	102.0	0.67468	-1.70899	-2.57002
14.0	0.00544	-22.64081	-23.50185	104.0	0.69294	-1.59303	-2.45407
16.0	0.00417	-23.79734	-24.65838	106.0	0.71581	-1.45200	-2.31303
18.0	0.00330	-24.81589	-25.67693	108.0	0.74299	-1.29018	-2.15121
20.0	0.00268	-25.72545	-26.58649	110.0	0.77408	-1.11214	-1.97317
22.0	0.00221	-26.54673	-27.40776	112.0	0.80863	-0.92250	-1.78353
24.0	0.00186	-27.29504	-28.15607	114.0	0.84610	-0.72580	-1.58683
26.0	0.00159	-27.98207	-28.84310	116.0	0.88586	-0.52636	-1.38740
28.0	0.00138	-28.61692	-29.47795	118.0	0.92720	-0.32827	-1.18930
30.0	0.00120	-29.20685	-30.06788	120.0	0.96932	-0.13531	-0.99634
32.0	0.00106	-29.75775	-30.61878	122.0	1.01135	0.04902	-0.81201
34.0	0.00094	-30.27448	-31.13551	124.0	1.05232	0.22148	-0.63956
36.0	0.00084	-30.76111	-31.62214	126.0	1.09120	0.37906	-0.48197
38.0	0.00075	-31.22109	-32.08212	128.0	1.12692	0.51894	-0.34210
40.0	0.00068	-31.65739	-32.51842	130.0	1.15836	0.63843	-0.22260
42.0	0.00062	-32.07261	-32.93364	132.0	1.18439	0.73494	-0.12609
44.0	0.00057	-32.46903	-33.33007	134.0	1.20391	0.80593	-0.05510
46.0	0.00052	-32.84870	-33.70974	136.0	1.21586	0.84883	-0.01220
48.0	0.00048	-33.21348	-34.07452	138.0	1.21928	0.86103	0.0
50.0	0.00044	-33.56508	-34.42611	140.0	1.21333	0.83979	-0.02124
52.0	0.00041	-33.90509	-34.76613	142.0	1.19735	0.78220	-0.07883
54.0	0.00038	-34.23504	-35.09607	144.0	1.17087	0.68509	-0.17594
56.0	0.00035	-34.55638	-35.41741	146.0	1.13370	0.54497	-0.31607
58.0	0.00033	-34.87057	-35.73160	148.0	1.08589	0.35788	-0.50316
60.0	0.00030	-35.17904	-36.04008	150.0	1.02785	0.11930	-0.74173
62.0	0.00028	-35.48329	-36.34432	152.0	0.96028	-0.17604	-1.03707
64.0	0.00026	-35.78486	-36.64589	154.0	0.88421	-0.53442	-1.39546
66.0	0.00025	-36.08539	-36.94643	156.0	0.80104	-0.96348	-1.82451
68.0	0.00023	-36.38670	-37.24773	158.0	0.71242	-1.47265	-2.33368
70.0	0.00021	-36.69077	-37.55180	160.0	0.62031	-2.07391	-2.93494
72.0	0.00020	-36.99987	-37.86090	162.0	0.52688	-2.78287	-3.64390
74.0	0.00019	-37.31664	-38.17768	164.0	0.43446	-3.62048	-4.48151
76.0	0.00017	-37.64420	-38.50523	166.0	0.34547	-4.61594	-5.47697
78.0	0.00016	-37.98631	-38.84735	168.0	0.26231	-5.81185	-6.67288
80.0	0.00015	-38.34765	-39.20868	170.0	0.18733	-7.27404	-8.13507
82.0	0.00013	-38.73414	-39.59517	172.0	0.12267	-9.11260	-9.97363
84.0	0.00012	-39.15351	-40.01454	174.0	0.07025	-11.53370	-12.39473
86.0	0.00011	-39.61614	-40.47718	176.0	0.03162	-14.99995	-15.86098
88.0	0.00010	-40.13656	-40.99760	178.0	0.00797	-20.98716	-21.84820
90.0	0.00008	-40.73596	-41.59699	180.	0.00000	-INFINITY	-INFINITY

Tables A1-14–A1-25; $h/\lambda = 0.10,\ 0 \leq ka \leq 3.0$

TABLE A1-14

INTEGRAL EQUATION METHOD
EPSILON= 0.0 (NORMALIZED GROUNDPLANE RADIUS)
MONOPOLE LENGTH= 0.100000(WAVELENGTH)
INFINITELY THIN MONOPOLE RADIUS
RADIATION RESISTANCE = 2.100720 (OHMS)

ELEVATION ANGLE θ (DEG)	DIRECTIVITY (NUMERIC)	DIRECTIVITY (DBI)	RELATIVE POWER (DB)	ELEVATION ANGLE (DEG)	DIRECTIVITY (NUMERIC)	DIRECTIVITY (DBI)	RELATIVE POWER (DB)
0.	0.00000	-INFINITY	-INFINITY				
2.0	0.00179	-27.45943	-29.23947	92.0	1.50475	1.77464	-0.00541
4.0	0.00717	-21.44377	-23.22382	94.0	1.49913	1.75840	-0.02165
6.0	0.01611	-17.93019	-19.71023	96.0	1.48980	1.73129	-0.04876
8.0	0.02856	-15.44296	-17.22301	98.0	1.47681	1.69324	-0.08680
10.0	0.04447	-13.51963	-15.29967	100.0	1.46022	1.64418	-0.13586
12.0	0.06376	-11.95420	-13.73424	102.0	1.44012	1.58399	-0.19606
14.0	0.08636	-10.63680	-12.41685	104.0	1.41661	1.51252	-0.26753
16.0	0.11215	-9.50187	-11.28191	106.0	1.38982	1.42960	-0.35045
18.0	0.14102	-8.50711	-10.28715	108.0	1.35988	1.33502	-0.44503
20.0	0.17284	-7.62366	-9.40370	110.0	1.32695	1.22855	-0.55150
22.0	0.20745	-6.83094	-8.61098	112.0	1.29119	1.10991	-0.67014
24.0	0.24469	-6.11377	-7.89381	114.0	1.25279	0.97879	-0.80126
26.0	0.28441	-5.46063	-7.24067	116.0	1.21194	0.83482	-0.94522
28.0	0.32639	-4.86257	-6.64262	118.0	1.16886	0.67761	-1.10244
30.0	0.37047	-4.31251	-6.09256	120.0	1.12375	0.50669	-1.27336
32.0	0.41641	-3.80476	-5.58481	122.0	1.07684	0.32153	-1.45852
34.0	0.46402	-3.33466	-5.11471	124.0	1.02838	0.12154	-1.65850
36.0	0.51305	-2.89837	-4.67842	126.0	0.97860	-0.09393	-1.87398
38.0	0.56329	-2.49268	-4.27273	128.0	0.92776	-0.32566	-2.10571
40.0	0.61448	-2.11490	-3.89494	130.0	0.87609	-0.57450	-2.35455
42.0	0.66639	-1.76271	-3.54276	132.0	0.82387	-0.84143	-2.62147
44.0	0.71876	-1.43416	-3.21421	134.0	0.77134	-1.12756	-2.90760
46.0	0.77134	-1.12756	-2.90760	136.0	0.71876	-1.43416	-3.21421
48.0	0.82387	-0.84143	-2.62147	138.0	0.66639	-1.76271	-3.54276
50.0	0.87609	-0.57450	-2.35455	140.0	0.61448	-2.11490	-3.89494
52.0	0.92776	-0.32566	-2.10571	142.0	0.56329	-2.49268	-4.27273
54.0	0.97860	-0.09393	-1.87398	144.0	0.51305	-2.89837	-4.67842
56.0	1.02838	0.12154	-1.65850	146.0	0.46402	-3.33466	-5.11471
58.0	1.07684	0.32153	-1.45852	148.0	0.41641	-3.80476	-5.58481
60.0	1.12375	0.50669	-1.27336	150.0	0.37047	-4.31251	-6.09256
62.0	1.16886	0.67761	-1.10244	152.0	0.32639	-4.86257	-6.64262
64.0	1.21194	0.83482	-0.94522	154.0	0.28441	-5.46063	-7.24067
66.0	1.25279	0.97879	-0.80126	156.0	0.24469	-6.11377	-7.89381
68.0	1.29119	1.10991	-0.67014	158.0	0.20745	-6.83094	-8.61098
70.0	1.32695	1.22855	-0.55150	160.0	0.17284	-7.62366	-9.40370
72.0	1.35988	1.33502	-0.44503	162.0	0.14102	-8.50711	-10.28715
74.0	1.38982	1.42960	-0.35045	164.0	0.11215	-9.50187	-11.28191
76.0	1.41661	1.51252	-0.26753	166.0	0.08636	-10.63680	-12.41685
78.0	1.44012	1.58399	-0.19606	168.0	0.06376	-11.95420	-13.73424
80.0	1.46022	1.64418	-0.13586	170.0	0.04447	-13.51963	-15.29967
82.0	1.47681	1.69324	-0.08680	172.0	0.02856	-15.44296	-17.22301
84.0	1.48980	1.73129	-0.04876	174.0	0.01611	-17.93019	-19.71023
86.0	1.49913	1.75840	-0.02165	176.0	0.00717	-21.44377	-23.22382
88.0	1.50475	1.77464	-0.00541	178.0	0.00179	-27.45943	-29.23947
90.0	1.50662	1.78005	0.0	180.	0.00000	-INFINITY	-INFINITY

TABLE A1-15

INTEGRAL EQUATION METHOD
EPSILON= 0.2500(NORMALIZED GROUNDPLANE RADIUS)
MONOPOLE LENGTH= 0.100000(WAVELENGTH)
INFINITELY THIN MONOPOLE RADIUS
RADIATION RESISTANCE = 2.105873 (OHMS)

ELEVATION ANGLE (DEG)	DIRECTIVITY (NUMERIC)	DIRECTIVITY (DBI)	RELATIVE POWER (DB)	ELEVATION ANGLE (DEG)	DIRECTIVITY (NUMERIC)	DIRECTIVITY (DBI)	RELATIVE POWER (DB)
0.	0.00000	-INFINITY	-INFINITY				
2.0	0.00181	-27.41673	-29.18613	92.0	1.50109	1.76406	-0.00534
4.0	0.00724	-21.40126	-23.17067	94.0	1.49555	1.74802	-0.02139
6.0	0.01626	-17.88800	-19.65741	96.0	1.48636	1.72123	-0.04818
8.0	0.02883	-15.40123	-17.17064	98.0	1.47355	1.68364	-0.08577
10.0	0.04489	-13.47847	-15.24788	100.0	1.45719	1.63515	-0.13425
12.0	0.06436	-11.91374	-13.68315	102.0	1.43736	1.57566	-0.19375
14.0	0.08715	-10.59716	-12.36657	104.0	1.41417	1.50500	-0.26441
16.0	0.11316	-9.46316	-11.23257	106.0	1.38772	1.42301	-0.34640
18.0	0.14225	-8.46944	-10.23885	108.0	1.35815	1.32948	-0.43993
20.0	0.17430	-7.58714	-9.35654	110.0	1.32561	1.22415	-0.54526
22.0	0.20914	-6.79567	-8.56507	112.0	1.29026	1.10676	-0.66265
24.0	0.24661	-6.07984	-7.84925	114.0	1.25227	0.97698	-0.79243
26.0	0.28654	-5.42813	-7.19753	116.0	1.21184	0.83444	-0.93496
28.0	0.32873	-4.83158	-6.60098	118.0	1.16916	0.67874	-1.09067
30.0	0.37298	-4.28310	-6.05251	120.0	1.12445	0.50939	-1.26002
32.0	0.41908	-3.77699	-5.54640	122.0	1.07792	0.32588	-1.44353
34.0	0.46681	-3.30860	-5.07800	124.0	1.02982	0.12760	-1.64181
36.0	0.51593	-2.87406	-4.64347	126.0	0.98036	-0.08613	-1.85554
38.0	0.56622	-2.47016	-4.23957	128.0	0.92981	-0.31607	-2.08547
40.0	0.61742	-2.09420	-3.86361	130.0	0.87840	-0.56308	-2.33249
42.0	0.66929	-1.74386	-3.51327	132.0	0.82639	-0.82816	-2.59757
44.0	0.72158	-1.41717	-3.18658	134.0	0.77403	-1.11243	-2.88184
46.0	0.77403	-1.11243	-2.88184	136.0	0.72158	-1.41718	-3.18658
48.0	0.82639	-0.82816	-2.59757	138.0	0.66929	-1.74387	-3.51327
50.0	0.87840	-0.56308	-2.33249	140.0	0.61742	-2.09421	-3.86361
52.0	0.92981	-0.31606	-2.08547	142.0	0.56622	-2.47017	-4.23958
54.0	0.98036	-0.08613	-1.85553	144.0	0.51593	-2.87406	-4.64347
56.0	1.02982	0.12760	-1.64181	146.0	0.46681	-3.30860	-5.07801
58.0	1.07792	0.32588	-1.44353	148.0	0.41908	-3.77700	-5.54641
60.0	1.12445	0.50940	-1.26001	150.0	0.37298	-4.28311	-6.05251
62.0	1.16916	0.67874	-1.09067	152.0	0.32873	-4.83158	-6.60099
64.0	1.21184	0.83445	-0.93496	154.0	0.28654	-5.42813	-7.19754
66.0	1.25227	0.97698	-0.79243	156.0	0.24661	-6.07985	-7.84925
68.0	1.29026	1.10676	-0.66264	158.0	0.20914	-6.79567	-8.56508
70.0	1.32561	1.22415	-0.54525	160.0	0.17430	-7.58714	-9.35655
72.0	1.35815	1.32948	-0.43993	162.0	0.14225	-8.46945	-10.23885
74.0	1.38772	1.42301	-0.34639	164.0	0.11316	-9.46316	-11.23257
76.0	1.41417	1.50500	-0.26440	166.0	0.08715	-10.59717	-12.36657
78.0	1.43736	1.57566	-0.19375	168.0	0.06436	-11.91374	-13.68315
80.0	1.45719	1.63516	-0.13425	170.0	0.04489	-13.47848	-15.24788
82.0	1.47355	1.68364	-0.08577	172.0	0.02883	-15.40123	-17.17064
84.0	1.48636	1.72123	-0.04818	174.0	0.01626	-17.88801	-19.65742
86.0	1.49555	1.74802	-0.02139	176.0	0.00724	-21.40127	-23.17068
88.0	1.50109	1.76406	-0.00534	178.0	0.00181	-27.41673	-29.18614
90.0	1.50294	1.76941	0.0	180.	0.00000	-INFINITY	-INFINITY

TABLE A1-16

INTEGRAL EQUATION METHOD
EPSILON= 0.5000(NORMALIZED GROUNDPLANE RADIUS)
MONOPOLE LENGTH= 0.100000(WAVELENGTH)
INFINITELY THIN MONOPOLE RADIUS
RADIATION RESISTANCE = 2.120811 (OHMS)

ELEVATION ANGLE (DEG)	DIRECTIVITY (NUMERIC)	DIRECTIVITY (DBI)	RELATIVE POWER (DB)	ELEVATION ANGLE (DEG)	DIRECTIVITY (NUMERIC)	DIRECTIVITY (DBI)	RELATIVE POWER (DB)
0.	0.00000	-INFINITY	-INFINITY				
2.0	0.00186	-27.29481	-29.03352	92.0	1.49058	1.73355	-0.00516
4.0	0.00745	-21.27990	-23.01861	94.0	1.48527	1.71806	-0.02065
6.0	0.01672	-17.76756	-19.50627	96.0	1.47645	1.69220	-0.04651
8.0	0.02963	-15.28207	-17.02078	98.0	1.46417	1.65590	-0.08281
10.0	0.04612	-13.36095	-15.09966	100.0	1.44846	1.60907	-0.12964
12.0	0.06610	-11.79820	-13.53691	102.0	1.42941	1.55158	-0.18713
14.0	0.08946	-10.48395	-12.22266	104.0	1.40711	1.48327	-0.25544
16.0	0.11608	-9.35260	-11.09131	106.0	1.38164	1.40395	-0.33476
18.0	0.14582	-8.36185	-10.10056	108.0	1.35313	1.31341	-0.42530
20.0	0.17853	-7.48281	-9.22152	110.0	1.32171	1.21137	-0.52733
22.0	0.21405	-6.69490	-8.43360	112.0	1.28753	1.09756	-0.64115
24.0	0.25218	-5.98289	-7.72160	114.0	1.25073	0.97163	-0.76708
26.0	0.29274	-5.33524	-7.07395	116.0	1.21149	0.83318	-0.90552
28.0	0.33551	-4.74299	-6.48170	118.0	1.16998	0.68180	-1.05691
30.0	0.38028	-4.19902	-5.93773	120.0	1.12642	0.51699	-1.22172
32.0	0.42681	-3.69761	-5.43632	122.0	1.08098	0.33819	-1.40052
34.0	0.47489	-3.23407	-4.97278	124.0	1.03390	0.14478	-1.59392
36.0	0.52426	-2.80453	-4.54324	126.0	0.98539	-0.06393	-1.80264
38.0	0.57468	-2.40575	-4.14446	128.0	0.93568	-0.28874	-2.02745
40.0	0.62589	-2.03500	-3.77371	130.0	0.88501	-0.53054	-2.26925
42.0	0.67765	-1.68993	-3.42864	132.0	0.83362	-0.79034	-2.52905
44.0	0.72970	-1.36856	-3.10727	134.0	0.78176	-1.06929	-2.80800
46.0	0.78178	-1.06915	-2.80786	136.0	0.72967	-1.36871	-3.10742
48.0	0.83364	-0.79020	-2.52891	138.0	0.67763	-1.69009	-3.42880
50.0	0.88503	-0.53040	-2.26911	140.0	0.62587	-2.03516	-3.77386
52.0	0.93571	-0.28861	-2.02732	142.0	0.57466	-2.40592	-4.14462
54.0	0.98542	-0.06380	-1.80251	144.0	0.52424	-2.80470	-4.54341
56.0	1.03393	0.14490	-1.59381	146.0	0.47487	-3.23424	-4.97295
58.0	1.08101	0.33830	-1.40041	148.0	0.42680	-3.69778	-5.43649
60.0	1.12644	0.51709	-1.22162	150.0	0.38026	-4.19920	-5.93791
62.0	1.17001	0.68190	-1.05681	152.0	0.33549	-4.74317	-6.48188
64.0	1.21151	0.83328	-0.90543	154.0	0.29272	-5.33543	-7.07414
66.0	1.25075	0.97171	-0.76700	156.0	0.25217	-5.98307	-7.72178
68.0	1.28755	1.09764	-0.64107	158.0	0.21404	-6.69509	-8.43379
70.0	1.32174	1.21145	-0.52726	160.0	0.17853	-7.48301	-9.22172
72.0	1.35315	1.31347	-0.42524	162.0	0.14581	-8.36204	-10.10075
74.0	1.38166	1.40401	-0.33470	164.0	0.11607	-9.35280	-11.09151
76.0	1.40712	1.48332	-0.25539	166.0	0.08945	-10.48415	-12.22286
78.0	1.42943	1.55162	-0.18709	168.0	0.06609	-11.79840	-13.53711
80.0	1.44847	1.60911	-0.12960	170.0	0.04612	-13.36115	-15.09986
82.0	1.46418	1.65593	-0.08278	172.0	0.02963	-15.28227	-17.02098
84.0	1.47646	1.69222	-0.04649	174.0	0.01672	-17.76777	-19.50647
86.0	1.48528	1.71807	-0.02064	176.0	0.00745	-21.28010	-23.01881
88.0	1.49058	1.73356	-0.00515	178.0	0.00186	-27.29501	-29.03372
90.0	1.49235	1.73871	0.0	180.	0.00000	-INFINITY	-INFINITY

TABLE A1-17

INTEGRAL EQUATION METHOD
EPSILON= 0.7500(NORMALIZED GROUNDPLANE RADIUS)
MONOPOLE LENGTH= 0.100000(WAVELENGTH)
INFINITELY THIN MONOPOLE RADIUS
RADIATION RESISTANCE = 2.146484 (OHMS)

ELEVATION ANGLE (DEG)	DIRECTIVITY (NUMERIC)	DIRECTIVITY (DBI)	RELATIVE POWER (DB)	ELEVATION ANGLE (DEG)	DIRECTIVITY (NUMERIC)	DIRECTIVITY (DBI)	RELATIVE POWER (DB)
0.	0.00000	-INFINITY	-INFINITY				
2.0	0.00195	-27.09141	-28.77786	92.0	1.47285	1.68158	-0.00487
4.0	0.00780	-21.07742	-22.76387	94.0	1.46792	1.66702	-0.01943
6.0	0.01751	-17.56660	-19.25305	96.0	1.45973	1.64271	-0.04374
8.0	0.03102	-15.08323	-16.76968	98.0	1.44830	1.60859	-0.07787
10.0	0.04825	-13.16482	-14.85127	100.0	1.43368	1.56453	-0.12192
12.0	0.06910	-11.60536	-13.29181	102.0	1.41593	1.51041	-0.17604
14.0	0.09343	-10.29495	-11.98140	104.0	1.39510	1.44605	-0.24040
16.0	0.12112	-9.16799	-10.85444	106.0	1.37127	1.37124	-0.31521
18.0	0.15198	-8.18215	-9.86860	108.0	1.34454	1.28573	-0.40072
20.0	0.18584	-7.30853	-8.99498	110.0	1.31500	1.18925	-0.49721
22.0	0.22251	-6.52650	-8.21295	112.0	1.28276	1.08146	-0.60500
24.0	0.26177	-5.82083	-7.50728	114.0	1.24796	0.96199	-0.72446
26.0	0.30339	-5.17994	-6.86639	116.0	1.21072	0.83044	-0.85601
28.0	0.34715	-4.59483	-6.28128	118.0	1.17120	0.68632	-1.00013
30.0	0.39279	-4.05835	-5.74480	120.0	1.12957	0.52912	-1.15733
32.0	0.44007	-3.56475	-5.25120	122.0	1.08598	0.35824	-1.32822
34.0	0.48873	-3.10930	-4.79575	124.0	1.04064	0.17301	-1.51344
36.0	0.53851	-2.68809	-4.37454	126.0	0.99374	-0.02729	-1.71374
38.0	0.58914	-2.29785	-3.98430	128.0	0.94547	-0.24351	-1.92996
40.0	0.64036	-1.93579	-3.62224	130.0	0.89607	-0.47658	-2.16303
42.0	0.69190	-1.59954	-3.28599	132.0	0.84576	-0.72755	-2.41400
44.0	0.74352	-1.28705	-2.97350	134.0	0.79476	-0.99762	-2.68407
46.0	0.79496	-0.99657	-2.68302	136.0	0.74334	-1.28814	-2.97459
48.0	0.84595	-0.72653	-2.41298	138.0	0.69173	-1.60066	-3.28711
50.0	0.89627	-0.47560	-2.16205	140.0	0.64019	-1.93694	-3.62339
52.0	0.94568	-0.24257	-1.92902	142.0	0.58897	-2.29903	-3.98549
54.0	0.99394	-0.02639	-1.71284	144.0	0.53836	-2.68931	-4.37576
56.0	1.04085	0.17387	-1.51258	146.0	0.48859	-3.11054	-4.79699
58.0	1.08619	0.35905	-1.32740	148.0	0.43995	-3.56602	-5.25247
60.0	1.12977	0.52989	-1.15656	150.0	0.39268	-4.05964	-5.74609
62.0	1.17140	0.68705	-0.99940	152.0	0.34705	-4.59614	-6.28259
64.0	1.21091	0.83112	-0.85533	154.0	0.30330	-5.18127	-6.86772
66.0	1.24814	0.96263	-0.72383	156.0	0.26169	-5.82218	-7.50863
68.0	1.28293	1.08204	-0.60441	158.0	0.22244	-6.52787	-8.21432
70.0	1.31516	1.18978	-0.49667	160.0	0.18578	-7.30992	-8.99637
72.0	1.34469	1.28622	-0.40023	162.0	0.15193	-8.18355	-9.87000
74.0	1.37141	1.37167	-0.31478	164.0	0.12108	-9.16940	-10.85586
76.0	1.39522	1.44643	-0.24002	166.0	0.09340	-10.29638	-11.98283
78.0	1.41603	1.51074	-0.17571	168.0	0.06908	-11.60679	-13.29324
80.0	1.43377	1.56481	-0.12164	170.0	0.04824	-13.16626	-14.85271
82.0	1.44837	1.60881	-0.07765	172.0	0.03101	-15.08468	-16.77113
84.0	1.45978	1.64288	-0.04357	174.0	0.01751	-17.56805	-19.25450
86.0	1.46796	1.66713	-0.01932	176.0	0.00780	-21.07887	-22.76533
88.0	1.47287	1.68164	-0.00481	178.0	0.00195	-27.09287	-28.77932
90.0	1.47450	1.68645	0.0	180.	0.00000	-INFINITY	-INFINITY

TABLE A1-18

INTEGRAL EQUATION METHOD
EPSILON= 1.0000(NORMALIZED GROUNDPLANE RADIUS)
MONOPOLE LENGTH= 0.100000(WAVELENGTH)
INFINITELY THIN MONOPOLE RADIUS
RADIATION RESISTANCE = 2.185235 (OHMS)

ELEVATION ANGLE (DEG)	DIRECTIVITY (NUMERIC)	DIRECTIVITY (DBI)	RELATIVE POWER (DB)	ELEVATION ANGLE (DEG)	DIRECTIVITY (NUMERIC)	DIRECTIVITY (DBI)	RELATIVE POWER (DB)
0.	0.00000	-INFINITY	-INFINITY				
2.0	0.00209	-26.79818	-28.40693	92.0	1.44686	1.60426	-0.00449
4.0	0.00835	-20.78549	-22.39424	94.0	1.44245	1.59101	-0.01774
6.0	0.01872	-17.27683	-18.88558	96.0	1.43514	1.56894	-0.03980
8.0	0.03314	-14.79647	-16.40522	98.0	1.42494	1.53797	-0.07078
10.0	0.05150	-12.88191	-14.49066	100.0	1.41188	1.49797	-0.11078
12.0	0.07367	-11.32713	-12.93587	102.0	1.39597	1.44877	-0.15998
14.0	0.09949	-10.02219	-11.63094	104.0	1.37726	1.39017	-0.21857
16.0	0.12878	-8.90148	-10.51023	106.0	1.35580	1.32194	-0.28680
18.0	0.16134	-7.92264	-9.53139	108.0	1.33162	1.24380	-0.36495
20.0	0.19694	-7.05674	-8.66549	110.0	1.30479	1.15541	-0.45333
22.0	0.23534	-6.28311	-7.89186	112.0	1.27539	1.05642	-0.55233
24.0	0.27628	-5.58649	-7.19524	114.0	1.24348	0.94640	-0.66235
26.0	0.31950	-4.95526	-6.56400	116.0	1.20917	0.82489	-0.78386
28.0	0.36472	-4.38036	-5.98911	118.0	1.17256	0.69136	-0.91738
30.0	0.41166	-3.85462	-5.46337	120.0	1.13377	0.54524	-1.06351
32.0	0.46002	-3.37222	-4.98097	122.0	1.09292	0.38587	-1.22287
34.0	0.50952	-2.92839	-4.53714	124.0	1.05016	0.21254	-1.39621
36.0	0.55987	-2.51917	-4.12791	126.0	1.00564	0.02444	-1.58431
38.0	0.61077	-2.14121	-3.74996	128.0	0.95955	-0.17933	-1.78808
40.0	0.66196	-1.79168	-3.40043	130.0	0.91206	-0.39976	-2.00850
42.0	0.71315	-1.46816	-3.07691	132.0	0.86338	-0.63796	-2.24671
44.0	0.76409	-1.16853	-2.77728	134.0	0.81373	-0.89520	-2.50395
46.0	0.81452	-0.89097	-2.49972	136.0	0.76333	-1.17290	-2.78164
48.0	0.86420	-0.63387	-2.24262	138.0	0.71242	-1.47265	-3.08139
50.0	0.91289	-0.39581	-2.00455	140.0	0.66126	-1.79629	-3.40503
52.0	0.96039	-0.17553	-1.78428	142.0	0.61011	-2.14592	-3.75467
54.0	1.00649	0.02808	-1.58067	144.0	0.55924	-2.52398	-4.13273
56.0	1.05100	0.21602	-1.39273	146.0	0.50894	-2.93331	-4.54205
58.0	1.09375	0.38919	-1.21956	148.0	0.45949	-3.37723	-4.98597
60.0	1.13459	0.54838	-1.06037	150.0	0.41118	-3.85971	-5.46846
62.0	1.17336	0.69432	-0.91442	152.0	0.36429	-4.38553	-5.99428
64.0	1.20995	0.82766	-0.78108	154.0	0.31912	-4.96050	-6.56925
66.0	1.24422	0.94898	-0.65976	156.0	0.27594	-5.59180	-7.20055
68.0	1.27609	1.05881	-0.54994	158.0	0.23505	-6.28849	-7.89724
70.0	1.30545	1.15760	-0.45115	160.0	0.19669	-7.06217	-8.67092
72.0	1.33223	1.24578	-0.36297	162.0	0.16113	-7.92813	-9.53687
74.0	1.35635	1.32372	-0.28503	164.0	0.12862	-8.90701	-10.51576
76.0	1.37776	1.39173	-0.21701	166.0	0.09936	-10.02776	-11.63651
78.0	1.39640	1.45011	-0.15864	168.0	0.07357	-11.33273	-12.94147
80.0	1.41224	1.49909	-0.10966	170.0	0.05143	-12.88754	-14.49629
82.0	1.42524	1.53887	-0.06987	172.0	0.03310	-14.80213	-16.41087
84.0	1.43536	1.56962	-0.03913	174.0	0.01870	-17.28250	-18.89125
86.0	1.44260	1.59146	-0.01728	176.0	0.00833	-20.79117	-22.39992
88.0	1.44693	1.60449	-0.00426	178.0	0.00209	-26.80387	-28.41262
90.0	1.44835	1.60875	0.0	180.	0.00000	-INFINITY	-INFINITY

TABLE A1-19

INTEGRAL EQUATION METHOD
EPSILON= 1.2500(NORMALIZED GROUNDPLANE RADIUS)
MONOPOLE LENGTH= 0.100000(WAVELENGTH)
INFINITELY THIN MONOPOLE RADIUS
RADIATION RESISTANCE = 2.241091 (OHMS)

ELEVATION ANGLE θ (DEG)	DIRECTIVITY (NUMERIC)	DIRECTIVITY (DBI)	RELATIVE POWER (DB)	ELEVATION ANGLE (DEG)	DIRECTIVITY (NUMERIC)	DIRECTIVITY (DBI)	RELATIVE POWER (DB)
0.	0.00000	-INFINITY	-INFINITY				
2.0	0.00229	-26.40280	-27.90193	92.0	1.41094	1.49508	-0.00405
4.0	0.00914	-20.39182	-21.89095	94.0	1.40720	1.48357	-0.01557
6.0	0.02048	-16.88601	-18.38515	96.0	1.40105	1.46453	-0.03461
8.0	0.03623	-14.40963	-15.90877	98.0	1.39247	1.43786	-0.06128
10.0	0.05623	-12.50016	-13.99929	100.0	1.38147	1.40340	-0.09573
12.0	0.08032	-10.95154	-12.45068	102.0	1.36803	1.36097	-0.13816
14.0	0.10830	-9.65385	-11.15298	104.0	1.35217	1.31032	-0.18882
16.0	0.13991	-8.54141	-10.04054	106.0	1.33388	1.25116	-0.24798
18.0	0.17491	-7.57184	-9.07098	108.0	1.31315	1.18315	-0.31598
20.0	0.21300	-6.71618	-8.21531	110.0	1.29001	1.10592	-0.39321
22.0	0.25388	-5.95371	-7.45284	112.0	1.26446	1.01903	-0.48010
24.0	0.29723	-5.26911	-6.76825	114.0	1.23652	0.92200	-0.57713
26.0	0.34271	-4.65073	-6.14987	116.0	1.20623	0.81429	-0.68485
28.0	0.38999	-4.08947	-5.58860	118.0	1.17362	0.69529	-0.80384
30.0	0.43873	-3.57806	-5.07720	120.0	1.13877	0.56435	-0.93479
32.0	0.48858	-3.11065	-4.60978	122.0	1.10172	0.42073	-1.07841
34.0	0.53921	-2.68239	-4.18152	124.0	1.06258	0.26363	-1.23551
36.0	0.59030	-2.28925	-3.78839	126.0	1.02145	0.09216	-1.40698
38.0	0.64153	-1.92783	-3.42696	128.0	0.97844	-0.09466	-1.59380
40.0	0.69260	-1.59521	-3.09434	130.0	0.93370	-0.29792	-1.79706
42.0	0.74321	-1.28888	-2.78801	132.0	0.88740	-0.51882	-2.01795
44.0	0.79311	-1.00666	-2.50579	134.0	0.83971	-0.75870	-2.25783
46.0	0.84205	-0.74663	-2.24577	136.0	0.79085	-1.01906	-2.51820
48.0	0.88979	-0.50711	-2.00625	138.0	0.74104	-1.30160	-2.80073
50.0	0.93614	-0.28660	-1.78573	140.0	0.69052	-1.60823	-3.10736
52.0	0.98090	-0.08374	-1.58288	142.0	0.63957	-1.94113	-3.44027
54.0	1.02392	0.10266	-1.39648	144.0	0.58846	-2.30282	-3.80195
56.0	1.06505	0.27369	-1.22545	146.0	0.53750	-2.69620	-4.19533
58.0	1.10416	0.43033	-1.06881	148.0	0.48700	-3.12468	-4.62382
60.0	1.14116	0.57346	-0.92567	150.0	0.43729	-3.59230	-5.09144
62.0	1.17595	0.70390	-0.79523	152.0	0.38870	-4.10390	-5.60304
64.0	1.20848	0.82238	-0.67675	154.0	0.34156	-4.66535	-6.16448
66.0	1.23867	0.92955	-0.56958	156.0	0.29622	-5.28389	-6.78302
68.0	1.26649	1.02603	-0.47311	158.0	0.25301	-5.96862	-7.46776
70.0	1.29191	1.11234	-0.38680	160.0	0.21226	-6.73123	-8.23036
72.0	1.31491	1.18897	-0.31016	162.0	0.17430	-7.58701	-9.08615
74.0	1.33548	1.25637	-0.24276	164.0	0.13942	-8.55668	-10.05582
76.0	1.35360	1.31491	-0.18422	166.0	0.10791	-9.66921	-11.16834
78.0	1.36928	1.36493	-0.13420	168.0	0.08004	-10.96698	-12.46612
80.0	1.38252	1.40672	-0.09241	170.0	0.05603	-12.51566	-14.01480
82.0	1.39332	1.44052	-0.05861	172.0	0.03610	-14.42519	-15.92432
84.0	1.40169	1.46653	-0.03260	174.0	0.02041	-16.90161	-18.40075
86.0	1.40764	1.48491	-0.01423	176.0	0.00910	-20.40745	-21.90658
88.0	1.41116	1.49575	-0.00338	178.0	0.00228	-26.41844	-27.91758
90.0	1.41226	1.49913	0.0	180.	0.00000	-INFINITY	-INFINITY

TABLE A1-20

INTEGRAL EQUATION METHOD
EPSILON= 1.5000(NORMALIZED GROUNDPLANE RADIUS)
MONOPOLE LENGTH= 0.100000(WAVELENGTH)
INFINITELY THIN MONOPOLE RADIUS
RADIATION RESISTANCE = 2.320740 (OHMS)

ELEVATION ANGLE (DEG)	DIRECTIVITY (NUMERIC)	DIRECTIVITY (DBI)	RELATIVE POWER (DB)	ELEVATION ANGLE (DEG)	DIRECTIVITY (NUMERIC)	DIRECTIVITY (DBI)	RELATIVE POWER (DB)
0.	0.00000	-INFINITY	-INFINITY				
2.0	0.00258	-25.89000	-27.23747	92.0	1.36265	1.34383	-0.00363
4.0	0.01028	-19.88119	-21.22865	94.0	1.35973	1.33453	-0.01293
6.0	0.02302	-16.37897	-17.72644	96.0	1.35503	1.31948	-0.02799
8.0	0.04067	-13.90760	-15.25506	98.0	1.34851	1.29854	-0.04893
10.0	0.06303	-12.00453	-13.35200	100.0	1.34014	1.27151	-0.07595
12.0	0.08987	-10.46371	-11.81118	102.0	1.32989	1.23817	-0.10930
14.0	0.12092	-9.17516	-10.52262	104.0	1.31771	1.19819	-0.14927
16.0	0.15584	-8.07318	-9.42064	106.0	1.30354	1.15123	-0.19624
18.0	0.19430	-7.11535	-8.46281	108.0	1.28732	1.09686	-0.25060
20.0	0.23590	-6.27266	-7.62012	110.0	1.26900	1.03463	-0.31284
22.0	0.28026	-5.52435	-6.87182	112.0	1.24853	0.96399	-0.38348
24.0	0.32696	-4.85506	-6.20253	114.0	1.22585	0.88436	-0.46310
26.0	0.37557	-4.25307	-5.60053	116.0	1.20091	0.79510	-0.55236
28.0	0.42568	-3.70920	-5.05666	118.0	1.17368	0.69550	-0.65197
30.0	0.47685	-3.21614	-4.56361	120.0	1.14414	0.58477	-0.76269
32.0	0.52869	-2.76795	-4.11541	122.0	1.11226	0.46208	-0.88538
34.0	0.58080	-2.35972	-3.70718	124.0	1.07808	0.32650	-1.02097
36.0	0.63280	-1.98732	-3.33478	126.0	1.04160	0.17701	-1.17045
38.0	0.68434	-1.64726	-2.99473	128.0	1.00289	0.01254	-1.33493
40.0	0.73510	-1.33655	-2.68401	130.0	0.96202	-0.16814	-1.51561
42.0	0.78477	-1.05257	-2.40004	132.0	0.91911	-0.36634	-1.71380
44.0	0.83310	-0.79305	-2.14051	134.0	0.87428	-0.58351	-1.93097
46.0	0.87984	-0.55596	-1.90343	136.0	0.82770	-0.82127	-2.16873
48.0	0.92480	-0.33952	-1.68698	138.0	0.77958	-1.08142	-2.42888
50.0	0.96781	-0.14210	-1.48956	140.0	0.73013	-1.36597	-2.71344
52.0	1.00873	0.03775	-1.30972	142.0	0.67964	-1.67722	-3.02468
54.0	1.04745	0.20134	-1.14612	144.0	0.62838	-2.01777	-3.36523
56.0	1.08390	0.34988	-0.99758	146.0	0.57668	-2.39061	-3.73808
58.0	1.11801	0.48447	-0.86300	148.0	0.52490	-2.79926	-4.14672
60.0	1.14977	0.60610	-0.74136	150.0	0.47339	-3.24782	-4.59529
62.0	1.17916	0.71572	-0.63175	152.0	0.42255	-3.74122	-5.08868
64.0	1.20619	0.81416	-0.53331	154.0	0.37279	-4.28540	-5.63286
66.0	1.23089	0.90220	-0.44527	156.0	0.32451	-4.88767	-6.23513
68.0	1.25330	0.98055	-0.36692	158.0	0.27815	-5.55720	-6.90467
70.0	1.27346	1.04986	-0.29760	160.0	0.23411	-6.30573	-7.65319
72.0	1.29144	1.11073	-0.23674	162.0	0.19281	-7.14861	-8.49608
74.0	1.30728	1.16367	-0.18379	164.0	0.15465	-8.10661	-9.45407
76.0	1.32105	1.20918	-0.13829	166.0	0.11998	-9.20873	-10.55620
78.0	1.33280	1.24766	-0.09981	168.0	0.08918	-10.49741	-11.84487
80.0	1.34260	1.27948	-0.06799	170.0	0.06254	-12.03834	-13.38580
82.0	1.35050	1.30494	-0.04253	172.0	0.04035	-13.94148	-15.28895
84.0	1.35653	1.32430	-0.02316	174.0	0.02284	-16.41292	-17.76039
86.0	1.36074	1.33776	-0.00971	176.0	0.01020	-19.91518	-21.26264
88.0	1.36315	1.34545	-0.00202	178.0	0.00256	-25.92402	-27.27149
90.0	1.36379	1.34746	0.0	180.	0.00000	-INFINITY	-INFINITY

TABLE A1-21

INTEGRAL EQUATION METHOD
EPSILON= 1.7500(NORMALIZED GROUNDPLANE RADIUS)
MONOPOLE LENGTH= 0.100000(WAVELENGTH)
INFINITELY THIN MONOPOLE RADIUS
RADIATION RESISTANCE = 2.435146 (OHMS)

ELEVATION ANGLE (DEG)	DIRECTIVITY (NUMERIC)	DIRECTIVITY (DBI)	RELATIVE POWER (DB)	ELEVATION ANGLE (DEG)	DIRECTIVITY (NUMERIC)	DIRECTIVITY (DBI)	RELATIVE POWER (DB)
0.	0.00000	-INFINITY	-INFINITY				
2.0	0.00299	-25.24498	-26.38355	92.0	1.29873	1.13520	-0.00337
4.0	0.01192	-19.23878	-20.37735	94.0	1.29679	1.12869	-0.00988
6.0	0.02666	-15.74092	-16.87949	96.0	1.29385	1.11885	-0.01972
8.0	0.04704	-13.27562	-14.41419	98.0	1.28988	1.10550	-0.03306
10.0	0.07277	-11.38034	-12.51890	100.0	1.28481	1.08839	-0.05017
12.0	0.10354	-9.84898	-10.98754	102.0	1.27855	1.06719	-0.07137
14.0	0.13895	-8.57154	-9.71010	104.0	1.27101	1.04151	-0.09706
16.0	0.17855	-7.48229	-8.62086	106.0	1.26208	1.01086	-0.12771
18.0	0.22188	-6.53879	-7.67735	108.0	1.25162	0.97472	-0.16385
20.0	0.26841	-5.71195	-6.85052	110.0	1.23950	0.93248	-0.20609
22.0	0.31762	-4.98099	-6.11955	112.0	1.22559	0.88346	-0.25510
24.0	0.36894	-4.33047	-5.46904	114.0	1.20974	0.82694	-0.31163
26.0	0.42183	-3.74862	-4.88718	116.0	1.19182	0.76209	-0.37647
28.0	0.47575	-3.22620	-4.36476	118.0	1.17167	0.68806	-0.45050
30.0	0.53017	-2.75582	-3.89439	120.0	1.14919	0.60390	-0.53466
32.0	0.58459	-2.33146	-3.47003	122.0	1.12424	0.50861	-0.62996
34.0	0.63854	-1.94813	-3.08670	124.0	1.09675	0.40109	-0.73748
36.0	0.69157	-1.60161	-2.74018	126.0	1.06664	0.28017	-0.85839
38.0	0.74331	-1.28831	-2.42688	128.0	1.03386	0.14461	-0.99396
40.0	0.79339	-1.00514	-2.14370	130.0	0.99840	-0.00696	-1.14553
42.0	0.84152	-0.74936	-1.88793	132.0	0.96028	-0.17601	-1.31458
44.0	0.88744	-0.51859	-1.65716	134.0	0.91957	-0.36414	-1.50271
46.0	0.93096	-0.31069	-1.44926	136.0	0.87637	-0.57310	-1.71167
48.0	0.97191	-0.12372	-1.26229	138.0	0.83084	-0.80483	-1.94339
50.0	1.01020	0.04405	-1.09451	140.0	0.78317	-1.06146	-2.20003
52.0	1.04574	0.19423	-0.94434	142.0	0.73360	-1.34541	-2.48398
54.0	1.07852	0.32828	-0.81029	144.0	0.68243	-1.65940	-2.79797
56.0	1.10855	0.44755	-0.69102	146.0	0.63001	-2.00653	-3.14510
58.0	1.13587	0.55330	-0.58527	148.0	0.57671	-2.39041	-3.52898
60.0	1.16056	0.64668	-0.49189	150.0	0.52297	-2.81525	-3.95382
62.0	1.18271	0.72879	-0.40977	152.0	0.46924	-3.28605	-4.42462
64.0	1.20244	0.80064	-0.33793	154.0	0.41602	-3.80885	-4.94741
66.0	1.21988	0.86316	-0.27540	156.0	0.36383	-4.39102	-5.52959
68.0	1.23516	0.91724	-0.22133	158.0	0.31320	-5.04182	-6.18039
70.0	1.24844	0.96367	-0.17490	160.0	0.26467	-5.77302	-6.91159
72.0	1.25985	1.00320	-0.13537	162.0	0.21877	-6.60007	-7.73863
74.0	1.26955	1.03651	-0.10206	164.0	0.17605	-7.54375	-8.68232
76.0	1.27768	1.06421	-0.07436	166.0	0.13699	-8.63314	-9.77171
78.0	1.28436	1.08686	-0.05171	168.0	0.10208	-9.91070	-11.04927
80.0	1.28971	1.10492	-0.03364	170.0	0.07174	-11.44216	-12.58073
82.0	1.29385	1.11882	-0.01974	172.0	0.04637	-13.33752	-14.47609
84.0	1.29685	1.12890	-0.00967	174.0	0.02629	-15.80288	-16.94145
86.0	1.29880	1.13541	-0.00315	176.0	0.01175	-19.30078	-20.43935
88.0	1.29974	1.13857	0.0	178.0	0.00295	-25.30701	-26.44558
90.0	1.29971	1.13848	-0.00009	180.	0.00000	-INFINITY	-INFINITY

TABLE A1-22

INTEGRAL EQUATION METHOD
EPSILON= 2.0000(NORMALIZED GROUNDPLANE RADIUS)
MONOPOLE LENGTH= 0.100000(WAVELENGTH)
INFINITELY THIN MONOPOLE RADIUS
RADIATION RESISTANCE = 2.602246 (OHMS)

ELEVATION ANGLE (DEG)	DIRECTIVITY (NUMERIC)	DIRECTIVITY (DBI)	RELATIVE POWER (DB)	ELEVATION ANGLE (DEG)	DIRECTIVITY (NUMERIC)	DIRECTIVITY (DBI)	RELATIVE POWER (DB)
0.	0.00000	-INFINITY	-INFINITY				
2.0	0.00358	-24.45939	-25.32544	92.0	1.21539	0.84716	-0.01889
4.0	0.01427	-18.45624	-19.32230	94.0	1.21457	0.84422	-0.02183
6.0	0.03189	-14.96346	-15.82951	96.0	1.21375	0.84128	-0.02477
8.0	0.05617	-12.50525	-13.37131	98.0	1.21286	0.83809	-0.02796
10.0	0.08671	-10.61906	-11.48511	100.0	1.21180	0.83429	-0.03176
12.0	0.12306	-9.09877	-9.96482	102.0	1.21044	0.82943	-0.03662
14.0	0.16465	-7.83435	-8.70040	104.0	1.20863	0.82295	-0.04310
16.0	0.21086	-6.76004	-7.62609	106.0	1.20620	0.81420	-0.05185
18.0	0.26101	-5.83335	-6.69940	108.0	1.20294	0.80244	-0.06361
20.0	0.31440	-5.02518	-5.89123	110.0	1.19863	0.78686	-0.07919
22.0	0.37028	-4.31467	-5.18072	112.0	1.19304	0.76656	-0.09950
24.0	0.42792	-3.68636	-4.55241	114.0	1.18592	0.74056	-0.12549
26.0	0.48659	-3.12839	-3.99444	116.0	1.17702	0.70783	-0.15822
28.0	0.54557	-2.63147	-3.49752	118.0	1.16607	0.66726	-0.19879
30.0	0.60421	-2.18815	-3.05420	120.0	1.15284	0.61768	-0.24838
32.0	0.66186	-1.79232	-2.65838	122.0	1.13706	0.55783	-0.30822
34.0	0.71797	-1.43891	-2.30496	124.0	1.11852	0.48643	-0.37963
36.0	0.77204	-1.12361	-1.98966	126.0	1.09700	0.40207	-0.46399
38.0	0.82362	-0.84273	-1.70878	128.0	1.07233	0.30329	-0.56276
40.0	0.87236	-0.59306	-1.45911	130.0	1.04437	0.18855	-0.67750
42.0	0.91796	-0.37178	-1.23783	132.0	1.01302	0.05618	-0.80988
44.0	0.96021	-0.17635	-1.04240	134.0	0.97823	-0.09560	-0.96165
46.0	0.99896	-0.00450	-0.87055	136.0	0.94001	-0.26869	-1.13475
48.0	1.03415	0.14584	-0.72022	138.0	0.89842	-0.46520	-1.33125
50.0	1.06575	0.27656	-0.58949	140.0	0.85361	-0.68740	-1.55345
52.0	1.09381	0.38943	-0.47662	142.0	0.80578	-0.93784	-1.80390
54.0	1.11843	0.48610	-0.37995	144.0	0.75520	-1.21937	-2.08542
56.0	1.13975	0.56810	-0.29795	146.0	0.70223	-1.53520	-2.40126
58.0	1.15794	0.63687	-0.22918	148.0	0.64728	-1.88905	-2.75511
60.0	1.17322	0.69380	-0.17225	150.0	0.59085	-2.28523	-3.15128
62.0	1.18582	0.74019	-0.12586	152.0	0.53348	-2.72882	-3.59487
64.0	1.19598	0.77726	-0.08880	154.0	0.47578	-3.22595	-4.09200
66.0	1.20398	0.80618	-0.05987	156.0	0.41840	-3.78407	-4.65012
68.0	1.21006	0.82807	-0.03798	158.0	0.36204	-4.41249	-5.27854
70.0	1.21450	0.84396	-0.02209	160.0	0.30739	-5.12307	-5.98912
72.0	1.21754	0.85483	-0.01122	162.0	0.25519	-5.93129	-6.79734
74.0	1.21944	0.86160	-0.00446	164.0	0.20616	-6.85800	-7.72405
76.0	1.22042	0.86508	-0.00097	166.0	0.16098	-7.93231	-8.79836
78.0	1.22069	0.86605	0.0	168.0	0.12032	-9.19673	-10.06278
80.0	1.22044	0.86518	-0.00088	170.0	0.08478	-10.71701	-11.58306
82.0	1.21984	0.86304	-0.00301	172.0	0.05491	-12.60319	-13.46924
84.0	1.21903	0.86014	-0.00591	174.0	0.03118	-15.06138	-15.92743
86.0	1.21811	0.85686	-0.00919	176.0	0.01395	-18.55415	-19.42020
88.0	1.21717	0.85350	-0.01255	178.0	0.00350	-24.55729	-25.42334
90.0	1.21625	0.85025	-0.01581	180.	0.00000	-INFINITY	-INFINITY

TABLE A1-23

INTEGRAL EQUATION METHOD
EPSILON= 2.2500(NORMALIZED GROUNDPLANE RADIUS)
MONOPOLE LENGTH= 0.100000(WAVELENGTH)
INFINITELY THIN MONOPOLE RADIUS
RADIATION RESISTANCE = 2.851602 (OHMS)

ELEVATION ANGLE (DEG)	DIRECTIVITY (NUMERIC)	DIRECTIVITY (DBI)	RELATIVE POWER (DB)
0.	0.00000	-INFINITY	-INFINITY
2.0	0.00443	-23.53888	-24.28243
4.0	0.01762	-17.53915	-18.28271
6.0	0.03934	-14.05206	-14.79562
8.0	0.06915	-11.60181	-12.34537
10.0	0.10652	-9.72583	-10.46939
12.0	0.15073	-8.21798	-8.96154
14.0	0.20099	-6.96821	-7.71177
16.0	0.25640	-5.91074	-6.65430
18.0	0.31601	-5.00304	-5.74660
20.0	0.37879	-4.21598	-4.95953
22.0	0.44375	-3.52866	-4.27222
24.0	0.50985	-2.92557	-3.66913
26.0	0.57613	-2.39482	-3.13838
28.0	0.64164	-1.92705	-2.67061
30.0	0.70554	-1.51475	-2.25831
32.0	0.76705	-1.15176	-1.89532
34.0	0.82548	-0.83291	-1.57647
36.0	0.88027	-0.55384	-1.29740
38.0	0.93095	-0.31076	-1.05431
40.0	0.97716	-0.10036	-0.84392
42.0	1.01866	0.08029	-0.66327
44.0	1.05532	0.23383	-0.50973
46.0	1.08709	0.36266	-0.38089
48.0	1.11404	0.46901	-0.27454
50.0	1.13630	0.55493	-0.18863
52.0	1.15408	0.62235	-0.12120
54.0	1.16765	0.67311	-0.07044
56.0	1.17732	0.70896	-0.03460
58.0	1.18347	0.73158	-0.01198
60.0	1.18647	0.74258	-0.00097
62.0	1.18674	0.74356	0.0
64.0	1.18468	0.73602	-0.00754
66.0	1.18071	0.72145	-0.02211
68.0	1.17525	0.70129	-0.04227
70.0	1.16867	0.67692	-0.06664
72.0	1.16136	0.64966	-0.09389
74.0	1.15366	0.62079	-0.12276
76.0	1.14591	0.59150	-0.15206
78.0	1.13838	0.56288	-0.18067
80.0	1.13134	0.53595	-0.20761
82.0	1.12501	0.51158	-0.23198
84.0	1.11958	0.49055	-0.25301
86.0	1.11518	0.47346	-0.27010
88.0	1.11193	0.46079	-0.28277
90.0	1.10990	0.45284	-0.29072
92.0	1.10911	0.44975	-0.29381
94.0	1.10955	0.45149	-0.29207
96.0	1.11118	0.45784	-0.28571
98.0	1.11390	0.46844	-0.27511
100.0	1.11757	0.48276	-0.26080
102.0	1.12204	0.50009	-0.24346
104.0	1.12710	0.51963	-0.22393
106.0	1.13251	0.54041	-0.20315
108.0	1.13798	0.56136	-0.18219
110.0	1.14323	0.58133	-0.16223
112.0	1.14790	0.59904	-0.14451
114.0	1.15164	0.61319	-0.13037
116.0	1.15408	0.62236	-0.12120
118.0	1.15481	0.62510	-0.11846
120.0	1.15343	0.61991	-0.12365
122.0	1.14954	0.60523	-0.13832
124.0	1.14274	0.57946	-0.16410
126.0	1.13264	0.54093	-0.20263
128.0	1.11890	0.48793	-0.25563
130.0	1.10120	0.41866	-0.32490
132.0	1.07926	0.33125	-0.41230
134.0	1.05287	0.22374	-0.51982
136.0	1.02189	0.09402	-0.64954
138.0	0.98624	-0.06015	-0.80371
140.0	0.94597	-0.24122	-0.98477
142.0	0.90119	-0.45184	-1.19540
144.0	0.85212	-0.69500	-1.43855
146.0	0.79910	-0.97401	-1.71757
148.0	0.74256	-1.29269	-2.03625
150.0	0.68306	-1.65544	-2.39899
152.0	0.62124	-2.06742	-2.81098
154.0	0.55785	-2.53482	-3.27838
156.0	0.49372	-3.06519	-3.80874
158.0	0.42975	-3.66787	-4.41142
160.0	0.36688	-4.35477	-5.09833
162.0	0.30610	-5.14143	-5.88499
164.0	0.24839	-6.04874	-6.79230
166.0	0.19472	-7.10585	-7.84941
168.0	0.14604	-8.35529	-9.09885
170.0	0.10321	-9.86286	-10.60642
172.0	0.06701	-11.73860	-12.48216
174.0	0.03812	-14.18866	-14.93221
176.0	0.01708	-17.67561	-18.41916
178.0	0.00429	-23.67525	-24.41881
180.	0.00000	-INFINITY	-INFINITY

TABLE A1-24

INTEGRAL EQUATION METHOD
EPSILON= 2.5000(NORMALIZED GROUNDPLANE RADIUS)
MONOPOLE LENGTH= 0.100000(WAVELENGTH)
INFINITELY THIN MONOPOLE RADIUS
RADIATION RESISTANCE = 3.232568 (OHMS)

ELEVATION ANGLE (DEG)	DIRECTIVITY (NUMERIC)	DIRECTIVITY (DBI)	RELATIVE POWER (DB)	ELEVATION ANGLE (DEG)	DIRECTIVITY (NUMERIC)	DIRECTIVITY (DBI)	RELATIVE POWER (DB)
0.	0.00000	-INFINITY	-INFINITY				
2.0	0.00561	-22.50875	-23.39219	92.0	0.97835	-0.09507	-0.97852
4.0	0.02232	-16.51271	-17.39616	94.0	0.98017	-0.08699	-0.97044
6.0	0.04975	-13.03178	-13.91522	96.0	0.98450	-0.06783	-0.95127
8.0	0.08729	-10.59013	-11.47357	98.0	0.99125	-0.03818	-0.92162
10.0	0.13412	-8.72519	-9.60863	100.0	1.00025	0.00110	-0.88234
12.0	0.18920	-7.23080	-8.11424	102.0	1.01132	0.04889	-0.83455
14.0	0.25137	-5.99690	-6.88034	104.0	1.02421	0.10389	-0.77956
16.0	0.31932	-4.95768	-5.84113	106.0	1.03863	0.16460	-0.71884
18.0	0.39169	-4.07060	-4.95405	108.0	1.05424	0.22939	-0.65405
20.0	0.46704	-3.30649	-4.18993	110.0	1.07066	0.29652	-0.58692
22.0	0.54395	-2.64443	-3.52788	112.0	1.08747	0.36417	-0.51928
24.0	0.62103	-2.06888	-2.95232	114.0	1.10419	0.43044	-0.45300
26.0	0.69696	-1.56790	-2.45134	116.0	1.12032	0.49342	-0.39002
28.0	0.77053	-1.13210	-2.01554	118.0	1.13532	0.55117	-0.33227
30.0	0.84063	-0.75393	-1.63737	120.0	1.14861	0.60174	-0.28170
32.0	0.90632	-0.42717	-1.31061	122.0	1.15962	0.64317	-0.24028
34.0	0.96680	-0.14661	-1.03006	124.0	1.16774	0.67348	-0.20996
36.0	1.02145	0.09217	-0.79127	126.0	1.17238	0.69070	-0.19274
38.0	1.06980	0.29302	-0.59042	128.0	1.17296	0.69283	-0.19062
40.0	1.11156	0.45932	-0.42413	130.0	1.16891	0.67782	-0.20562
42.0	1.14659	0.59408	-0.28937	132.0	1.15974	0.64360	-0.23984
44.0	1.17491	0.70003	-0.18341	134.0	1.14499	0.58800	-0.29544
46.0	1.19665	0.77968	-0.10377	136.0	1.12429	0.50877	-0.37467
48.0	1.21209	0.83535	-0.04809	138.0	1.09737	0.40352	-0.47992
50.0	1.22159	0.86925	-0.01420	140.0	1.06407	0.26969	-0.61376
52.0	1.22559	0.88344	0.0	142.0	1.02435	0.10449	-0.77895
54.0	1.22460	0.87996	-0.00349	144.0	0.97834	-0.09512	-0.97857
56.0	1.21920	0.86074	-0.02270	146.0	0.92628	-0.33257	-1.21602
58.0	1.20996	0.82771	-0.05573	148.0	0.86861	-0.61173	-1.49518
60.0	1.19750	0.78276	-0.10068	150.0	0.80592	-0.93707	-1.82051
62.0	1.18243	0.72776	-0.15568	152.0	0.73896	-1.31381	-2.19726
64.0	1.16536	0.66460	-0.21884	154.0	0.66862	-1.74820	-2.63165
66.0	1.14687	0.59515	-0.28830	156.0	0.59596	-2.24781	-3.13126
68.0	1.12753	0.52126	-0.36218	158.0	0.52215	-2.82206	-3.70550
70.0	1.10785	0.44481	-0.43863	160.0	0.44845	-3.48287	-4.36631
72.0	1.08834	0.36764	-0.51580	162.0	0.37620	-4.24582	-5.12927
74.0	1.06944	0.29156	-0.59188	164.0	0.30677	-5.13184	-6.01529
76.0	1.05155	0.21832	-0.66513	166.0	0.24154	-6.17010	-7.05354
78.0	1.03505	0.14959	-0.73385	168.0	0.18184	-7.40316	-8.28660
80.0	1.02022	0.08695	-0.79649	170.0	0.12892	-8.89682	-9.78026
82.0	1.00736	0.03183	-0.85161	172.0	0.08392	-10.76117	-11.64461
84.0	0.99666	-0.01453	-0.89797	174.0	0.04784	-13.20234	-14.08579
86.0	0.98831	-0.05107	-0.93451	176.0	0.02146	-16.68294	-17.56638
88.0	0.98243	-0.07699	-0.96044	178.0	0.00540	-22.67877	-23.56222
90.0	0.97910	-0.09175	-0.97519	180.	0.00000	-INFINITY	-INFINITY

TABLE A1-25

INTEGRAL EQUATION METHOD
EPSILON= 2.7500(NORMALIZED GROUNDPLANE RADIUS)
MONOPOLE LENGTH= 0.100000(WAVELENGTH)
INFINITELY THIN MONOPOLE RADIUS
RADIATION RESISTANCE = 3.832907 (OHMS)

ELEVATION ANGLE (DEG)	DIRECTIVITY (NUMERIC)	DIRECTIVITY (DBI)	RELATIVE POWER (DB)	ELEVATION ANGLE (DEG)	DIRECTIVITY (NUMERIC)	DIRECTIVITY (DBI)	RELATIVE POWER (DB)
0.	0.00000	-INFINITY	-INFINITY				
2.0	0.33458	-4.75496	-6.01229	92.0	0.82283	-0.84689	-2.10422
4.0	0.08184	-10.87035	-12.12768	94.0	0.82389	-0.84132	-2.09866
6.0	0.03555	-14.49103	-15.74836	96.0	0.82886	-0.81521	-2.07254
8.0	0.01953	-17.09339	-18.35072	98.0	0.83765	-0.76940	-2.02673
10.0	0.01219	-19.14066	-20.39799	100.0	0.85011	-0.70525	-1.96258
12.0	0.00824	-20.83972	-22.09705	102.0	0.86606	-0.62453	-1.88186
14.0	0.00589	-22.30144	-23.55878	104.0	0.88524	-0.52937	-1.78671
16.0	0.00437	-23.59260	-24.84993	106.0	0.90737	-0.42217	-1.67950
18.0	0.00334	-24.75685	-26.01418	108.0	0.93207	-0.30550	-1.56283
20.0	0.00262	-25.82470	-27.08203	110.0	0.95895	-0.18206	-1.43939
22.0	0.00208	-26.81867	-28.07601	112.0	0.98751	-0.05460	-1.31193
24.0	0.00168	-27.75632	-29.01366	114.0	1.01722	0.07415	-1.18318
26.0	0.00136	-28.65205	-29.90938	116.0	1.04749	0.20148	-1.05585
28.0	0.00112	-29.51836	-30.77569	118.0	1.07764	0.32475	-0.93259
30.0	0.00092	-30.36685	-31.62418	120.0	1.10698	0.44139	-0.81595
32.0	0.00076	-31.20899	-32.46633	122.0	1.13473	0.54891	-0.70842
34.0	0.00062	-32.05707	-33.31441	124.0	1.16009	0.64491	-0.61242
36.0	0.00051	-32.92525	-34.18258	126.0	1.18223	0.72702	-0.53031
38.0	0.00041	-33.83118	-35.08851	128.0	1.20031	0.79293	-0.46440
40.0	0.00033	-34.79874	-36.05607	130.0	1.21348	0.84034	-0.41699
42.0	0.00026	-35.86305	-37.12038	132.0	1.22094	0.86696	-0.39038
44.0	0.00020	-37.08103	-38.33836	134.0	1.22192	0.87044	-0.38689
46.0	0.00014	-38.55739	-39.81472	136.0	1.21574	0.84839	-0.40894
48.0	0.00009	-40.52488	-41.78221	138.0	1.20180	0.79830	-0.45903
50.0	0.00004	-43.72276	-44.98009	140.0	1.17965	0.71752	-0.53982
52.0	1.33577	1.25733	0.0	142.0	1.14899	0.60317	-0.65417
54.0	1.31585	1.19205	-0.06528	144.0	1.10972	0.45212	-0.80521
56.0	1.29097	1.10916	-0.14817	146.0	1.06192	0.26091	-0.99642
58.0	1.26203	1.01071	-0.24662	148.0	1.00591	0.02560	-1.23173
60.0	1.22992	0.89877	-0.35856	150.0	0.94225	-0.25832	-1.51565
62.0	1.19549	0.77545	-0.48189	152.0	0.87174	-0.59613	-1.85346
64.0	1.15955	0.64289	-0.61444	154.0	0.79541	-0.99409	-2.25142
66.0	1.12288	0.50334	-0.75399	156.0	0.71452	-1.45984	-2.71717
68.0	1.08620	0.35911	-0.89822	158.0	0.63055	-2.00280	-3.26013
70.0	1.05017	0.21259	-1.04474	160.0	0.54514	-2.63494	-3.89227
72.0	1.01537	0.06626	-1.19108	162.0	0.46006	-3.37187	-4.62921
74.0	0.98235	-0.07734	-1.33468	164.0	0.37718	-4.23454	-5.49187
76.0	0.95156	-0.21562	-1.47295	166.0	0.29839	-5.25213	-6.50947
78.0	0.92342	-0.34599	-1.60332	168.0	0.22557	-6.46724	-7.72457
80.0	0.89828	-0.46590	-1.72323	170.0	0.16048	-7.94569	-9.20302
82.0	0.87641	-0.57291	-1.83025	172.0	0.10477	-9.79756	-11.05489
84.0	0.85807	-0.66479	-1.92212	174.0	0.05985	-12.22902	-13.48635
86.0	0.84342	-0.73954	-1.99687	176.0	0.02690	-15.70267	-16.96000
88.0	0.83262	-0.79553	-2.05286	178.0	0.00677	-21.69433	-22.95166
90.0	0.82574	-0.83155	-2.08889	180.	0.00000	-INFINITY	-INFINITY

Tables A1-26–A1-37; $h/\lambda = 0.025$, $0 \leqslant ka \leqslant 3.0$

TABLE A1-26

INTEGRAL EQUATION METHOD
EPSILON= 0.0 (NORMALIZED GROUNDPLANE RADIUS)
MONOPOLE LENGTH= 0.025000(WAVELENGTH)
INFINITELY THIN MONOPOLE RADIUS
RADIATION RESISTANCE = 0.123760 (OHMS)

ELEVATION ANGLE (DEG)	DIRECTIVITY (NUMERIC)	DIRECTIVITY (DBI)	RELATIVE POWER (DB)	ELEVATION ANGLE (DEG)	DIRECTIVITY (NUMERIC)	DIRECTIVITY (DBI)	RELATIVE POWER (DB)
0.	0.00000	-INFINITY	-INFINITY				
2.0	0.00182	-27.38746	-29.14957	92.0	1.49858	1.75680	-0.00530
4.0	0.00729	-21.37213	-23.13424	94.0	1.49310	1.74089	-0.02121
6.0	0.01637	-17.85910	-19.62120	96.0	1.48400	1.71433	-0.04778
8.0	0.02902	-15.37263	-17.13473	98.0	1.47131	1.67704	-0.08506
10.0	0.04518	-13.45027	-15.21237	100.0	1.45511	1.62895	-0.13315
12.0	0.06477	-11.88602	-13.64812	102.0	1.43547	1.56993	-0.19217
14.0	0.08770	-10.57000	-12.33210	104.0	1.41248	1.49984	-0.26227
16.0	0.11385	-9.43664	-11.19874	106.0	1.38627	1.41848	-0.34362
18.0	0.14310	-8.44364	-10.20574	108.0	1.35696	1.32566	-0.43644
20.0	0.17530	-7.56212	-9.32423	110.0	1.32468	1.22112	-0.54098
22.0	0.21030	-6.77151	-8.53361	112.0	1.28961	1.10458	-0.65752
24.0	0.24794	-6.05660	-7.81870	114.0	1.25191	0.97572	-0.78638
26.0	0.28801	-5.40587	-7.16797	116.0	1.21176	0.83416	-0.92794
28.0	0.33034	-4.81035	-6.57246	118.0	1.16936	0.67949	-1.08261
30.0	0.37472	-4.26296	-6.02507	120.0	1.12492	0.51123	-1.25088
32.0	0.42092	-3.75799	-5.52009	122.0	1.07866	0.32884	-1.43326
34.0	0.46873	-3.29076	-5.05286	124.0	1.03080	0.13173	-1.63038
36.0	0.51791	-2.85742	-4.61952	126.0	0.98157	-0.08080	-1.84290
38.0	0.56823	-2.45476	-4.21686	128.0	0.93121	-0.30951	-2.07161
40.0	0.61944	-2.08004	-3.84215	130.0	0.87998	-0.55528	-2.31738
42.0	0.67128	-1.73097	-3.49307	132.0	0.82811	-0.81909	-2.58120
44.0	0.72351	-1.40555	-3.16766	134.0	0.77587	-1.10209	-2.86419
46.0	0.77587	-1.10209	-2.86419	136.0	0.72351	-1.40555	-3.16766
48.0	0.82811	-0.81909	-2.58120	138.0	0.67128	-1.73097	-3.49307
50.0	0.87998	-0.55528	-2.31738	140.0	0.61944	-2.08004	-3.84215
52.0	0.93121	-0.30951	-2.07161	142.0	0.56823	-2.45476	-4.21686
54.0	0.98157	-0.08080	-1.84290	144.0	0.51791	-2.85742	-4.61952
56.0	1.03080	0.13173	-1.63038	146.0	0.46873	-3.29076	-5.05286
58.0	1.07866	0.32884	-1.43326	148.0	0.42092	-3.75799	-5.52009
60.0	1.12492	0.51123	-1.25088	150.0	0.37472	-4.26296	-6.02507
62.0	1.16936	0.67949	-1.08261	152.0	0.33034	-4.81035	-6.57246
64.0	1.21176	0.83416	-0.92794	154.0	0.28801	-5.40587	-7.16797
66.0	1.25191	0.97572	-0.78638	156.0	0.24794	-6.05660	-7.81870
68.0	1.28961	1.10458	-0.65752	158.0	0.21030	-6.77151	-8.53361
70.0	1.32468	1.22112	-0.54098	160.0	0.17530	-7.56212	-9.32423
72.0	1.35696	1.32566	-0.43644	162.0	0.14310	-8.44364	-10.20574
74.0	1.38627	1.41848	-0.34362	164.0	0.11385	-9.43664	-11.19874
76.0	1.41248	1.49984	-0.26227	166.0	0.08770	-10.57000	-12.33210
78.0	1.43547	1.56993	-0.19217	168.0	0.06477	-11.88602	-13.64812
80.0	1.45511	1.62895	-0.13315	170.0	0.04518	-13.45027	-15.21237
82.0	1.47131	1.67704	-0.08506	172.0	0.02902	-15.37263	-17.13473
84.0	1.48400	1.71433	-0.04778	174.0	0.01637	-17.85910	-19.62120
86.0	1.49310	1.74089	-0.02121	176.0	0.00729	-21.37213	-23.13424
88.0	1.49858	1.75680	-0.00530	178.0	0.00182	-27.38746	-29.14957
90.0	1.50041	1.76210	0.0	180.	0.00000	-INFINITY	-INFINITY

TABLE A1-27

INTEGRAL EQUATION METHOD
EPSILON= 0.2500(NORMALIZED GROUNDPLANE RADIUS)
MONOPOLE LENGTH= 0.025000(WAVELENGTH)
INFINITELY THIN MONOPOLE RADIUS
RADIATION RESISTANCE = 0.124029 (OHMS)

ELEVATION ANGLE (DEG)	DIRECTIVITY (NUMERIC)	DIRECTIVITY (DBI)	RELATIVE POWER (DB)	ELEVATION ANGLE (DEG)	DIRECTIVITY (NUMERIC)	DIRECTIVITY (DBI)	RELATIVE POWER (DB)
0.	0.00000	-INFINITY	-INFINITY				
2.0	0.00184	-27.34979	-29.10244	92.0	1.49534	1.74741	-0.00524
4.0	0.00735	-21.33463	-23.08729	94.0	1.48993	1.73167	-0.02098
6.0	0.01651	-17.82188	-19.57453	96.0	1.48095	1.70539	-0.04726
8.0	0.02927	-15.33581	-17.08847	98.0	1.46842	1.66851	-0.08415
10.0	0.04556	-13.41396	-15.16661	100.0	1.45242	1.62093	-0.13173
12.0	0.06531	-11.85032	-13.60298	102.0	1.43302	1.56252	-0.19013
14.0	0.08841	-10.53503	-12.28768	104.0	1.41031	1.49315	-0.25950
16.0	0.11475	-9.40249	-11.15514	106.0	1.38440	1.41262	-0.34003
18.0	0.14420	-8.41041	-10.16306	108.0	1,35541	1.32072	-0.43193
20.0	0.17661	-7.52991	-9.28256	110.0	1.32349	1.21720	-0.53546
22.0	0.21182	-6.74039	-8.49305	112.0	1.28877	1.10176	-0.65089
24.0	0.24965	-6.02667	-7.77932	114.0	1.25143	0.97408	-0.77857
26.0	0.28992	-5.37720	-7.12985	116.0	1.21165	0.83379	-0.91887
28.0	0.33243	-4.78301	-6.53567	118.0	1.16962	0.68045	-1.07221
30.0	0.37696	-4.23702	-5.98967	120.0	1.12553	0.51358	-1.23907
32.0	0.42330	-3.73349	-5.48615	122.0	1.07961	0.33265	-1.42000
34.0	0.47122	-3.26777	-5.02042	124.0	1.03206	0.13704	-1.61561
36.0	0.52048	-2.83598	-4.58863	126.0	0.98312	-0.07394	-1.82659
38.0	0.57083	-2.43490	-4.18755	128.0	0.93303	-0.30107	-2.05372
40.0	0.62204	-2.06179	-3.81445	130.0	0.88202	-0.54522	-2.29788
42.0	0.67385	-1.71435	-3.46700	132.0	0.83035	-0.80741	-2.56006
44.0	0.72601	-1.39058	-3.14323	134.0	0.77826	-1.08876	-2.84141
46.0	0.77826	-1.08876	-2.84141	136.0	0.72601	-1.39058	-3.14323
48.0	0.83035	-0.80741	-2.56006	138.0	0.67385	-1.71435	-3.46700
50.0	0.88202	-0.54522	-2.29787	140.0	0.62204	-2.06179	-3.81445
52.0	0.93303	-0.30107	-2.05372	142.0	0.57083	-2.43490	-4.18755
54.0	0.98312	-0.07394	-1.82659	144.0	0.52048	-2.83598	-4.58863
56.0	1.03206	0.13704	-1.61561	146.0	0.47122	-3.26777	-5.02042
58.0	1.07961	0.33265	-1.42000	148.0	0.42330	-3.73350	-5.48615
60.0	1.12553	0.51358	-1.23907	150.0	0.37696	-4.23702	-5.98967
62.0	1.16962	0.68045	-1.07221	152.0	0.33243	-4.78302	-6.53567
64.0	1.21165	0.83379	-0.91887	154.0	0.28992	-5.37720	-7.12985
66.0	1.25144	0.97408	-0.77857	156.0	0.24965	-6.02667	-7.77932
68.0	1.28877	1.10176	-0.65089	158.0	0.21182	-6.74039	-8.49305
70.0	1.32349	1.21720	-0.53546	160.0	0.17661	-7.52991	-9.28256
72.0	1.35541	1.32072	-0.43193	162.0	0.14420	-8.41041	-10.16306
74.0	1.38440	1.41262	-0.34003	164.0	0.11475	-9.40249	-11.15514
76.0	1.41031	1.49315	-0.25950	166.0	0.08841	-10.53503	-12.28768
78.0	1.43302	1.56252	-0.19013	168.0	0.06531	-11.85033	-13.60298
80.0	1.45242	1.62093	-0.13173	170.0	0.04556	-13.41396	-15.16661
82.0	1.46842	1.66851	-0.08414	172.0	0.02927	-15.33582	-17.08847
84.0	1.48095	1.70539	-0.04726	174.0	0.01651	-17.82188	-19.57454
86.0	1.48993	1.73167	-0.02098	176.0	0.00735	-21.33463	-23.08729
88.0	1.49534	1.74741	-0.00524	178.0	0.00184	-27.34979	-29.10244
90.0	1.49715	1.75265	0.0	180.	0.00000	-INFINITY	-INFINITY

TABLE A1-28

INTEGRAL EQUATION METHOD
EPSILON= 0.5000(NORMALIZED GROUNDPLANE RADIUS)
MONOPOLE LENGTH= 0.025000(WAVELENGTH)
INFINITELY THIN MONOPOLE RADIUS
RADIATION RESISTANCE = 0.124796 (OHMS)

ELEVATION ANGLE (DEG)	DIRECTIVITY (NUMERIC)	DIRECTIVITY (DBI)	RELATIVE POWER (DB)	ELEVATION ANGLE (DEG)	DIRECTIVITY (NUMERIC)	DIRECTIVITY (DBI)	RELATIVE POWER (DB)
0.	0.00000	-INFINITY	-INFINITY				
2.0	0.00189	-27.24403	-28.96992	92.0	1.48621	1.72081	-0.00508
4.0	0.00753	-21.22935	-22.95524	94.0	1.48100	1.70555	-0.02034
6.0	0.01691	-17.71740	-19.44329	96.0	1.47234	1.68008	-0.04581
8.0	0.02997	-15.23245	-16.95834	98.0	1.46027	1.64432	-0.08157
10.0	0.04664	-13.31202	-15.03791	100.0	1.44483	1.59818	-0.12771
12.0	0.06683	-11.75011	-13.47600	102.0	1.42611	1.54152	-0.18437
14.0	0.09043	-10.43683	-12.16272	104.0	1.40417	1.47418	-0.25171
16.0	0.11731	-9.30660	-11.03249	106.0	1.37911	1.39598	-0.32991
18.0	0.14733	-8.31709	-10.04298	108.0	1.35104	1.30668	-0.41921
20.0	0.18033	-7.43943	-9.16532	110.0	1.32008	1.20602	-0.51987
22.0	0.21612	-6.65301	-8.37890	112.0	1.28638	1.09369	-0.63220
24.0	0.25453	-5.94260	-7.66849	114.0	1.25007	0.96935	-0.75654
26.0	0.29535	-5.29666	-7.02255	116.0	1.21133	0.83261	-0.89328
28.0	0.33836	-4.70621	-6.43210	118.0	1.17032	0.68303	-1.04286
30.0	0.38334	-4.16414	-5.89003	120.0	1.12722	0.52010	-1.20579
32.0	0.43006	-3.66469	-5.39058	122.0	1.08225	0.34327	-1.38262
34.0	0.47828	-3.20318	-4.92907	124.0	1.03559	0.15189	-1.57400
36.0	0.52775	-2.77574	-4.50163	126.0	0.98748	-0.05473	-1.78062
38.0	0.57822	-2.37910	-4.10499	128.0	0.93812	-0.27740	-2.00329
40.0	0.62943	-2.01052	-3.73641	130.0	0.88776	-0.51702	-2.24291
42.0	0.68114	-1.66766	-3.39355	132.0	0.83664	-0.77462	-2.50051
44.0	0.73308	-1.34850	-3.07439	134.0	0.78499	-1.05135	-2.77724
46.0	0.78500	-1.05132	-2.77721	136.0	0.73307	-1.34854	-3.07443
48.0	0.83665	-0.77458	-2.50048	138.0	0.68113	-1.66770	-3.39359
50.0	0.88777	-0.51699	-2.24288	140.0	0.62942	-2.01056	-3.73645
52.0	0.93813	-0.27737	-2.00326	142.0	0.57821	-2.37914	-4.10503
54.0	0.98748	-0.05470	-1.78059	144.0	0.52774	-2.77578	-4.50167
56.0	1.03560	0.15192	-1.57397	146.0	0.47827	-3.20323	-4.92912
58.0	1.08225	0.34329	-1.38260	148.0	0.43006	-3.66473	-5.39062
60.0	1.12723	0.52013	-1.20576	150.0	0.38334	-4.16418	-5.89007
62.0	1.17032	0.68305	-1.04284	152.0	0.33836	-4.70626	-6.43215
64.0	1.21133	0.83264	-0.89325	154.0	0.29534	-5.29671	-7.02260
66.0	1.25008	0.96937	-0.75652	156.0	0.25453	-5.94265	-7.66854
68.0	1.28638	1.09371	-0.63218	158.0	0.21612	-6.65305	-8.37894
70.0	1.32009	1.20603	-0.51986	160.0	0.18032	-7.43948	-9.16537
72.0	1.35104	1.30669	-0.41920	162.0	0.14733	-8.31714	-10.04303
74.0	1.37911	1.39599	-0.32990	164.0	0.11731	-9.30665	-11.03254
76.0	1.40417	1.47420	-0.25169	166.0	0.09043	-10.43688	-12.16277
78.0	1.42611	1.54153	-0.18436	168.0	0.06683	-11.75016	-13.47605
80.0	1.44484	1.59819	-0.12770	170.0	0.04664	-13.31207	-15.03796
82.0	1.46027	1.64433	-0.08156	172.0	0.02997	-15.23250	-16.95839
84.0	1.47234	1.68009	-0.04580	174.0	0.01691	-17.71745	-19.44334
86.0	1.48100	1.70556	-0.02033	176.0	0.00753	-21.22940	-22.95529
88.0	1.48621	1.72081	-0.00508	178.0	0.00189	-27.24408	-28.96997
90.0	1.48795	1.72589	0.0	180.	0.00000	-INFINITY	-INFINITY

TABLE A1-29

INTEGRAL EQUATION METHOD
EPSILON= 0.7500(NORMALIZED GROUNDPLANE RADIUS)
MONOPOLE LENGTH= 0.025000(WAVELENGTH)
INFINITELY THIN MONOPOLE RADIUS
RADIATION RESISTANCE = 0.126059 (OHMS)

ELEVATION ANGLE (DEG)	DIRECTIVITY (NUMERIC)	DIRECTIVITY (DBI)	RELATIVE POWER (DB)	ELEVATION ANGLE (DEG)	DIRECTIVITY (NUMERIC)	DIRECTIVITY (DBI)	RELATIVE POWER (DB)
0.	0.00000	-INFINITY	-INFINITY				
2.0	0.00196	-27.07383	-28.75599	92.0	1.47141	1.67733	-0.00482
4.0	0.00783	-21.05993	-22.74208	94.0	1.46651	1.66286	-0.01929
6.0	0.01758	-17.54926	-19.23141	96.0	1.45838	1.63869	-0.04346
8.0	0.03115	-15.06609	-16.74825	98.0	1.44702	1.60476	-0.07740
10.0	0.04844	-13.14795	-14.83010	100.0	1.43250	1.56094	-0.12121
12.0	0.06936	-11.58881	-13.27096	102.0	1.41485	1.50711	-0.17505
14.0	0.09378	-10.27877	-11.96093	104.0	1.39414	1.44307	-0.23908
16.0	0.12156	-9.15224	-10.83439	106.0	1.37045	1.36864	-0.31352
18.0	0.15252	-8.16687	-9.84903	108.0	1.34386	1.28355	-0.39861
20.0	0.18648	-7.29377	-8.97592	110.0	1.31447	1.18752	-0.49463
22.0	0.22324	-6.51230	-8.19445	112.0	1.28240	1.08023	-0.60192
24.0	0.26259	-5.80722	-7.48938	114.0	1.24776	0.96130	-0.72085
26.0	0.30430	-5.16697	-6.84912	116.0	1.21069	0.83033	-0.85183
28.0	0.34814	-4.58252	-6.26468	118.0	1.17134	0.68682	-0.99533
30.0	0.39385	-4.04674	-5.72889	120.0	1.12986	0.53026	-1.15189
32.0	0.44118	-3.55386	-5.23601	122.0	1.08644	0.36005	-1.32210
34.0	0.48988	-3.09915	-4.78130	124.0	1.04125	0.17553	-1.50662
36.0	0.53967	-2.67869	-4.36085	126.0	0.99448	-0.02405	-1.70620
38.0	0.59031	-2.28921	-3.97137	128.0	0.94634	-0.23952	-1.92167
40.0	0.64152	-1.92793	-3.61008	130.0	0.89705	-0.47182	-2.15398
42.0	0.69303	-1.59245	-3.27461	132.0	0.84683	-0.72202	-2.40417
44.0	0.74460	-1.28075	-2.96290	134.0	0.79592	-0.99130	-2.67345
46.0	0.79597	-0.99103	-2.67318	136.0	0.74456	-1.28102	-2.96317
48.0	0.84688	-0.72176	-2.40391	138.0	0.69299	-1.59274	-3.27489
50.0	0.89710	-0.47158	-2.15373	140.0	0.64147	-1.92822	-3.61037
52.0	0.94639	-0.23928	-1.92144	142.0	0.59027	-2.28951	-3.97167
54.0	0.99453	-0.02382	-1.70597	144.0	0.53963	-2.67900	-4.36115
56.0	1.04130	0.17575	-1.50641	146.0	0.48984	-3.09946	-4.78161
58.0	1.08649	0.36026	-1.32190	148.0	0.44115	-3.55418	-5.23633
60.0	1.12991	0.53046	-1.15170	150.0	0.39382	-4.04707	-5.72922
62.0	1.17139	0.68700	-0.99515	152.0	0.34811	-4.58286	-6.26501
64.0	1.21074	0.83050	-0.85166	154.0	0.30428	-5.16731	-6.84946
66.0	1.24780	0.96146	-0.72069	156.0	0.26257	-5.80757	-7.48972
68.0	1.28244	1.08038	-0.60178	158.0	0.22322	-6.51265	-8.19480
70.0	1.31451	1.18765	-0.49450	160.0	0.18646	-7.29412	-8.97627
72.0	1.34390	1.28367	-0.39849	162.0	0.15250	-8.16723	-9.84938
74.0	1.37049	1.36874	-0.31341	164.0	0.12155	-9.15260	-10.83475
76.0	1.39417	1.44317	-0.23899	166.0	0.09377	-10.27914	-11.96129
78.0	1.41488	1.50719	-0.17497	168.0	0.06936	-11.58917	-13.27133
80.0	1.43252	1.56101	-0.12114	170.0	0.04844	-13.14832	-14.83047
82.0	1.44704	1.60481	-0.07734	172.0	0.03114	-15.06646	-16.74862
84.0	1.45839	1.63874	-0.04342	174.0	0.01758	-17.54963	-19.23179
86.0	1.46652	1.66289	-0.01927	176.0	0.00783	-21.06030	-22.74246
88.0	1.47141	1.67734	-0.00481	178.0	0.00196	-27.07420	-28.75636
90.0	1.47304	1.68215	0.0	180.	0.00000	-INFINITY	-INFINITY

TABLE A1-30

INTEGRAL EQUATION METHOD
EPSILON= 1.0000(NORMALIZED GROUNDPLANE RADIUS)
MONOPOLE LENGTH= 0.025000(WAVELENGTH)
INFINITELY THIN MONOPOLE RADIUS
RADIATION RESISTANCE = 0.127888 (OHMS)

ELEVATION ANGLE (DEG)	DIRECTIVITY (NUMERIC)	DIRECTIVITY (DBI)	RELATIVE POWER (DB)	ELEVATION ANGLE (DEG)	DIRECTIVITY (NUMERIC)	DIRECTIVITY (DBI)	RELATIVE POWER (DB)
0.	0.00000	-INFINITY	-INFINITY				
2.0	0.00207	-26.83564	-28.45522	92.0	1.45048	1.61512	-0.00447
4.0	0.00827	-20.82281	-22.44239	94.0	1.44602	1.60174	-0.01784
6.0	0.01856	-17.31393	-18.93351	96.0	1.43860	1.57941	-0.04017
8.0	0.03286	-14.83325	-16.45284	98.0	1.42825	1.54805	-0.07154
10.0	0.05107	-12.91829	-14.53787	100.0	1.41498	1.50751	-0.11207
12.0	0.07306	-11.36301	-12.98259	102.0	1.39883	1.45766	-0.16192
14.0	0.09868	-10.05749	-11.67708	104.0	1.37984	1.39829	-0.22129
16.0	0.12776	-8.93611	-10.55570	106.0	1.35805	1.32917	-0.29041
18.0	0.16008	-7.95652	-9.57611	108.0	1.33353	1.25003	-0.36956
20.0	0.19544	-7.08978	-8.70936	110.0	1.30634	1.16055	-0.45904
22.0	0.23360	-6.31523	-7.93481	112.0	1.27655	1.06036	-0.55922
24.0	0.27431	-5.61760	-7.23718	114.0	1.24425	0.94907	-0.67051
26.0	0.31730	-4.98528	-6.60487	116.0	1.20954	0.82622	-0.79337
28.0	0.36231	-4.40923	-6.02881	118.0	1.17254	0.69128	-0.92830
30.0	0.40905	-3.88225	-5.50184	120.0	1.13336	0.54370	-1.07589
32.0	0.45724	-3.39855	-5.01814	122.0	1.09215	0.38282	-1.23677
34.0	0.50660	-2.95335	-4.57294	124.0	1.04905	0.20794	-1.41164
36.0	0.55684	-2.54270	-4.16228	126.0	1.00422	0.01827	-1.60131
38.0	0.60768	-2.16325	-3.78284	128.0	0.95784	-0.18708	-1.80666
40.0	0.65884	-1.81219	-3.43178	130.0	0.91010	-0.40909	-2.02868
42.0	0.71005	-1.48709	-3.10667	132.0	0.86121	-0.64888	-2.26847
44.0	0.76105	-1.18584	-2.80543	134.0	0.81139	-0.90770	-2.52729
46.0	0.81159	-0.90663	-2.52621	136.0	0.76086	-1.18695	-2.80654
48.0	0.86142	-0.64785	-2.26743	138.0	0.70987	-1.48823	-3.10782
50.0	0.91031	-0.40809	-2.02768	140.0	0.65866	-1.81337	-3.43295
52.0	0.95805	-0.18611	-1.80570	142.0	0.60751	-2.16446	-3.78405
54.0	1.00443	0.01919	-1.60039	144.0	0.55668	-2.54393	-4.16352
56.0	1.04926	0.20882	-1.41076	146.0	0.50645	-2.95461	-4.57420
58.0	1.09236	0.38366	-1.23593	148.0	0.45711	-3.39984	-5.01942
60.0	1.13357	0.54449	-1.07510	150.0	0.40892	-3.88356	-5.50315
62.0	1.17274	0.69203	-0.92756	152.0	0.36220	-4.41056	-6.03015
64.0	1.20974	0.82691	-0.79267	154.0	0.31720	-4.98663	-6.60622
66.0	1.24443	0.94972	-0.66987	156.0	0.27422	-5.61897	-7.23856
68.0	1.27672	1.06096	-0.55862	158.0	0.23353	-6.31661	-7.93620
70.0	1.30650	1.16109	-0.45849	160.0	0.19538	-7.09118	-8.71077
72.0	1.33368	1.25052	-0.36906	162.0	0.16003	-7.95794	-9.57752
74.0	1.35819	1.32962	-0.28997	164.0	0.12772	-8.93754	-10.55713
76.0	1.37997	1.39868	-0.22090	166.0	0.09865	-10.05893	-11.67852
78.0	1.39894	1.45800	-0.16159	168.0	0.07304	-11.36446	-12.98404
80.0	1.41507	1.50779	-0.11179	170.0	0.05105	-12.91975	-14.53933
82.0	1.42832	1.54827	-0.07132	172.0	0.03285	-14.83472	-16.45430
84.0	1.43866	1.57958	-0.04000	174.0	0.01855	-17.31540	-18.93498
86.0	1.44606	1.60185	-0.01773	176.0	0.00827	-20.82428	-22.44387
88.0	1.45050	1.61517	-0.00441	178.0	0.00207	-26.83711	-28.45670
90.0	1.45197	1.61959	0.0	180.	0.00000	-INFINITY	-INFINITY

TABLE A1-31

INTEGRAL EQUATION METHOD
EPSILON= 1.2500(NORMALIZED GROUNDPLANE RADIUS)
MONOPOLE LENGTH= 0.025000(WAVELENGTH)
INFINITELY THIN MONOPOLE RADIUS
RADIATION RESISTANCE = 0.130457 (OHMS)

ELEVATION ANGLE θ (DEG)	DIRECTIVITY (NUMERIC)	DIRECTIVITY (DBI)	RELATIVE POWER (DB)	ELEVATION ANGLE (DEG)	DIRECTIVITY (NUMERIC)	DIRECTIVITY (DBI)	RELATIVE POWER (DB)
0.	0.00000	-INFINITY	-INFINITY				
2.0	0.00223	-26.51682	-28.05005	92.0	1.42207	1.52920	-0.00402
4.0	0.00890	-20.50543	-22.03865	94.0	1.41818	1.51730	-0.01592
6.0	0.01996	-16.99892	-18.53214	96.0	1.41171	1.49745	-0.03577
8.0	0.03531	-14.52156	-16.05478	98.0	1.40267	1.46955	-0.06367
10.0	0.05482	-12.61082	-14.14404	100.0	1.39106	1.43347	-0.09975
12.0	0.07833	-11.06068	-12.59390	102.0	1.37690	1.38901	-0.14421
14.0	0.10565	-9.76118	-11.29440	104.0	1.36018	1.33597	-0.19726
16.0	0.13656	-8.64667	-10.17989	106.0	1.34093	1.27406	-0.25917
18.0	0.17081	-7.67476	-9.20798	108.0	1.31916	1.20297	-0.33025
20.0	0.20814	-6.81649	-8.34971	110.0	1.29490	1.12236	-0.41086
22.0	0.24825	-6.05117	-7.58439	112.0	1.26818	1.03180	-0.50142
24.0	0.29084	-5.36347	-6.89670	114.0	1.23904	0.93085	-0.60237
26.0	0.33560	-4.74175	-6.27497	116.0	1.20753	0.81899	-0.71423
28.0	0.38222	-4.17691	-5.71013	118.0	1.17372	0.69564	-0.83758
30.0	0.43036	-3.66171	-5.19493	120.0	1.13768	0.56019	-0.97303
32.0	0.47970	-3.19029	-4.72351	122.0	1.09949	0.41192	-1.12130
34.0	0.52993	-2.75783	-4.29105	124.0	1.05927	0.25007	-1.28315
36.0	0.58072	-2.36030	-3.89352	126.0	1.01713	0.07376	-1.45946
38.0	0.63178	-1.99432	-3.52754	128.0	0.97321	-0.11794	-1.65116
40.0	0.68281	-1.65698	-3.19020	130.0	0.92766	-0.32610	-1.85932
42.0	0.73353	-1.34580	-2.87902	132.0	0.88066	-0.55190	-2.08512
44.0	0.78368	-1.05862	-2.59184	134.0	0.83241	-0.79664	-2.32986
46.0	0.83300	-0.79354	-2.32676	136.0	0.78310	-1.06182	-2.59504
48.0	0.88127	-0.54889	-2.08211	138.0	0.73298	-1.34909	-2.88231
50.0	0.92828	-0.32320	-1.85642	140.0	0.68228	-1.66035	-3.19357
52.0	0.97383	-0.11515	-1.64837	142.0	0.63128	-1.99777	-3.53099
54.0	1.01776	0.07644	-1.45678	144.0	0.58025	-2.36383	-3.89705
56.0	1.05989	0.25263	-1.28060	146.0	0.52949	-2.76143	-4.29465
58.0	1.10011	0.41436	-1.11886	148.0	0.47930	-3.19395	-4.72718
60.0	1.13828	0.56250	-0.97072	150.0	0.42999	-3.66544	-5.19866
62.0	1.17431	0.69782	-0.83540	152.0	0.38188	-4.18069	-5.71391
64.0	1.20810	0.82103	-0.71219	154.0	0.33531	-4.74559	-6.27881
66.0	1.23958	0.93275	-0.60047	156.0	0.29058	-5.36736	-6.90058
68.0	1.26869	1.03356	-0.49966	158.0	0.24802	-6.05510	-7.58832
70.0	1.29538	1.12397	-0.40925	160.0	0.20795	-6.82046	-8.35368
72.0	1.31960	1.20443	-0.32879	162.0	0.17066	-7.67877	-9.21199
74.0	1.34133	1.27536	-0.25786	164.0	0.13644	-8.65070	-10.18392
76.0	1.36054	1.33711	-0.19611	166.0	0.10555	-9.76524	-11.29846
78.0	1.37721	1.39000	-0.14322	168.0	0.07826	-11.06477	-12.59799
80.0	1.39133	1.43430	-0.09893	170.0	0.05477	-12.61493	-14.14815
82.0	1.40288	1.47022	-0.06300	172.0	0.03527	-14.52568	-16.05890
84.0	1.41187	1.49795	-0.03527	174.0	0.01994	-17.00306	-18.53628
86.0	1.41828	1.51763	-0.01559	176.0	0.00889	-20.50957	-22.04279
88.0	1.42212	1.52937	-0.00385	178.0	0.00223	-26.52098	-28.05420
90.0	1.42338	1.53322	0.0	180.	0.00000	-INFINITY	-INFINITY

TABLE A1-32

INTEGRAL EQUATION METHOD
EPSILON= 1.5000(NORMALIZED GROUNDPLANE RADIUS)
MONOPOLE LENGTH= 0.025000(WAVELENGTH)
INFINITELY THIN MONOPOLE RADIUS
RADIATION RESISTANCE = 0.134091 (OHMS)

ELEVATION ANGLE (DEG)	DIRECTIVITY (NUMERIC)	DIRECTIVITY (DBI)	RELATIVE POWER (DB)	ELEVATION ANGLE (DEG)	DIRECTIVITY (NUMERIC)	DIRECTIVITY (DBI)	RELATIVE POWER (DB)
0.	0.00000	-INFINITY	-INFINITY				
2.0	0.00246	-26.09575	-27.50963	92.0	1.38371	1.41044	-0.00344
4.0	0.00980	-20.08621	-21.50010	94.0	1.38054	1.40050	-0.01339
6.0	0.02196	-16.58281	-17.99669	96.0	1.37530	1.38398	-0.02991
8.0	0.03882	-14.10977	-15.52365	98.0	1.36797	1.36076	-0.05313
10.0	0.06019	-12.20456	-13.61845	100.0	1.35853	1.33068	-0.08321
12.0	0.08588	-10.66113	-12.07501	102.0	1.34695	1.29352	-0.12037
14.0	0.11562	-9.36949	-10.78337	104.0	1.33322	1.24901	-0.16487
16.0	0.14914	-8.26396	-9.67785	106.0	1.31730	1.19685	-0.21704
18.0	0.18612	-7.30212	-8.71601	108.0	1.29917	1.13665	-0.27723
20.0	0.22621	-6.45496	-7.86885	110.0	1.27879	1.06801	-0.34588
22.0	0.26905	-5.70173	-7.11562	112.0	1.25616	0.99044	-0.42344
24.0	0.31426	-5.02708	-6.44096	114.0	1.23124	0.90343	-0.51046
26.0	0.36147	-4.41928	-5.83317	116.0	1.20403	0.80638	-0.60751
28.0	0.41028	-3.86919	-5.28307	118.0	1.17453	0.69863	-0.71525
30.0	0.46031	-3.36949	-4.78338	120.0	1.14274	0.57948	-0.83440
32.0	0.51118	-2.91427	-4.32815	122.0	1.10870	0.44814	-0.96574
34.0	0.56252	-2.49862	-3.91250	124.0	1.07244	0.30374	-1.11014
36.0	0.61398	-2.11844	-3.53233	126.0	1.03403	0.14534	-1.26855
38.0	0.66523	-1.77028	-3.18416	128.0	0.99355	-0.02812	-1.44200
40.0	0.71596	-1.45114	-2.86502	130.0	0.95109	-0.21778	-1.63166
42.0	0.76587	-1.15846	-2.57234	132.0	0.90680	-0.42491	-1.83879
44.0	0.81471	-0.88998	-2.30386	134.0	0.86081	-0.65091	-2.06480
46.0	0.86224	-0.64372	-2.05760	136.0	0.81332	-0.89737	-2.31126
48.0	0.90826	-0.41792	-1.83180	138.0	0.76453	-1.16604	-2.57993
50.0	0.95257	-0.21102	-1.62490	140.0	0.71468	-1.45890	-2.87278
52.0	0.99504	-0.02160	-1.43548	142.0	0.66402	-1.77820	-3.19209
54.0	1.03553	0.15161	-1.26227	144.0	0.61284	-2.12652	-3.54041
56.0	1.07393	0.30975	-1.10413	146.0	0.56146	-2.50684	-3.92072
58.0	1.11017	0.45388	-0.96001	148.0	0.51020	-2.92262	-4.33650
60.0	1.14418	0.58493	-0.82895	150.0	0.45941	-3.37796	-4.79185
62.0	1.17592	0.70378	-0.71010	152.0	0.40947	-3.87777	-5.29165
64.0	1.20537	0.81122	-0.60267	154.0	0.36075	-4.42797	-5.84185
66.0	1.23252	0.90795	-0.50594	156.0	0.31363	-5.03586	-6.44974
68.0	1.25737	0.99463	-0.41926	158.0	0.26850	-5.71060	-7.12448
70.0	1.27993	1.07185	-0.34204	160.0	0.22574	-6.46390	-7.87778
72.0	1.30021	1.14014	-0.27375	162.0	0.18573	-7.31113	-8.72501
74.0	1.31825	1.19997	-0.21391	164.0	0.14883	-8.27303	-9.68691
76.0	1.33406	1.25177	-0.16212	166.0	0.11538	-9.37861	-10.79249
78.0	1.34769	1.29589	-0.11799	168.0	0.08570	-10.67029	-12.08417
80.0	1.35915	1.33267	-0.08122	170.0	0.06007	-12.21376	-13.62765
82.0	1.36847	1.36236	-0.05153	172.0	0.03873	-14.11900	-15.53288
84.0	1.37568	1.38518	-0.02871	174.0	0.02192	-16.59206	-18.00595
86.0	1.38080	1.40130	-0.01258	176.0	0.00978	-20.09548	-21.50937
88.0	1.38384	1.41085	-0.00304	178.0	0.00245	-26.10503	-27.51891
90.0	1.38480	1.41388	0.0	180.	0.00000	-INFINITY	-INFINITY

TABLE A1-33

INTEGRAL EQUATION METHOD
EPSILON= 1.7500(NORMALIZED GROUNDPLANE RADIUS)
MONOPOLE LENGTH= 0.025000(WAVELENGTH)
INFINITELY THIN MONOPOLE RADIUS
RADIATION RESISTANCE = 0.139354 (OHMS)

ELEVATION ANGLE (DEG)	DIRECTIVITY (NUMERIC)	DIRECTIVITY (DBI)	RELATIVE POWER (DB)	ELEVATION ANGLE (DEG)	DIRECTIVITY (NUMERIC)	DIRECTIVITY (DBI)	RELATIVE POWER (DB)
0.	0.00000	-INFINITY	-INFINITY				
2.0	0.00279	-25.54439	-26.79107	92.0	1.33167	1.24398	-0.00270
4.0	0.01112	-19.53725	-20.78394	94.0	1.32944	1.23668	-0.01001
6.0	0.02490	-16.03782	-17.28451	96.0	1.32577	1.22469	-0.02200
8.0	0.04395	-13.57033	-14.81702	98.0	1.32065	1.20787	-0.03882
10.0	0.06804	-11.67223	-12.91892	100.0	1.31401	1.18600	-0.06069
12.0	0.09689	-10.13742	-11.38410	102.0	1.30582	1.15882	-0.08787
14.0	0.13014	-8.85589	-10.10258	104.0	1.29598	1.12600	-0.12069
16.0	0.16742	-7.76193	-9.00862	106.0	1.28444	1.08715	-0.15954
18.0	0.20830	-6.81307	-8.05976	108.0	1.27111	1.04182	-0.20487
20.0	0.25233	-5.98024	-7.22693	110.0	1.25589	0.98950	-0.25718
22.0	0.29904	-5.24266	-6.48934	112.0	1.23869	0.92963	-0.31706
24.0	0.34795	-4.58488	-5.83157	114.0	1.21943	0.86155	-0.38513
26.0	0.39855	-3.99514	-5.24182	116.0	1.19801	0.78459	-0.46210
28.0	0.45038	-3.46420	-4.71088	118.0	1.17435	0.69796	-0.54872
30.0	0.50296	-2.98468	-4.23137	120.0	1.14838	0.60084	-0.64584
32.0	0.55583	-2.55058	-3.79726	122.0	1.12004	0.49232	-0.75437
34.0	0.60857	-2.15690	-3.40358	124.0	1.08928	0.37140	-0.87529
36.0	0.66078	-1.79945	-3.04614	126.0	1.05609	0.23700	-1.00968
38.0	0.71209	-1.47467	-2.72136	128.0	1.02046	0.08797	-1.15871
40.0	0.76217	-1.17948	-2.42617	130.0	0.98243	-0.07698	-1.32366
42.0	0.81074	-0.91119	-2.15788	132.0	0.94205	-0.25924	-1.50593
44.0	0.85754	-0.66744	-1.91412	134.0	0.89942	-0.46035	-1.70704
46.0	0.90238	-0.44613	-1.69281	136.0	0.85467	-0.68200	-1.92869
48.0	0.94507	-0.24538	-1.49207	138.0	0.80797	-0.92607	-2.17276
50.0	0.98548	-0.06351	-1.31020	140.0	0.75951	-1.19465	-2.44134
52.0	1.02353	0.10101	-1.14567	142.0	0.70956	-1.49011	-2.73680
54.0	1.05915	0.24959	-0.99709	144.0	0.65839	-1.81514	-3.06182
56.0	1.09232	0.38350	-0.86319	146.0	0.60634	-2.17280	-3.41949
58.0	1.12303	0.50391	-0.74278	148.0	0.55377	-2.56669	-3.81337
60.0	1.15130	0.61189	-0.63479	150.0	0.50107	-3.00098	-4.24766
62.0	1.17718	0.70844	-0.53824	152.0	0.44868	-3.48066	-4.72734
64.0	1.20073	0.79447	-0.45222	154.0	0.39703	-4.01174	-5.25843
66.0	1.22203	0.87080	-0.37588	156.0	0.34661	-4.60162	-5.84831
68.0	1.24114	0.93822	-0.30847	158.0	0.29788	-5.25952	-6.50620
70.0	1.25817	0.99741	-0.24928	160.0	0.25135	-5.99721	-7.24390
72.0	1.27321	1.04902	-0.19767	162.0	0.20749	-6.83013	-8.07681
74.0	1.28636	1.09361	-0.15308	164.0	0.16676	-7.77907	-9.02576
76.0	1.29769	1.13170	-0.11498	166.0	0.12963	-8.87310	-10.11979
78.0	1.30730	1.16375	-0.08294	168.0	0.09650	-10.15468	-11.40137
80.0	1.31527	1.19013	-0.05655	170.0	0.06777	-11.68954	-12.93623
82.0	1.32166	1.21119	-0.03549	172.0	0.04378	-13.58769	-14.83437
84.0	1.32654	1.22720	-0.01949	174.0	0.02480	-16.05521	-17.30189
86.0	1.32995	1.23836	-0.00833	176.0	0.01108	-19.55465	-20.80134
88.0	1.33193	1.24482	-0.00187	178.0	0.00278	-25.56180	-26.80849
90.0	1.33250	1.24669	0.0	180.	0.00000	-INFINITY	-INFINITY

TABLE A1-34

INTEGRAL EQUATION METHOD
EPSILON= 2.0000(NORMALIZED GROUNDPLANE RADIUS)
MONOPOLE LENGTH= 0.025000(WAVELENGTH)
INFINITELY THIN MONOPOLE RADIUS
RADIATION RESISTANCE = 0.147191 (OHMS)

ELEVATION ANGLE (DEG)	DIRECTIVITY (NUMERIC)	DIRECTIVITY (DBI)	RELATIVE POWER (DB)	ELEVATION ANGLE (DEG)	DIRECTIVITY (NUMERIC)	DIRECTIVITY (DBI)	RELATIVE POWER (DB)
0.	0.00000	-INFINITY	-INFINITY				
2.0	0.00328	-24.83551	-25.84458	92.0	1.26105	1.00733	-0.00173
4.0	0.01309	-18.83135	-19.84042	94.0	1.25999	1.00368	-0.00539
6.0	0.02926	-15.33688	-16.34595	96.0	1.25834	0.99798	-0.01109
8.0	0.05157	-12.87631	-13.88538	98.0	1.25604	0.99005	-0.01902
10.0	0.07967	-10.98706	-11.99613	100.0	1.25303	0.97960	-0.02946
12.0	0.11316	-9.46302	-10.47209	102.0	1.24919	0.96629	-0.04277
14.0	0.15156	-8.19414	-9.20321	104.0	1.24443	0.94969	-0.05938
16.0	0.19433	-7.11466	-8.12373	106.0	1.23859	0.92929	-0.07978
18.0	0.24088	-6.18207	-7.19114	108.0	1.23155	0.90451	-0.10456
20.0	0.29059	-5.36726	-6.37633	110.0	1.22312	0.87469	-0.13438
22.0	0.34282	-4.64936	-5.65842	112.0	1.21314	0.83912	-0.16995
24.0	0.39693	-4.01288	-5.02195	114.0	1.20143	0.79699	-0.21208
26.0	0.45228	-3.44597	-4.45504	116.0	1.18780	0.74744	-0.26163
28.0	0.50824	-2.93932	-3.94839	118.0	1.17207	0.68953	-0.31953
30.0	0.56423	-2.48547	-3.49454	120.0	1.15405	0.62225	-0.38682
32.0	0.61968	-2.07830	-3.08737	122.0	1.13358	0.54451	-0.46456
34.0	0.67410	-1.71273	-2.72180	124.0	1.11049	0.45514	-0.55393
36.0	0.72703	-1.38447	-2.39354	126.0	1.08465	0.35288	-0.65618
38.0	0.77807	-1.08982	-2.09889	128.0	1.05594	0.23640	-0.77267
40.0	0.82688	-0.82559	-1.83466	130.0	1.02429	0.10424	-0.90483
42.0	0.87318	-0.58896	-1.59802	132.0	0.98966	-0.04516	-1.05423
44.0	0.91676	-0.37743	-1.38650	134.0	0.95203	-0.21350	-1.22257
46.0	0.95747	-0.18877	-1.19783	136.0	0.91146	-0.40262	-1.41169
48.0	0.99519	-0.02094	-1.03001	138.0	0.86805	-0.61457	-1.62363
50.0	1.02989	0.12789	-0.88117	140.0	0.82195	-0.85157	-1.86063
52.0	1.06156	0.25943	-0.74964	142.0	0.77337	-1.11613	-2.12520
54.0	1.09024	0.37523	-0.63384	144.0	0.72259	-1.41107	-2.42013
56.0	1.11603	0.47674	-0.53233	146.0	0.66995	-1.73959	-2.74865
58.0	1.13902	0.56530	-0.44376	148.0	0.61583	-2.10538	-3.11444
60.0	1.15936	0.64217	-0.36689	150.0	0.56069	-2.51274	-3.52181
62.0	1.17720	0.70852	-0.30055	152.0	0.50504	-2.96676	-3.97583
64.0	1.19273	0.76542	-0.24365	154.0	0.44941	-3.47355	-4.48262
66.0	1.20612	0.81390	-0.19517	156.0	0.39440	-4.04059	-5.04965
68.0	1.21756	0.85489	-0.15418	158.0	0.34063	-4.67717	-5.68624
70.0	1.22723	0.88926	-0.11980	160.0	0.28872	-5.39516	-6.40423
72.0	1.23533	0.91782	-0.09125	162.0	0.23933	-6.21005	-7.21912
74.0	1.24202	0.94129	-0.06778	164.0	0.19308	-7.14270	-8.15177
76.0	1.24747	0.96032	-0.04875	166.0	0.15058	-8.22223	-9.23130
78.0	1.25184	0.97550	-0.03357	168.0	0.11243	-9.49115	-10.50021
80.0	1.25526	0.98734	-0.02172	170.0	0.07915	-11.01522	-12.02429
82.0	1.25785	0.99629	-0.01278	172.0	0.05123	-12.90450	-13.91356
84.0	1.25970	1.00269	-0.00638	174.0	0.02907	-15.36509	-16.37415
86.0	1.26091	1.00683	-0.00224	176.0	0.01300	-18.85957	-19.86864
88.0	1.26151	1.00891	-0.00015	178.0	0.00326	-24.86374	-25.87280
90.0	1.26156	1.00907	0.0	180.	0.00000	-INFINITY	-INFINITY

TABLE A1-35

INTEGRAL EQUATION METHOD
EPSILON= 2.2500(NORMALIZED GROUNDPLANE RADIUS)
MONOPOLE LENGTH= 0.025000(WAVELENGTH)
INFINITELY THIN MONOPOLE RADIUS
RADIATION RESISTANCE = 0.159189 (OHMS)

ELEVATION ANGLE (DEG)	DIRECTIVITY (NUMERIC)	DIRECTIVITY (DBI)	RELATIVE POWER (DB)	ELEVATION ANGLE (DEG)	DIRECTIVITY (NUMERIC)	DIRECTIVITY (DBI)	RELATIVE POWER (DB)
0.	0.00000	-INFINITY	-INFINITY				
2.0	0.00402	-23.95498	-24.68588	92.0	1.16637	0.66835	-0.06255
4.0	0.01602	-17.95436	-18.68527	94.0	1.16676	0.66983	-0.06107
6.0	0.03576	-14.46579	-15.19669	96.0	1.16762	0.67303	-0.05787
8.0	0.06290	-12.01346	-12.74436	98.0	1.16887	0.67767	-0.05323
10.0	0.09694	-10.13477	-10.86567	100.0	1.17040	0.68336	-0.04754
12.0	0.13729	-8.62357	-9.35447	102.0	1.17208	0.68959	-0.04131
14.0	0.18324	-7.36981	-8.10071	104.0	1.17375	0.69575	-0.03515
16.0	0.23401	-6.30767	-7.03857	106.0	1.17521	0.70114	-0.02976
18.0	0.28876	-5.39461	-6.12551	108.0	1.17624	0.70496	-0.02594
20.0	0.34662	-4.60145	-5.33235	110.0	1.17660	0.70631	-0.02459
22.0	0.40670	-3.90729	-4.63819	112.0	1.17604	0.70421	-0.02669
24.0	0.46811	-3.29656	-4.02746	114.0	1.17425	0.69762	-0.03328
26.0	0.52999	-2.75735	-3.48825	116.0	1.17095	0.68539	-0.04551
28.0	0.59152	-2.28027	-3.01118	118.0	1.16582	0.66633	-0.06457
30.0	0.65196	-1.85777	-2.58867	120.0	1.15855	0.63915	-0.09175
32.0	0.71062	-1.48365	-2.21455	122.0	1.14882	0.60251	-0.12839
34.0	0.76688	-1.15272	-1.88362	124.0	1.13631	0.55499	-0.17592
36.0	0.82024	-0.86057	-1.59147	126.0	1.12075	0.49508	-0.23582
38.0	0.87028	-0.60339	-1.33429	128.0	1.10184	0.42120	-0.30970
40.0	0.91667	-0.37786	-1.10876	130.0	1.07937	0.33169	-0.39921
42.0	0.95918	-0.18102	-0.91192	132.0	1.05311	0.22475	-0.50615
44.0	0.99765	-0.01022	-0.74112	134.0	1.02294	0.09850	-0.63240
46.0	1.03203	0.13694	-0.59396	136.0	0.98876	-0.04910	-0.78000
48.0	1.06235	0.26266	-0.46824	138.0	0.95054	-0.22028	-0.95118
50.0	1.08867	0.36897	-0.36193	140.0	0.90836	-0.41742	-1.14832
52.0	1.11116	0.45776	-0.27314	142.0	0.86234	-0.64320	-1.37410
54.0	1.13001	0.53081	-0.20009	144.0	0.81272	-0.90057	-1.63147
56.0	1.14546	0.58978	-0.14112	146.0	0.75982	-1.19287	-1.92377
58.0	1.15778	0.63625	-0.09465	148.0	0.70406	-1.52391	-2.25481
60.0	1.16727	0.67170	-0.05920	150.0	0.64594	-1.89810	-2.62900
62.0	1.17424	0.69756	-0.03334	152.0	0.58605	-2.32065	-3.05155
64.0	1.17901	0.71517	-0.01573	154.0	0.52508	-2.79775	-3.52865
66.0	1.18190	0.72581	-0.00509	156.0	0.46377	-3.33696	-4.06786
68.0	1.18322	0.73067	-0.00023	158.0	0.40293	-3.94767	-4.67857
70.0	1.18329	0.73090	0.0	160.0	0.34341	-4.64181	-5.37271
72.0	1.18237	0.72755	-0.00335	162.0	0.28609	-5.43493	-6.16583
74.0	1.18076	0.72160	-0.00930	164.0	0.23185	-6.34796	-7.07886
76.0	1.17868	0.71396	-0.01694	166.0	0.18155	-7.41006	-8.14096
78.0	1.17637	0.70542	-0.02548	168.0	0.13603	-8.66378	-9.39468
80.0	1.17401	0.69672	-0.03418	170.0	0.09605	-10.17494	-10.90584
82.0	1.17178	0.68848	-0.04242	172.0	0.06232	-12.05360	-12.78450
84.0	1.16982	0.68120	-0.04970	174.0	0.03543	-14.50591	-15.23681
86.0	1.16824	0.67531	-0.05559	176.0	0.01587	-17.99447	-18.72537
88.0	1.16711	0.67110	-0.05980	178.0	0.00399	-23.99507	-24.72597
90.0	1.16648	0.66876	-0.06214	180.	0.00000	-INFINITY	-INFINITY

TABLE A1-36

INTEGRAL EQUATION METHOD
EPSILON= 2.5000(NORMALIZED GROUNDPLANE RADIUS)
MONOPOLE LENGTH= 0.025000(WAVELENGTH)
INFINITELY THIN MONOPOLE RADIUS
RADIATION RESISTANCE = 0.178038 (OHMS)

ELEVATION ANGLE (DEG)	DIRECTIVITY (NUMERIC)	DIRECTIVITY (DBI)	RELATIVE POWER (DB)	ELEVATION ANGLE (DEG)	DIRECTIVITY (NUMERIC)	DIRECTIVITY (DBI)	RELATIVE POWER (DB)
0.	0.00000	-INFINITY	-INFINITY				
2.0	0.00511	-22.91583	-23.63294	92.0	1.04333	0.18423	-0.53288
4.0	0.02033	-16.91920	-17.63631	94.0	1.04546	0.19309	-0.52402
6.0	0.04532	-13.43726	-14.15437	96.0	1.04932	0.20907	-0.50805
8.0	0.07954	-10.99419	-11.71130	98.0	1.05480	0.23171	-0.48540
10.0	0.12225	-9.12739	-9.84450	100.0	1.06180	0.26041	-0.45670
12.0	0.17256	-7.63068	-8.34780	102.0	1.07013	0.29437	-0.42274
14.0	0.22940	-6.39400	-7.11111	104.0	1.07960	0.33265	-0.38447
16.0	0.29164	-5.35148	-6.06859	106.0	1.08997	0.37416	-0.34295
18.0	0.35805	-4.46056	-5.17767	108.0	1.10096	0.41772	-0.29939
20.0	0.42736	-3.69203	-4.40914	110.0	1.11225	0.46202	-0.25509
22.0	0.49832	-3.02493	-3.74204	112.0	1.12348	0.50564	-0.21147
24.0	0.56968	-2.44365	-3.16077	114.0	1.13426	0.54712	-0.16999
26.0	0.64029	-1.93623	-2.65334	116.0	1.14417	0.58490	-0.13221
28.0	0.70905	-1.49321	-2.21033	118.0	1.15276	0.61738	-0.09974
30.0	0.77500	-1.10698	-1.82409	120.0	1.15955	0.64288	-0.07423
32.0	0.83729	-0.77126	-1.48837	122.0	1.16404	0.65970	-0.05742
34.0	0.89520	-0.48078	-1.19789	124.0	1.16575	0.66606	-0.05105
36.0	0.94819	-0.23103	-0.94814	126.0	1.16417	0.66017	-0.05695
38.0	0.99584	-0.01811	-0.73522	128.0	1.15881	0.64013	-0.07699
40.0	1.03787	0.16142	-0.55569	130.0	1.14921	0.60400	-0.11311
42.0	1.07415	0.31066	-0.40645	132.0	1.13495	0.54977	-0.16734
44.0	1.10468	0.43238	-0.28473	134.0	1.11566	0.47531	-0.24180
46.0	1.12957	0.52915	-0.18797	136.0	1.09104	0.37839	-0.33872
48.0	1.14904	0.60334	-0.11378	138.0	1.06087	0.25660	-0.46051
50.0	1.16337	0.65717	-0.05994	140.0	1.02504	0.10739	-0.60972
52.0	1.17294	0.69276	-0.02436	142.0	0.98354	-0.07206	-0.78917
54.0	1.17817	0.71209	-0.00502	144.0	0.93652	-0.28484	-1.00195
56.0	1.17954	0.71711	0.0	146.0	0.88422	-0.53440	-1.25151
58.0	1.17752	0.70967	-0.00745	148.0	0.82705	-0.82466	-1.54178
60.0	1.17262	0.69156	-0.02556	150.0	0.76557	-1.16014	-1.87726
62.0	1.16534	0.66453	-0.05258	152.0	0.70047	-1.54611	-2.26323
64.0	1.15619	0.63029	-0.08682	154.0	0.63258	-1.98886	-2.70598
66.0	1.14564	0.59048	-0.12663	156.0	0.56286	-2.49601	-3.21313
68.0	1.13415	0.54671	-0.17041	158.0	0.49238	-3.07702	-3.79413
70.0	1.12215	0.50051	-0.21660	160.0	0.42229	-3.74386	-4.46097
72.0	1.11003	0.45337	-0.26375	162.0	0.35382	-4.51215	-5.22926
74.0	1.09817	0.40668	-0.31043	164.0	0.28821	-5.40284	-6.11995
76.0	1.08687	0.36178	-0.35533	166.0	0.22672	-6.44514	-7.16225
78.0	1.07643	0.31987	-0.39725	168.0	0.17054	-7.68164	-8.39875
80.0	1.06710	0.28204	-0.43507	170.0	0.12083	-9.17818	-9.89529
82.0	1.05908	0.24928	-0.46783	172.0	0.07862	-11.04484	-11.76196
84.0	1.05254	0.22240	-0.49472	174.0	0.04479	-13.48781	-14.20492
86.0	1.04762	0.20205	-0.51506	176.0	0.02009	-16.96967	-17.68678
88.0	1.04442	0.18874	-0.52837	178.0	0.00505	-22.96626	-23.68337
90.0	1.04298	0.18276	-0.53435	180.	0.00000	-INFINITY	-INFINITY

TABLE A1-37

INTEGRAL EQUATION METHOD
EPSILON= 2.7500(NORMALIZED GROUNDPLANE RADIUS)
MONOPOLE LENGTH= 0.025000(WAVELENGTH)
INFINITELY THIN MONOPOLE RADIUS
RADIATION RESISTANCE = 0.208486 (OHMS)

ELEVATION ANGLE θ (DEG)	DIRECTIVITY (NUMERIC)	DIRECTIVITY (DBI)	RELATIVE POWER (DB)	ELEVATION ANGLE (DEG)	DIRECTIVITY (NUMERIC)	DIRECTIVITY (DBI)	RELATIVE POWER (DB)
0.	0.00000	-INFINITY	-INFINITY				
2.0	1.39465	1.44466	0.0	92.0	0.88914	-0.51032	-1.95497
4.0	0.34114	-4.67071	-6.11536	94.0	0.89079	-0.50225	-1.94691
6.0	0.14821	-8.29136	-9.73601	96.0	0.89557	-0.47900	-1.92365
8.0	0.08140	-10.89369	-12.33834	98.0	0.90340	-0.44119	-1.88584
10.0	0.05080	-12.94094	-14.38559	100.0	0.91415	-0.38980	-1.83446
12.0	0.03436	-14.63996	-16.08462	102.0	0.92766	-0.32610	-1.77075
14.0	0.02454	-16.10166	-17.54632	104.0	0.94372	-0.25158	-1.69624
16.0	0.01823	-17.39279	-18.83744	106.0	0.96206	-0.16798	-1.61263
18.0	0.01394	-18.55701	-20.00167	108.0	0.98239	-0.07717	-1.52183
20.0	0.01090	-19.62483	-21.06949	110.0	1.00434	0.01880	-1.42585
22.0	0.00867	-20.61878	-22.06344	112.0	1.02751	0.11784	-1.32681
24.0	0.00699	-21.55640	-23.00106	114.0	1.05142	0.21776	-1.22689
26.0	0.00569	-22.45210	-23.89675	116.0	1.07556	0.31635	-1.12830
28.0	0.00466	-23.31838	-24.76304	118.0	1.09936	0.41138	-1.03327
30.0	0.00383	-24.16684	-25.61150	120.0	1.12218	0.50062	-0.94403
32.0	0.00316	-25.00896	-26.45362	122.0	1.14336	0.58183	-0.86282
34.0	0.00260	-25.85701	-27.30167	124.0	1.16220	0.65280	-0.79186
36.0	0.00213	-26.72516	-28.16981	126.0	1.17795	0.71128	-0.73338
38.0	0.00173	-27.63106	-29.07572	128.0	1.18988	0.75504	-0.68962
40.0	0.00138	-28.59860	-30.04325	130.0	1.19724	0.78182	-0.66283
42.0	0.00108	-29.66287	-31.10753	132.0	1.19931	0.78933	-0.65533
44.0	0.00082	-30.88082	-32.32548	134.0	1.19542	0.77520	-0.66945
46.0	0.00058	-32.35716	-33.80181	136.0	1.18495	0.73699	-0.70766
48.0	0.00037	-34.32462	-35.76927	138.0	1.16738	0.67213	-0.77252
50.0	0.00018	-37.52247	-38.96713	140.0	1.14232	0.57790	-0.86676
52.0	1.27923	1.06948	-0.37518	142.0	1.10952	0.45134	-0.99331
54.0	1.26582	1.02373	-0.42092	144.0	1.06887	0.28926	-1.15539
56.0	1.24803	0.96224	-0.48242	146.0	1.02049	0.08810	-1.35655
58.0	1.22662	0.88709	-0.55757	148.0	0.96468	-0.15615	-1.60081
60.0	1.20236	0.80035	-0.64431	150.0	0.90197	-0.44808	-1.89274
62.0	1.17601	0.70411	-0.74055	152.0	0.83310	-0.79304	-2.23770
64.0	1.14827	0.60044	-0.84422	154.0	0.75904	-1.19738	-2.64203
66.0	1.11981	0.49143	-0.95322	156.0	0.68096	-1.66876	-3.11342
68.0	1.09124	0.37919	-1.06547	158.0	0.60025	-2.21670	-3.66135
70.0	1.06312	0.26581	-1.17884	160.0	0.51842	-2.85321	-4.29787
72.0	1.03595	0.15340	-1.29126	162.0	0.43712	-3.59395	-5.03860
74.0	1.01019	0.04402	-1.40063	164.0	0.35810	-4.45990	-5.90456
76.0	0.98622	-0.06028	-1.50494	166.0	0.28312	-5.48031	-6.92496
78.0	0.96437	-0.15756	-1.60222	168.0	0.21391	-6.69778	-8.14244
80.0	0.94494	-0.24598	-1.69063	170.0	0.15212	-8.17818	-9.62284
82.0	0.92815	-0.32381	-1.76847	172.0	0.09927	-10.03162	-11.47628
84.0	0.91420	-0.38957	-1.83422	174.0	0.05670	-12.46428	-13.90894
86.0	0.90324	-0.44195	-1.88661	176.0	0.02548	-15.93878	-17.38343
88.0	0.89537	-0.47996	-1.92461	178.0	0.00641	-21.93094	-23.37560
90.0	0.89066	-0.50287	-1.94753	180.	0.00000	-INFINITY	-INFINITY

Tables A1-38–A1-49; $h/\lambda = 0.01$, $0 \leqslant ka \leqslant 3.0$

TABLE A1-38

INTEGRAL EQUATION METHOD
EPSILON= 0.0 (NORMALIZED GROUNDPLANE RADIUS)
MONOPOLE LENGTH= 0.010000(WAVELENGTH)
INFINITELY THIN MONOPOLE RADIUS
RADIATION RESISTANCE = 0.019738 (OHMS)

ELEVATION ANGLE (DEG)	DIRECTIVITY (NUMERIC)	DIRECTIVITY (DBI)	RELATIVE POWER (DB)	ELEVATION ANGLE (DEG)	DIRECTIVITY (NUMERIC)	DIRECTIVITY (DBI)	RELATIVE POWER (DB)
0.	0.00000	-INFINITY	-INFINITY				
2.0	0.00183	-27.38347	-29.14457	92.0	1.49824	1.75581	-0.00529
4.0	0.00730	-21.36815	-23.12926	94.0	1.49276	1.73991	-0.02119
6.0	0.01639	-17.85515	-19.61625	96.0	1.48367	1.71338	-0.04772
8.0	0.02905	-15.36873	-17.12983	98.0	1.47100	1.67614	-0.08496
10.0	0.04522	-13.44642	-15.20752	100.0	1.45482	1.62810	-0.13300
12.0	0.06483	-11.88223	-13.64333	102.0	1.43521	1.56915	-0.19195
14.0	0.08778	-10.56629	-12.32739	104.0	1.41225	1.49913	-0.26197
16.0	0.11395	-9.43302	-11.19412	106.0	1.38607	1.41786	-0.34324
18.0	0.14322	-8.44011	-10.20121	108.0	1.35679	1.32514	-0.43596
20.0	0.17544	-7.55870	-9.31981	110.0	1.32456	1.22071	-0.54040
22.0	0.21046	-6.76821	-8.52931	112.0	1.28952	1.10429	-0.65682
24.0	0.24812	-6.05343	-7.81453	114.0	1.25186	0.97555	-0.78555
26.0	0.28822	-5.40283	-7.16393	116.0	1.21175	0.83412	-0.92698
28.0	0.33056	-4.80745	-6.56856	118.0	1.16939	0.67959	-1.08151
30.0	0.37495	-4.26021	-6.02131	120.0	1.12499	0.51148	-1.24963
32.0	0.42117	-3.75539	-5.51649	122.0	1.07876	0.32925	-1.43186
34.0	0.46899	-3.28832	-5.04942	124.0	1.03093	0.13229	-1.62881
36.0	0.51819	-2.85515	-4.61625	126.0	0.98173	-0.08007	-1.84118
38.0	0.56851	-2.45265	-4.21375	128.0	0.93140	-0.30862	-2.06972
40.0	0.61971	-2.07811	-3.83921	130.0	0.88019	-0.55421	-2.31531
42.0	0.67155	-1.72921	-3.49031	132.0	0.82835	-0.81785	-2.57896
44.0	0.72377	-1.40396	-3.16507	134.0	0.77613	-1.10067	-2.86178
46.0	0.77613	-1.10067	-2.86178	136.0	0.72377	-1.40396	-3.16507
48.0	0.82835	-0.81785	-2.57896	138.0	0.67155	-1.72921	-3.49031
50.0	0.88019	-0.55421	-2.31531	140.0	0.61971	-2.07811	-3.83921
52.0	0.93140	-0.30862	-2.06972	142.0	0.56851	-2.45265	-4.21375
54.0	0.98173	-0.08007	-1.84118	144.0	0.51819	-2.85515	-4.61625
56.0	1.03093	0.13229	-1.62881	146.0	0.46899	-3.28832	-5.04942
58.0	1.07876	0.32925	-1.43186	148.0	0.42117	-3.75539	-5.51649
60.0	1.12499	0.51148	-1.24963	150.0	0.37495	-4.26021	-6.02131
62.0	1.16939	0.67959	-1.08151	152.0	0.33056	-4.80745	-6.56856
64.0	1.21175	0.83412	-0.92698	154.0	0.28822	-5.40283	-7.16393
66.0	1.25186	0.97555	-0.78555	156.0	0.24812	-6.05343	-7.81453
68.0	1.28952	1.10429	-0.65682	158.0	0.21046	-6.76821	-8.52931
70.0	1.32456	1.22071	-0.54040	160.0	0.17544	-7.55870	-9.31981
72.0	1.35679	1.32514	-0.43596	162.0	0.14322	-8.44011	-10.20121
74.0	1.38607	1.41786	-0.34324	164.0	0.11395	-9.43302	-11.19412
76.0	1.41225	1.49913	-0.26197	166.0	0.08778	-10.56629	-12.32739
78.0	1.43521	1.56915	-0.19195	168.0	0.06483	-11.88223	-13.64333
80.0	1.45482	1.62810	-0.13300	170.0	0.04522	-13.44642	-15.20752
82.0	1.47100	1.67614	-0.08496	172.0	0.02905	-15.36873	-17.12983
84.0	1.48367	1.71338	-0.04772	174.0	0.01639	-17.85515	-19.61625
86.0	1.49276	1.73991	-0.02119	176.0	0.00730	-21.36815	-23.12926
88.0	1.49824	1.75581	-0.00529	178.0	0.00183	-27.38347	-29.14457
90.0	1.50007	1.76110	0.0	180.	0.00000	-INFINITY	-INFINITY

TABLE A1-39

INTEGRAL EQUATION METHOD
EPSILON= 0.2500(NORMALIZED GROUNDPLANE RADIUS)
MONOPOLE LENGTH= 0.010000(WAVELENGTH)
INFINITELY THIN MONOPOLE RADIUS
RADIATION RESISTANCE = 0.019769 (OHMS)

ELEVATION ANGLE (DEG)	DIRECTIVITY (NUMERIC)	DIRECTIVITY (DBI)	RELATIVE POWER (DB)	ELEVATION ANGLE (DEG)	DIRECTIVITY (NUMERIC)	DIRECTIVITY (DBI)	RELATIVE POWER (DB)
0.	0.00000	-INFINITY	-INFINITY				
2.0	0.00184	-27.35570	-29.10984	92.0	1.49585	1.74889	-0.00525
4.0	0.00734	-21.34052	-23.09466	94.0	1.49043	1.73312	-0.02102
6.0	0.01649	-17.82772	-19.58186	96.0	1.48143	1.70680	-0.04734
8.0	0.02923	-15.34159	-17.09574	98.0	1.46888	1.66985	-0.08429
10.0	0.04550	-13.41966	-15.17380	100.0	1.45284	1.62219	-0.13195
12.0	0.06522	-11.85593	-13.61007	102.0	1.43341	1.56369	-0.19045
14.0	0.08830	-10.54052	-12.29466	104.0	1.41065	1.49420	-0.25994
16.0	0.11461	-9.40785	-11.16199	106.0	1.38470	1.41355	-0.34060
18.0	0.14402	-8.41562	-10.16977	108.0	1.35566	1.32150	-0.43264
20.0	0.17640	-7.53496	-9.28911	110.0	1.32368	1.21782	-0.53633
22.0	0.21158	-6.74528	-8.49942	112.0	1.28890	1.10221	-0.65194
24.0	0.24938	-6.03137	-7.78551	114.0	1.25151	0.97434	-0.77980
26.0	0.28962	-5.38170	-7.13584	116.0	1.21167	0.83385	-0.92030
28.0	0.33210	-4.78731	-6.54145	118.0	1.16958	0.68030	-1.07385
30.0	0.37661	-4.24110	-5.99524	120.0	1.12544	0.51321	-1.24093
32.0	0.42293	-3.73734	-5.49149	122.0	1.07946	0.33205	-1.42209
34.0	0.47083	-3.27138	-5.02553	124.0	1.03186	0.13620	-1.61794
36.0	0.52007	-2.83935	-4.59349	126.0	0.98287	-0.07502	-1.82916
38.0	0.57042	-2.43802	-4.19216	128.0	0.93274	-0.30240	-2.05654
40.0	0.62163	-2.06466	-3.81881	130.0	0.88170	-0.54681	-2.30095
42.0	0.67345	-1.71696	-3.47111	132.0	0.82999	-0.80925	-2.56339
44.0	0.72562	-1.39293	-3.14708	134.0	0.77788	-1.09086	-2.84500
46.0	0.77788	-1.09086	-2.84500	136.0	0.72562	-1.39293	-3.14708
48.0	0.82999	-0.80925	-2.56339	138.0	0.67345	-1.71696	-3.47111
50.0	0.88170	-0.54680	-2.30095	140.0	0.62163	-2.06466	-3.81881
52.0	0.93274	-0.30240	-2.05654	142.0	0.57042	-2.43802	-4.19216
54.0	0.98287	-0.07502	-1.82916	144.0	0.52007	-2.83935	-4.59349
56.0	1.03186	0.13620	-1.61794	146.0	0.47083	-3.27138	-5.02553
58.0	1.07946	0.33205	-1.42209	148.0	0.42293	-3.73734	-5.49149
60.0	1.12544	0.51321	-1.24093	150.0	0.37661	-4.24110	-5.99524
62.0	1.16958	0.68030	-1.07385	152.0	0.33210	-4.78731	-6.54145
64.0	1.21167	0.83385	-0.92030	154.0	0.28962	-5.38170	-7.13584
66.0	1.25151	0.97434	-0.77980	156.0	0.24938	-6.03137	-7.78551
68.0	1.28890	1.10221	-0.65194	158.0	0.21158	-6.74528	-8.49942
70.0	1.32368	1.21782	-0.53633	160.0	0.17640	-7.53497	-9.28911
72.0	1.35566	1.32150	-0.43264	162.0	0.14402	-8.41563	-10.16977
74.0	1.38470	1.41355	-0.34060	164.0	0.11461	-9.40785	-11.16199
76.0	1.41065	1.49420	-0.25994	166.0	0.08830	-10.54052	-12.29466
78.0	1.43341	1.56369	-0.19045	168.0	0.06522	-11.85593	-13.61007
80.0	1.45284	1.62219	-0.13195	170.0	0.04550	-13.41966	-15.17380
82.0	1.46888	1.66985	-0.08429	172.0	0.02923	-15.34159	-17.09574
84.0	1.48143	1.70680	-0.04734	174.0	0.01649	-17.82772	-19.58187
86.0	1.49043	1.73312	-0.02102	176.0	0.00734	-21.34052	-23.09466
88.0	1.49585	1.74889	-0.00525	178.0	0.00184	-27.35570	-29.10985
90.0	1.49766	1.75414	0.0	180.	0.00000	-INFINITY	-INFINITY

TABLE A1-40

INTEGRAL EQUATION METHOD
EPSILON= 0.5000(NORMALIZED GROUNDPLANE RADIUS)
MONOPOLE LENGTH= 0.010000(WAVELENGTH)
INFINITELY THIN MONOPOLE RADIUS
RADIATION RESISTANCE = 0.019825 (OHMS)

ELEVATION ANGLE (DEG)	DIRECTIVITY (NUMERIC)	DIRECTIVITY (DBI)	RELATIVE POWER (DB)	ELEVATION ANGLE (DEG)	DIRECTIVITY (NUMERIC)	DIRECTIVITY (DBI)	RELATIVE POWER (DB)
0.	0.00000	-INFINITY	-INFINITY				
2.0	0.00186	-27.30677	-29.04865	92.0	1.49166	1.73670	-0.00518
4.0	0.00743	-21.29182	-23.03369	94.0	1.48633	1.72115	-0.02072
6.0	0.01667	-17.77939	-19.52127	96.0	1.47747	1.69520	-0.04668
8.0	0.02955	-15.29378	-17.03566	98.0	1.46513	1.65877	-0.08311
10.0	0.04600	-13.37251	-15.11439	100.0	1.44936	1.61176	-0.13011
12.0	0.06592	-11.80958	-13.55146	102.0	1.43023	1.55406	-0.18781
14.0	0.08923	-10.49511	-12.23699	104.0	1.40783	1.48551	-0.25637
16.0	0.11578	-9.36352	-11.10539	106.0	1.38226	1.40591	-0.33596
18.0	0.14546	-8.37249	-10.11437	108.0	1.35365	1.31506	-0.42682
20.0	0.17811	-7.49315	-9.23503	110.0	1.32211	1.21269	-0.52919
22.0	0.21356	-6.70490	-8.44678	112.0	1.28781	1.09850	-0.64337
24.0	0.25162	-5.99253	-7.73441	114.0	1.25088	0.97217	-0.76970
26.0	0.29211	-5.34451	-7.08638	116.0	1.21152	0.83331	-0.90856
28.0	0.33482	-4.75185	-6.49372	118.0	1.16990	0.68149	-1.06039
30.0	0.37954	-4.20745	-5.94933	120.0	1.12621	0.51621	-1.22567
32.0	0.42603	-3.70559	-5.44747	122.0	1.08067	0.33693	-1.40494
34.0	0.47407	-3.24159	-4.98346	124.0	1.03348	0.14304	-1.59884
36.0	0.52341	-2.81157	-4.55344	126.0	0.98488	-0.06618	-1.80806
38.0	0.57381	-2.41229	-4.15417	128.0	0.93508	-0.29150	-2.03338
40.0	0.62502	-2.04103	-3.78290	130.0	0.88434	-0.53382	-2.27570
42.0	0.67679	-1.69545	-3.43732	132.0	0.83289	-0.79415	-2.53602
44.0	0.72886	-1.37355	-3.11543	134.0	0.78098	-1.07363	-2.81550
46.0	0.78098	-1.07361	-2.81549	136.0	0.72886	-1.37357	-3.11544
48.0	0.83289	-0.79414	-2.53601	138.0	0.67679	-1.69546	-3.43734
50.0	0.88434	-0.53381	-2.27569	140.0	0.62502	-2.04104	-3.78292
52.0	0.93508	-0.29149	-2.03337	142.0	0.57381	-2.41231	-4.15418
54.0	0.98488	-0.06617	-1.80804	144.0	0.52341	-2.81158	-4.55346
56.0	1.03349	0.14305	-1.59883	146.0	0.47407	-3.24160	-4.98348
58.0	1.08067	0.33694	-1.40493	148.0	0.42603	-3.70561	-5.44748
60.0	1.12622	0.51622	-1.22565	150.0	0.37954	-4.20747	-5.94935
62.0	1.16990	0.68150	-1.06038	152.0	0.33482	-4.75187	-6.49374
64.0	1.21152	0.83332	-0.90856	154.0	0.29211	-5.34453	-7.08640
66.0	1.25089	0.97218	-0.76969	156.0	0.25162	-5.99255	-7.73443
68.0	1.28781	1.09851	-0.64336	158.0	0.21355	-6.70492	-8.44680
70.0	1.32212	1.21269	-0.52918	160.0	0.17811	-7.49317	-9.23505
72.0	1.35365	1.31506	-0.42681	162.0	0.14546	-8.37251	-10.11439
74.0	1.38227	1.40592	-0.33596	164.0	0.11578	-9.36354	-11.10541
76.0	1.40783	1.48551	-0.25636	166.0	0.08922	-10.49514	-12.23701
78.0	1.43023	1.55407	-0.18781	168.0	0.06592	-11.80960	-13.55148
80.0	1.44936	1.61177	-0.13011	170.0	0.04600	-13.37253	-15.11441
82.0	1.46513	1.65877	-0.08311	172.0	0.02955	-15.29380	-17.03568
84.0	1.47747	1.69520	-0.04668	174.0	0.01667	-17.77941	-19.52129
86.0	1.48633	1.72115	-0.02072	176.0	0.00743	-21.29184	-23.03371
88.0	1.49166	1.73670	-0.00518	178.0	0.00186	-27.30680	-29.04867
90.0	1.49344	1.74188	0.0	180.	0.00000	-INFINITY	-INFINITY

TABLE A1-41

INTEGRAL EQUATION METHOD
EPSILON= 0.7500(NORMALIZED GROUNDPLANE RADIUS)
MONOPOLE LENGTH= 0.010000(WAVELENGTH)
INFINITELY THIN MONOPOLE RADIUS
RADIATION RESISTANCE = 0.019871 (OHMS)

ELEVATION ANGLE (DEG)	DIRECTIVITY (NUMERIC)	DIRECTIVITY (DBI)	RELATIVE POWER (DB)	ELEVATION ANGLE (DEG)	DIRECTIVITY (NUMERIC)	DIRECTIVITY (DBI)	RELATIVE POWER (DB)
0.	0.00000	-INFINITY	-INFINITY				
2.0	0.00188	-27.26650	-28.99837	92.0	1.48825	1.72675	-0.00512
4.0	0.00750	-21.25172	-22.98360	94.0	1.48299	1.71138	-0.02049
6.0	0.01683	-17.73961	-19.47148	96.0	1.47425	1.68572	-0.04615
8.0	0.02982	-15.25443	-16.98631	98.0	1.46208	1.64971	-0.08216
10.0	0.04641	-13.33372	-15.06559	100.0	1.44652	1.60323	-0.12864
12.0	0.06650	-11.77146	-13.50333	102.0	1.42764	1.54618	-0.18570
14.0	0.09000	-10.45777	-12.18965	104.0	1.40552	1.47837	-0.25350
16.0	0.11676	-9.32707	-11.05894	106.0	1.38027	1.39964	-0.33223
18.0	0.14665	-8.33704	-10.06891	108.0	1.35199	1.30975	-0.42212
20.0	0.17952	-7.45880	-9.19067	110.0	1.32082	1.20844	-0.52343
22.0	0.21519	-6.67174	-8.40361	112.0	1.28689	1.09541	-0.63646
24.0	0.25347	-5.96066	-7.69253	114.0	1.25035	0.97032	-0.76155
26.0	0.29417	-5.31399	-7.04586	116.0	1.21137	0.83278	-0.89909
28.0	0.33707	-4.72277	-6.45464	118.0	1.17013	0.68236	-1.04951
30.0	0.38196	-4.17988	-5.91175	120.0	1.12682	0.51855	-1.21332
32.0	0.42859	-3.67958	-5.41145	122.0	1.08164	0.34081	-1.39106
34.0	0.47674	-3.21719	-4.94906	124.0	1.03478	0.14850	-1.58337
36.0	0.52616	-2.78883	-4.52071	126.0	0.98649	-0.05908	-1.79095
38.0	0.57660	-2.39126	-4.12313	128.0	0.93697	-0.28273	-2.01460
40.0	0.62781	-2.02172	-3.75359	130.0	0.88647	-0.52334	-2.25521
42.0	0.67953	-1.67788	-3.40975	132.0	0.83523	-0.78194	-2.51381
44.0	0.73152	-1.35774	-3.08961	134.0	0.78349	-1.05967	-2.79154
46.0	0.78351	-1.05956	-2.79143	136.0	0.73150	-1.35785	-3.08972
48.0	0.83525	-0.78183	-2.51370	138.0	0.67952	-1.67800	-3.40987
50.0	0.88649	-0.52324	-2.25511	140.0	0.62779	-2.02184	-3.75371
52.0	0.93699	-0.28263	-2.01450	142.0	0.57658	-2.39138	-4.12325
54.0	0.98651	-0.05899	-1.79086	144.0	0.52614	-2.78896	-4.52083
56.0	1.03481	0.14859	-1.58328	146.0	0.47672	-3.21732	-4.94920
58.0	1.08166	0.34089	-1.39098	148.0	0.42858	-3.67971	-5.41159
60.0	1.12684	0.51863	-1.21324	150.0	0.38194	-4.18001	-5.91188
62.0	1.17015	0.68243	-1.04944	152.0	0.33706	-4.72291	-6.45478
64.0	1.21139	0.83285	-0.89902	154.0	0.29416	-5.31413	-7.04600
66.0	1.25037	0.97038	-0.76149	156.0	0.25347	-5.96080	-7.69267
68.0	1.28691	1.09547	-0.63640	158.0	0.21518	-6.67189	-8.40376
70.0	1.32084	1.20849	-0.52338	160.0	0.17952	-7.45895	-9.19082
72.0	1.35201	1.30980	-0.42207	162.0	0.14665	-8.33719	-10.06906
74.0	1.38028	1.39968	-0.33219	164.0	0.11676	-9.32722	-11.05909
76.0	1.40553	1.47841	-0.25346	166.0	0.08999	-10.45793	-12.18980
78.0	1.42765	1.54621	-0.18566	168.0	0.06650	-11.77161	-13.50348
80.0	1.44653	1.60326	-0.12861	170.0	0.04641	-13.33387	-15.06574
82.0	1.46209	1.64973	-0.08214	172.0	0.02982	-15.25459	-16.98646
84.0	1.47426	1.68574	-0.04613	174.0	0.01683	-17.73977	-19.47164
86.0	1.48299	1.71139	-0.02048	176.0	0.00750	-21.25188	-22.98375
88.0	1.48825	1.72676	-0.00511	178.0	0.00188	-27.26666	-28.99853
90.0	1.49000	1.73187	0.0	180.	0.00000	-INFINITY	-INFINITY

TABLE A1-42

INTEGRAL EQUATION METHOD
EPSILON= 1.0000(NORMALIZED GROUNDPLANE RADIUS)
MONOPOLE LENGTH= 0.010000(WAVELENGTH)
INFINITELY THIN MONOPOLE RADIUS
RADIATION RESISTANCE = 0.019898 (OHMS)

ELEVATION ANGLE (DEG)	DIRECTIVITY (NUMERIC)	DIRECTIVITY (DBI)	RELATIVE POWER (DB)	ELEVATION ANGLE (DEG)	DIRECTIVITY (NUMERIC)	DIRECTIVITY (DBI)	RELATIVE POWER (DB)
0.	0.00000	-INFINITY	-INFINITY				
2.0	0.00189	-27.24179	-28.96770	92.0	1.48622	1.72082	-0.00509
4.0	0.00754	-21.22714	-22.95305	94.0	1.48100	1.70555	-0.02036
6.0	0.01692	-17.71522	-19.44113	96.0	1.47233	1.68005	-0.04586
8.0	0.02999	-15.23031	-16.95622	98.0	1.46025	1.64427	-0.08164
10.0	0.04667	-13.30994	-15.03585	100.0	1.44481	1.59810	-0.12781
12.0	0.06686	-11.74810	-13.47401	102.0	1.42607	1.54141	-0.18450
14.0	0.09047	-10.43491	-12.16082	104.0	1.40412	1.47405	-0.25186
16.0	0.11736	-9.30476	-11.03067	106.0	1.37905	1.39581	-0.33010
18.0	0.14739	-8.31536	-10.04127	108.0	1.35098	1.30648	-0.41943
20.0	0.18039	-7.43780	-9.16371	110.0	1.32002	1.20579	-0.52012
22.0	0.21620	-6.65148	-8.37739	112.0	1.28630	1.09344	-0.63247
24.0	0.25461	-5.94119	-7.66710	114.0	1.24999	0.96908	-0.75683
26.0	0.29544	-5.29537	-7.02128	116.0	1.21125	0.83232	-0.89359
28.0	0.33845	-4.70504	-6.43095	118.0	1.17023	0.68272	-1.04319
30.0	0.38344	-4.16308	-5.88899	120.0	1.12714	0.51979	-1.20612
32.0	0.43016	-3.66375	-5.38966	122.0	1.08217	0.34295	-1.38296
34.0	0.47837	-3.20236	-4.92827	124.0	1.03552	0.15158	-1.57433
36.0	0.52783	-2.77502	-4.50093	126.0	0.98741	-0.05502	-1.78093
38.0	0.57830	-2.37848	-4.10439	128.0	0.93806	-0.27767	-2.00358
40.0	0.62951	-2.01000	-3.73591	130.0	0.88772	-0.51726	-2.24317
42.0	0.68120	-1.66723	-3.39314	132.0	0.83660	-0.77482	-2.50073
44.0	0.73314	-1.34816	-3.07407	134.0	0.78496	-1.05150	-2.77741
46.0	0.78505	-1.05105	-2.77696	136.0	0.73306	-1.34863	-3.07454
48.0	0.83668	-0.77438	-2.50029	138.0	0.68113	-1.66772	-3.39363
50.0	0.88780	-0.51684	-2.24275	140.0	0.62943	-2.01051	-3.73642
52.0	0.93815	-0.27727	-2.00318	142.0	0.57823	-2.37900	-4.10491
54.0	0.98750	-0.05464	-1.78055	144.0	0.52777	-2.77555	-4.50146
56.0	1.03561	0.15195	-1.57396	146.0	0.47831	-3.20290	-4.92881
58.0	1.08225	0.34330	-1.38261	148.0	0.43010	-3.66431	-5.39022
60.0	1.12723	0.52011	-1.20580	150.0	0.38338	-4.16365	-5.88956
62.0	1.17032	0.68303	-1.04288	152.0	0.33841	-4.70562	-6.43153
64.0	1.21132	0.83261	-0.89330	154.0	0.29540	-5.29597	-7.02188
66.0	1.25007	0.96934	-0.75657	156.0	0.25458	-5.94180	-7.66771
68.0	1.28638	1.09368	-0.63223	158.0	0.21617	-6.65210	-8.37801
70.0	1.32008	1.20601	-0.51990	160.0	0.18037	-7.43842	-9.16433
72.0	1.35104	1.30668	-0.41923	162.0	0.14737	-8.31599	-10.04190
74.0	1.37911	1.39599	-0.32992	164.0	0.11734	-9.30540	-11.03131
76.0	1.40417	1.47421	-0.25170	166.0	0.09046	-10.43555	-12.16146
78.0	1.42612	1.54155	-0.18436	168.0	0.06685	-11.74875	-13.47466
80.0	1.44485	1.59822	-0.12770	170.0	0.04666	-13.31060	-15.03651
82.0	1.46028	1.64436	-0.08155	172.0	0.02998	-15.23097	-16.95688
84.0	1.47235	1.68012	-0.04579	174.0	0.01692	-17.71588	-19.44179
86.0	1.48101	1.70559	-0.02032	176.0	0.00754	-21.22780	-22.95371
88.0	1.48622	1.72084	-0.00507	178.0	0.00189	-27.24246	-28.96837
90.0	1.48796	1.72591	0.0	180.	0.00000	-INFINITY	-INFINITY

TABLE A1-43

INTEGRAL EQUATION METHOD
EPSILON= 1.2500(NORMALIZED GROUNDPLANE RADIUS)
MONOPOLE LENGTH= 0.010000(WAVELENGTH)
INFINITELY THIN MONOPOLE RADIUS
RADIATION RESISTANCE = 0.019916 (OHMS)

ELEVATION ANGLE (DEG)	DIRECTIVITY (NUMERIC)	DIRECTIVITY (DBI)	RELATIVE POWER (DB)	ELEVATION ANGLE (DEG)	DIRECTIVITY (NUMERIC)	DIRECTIVITY (DBI)	RELATIVE POWER (DB)
0.	0.00000	-INFINITY	-INFINITY				
2.0	0.00189	-27.22511	-28.94725	92.0	1.48493	1.71705	-0.00509
4.0	0.00757	-21.21054	-22.93268	94.0	1.47973	1.70182	-0.02032
6.0	0.01699	-17.69875	-19.42089	96.0	1.47110	1.67641	-0.04573
8.0	0.03010	-15.21403	-16.93617	98.0	1.45907	1.64076	-0.08138
10.0	0.04684	-13.29390	-15.01604	100.0	1.44370	1.59476	-0.12738
12.0	0.06711	-11.73235	-13.45449	102.0	1.42505	1.53829	-0.18385
14.0	0.09079	-10.41949	-12.14163	104.0	1.40319	1.47118	-0.25097
16.0	0.11777	-9.28973	-11.01187	106.0	1.37824	1.39323	-0.32891
18.0	0.14789	-8.30075	-10.02289	108.0	1.35028	1.30424	-0.41791
20.0	0.18098	-7.42366	-9.14580	110.0	1.31945	1.20392	-0.51823
22.0	0.21688	-6.63785	-8.35999	112.0	1.28587	1.09197	-0.63017
24.0	0.25538	-5.92810	-7.65024	114.0	1.24970	0.96805	-0.75409
26.0	0.29629	-5.28285	-7.00500	116.0	1.21109	0.83177	-0.89038
28.0	0.33938	-4.69313	-6.41527	118.0	1.17022	0.68267	-1.03947
30.0	0.38443	-4.15179	-5.87394	120.0	1.12727	0.52027	-1.20187
32.0	0.43121	-3.65312	-5.37526	122.0	1.08243	0.34399	-1.37815
34.0	0.47947	-3.19240	-4.91454	124.0	1.03591	0.15321	-1.56893
36.0	0.52896	-2.76574	-4.48789	126.0	0.98792	-0.05279	-1.77493
38.0	0.57944	-2.36991	-4.09205	128.0	0.93868	-0.27481	-1.99695
40.0	0.63065	-2.00213	-3.72427	130.0	0.88843	-0.51376	-2.23590
42.0	0.68233	-1.66007	-3.38221	132.0	0.83740	-0.77066	-2.49280
44.0	0.73423	-1.34171	-3.06385	134.0	0.78584	-1.04668	-2.76882
46.0	0.78609	-1.04530	-2.76744	136.0	0.73398	-1.34313	-3.06528
48.0	0.83766	-0.76934	-2.49148	138.0	0.68210	-1.66155	-3.38369
50.0	0.88869	-0.51249	-2.23463	140.0	0.63043	-2.00366	-3.72580
52.0	0.93894	-0.27360	-1.99574	142.0	0.57923	-2.37148	-4.09362
54.0	0.98818	-0.05164	-1.77378	144.0	0.52877	-2.76736	-4.48951
56.0	1.03617	0.15431	-1.56784	146.0	0.47929	-3.19406	-4.91620
58.0	1.08269	0.34503	-1.37711	148.0	0.43104	-3.65482	-5.37696
60.0	1.12752	0.52124	-1.20090	150.0	0.38428	-4.15354	-5.87568
62.0	1.17047	0.68359	-1.03856	152.0	0.33924	-4.69491	-6.41705
64.0	1.21133	0.83262	-0.88953	154.0	0.29616	-5.28467	-7.00681
66.0	1.24992	0.96884	-0.75330	156.0	0.25527	-5.92995	-7.65209
68.0	1.28608	1.09269	-0.62945	158.0	0.21678	-6.63973	-8.36187
70.0	1.31965	1.20458	-0.51757	160.0	0.18090	-7.42557	-9.14771
72.0	1.35046	1.30483	-0.41731	162.0	0.14782	-8.30268	-10.02482
74.0	1.37840	1.39377	-0.32838	164.0	0.11771	-9.29169	-11.01383
76.0	1.40334	1.47164	-0.25050	166.0	0.09075	-10.42147	-12.14361
78.0	1.42518	1.53869	-0.18345	168.0	0.06708	-11.73434	-13.45649
80.0	1.44381	1.59510	-0.12705	170.0	0.04682	-13.29591	-15.01805
82.0	1.45916	1.64103	-0.08111	172.0	0.03009	-15.21606	-16.93820
84.0	1.47117	1.67661	-0.04553	174.0	0.01698	-17.70079	-19.42293
86.0	1.47977	1.70195	-0.02019	176.0	0.00756	-21.21258	-22.93472
88.0	1.48495	1.71711	-0.00503	178.0	0.00189	-27.22715	-28.94930
90.0	1.48667	1.72214	0.0	180.	0.00000	-INFINITY	-INFINITY

TABLE A1-44

INTEGRAL EQUATION METHOD
EPSILON= 1.5000(NORMALIZED GROUNDPLANE RADIUS)
MONOPOLE LENGTH= 0.010000(WAVELENGTH)
INFINITELY THIN MONOPOLE RADIUS
RADIATION RESISTANCE = 0.019940 (OHMS)

ELEVATION ANGLE (DEG)	DIRECTIVITY (NUMERIC)	DIRECTIVITY (DBI)	RELATIVE POWER (DB)	ELEVATION ANGLE (DEG)	DIRECTIVITY (NUMERIC)	DIRECTIVITY (DBI)	RELATIVE POWER (DB)
0.	0.00000	-INFINITY	-INFINITY				
2.0	0.00191	-27.19992	-28.91674	92.0	1.48310	1.71171	-0.00511
4.0	0.00761	-21.18547	-22.90229	94.0	1.47792	1.69652	-0.02030
6.0	0.01708	-17.67390	-19.39072	96.0	1.46934	1.67122	-0.04560
8.0	0.03027	-15.18947	-16.90629	98.0	1.45738	1.63573	-0.08109
10.0	0.04710	-13.26971	-14.98653	100.0	1.44210	1.58994	-0.12688
12.0	0.06747	-11.70861	-13.42543	102.0	1.42356	1.53375	-0.18307
14.0	0.09128	-10.39628	-12.11310	104.0	1.40183	1.46697	-0.24985
16.0	0.11838	-9.26711	-10.98393	106.0	1.37702	1.38941	-0.32741
18.0	0.14863	-8.27880	-9.99562	108.0	1.34923	1.30086	-0.41596
20.0	0.18187	-7.40243	-9.11925	110.0	1.31857	1.20103	-0.51579
22.0	0.21790	-6.61741	-8.33423	112.0	1.28518	1.08964	-0.62718
24.0	0.25654	-5.90850	-7.62532	114.0	1.24920	0.96632	-0.75050
26.0	0.29757	-5.26414	-6.98096	116.0	1.21079	0.83068	-0.88614
28.0	0.34077	-4.67534	-6.39216	118.0	1.17011	0.68228	-1.03454
30.0	0.38593	-4.13497	-5.85179	120.0	1.12736	0.52062	-1.19620
32.0	0.43278	-3.63728	-5.35410	122.0	1.08271	0.34512	-1.37170
34.0	0.48111	-3.17758	-4.89440	124.0	1.03637	0.15516	-1.56166
36.0	0.53064	-2.75197	-4.46879	126.0	0.98856	-0.04999	-1.76681
38.0	0.58114	-2.35719	-4.07401	128.0	0.93948	-0.27113	-1.98795
40.0	0.63234	-1.99047	-3.70729	130.0	0.88937	-0.50916	-2.22598
42.0	0.68400	-1.64947	-3.36629	132.0	0.83847	-0.76513	-2.48195
44.0	0.73584	-1.33217	-3.04899	134.0	0.78701	-1.04019	-2.75701
46.0	0.78762	-1.03681	-2.75363	136.0	0.73524	-1.33568	-3.05250
48.0	0.83910	-0.76189	-2.47871	138.0	0.68342	-1.65311	-3.36993
50.0	0.89001	-0.50606	-2.22288	140.0	0.63179	-1.99424	-3.71106
52.0	0.94012	-0.26817	-1.98499	142.0	0.58062	-2.36108	-4.07790
54.0	0.98920	-0.04717	-1.76399	144.0	0.53015	-2.75598	-4.47280
56.0	1.03701	0.15783	-1.55899	146.0	0.48065	-3.18171	-4.89853
58.0	1.08334	0.34764	-1.36918	148.0	0.43236	-3.64152	-5.35834
60.0	1.12797	0.52299	-1.19383	150.0	0.38554	-4.13931	-5.85613
62.0	1.17071	0.68450	-1.03232	152.0	0.34043	-4.67977	-6.39659
64.0	1.21136	0.83275	-0.88407	154.0	0.29726	-5.26867	-6.98549
66.0	1.24975	0.96823	-0.74859	156.0	0.25626	-5.91311	-7.62993
68.0	1.28570	1.09139	-0.62543	158.0	0.21767	-6.62211	-8.33893
70.0	1.31906	1.20263	-0.51419	160.0	0.18167	-7.40721	-9.12403
72.0	1.34968	1.30230	-0.41452	162.0	0.14847	-8.28364	-10.00046
74.0	1.37743	1.39070	-0.32612	164.0	0.11825	-9.27202	-10.98884
76.0	1.40220	1.46809	-0.24873	166.0	0.09118	-10.40124	-12.11806
78.0	1.42387	1.53471	-0.18211	168.0	0.06740	-11.71362	-13.43044
80.0	1.44236	1.59075	-0.12607	170.0	0.04705	-13.27476	-14.99158
82.0	1.45759	1.63637	-0.08045	172.0	0.03024	-15.19455	-16.91137
84.0	1.46950	1.67170	-0.04512	174.0	0.01706	-17.67900	-19.39582
86.0	1.47803	1.69684	-0.01998	176.0	0.00760	-21.19059	-22.90741
88.0	1.48316	1.71187	-0.00495	178.0	0.00190	-27.20505	-28.92187
90.0	1.48485	1.71682	0.0	180.	0.00000	-INFINITY	-INFINITY

TABLE A1-45

INTEGRAL EQUATION METHOD
EPSILON= 1.7500(NORMALIZED GROUNDPLANE RADIUS)
MONOPOLE LENGTH= 0.010000(WAVELENGTH)
INFINITELY THIN MONOPOLE RADIUS
RADIATION RESISTANCE = 0.019997 (OHMS)

ELEVATION ANGLE (DEG)	DIRECTIVITY (NUMERIC)	DIRECTIVITY (DBI)	RELATIVE POWER (DB)	ELEVATION ANGLE (DEG)	DIRECTIVITY (NUMERIC)	DIRECTIVITY (DBI)	RELATIVE POWER (DB)
0.	0.00000	-INFINITY	-INFINITY				
2.0	0.00193	-27.13917	-28.84367	92.0	1.47889	1.69937	-0.00513
4.0	0.00772	-21.12505	-22.82955	94.0	1.47377	1.68429	-0.02021
6.0	0.01732	-17.61402	-19.31852	96.0	1.46529	1.65923	-0.04527
8.0	0.03069	-15.13034	-16.83484	98.0	1.45349	1.62411	-0.08039
10.0	0.04774	-13.21154	-14.91604	100.0	1.43841	1.57883	-0.12567
12.0	0.06837	-11.65159	-13.35609	102.0	1.42012	1.52327	-0.18123
14.0	0.09246	-10.34060	-12.04510	104.0	1.39870	1.45725	-0.24725
16.0	0.11987	-9.21295	-10.91745	106.0	1.37423	1.38059	-0.32391
18.0	0.15044	-8.22632	-9.93082	108.0	1.34681	1.29305	-0.41145
20.0	0.18400	-7.35180	-9.05630	110.0	1.31655	1.19437	-0.51013
22.0	0.22036	-6.56876	-8.27326	112.0	1.28358	1.08424	-0.62026
24.0	0.25930	-5.86195	-7.56645	114.0	1.24805	0.96230	-0.74220
26.0	0.30062	-5.21981	-6.92431	116.0	1.21009	0.82817	-0.87633
28.0	0.34409	-4.63333	-6.33783	118.0	1.16987	0.68139	-1.02311
30.0	0.38946	-4.09536	-5.79986	120.0	1.12757	0.52145	-1.18305
32.0	0.43650	-3.60014	-5.30464	122.0	1.08337	0.34778	-1.35672
34.0	0.48496	-3.14296	-4.84746	124.0	1.03747	0.15974	-1.54476
36.0	0.53458	-2.71990	-4.42440	126.0	0.99006	-0.04338	-1.74788
38.0	0.58510	-2.32768	-4.03218	128.0	0.94137	-0.26242	-1.96692
40.0	0.63628	-1.96355	-3.66805	130.0	0.89161	-0.49827	-2.20277
42.0	0.68784	-1.62512	-3.32962	132.0	0.84101	-0.75199	-2.45649
44.0	0.73954	-1.31037	-3.01487	134.0	0.78982	-1.02474	-2.72924
46.0	0.79113	-1.01753	-2.72203	136.0	0.73827	-1.31787	-3.02237
48.0	0.84235	-0.74508	-2.44958	138.0	0.68661	-1.63292	-3.33742
50.0	0.89296	-0.49167	-2.19617	140.0	0.63509	-1.97162	-3.67612
52.0	0.94273	-0.25613	-1.96063	142.0	0.58398	-2.33604	-4.04054
54.0	0.99142	-0.03741	-1.74191	144.0	0.53352	-2.72851	-4.43301
56.0	1.03882	0.16540	-1.53910	146.0	0.48397	-3.15183	-4.85633
58.0	1.08470	0.35311	-1.35139	148.0	0.43559	-3.60926	-5.31376
60.0	1.12887	0.52645	-1.17805	150.0	0.38862	-4.10472	-5.80922
62.0	1.17113	0.68607	-1.01843	152.0	0.34333	-4.64291	-6.34741
64.0	1.21130	0.83252	-0.87198	154.0	0.29994	-5.22960	-6.93410
66.0	1.24920	0.96632	-0.73818	156.0	0.25871	-5.87194	-7.57644
68.0	1.28467	1.08792	-0.61658	158.0	0.21984	-6.57893	-8.28343
70.0	1.31757	1.19772	-0.50678	160.0	0.18356	-7.36215	-9.06665
72.0	1.34774	1.29607	-0.40843	162.0	0.15008	-8.23683	-9.94133
74.0	1.37508	1.38327	-0.32123	164.0	0.11957	-9.22360	-10.92810
76.0	1.39946	1.45960	-0.24490	166.0	0.09223	-10.35137	-12.05587
78.0	1.42078	1.52528	-0.17922	168.0	0.06820	-11.66247	-13.36697
80.0	1.43897	1.58051	-0.12399	170.0	0.04762	-13.22252	-14.92702
82.0	1.45394	1.62545	-0.07905	172.0	0.03061	-15.14140	-16.84590
84.0	1.46563	1.66024	-0.04426	174.0	0.01728	-17.62513	-19.32963
86.0	1.47400	1.68496	-0.01954	176.0	0.00770	-21.13621	-22.84071
88.0	1.47901	1.69970	-0.00480	178.0	0.00193	-27.15036	-28.85486
90.0	1.48064	1.70450	0.0	180.	0.00000	-INFINITY	-INFINITY

TABLE A1-46

INTEGRAL EQUATION METHOD
EPSILON= 2.0000(NORMALIZED GROUNDPLANE RADIUS)
MONOPOLE LENGTH= 0.010000(WAVELENGTH)
INFINITELY THIN MONOPOLE RADIUS
RADIATION RESISTANCE = 0.020129 (OHMS)

ELEVATION ANGLE (DEG)	DIRECTIVITY (NUMERIC)	DIRECTIVITY (DBI)	RELATIVE POWER (DB)	ELEVATION ANGLE (DEG)	DIRECTIVITY (NUMERIC)	DIRECTIVITY (DBI)	RELATIVE POWER (DB)
0.	0.00000	-INFINITY	-INFINITY				
2.0	0.00200	-26.99127	-28.66703	92.0	1.46914	1.67062	-0.00514
4.0	0.00798	-20.97800	-22.65376	94.0	1.46414	1.65584	-0.01992
6.0	0.01791	-17.46837	-19.14413	96.0	1.45591	1.63135	-0.04441
8.0	0.03172	-14.98665	-16.66241	98.0	1.44448	1.59711	-0.07865
10.0	0.04931	-13.07033	-14.74609	100.0	1.42988	1.55299	-0.12277
12.0	0.07058	-11.51337	-13.18913	102.0	1.41217	1.49888	-0.17688
14.0	0.09537	-10.20584	-11.88160	104.0	1.39143	1.43462	-0.24114
16.0	0.12353	-9.08212	-10.75788	106.0	1.36773	1.36000	-0.31576
18.0	0.15489	-8.09983	-9.77559	108.0	1.34116	1.27480	-0.40096
20.0	0.18923	-7.23003	-8.90579	110.0	1.31182	1.17875	-0.49701
22.0	0.22636	-6.45206	-8.12782	112.0	1.27983	1.07154	-0.60422
24.0	0.26603	-5.75063	-7.42639	114.0	1.24532	0.95282	-0.72294
26.0	0.30803	-5.11413	-6.78989	116.0	1.20842	0.82218	-0.85358
28.0	0.35209	-4.53352	-6.20928	118.0	1.16928	0.67918	-0.99658
30.0	0.39796	-4.00160	-5.67736	120.0	1.12805	0.52330	-1.15246
32.0	0.44539	-3.51256	-5.18832	122.0	1.08492	0.35396	-1.32180
34.0	0.49412	-3.06167	-4.73743	124.0	1.04004	0.17051	-1.50525
36.0	0.54388	-2.64495	-4.32071	126.0	0.99362	-0.02779	-1.70355
38.0	0.59442	-2.25910	-3.93486	128.0	0.94585	-0.24176	-1.91752
40.0	0.64546	-1.90130	-3.57706	130.0	0.89695	-0.47233	-2.14809
42.0	0.69676	-1.56917	-3.24493	132.0	0.84712	-0.72057	-2.39633
44.0	0.74806	-1.26062	-2.93638	134.0	0.79659	-0.98768	-2.66344
46.0	0.79912	-0.97387	-2.64963	136.0	0.74559	-1.27501	-2.95077
48.0	0.84970	-0.70735	-2.38311	138.0	0.69436	-1.58413	-3.25989
50.0	0.89956	-0.45971	-2.13547	140.0	0.64316	-1.91682	-3.59258
52.0	0.94847	-0.22975	-1.90551	142.0	0.59222	-2.27515	-3.95091
54.0	0.99623	-0.01639	-1.69215	144.0	0.54181	-2.66153	-4.33729
56.0	1.04263	0.18128	-1.49448	146.0	0.49218	-3.07875	-4.75451
58.0	1.08745	0.36411	-1.31165	148.0	0.44360	-3.53013	-5.20589
60.0	1.13053	0.53282	-1.14294	150.0	0.39631	-4.01963	-5.69539
62.0	1.17167	0.68807	-0.98769	152.0	0.35059	-4.55199	-6.22775
64.0	1.21072	0.83044	-0.84532	154.0	0.30669	-5.13302	-6.80878
66.0	1.24751	0.96043	-0.71533	156.0	0.26486	-5.76991	-7.44567
68.0	1.28189	1.07852	-0.59724	158.0	0.22534	-6.47171	-8.14747
70.0	1.31374	1.18509	-0.49067	160.0	0.18836	-7.25002	-8.92578
72.0	1.34292	1.28050	-0.39526	162.0	0.15417	-8.12013	-9.79589
74.0	1.36933	1.36507	-0.31069	164.0	0.12295	-9.10270	-10.77846
76.0	1.39285	1.43905	-0.23671	166.0	0.09491	-10.22667	-11.90243
78.0	1.41341	1.50268	-0.17308	168.0	0.07024	-11.53441	-13.21017
80.0	1.43092	1.55616	-0.11960	170.0	0.04907	-13.09156	-14.76732
82.0	1.44532	1.59964	-0.07612	172.0	0.03156	-15.00804	-16.68380
84.0	1.45655	1.63325	-0.04251	174.0	0.01782	-17.48988	-19.16564
86.0	1.46457	1.65710	-0.01866	176.0	0.00794	-20.99959	-22.67535
88.0	1.46935	1.67126	-0.00450	178.0	0.00199	-27.01291	-28.68867
90.0	1.47088	1.67576	0.0	180.	0.00000	-INFINITY	-INFINITY

TABLE A1-47

INTEGRAL EQUATION METHOD
EPSILON= 2.2500(NORMALIZED GROUNDPLANE RADIUS)
MONOPOLE LENGTH= 0.010000(WAVELENGTH)
INFINITELY THIN MONOPOLE RADIUS
RADIATION RESISTANCE = 0.020430 (OHMS)

ELEVATION ANGLE (DEG)	DIRECTIVITY (NUMERIC)	DIRECTIVITY (DBI)	RELATIVE POWER (DB)	ELEVATION ANGLE (DEG)	DIRECTIVITY (NUMERIC)	DIRECTIVITY (DBI)	RELATIVE POWER (DB)
0.	0.00000	-INFINITY	-INFINITY				
2.0	0.00216	-26.65013	-28.26161	92.0	1.44757	1.60641	-0.00507
4.0	0.00863	-20.63892	-22.25040	94.0	1.44288	1.59230	-0.01918
6.0	0.01935	-17.13271	-18.74418	96.0	1.43519	1.56910	-0.04237
8.0	0.03423	-14.65572	-16.26719	98.0	1.42454	1.53675	-0.07472
10.0	0.05314	-12.74540	-14.35688	100.0	1.41096	1.49516	-0.11631
12.0	0.07593	-11.19568	-12.80715	102.0	1.39450	1.44420	-0.16728
14.0	0.10241	-9.89654	-11.50802	104.0	1.37522	1.38371	-0.22776
16.0	0.13236	-8.78229	-10.39376	106.0	1.35317	1.31352	-0.29796
18.0	0.16556	-7.81047	-9.42194	108.0	1.32843	1.23339	-0.37809
20.0	0.20174	-6.95205	-8.56352	110.0	1.30109	1.14306	-0.46841
22.0	0.24064	-6.18627	-7.79774	112.0	1.27123	1.04225	-0.56923
24.0	0.28199	-5.49774	-7.10921	114.0	1.23896	0.93058	-0.68089
26.0	0.32548	-4.87475	-6.48622	116.0	1.20439	0.80768	-0.80380
28.0	0.37084	-4.30813	-5.91961	118.0	1.16764	0.67308	-0.93840
30.0	0.41777	-3.79060	-5.40208	120.0	1.12882	0.52626	-1.08522
32.0	0.46599	-3.31625	-4.92772	122.0	1.08809	0.36664	-1.24484
34.0	0.51520	-2.88021	-4.49168	124.0	1.04557	0.19355	-1.41793
36.0	0.56514	-2.47842	-4.08990	126.0	1.00144	0.00624	-1.60524
38.0	0.61553	-2.10749	-3.71896	128.0	0.95584	-0.19615	-1.80763
40.0	0.66612	-1.76450	-3.37597	130.0	0.90895	-0.41458	-2.02606
42.0	0.71664	-1.44697	-3.05845	132.0	0.86096	-0.65015	-2.26163
44.0	0.76687	-1.15276	-2.76423	134.0	0.81206	-0.90411	-2.51559
46.0	0.81658	-0.88000	-2.49147	136.0	0.76245	-1.17787	-2.78935
48.0	0.86555	-0.62706	-2.23853	138.0	0.71235	-1.47306	-3.08454
50.0	0.91358	-0.39253	-2.00401	140.0	0.66198	-1.79154	-3.40301
52.0	0.96047	-0.17516	-1.78663	142.0	0.61158	-2.13545	-3.74692
54.0	1.00604	0.02615	-1.58532	144.0	0.56140	-2.50727	-4.11874
56.0	1.05012	0.21238	-1.39909	146.0	0.51169	-2.90990	-4.52137
58.0	1.09254	0.38437	-1.22710	148.0	0.46273	-3.34676	-4.95823
60.0	1.13316	0.54289	-1.06858	150.0	0.41477	-3.82188	-5.43336
62.0	1.17182	0.68860	-0.92287	152.0	0.36812	-4.34014	-5.95162
64.0	1.20840	0.82209	-0.78938	154.0	0.32304	-4.90744	-6.51892
66.0	1.24276	0.94388	-0.66759	156.0	0.27983	-5.53108	-7.14255
68.0	1.27480	1.05443	-0.55704	158.0	0.23877	-6.22020	-7.83167
70.0	1.30441	1.15413	-0.45734	160.0	0.20015	-6.98653	-8.59800
72.0	1.33148	1.24334	-0.36813	162.0	0.16423	-7.84545	-9.45692
74.0	1.35593	1.32236	-0.28912	164.0	0.13129	-8.81771	-10.42919
76.0	1.37767	1.39144	-0.22003	166.0	0.10157	-9.93236	-11.54384
78.0	1.39663	1.45082	-0.16066	168.0	0.07530	-11.23184	-12.84332
80.0	1.41276	1.50067	-0.11080	170.0	0.05270	-12.78186	-14.39334
82.0	1.42599	1.54116	-0.07031	172.0	0.03394	-14.69242	-16.30389
84.0	1.43628	1.57241	-0.03907	174.0	0.01919	-17.16959	-18.78107
86.0	1.44361	1.59450	-0.01698	176.0	0.00856	-20.67594	-22.28742
88.0	1.44794	1.60751	-0.00397	178.0	0.00214	-26.68723	-28.29871
90.0	1.44926	1.61147	0.0	180.	0.00000	-INFINITY	-INFINITY

TABLE A1-48

INTEGRAL EQUATION METHOD
EPSILON= 2.5000(NORMALIZED GROUNDPLANE RADIUS)
MONOPOLE LENGTH= 0.010000(WAVELENGTH)
INFINITELY THIN MONOPOLE RADIUS
RADIATION RESISTANCE = 0.021093 (OHMS)

ELEVATION ANGLE (DEG)	DIRECTIVITY (NUMERIC)	DIRECTIVITY (DBI)	RELATIVE POWER (DB)	ELEVATION ANGLE (DEG)	DIRECTIVITY (NUMERIC)	DIRECTIVITY (DBI)	RELATIVE POWER (DB)
0.	0.00000	-INFINITY	-INFINITY				
2.0	0.00255	-25.93310	-27.40572	92.0	1.40210	1.46780	-0.00481
4.0	0.01017	-19.92621	-21.39883	94.0	1.39803	1.45517	-0.01745
6.0	0.02277	-16.42717	-17.89978	96.0	1.39145	1.43467	-0.03795
8.0	0.04018	-13.96013	-15.43274	98.0	1.38237	1.40626	-0.06636
10.0	0.06219	-12.06246	-13.53507	100.0	1.37084	1.36986	-0.10275
12.0	0.08855	-10.52797	-12.00059	102.0	1.35687	1.32538	-0.14724
14.0	0.11894	-9.24654	-10.71916	104.0	1.34050	1.27268	-0.19993
16.0	0.15303	-8.15231	-9.62492	106.0	1.32179	1.21162	-0.26100
18.0	0.19043	-7.20265	-8.67527	108.0	1.30077	1.14199	-0.33062
20.0	0.23076	-6.36837	-7.84098	110.0	1.27749	1.06357	-0.40904
22.0	0.27362	-5.62849	-7.10110	112.0	1.25201	0.97609	-0.49653
24.0	0.31861	-4.96742	-6.44004	114.0	1.22439	0.87921	-0.59340
26.0	0.36532	-4.37323	-5.84584	116.0	1.19469	0.77257	-0.70005
28.0	0.41338	-3.83652	-5.30914	118.0	1.16298	0.65571	-0.81690
30.0	0.46240	-3.34978	-4.82240	120.0	1.12931	0.52813	-0.94449
32.0	0.51205	-2.90686	-4.37948	122.0	1.09376	0.38922	-1.08340
34.0	0.56200	-2.50267	-3.97528	124.0	1.05640	0.23830	-1.23432
36.0	0.61194	-2.13293	-3.60554	126.0	1.01732	0.07457	-1.39804
38.0	0.66161	-1.79400	-3.26662	128.0	0.97659	-0.10287	-1.57548
40.0	0.71076	-1.48278	-2.95540	130.0	0.93432	-0.29506	-1.76767
42.0	0.75917	-1.19659	-2.66921	132.0	0.89060	-0.50318	-1.97580
44.0	0.80666	-0.93310	-2.40571	134.0	0.84555	-0.72860	-2.20122
46.0	0.85305	-0.69027	-2.16289	136.0	0.79931	-0.97286	-2.44548
48.0	0.89818	-0.46634	-1.93896	138.0	0.75202	-1.23773	-2.71035
50.0	0.94194	-0.25976	-1.73238	140.0	0.70384	-1.52524	-2.99785
52.0	0.98420	-0.06916	-1.54178	142.0	0.65498	-1.83770	-3.31031
54.0	1.02486	0.10665	-1.36597	144.0	0.60565	-2.17780	-3.65042
56.0	1.06383	0.26870	-1.20391	146.0	0.55608	-2.54865	-4.02127
58.0	1.10101	0.41793	-1.05469	148.0	0.50654	-2.95388	-4.42650
60.0	1.13635	0.55511	-0.91751	150.0	0.45732	-3.39777	-4.87038
62.0	1.16975	0.68094	-0.79167	152.0	0.40875	-3.88540	-5.35801
64.0	1.20117	0.79603	-0.67659	154.0	0.36117	-4.42293	-5.89554
66.0	1.23052	0.90089	-0.57173	156.0	0.31493	-5.01787	-6.49049
68.0	1.25776	0.99597	-0.47664	158.0	0.27042	-5.67962	-7.15223
70.0	1.28282	1.08165	-0.39096	160.0	0.22803	-6.42011	-7.89273
72.0	1.30565	1.15827	-0.31435	162.0	0.18815	-7.25494	-8.72756
74.0	1.32620	1.22608	-0.24653	164.0	0.15118	-8.20507	-9.67769
76.0	1.34442	1.28534	-0.18728	166.0	0.11750	-9.29973	-10.77234
78.0	1.36026	1.33622	-0.13639	168.0	0.08747	-10.58152	-12.05413
80.0	1.37369	1.37889	-0.09372	170.0	0.06143	-12.11630	-13.58892
82.0	1.38468	1.41348	-0.05913	172.0	0.03968	-14.01421	-15.48683
84.0	1.39318	1.44008	-0.03253	174.0	0.02248	-16.48144	-17.95406
86.0	1.39919	1.45877	-0.01384	176.0	0.01004	-19.98062	-21.45323
88.0	1.40269	1.46961	-0.00301	178.0	0.00252	-25.98759	-27.46020
90.0	1.40366	1.47262	0.0	180.	0.00000	-INFINITY	-INFINITY

TABLE A1-49

INTEGRAL EQUATION METHOD
EPSILON= 2.7500(NORMALIZED GROUNDPLANE RADIUS)
MONOPOLE LENGTH= 0.010000(WAVELENGTH)
INFINITELY THIN MONOPOLE RADIUS
RADIATION RESISTANCE = 0.022546 (OHMS)

ELEVATION ANGLE (DEG)	DIRECTIVITY (NUMERIC)	DIRECTIVITY (DBI)	RELATIVE POWER (DB)	ELEVATION ANGLE (DEG)	DIRECTIVITY (NUMERIC)	DIRECTIVITY (DBI)	RELATIVE POWER (DB)
0.	0.00000	-INFINITY	-INFINITY				
2.0	5.13099	7.10202	0.0	92.0	1.30845	1.16759	-5.93443
4.0	1.25506	0.98666	-6.11536	94.0	1.30203	1.14622	-5.95580
6.0	0.54526	-2.63399	-9.73601	96.0	1.29399	1.11932	-5.98269
8.0	0.29948	-5.23632	-12.33834	98.0	1.28436	1.08687	-6.01514
10.0	0.18691	-7.28357	-14.38559	100.0	1.27316	1.04884	-6.05318
12.0	0.12640	-8.98260	-16.08461	102.0	1.26043	1.00517	-6.09684
14.0	0.09028	-10.44429	-17.54631	104.0	1.24618	0.95582	-6.14619
16.0	0.06706	-11.73542	-18.83743	106.0	1.23047	0.90071	-6.20131
18.0	0.05129	-12.89964	-20.00165	108.0	1.21331	0.83973	-6.26229
20.0	0.04011	-13.96746	-21.06947	110.0	1.19474	0.77274	-6.32928
22.0	0.03191	-14.96141	-22.06342	112.0	1.17478	0.69955	-6.40247
24.0	0.02571	-15.89903	-23.00104	114.0	1.15343	0.61993	-6.48209
26.0	0.02092	-16.79472	-23.89674	116.0	1.13072	0.53356	-6.56846
28.0	0.01714	-17.66100	-24.76302	118.0	1.10664	0.44005	-6.66197
30.0	0.01409	-18.50946	-25.61147	120.0	1.08116	0.33892	-6.76310
32.0	0.01161	-19.35158	-26.45359	122.0	1.05428	0.22958	-6.87244
34.0	0.00955	-20.19963	-27.30164	124.0	1.02596	0.11130	-6.99072
36.0	0.00782	-21.06777	-28.16979	126.0	0.99615	-0.01677	-7.11878
38.0	0.00635	-21.97367	-29.07569	128.0	0.96480	-0.15564	-7.25766
40.0	0.00508	-22.94121	-30.04322	130.0	0.93186	-0.30651	-7.40853
42.0	0.00398	-24.00548	-31.10750	132.0	0.89727	-0.47076	-7.57277
44.0	0.00300	-25.22343	-32.32545	134.0	0.86100	-0.64997	-7.75199
46.0	0.00214	-26.69976	-33.80178	136.0	0.82300	-0.84599	-7.94800
48.0	0.00136	-28.66722	-35.76924	138.0	0.78327	-1.06088	-8.16290
50.0	0.00065	-31.86507	-38.96709	140.0	0.74182	-1.29704	-8.39906
52.0	1.07933	0.33153	-6.77048	142.0	0.69869	-1.55716	-8.65918
54.0	1.10870	0.44816	-6.65386	144.0	0.65398	-1.84433	-8.94635
56.0	1.13582	0.55309	-6.54892	146.0	0.60784	-2.16208	-9.26410
58.0	1.16077	0.64746	-6.45456	148.0	0.56047	-2.51447	-9.61649
60.0	1.18365	0.73222	-6.36979	150.0	0.51213	-2.90621	-10.00822
62.0	1.20454	0.80821	-6.29381	152.0	0.46315	-3.34281	-10.44483
64.0	1.22352	0.87612	-6.22590	154.0	0.41392	-3.83085	-10.93286
66.0	1.24066	0.93653	-6.16548	156.0	0.36490	-4.37825	-11.48026
68.0	1.25601	0.98995	-6.11207	158.0	0.31661	-4.99478	-12.09679
70.0	1.26962	1.03674	-6.06527	160.0	0.26960	-5.69274	-12.79476
72.0	1.28151	1.07724	-6.02478	162.0	0.22449	-6.48807	-13.59009
74.0	1.29172	1.11168	-5.99034	164.0	0.18189	-7.40203	-14.50404
76.0	1.30024	1.14024	-5.96177	166.0	0.14243	-8.46411	-15.56613
78.0	1.30710	1.16308	-5.93893	168.0	0.10673	-9.71718	-16.81920
80.0	1.31229	1.18029	-5.92173	170.0	0.07538	-11.22731	-18.32933
82.0	1.31581	1.19193	-5.91009	172.0	0.04892	-13.10481	-20.20682
84.0	1.31766	1.19805	-5.90397	174.0	0.02782	-15.55600	-22.65802
86.0	1.31785	1.19867	-5.90335	176.0	0.01246	-19.04365	-26.14567
88.0	1.31638	1.19380	-5.90821	178.0	0.00313	-25.04367	-32.14568
90.0	1.31324	1.18344	-5.91857	180.	0.00000	-INFINITY	-INFINITY

A.2 METHOD OF MOMENTS

Tables A2-1–A2-57; $0 < ka \leqslant 14.0$, $h/\lambda = 0.25$, $b/\lambda = 10^{-6}$, $N = 1$

TABLE A2-1

EPSLN= 0.2500(NORMALIZED GROUNDPLANE RADIUS)
MONOPOLE LENGTH= 0.250000(WAVELENGTH)
MONOPOLE RADIUS= 0.1000E-05(WAVELENGTH)
NUMBER OF MONOPOLE SEGMENTS= 1
NUMBER OF DISK SEGMENTS= 3
RADIATION RESISTANCE DETERMINED BY PATTERN= 0.194895E+02 OHMS

ELEVATION ANGLE (DEG)	DIRECTIVITY (NUMERIC)	DIRECTIVITY (DBI)	RELATIVE POWER (DB)	ELEVATION ANGLE (DEG)	DIRECTIVITY (NUMERIC)	DIRECTIVITY (DBI)	RELATIVE POWER (DB)
0.	0.00000	-INFINITY	-INFINITY				
2.0	0.00164	-27.84266	-29.71588	92.0	1.53717	1.86723	-0.00598
4.0	0.00657	-21.82525	-23.69846	94.0	1.53083	1.84927	-0.02395
6.0	0.01476	-18.30874	-20.18196	96.0	1.52030	1.81930	-0.05392
8.0	0.02620	-15.81745	-17.69066	98.0	1.50566	1.77727	-0.09594
10.0	0.04084	-13.88892	-15.76213	100.0	1.48700	1.72312	-0.15010
12.0	0.05865	-12.31719	-14.19040	102.0	1.46445	1.65675	-0.21647
14.0	0.07957	-10.99243	-12.86564	104.0	1.43815	1.57805	-0.29517
16.0	0.10354	-9.84909	-11.72230	106.0	1.40828	1.48688	-0.38634
18.0	0.13047	-8.84494	-10.71815	108.0	1.37502	1.38307	-0.49014
20.0	0.16028	-7.95115	-9.82437	110.0	1.33858	1.26644	-0.60677
22.0	0.19288	-7.14721	-9.02042	112.0	1.29920	1.13675	-0.73646
24.0	0.22814	-6.41797	-8.29119	114.0	1.25711	0.99374	-0.87947
26.0	0.26595	-5.75199	-7.62521	116.0	1.21258	0.83712	-1.03610
28.0	0.30617	-5.14038	-7.01360	118.0	1.16587	0.66652	-1.20669
30.0	0.34865	-4.57613	-6.44934	120.0	1.11726	0.48156	-1.39166
32.0	0.39322	-4.05361	-5.92682	122.0	1.06703	0.28177	-1.59144
34.0	0.43972	-3.56823	-5.44145	124.0	1.01546	0.06665	-1.80657
36.0	0.48795	-3.11624	-4.98946	126.0	0.96285	-0.16441	-2.03763
38.0	0.53771	-2.69451	-4.56772	128.0	0.90947	-0.41210	-2.28531
40.0	0.58879	-2.30040	-4.17361	130.0	0.85562	-0.67718	-2.55040
42.0	0.64096	-1.93171	-3.80492	132.0	0.80157	-0.96057	-2.83378
44.0	0.69398	-1.58654	-3.45975	134.0	0.74760	-1.26329	-3.13651
46.0	0.74760	-1.26329	-3.13650	136.0	0.69398	-1.58655	-3.45976
48.0	0.80157	-0.96056	-2.83378	138.0	0.64096	-1.93171	-3.80493
50.0	0.85562	-0.67718	-2.55039	140.0	0.58879	-2.30041	-4.17362
52.0	0.90948	-0.41209	-2.28531	142.0	0.53771	-2.69451	-4.56773
54.0	0.96285	-0.16441	-2.03762	144.0	0.48795	-3.11625	-4.98946
56.0	1.01547	0.06665	-1.80656	146.0	0.43972	-3.56824	-5.44145
58.0	1.06703	0.28178	-1.59144	148.0	0.39322	-4.05361	-5.92683
60.0	1.11727	0.48156	-1.39165	150.0	0.34865	-4.57614	-6.44935
62.0	1.16588	0.66652	-1.20669	152.0	0.30617	-5.14039	-7.01361
64.0	1.21258	0.83712	-1.03609	154.0	0.26595	-5.75200	-7.62522
66.0	1.25711	0.99375	-0.87947	156.0	0.22814	-6.41798	-8.29120
68.0	1.29920	1.13676	-0.73646	158.0	0.19288	-7.14722	-9.02043
70.0	1.33858	1.26644	-0.60677	160.0	0.16028	-7.95116	-9.82438
72.0	1.37502	1.38308	-0.49014	162.0	0.13047	-8.84495	-10.71816
74.0	1.40828	1.48688	-0.38633	164.0	0.10354	-9.84910	-11.72231
76.0	1.43815	1.57805	-0.29516	166.0	0.07957	-10.99243	-12.86565
78.0	1.46445	1.65675	-0.21646	168.0	0.05865	-12.31720	-14.19041
80.0	1.48700	1.72312	-0.15010	170.0	0.04084	-13.88892	-15.76214
82.0	1.50566	1.77727	-0.09594	172.0	0.02620	-15.81746	-17.69067
84.0	1.52030	1.81930	-0.05392	174.0	0.01476	-18.30875	-20.18197
86.0	1.53083	1.84927	-0.02395	176.0	0.00657	-21.82526	-23.69847
88.0	1.53717	1.86723	-0.00598	178.0	0.00164	-27.84267	-29.71589
90.0	1.53929	1.87321	0.00000	180.	0.00000	-INFINITY	-INFINITY

TABLE A2-2

EPSLN= 0.5000(NORMALIZED GROUNDPLANE RADIUS)
MONOPOLE LENGTH= 0.250000(WAVELENGTH)
MONOPOLE RADIUS= 0.1000E-05(WAVELENGTH)
NUMBER OF MONOPOLE SEGMENTS= 1
NUMBER OF DISK SEGMENTS= 7
RADIATION RESISTANCE DETERMINED BY PATTERN= 0.196224E+02 OHMS

ELEVATION ANGLE (DEG)	DIRECTIVITY (NUMERIC)	DIRECTIVITY (DBI)	RELATIVE POWER (DB)	ELEVATION ANGLE (DEG)	DIRECTIVITY (NUMERIC)	DIRECTIVITY (DBI)	RELATIVE POWER (DB)
0.	0.00000	-INFINITY	-INFINITY				
2.0	0.00169	-27.72040	-29.56410	92.0	1.52682	1.83789	-0.00581
4.0	0.00676	-21.70354	-23.54724	94.0	1.52071	1.82047	-0.02323
6.0	0.01518	-18.18796	-20.03166	96.0	1.51057	1.79141	-0.05229
8.0	0.02693	-15.69795	-17.54165	98.0	1.49646	1.75065	-0.09304
10.0	0.04197	-13.77107	-15.61476	100.0	1.47847	1.69812	-0.14557
12.0	0.06024	-12.20133	-14.04503	102.0	1.45671	1.63373	-0.20997
14.0	0.08168	-10.87890	-12.72260	104.0	1.43131	1.55733	-0.28636
16.0	0.10621	-9.73822	-11.58192	106.0	1.40243	1.46880	-0.37490
18.0	0.13375	-8.73704	-10.58074	108.0	1.37023	1.36794	-0.47576
20.0	0.16419	-7.84653	-9.69023	110.0	1.33492	1.25455	-0.58915
22.0	0.19742	-7.04614	-8.88984	112.0	1.29670	1.12839	-0.71531
24.0	0.23331	-6.32073	-8.16442	114.0	1.25579	0.98918	-0.85452
26.0	0.27172	-5.65882	-7.50251	116.0	1.21244	0.83659	-1.00711
28.0	0.31250	-5.05150	-6.89520	118.0	1.16688	0.67027	-1.17343
30.0	0.35549	-4.49175	-6.33545	120.0	1.11938	0.48978	-1.35392
32.0	0.40051	-3.97391	-5.81761	122.0	1.07020	0.29466	-1.54904
34.0	0.44736	-3.49339	-5.33708	124.0	1.01961	0.08436	-1.75934
36.0	0.49586	-3.04638	-4.89008	126.0	0.96789	-0.14174	-1.98544
38.0	0.54579	-2.62975	-4.47344	128.0	0.91530	-0.38434	-2.22804
40.0	0.59692	-2.24083	-4.08453	130.0	0.86213	-0.64426	-2.48795
42.0	0.64903	-1.87738	-3.72108	132.0	0.80865	-0.92241	-2.76610
44.0	0.70186	-1.53751	-3.38120	134.0	0.75512	-1.21985	-3.06355
46.0	0.75517	-1.21955	-3.06325	136.0	0.70181	-1.53781	-3.38151
48.0	0.80870	-0.92212	-2.76581	138.0	0.64898	-1.87770	-3.72140
50.0	0.86219	-0.64398	-2.48767	140.0	0.59688	-2.24115	-4.08485
52.0	0.91536	-0.38407	-2.22777	142.0	0.54575	-2.63008	-4.47378
54.0	0.96795	-0.14148	-1.98518	144.0	0.49582	-3.04672	-4.89042
56.0	1.01967	0.08461	-1.75909	146.0	0.44733	-3.49373	-5.33743
58.0	1.07026	0.29489	-1.54880	148.0	0.40047	-3.97427	-5.81796
60.0	1.11944	0.49000	-1.35370	150.0	0.35546	-4.49211	-6.33581
62.0	1.16694	0.67047	-1.17322	152.0	0.31247	-5.05187	-6.89557
64.0	1.21249	0.83679	-1.00691	154.0	0.27169	-5.65919	-7.50288
66.0	1.25585	0.98936	-0.85434	156.0	0.23329	-6.32110	-8.16480
68.0	1.29675	1.12856	-0.71514	158.0	0.19740	-7.04652	-8.89022
70.0	1.33497	1.25471	-0.58899	160.0	0.16418	-7.84691	-9.69061
72.0	1.37028	1.36808	-0.47562	162.0	0.13374	-8.73743	-10.58113
74.0	1.40247	1.46892	-0.37477	164.0	0.10620	-9.73861	-11.58231
76.0	1.43135	1.55744	-0.28625	166.0	0.08167	-10.87929	-12.72299
78.0	1.45674	1.63382	-0.20988	168.0	0.06023	-12.20173	-14.04542
80.0	1.47850	1.69820	-0.14549	170.0	0.04196	-13.77146	-15.61516
82.0	1.49648	1.75072	-0.09298	172.0	0.02693	-15.69835	-17.54205
84.0	1.51059	1.79146	-0.05224	174.0	0.01518	-18.18836	-20.03206
86.0	1.52072	1.82050	-0.02319	176.0	0.00675	-21.70394	-23.54764
88.0	1.52683	1.83791	-0.00579	178.0	0.00169	-27.72080	-29.56450
90.0	1.52887	1.84370	0.00000	180.	0.00000	-INFINITY	-INFINITY

TABLE A2-3

EPSLN= 0. 7500(NORMALIZED GROUNDPLANE RADIUS)
MONOPOLE LENGTH= 0. 250000(WAVELENGTH)
MONOPOLE RADIUS= 0. 1000E-05(WAVELENGTH)
NUMBER OF MONOPOLE SEGMENTS= 1
NUMBER OF DISK SEGMENTS= 16
RADIATION RESISTANCE DETERMINED BY PATTERN= 0. 198576E+02 OHMS

ELEVATION ANGLE (DEG)	DIRECTIVITY (NUMERIC)	DIRECTIVITY (DBI)	RELATIVE POWER (DB)	ELEVATION ANGLE (DEG)	DIRECTIVITY (NUMERIC)	DIRECTIVITY (DBI)	RELATIVE POWER (DB)
0.	0. 00000	-INFINITY	-INFINITY				
2. 0	0. 00177	-27. 50945	-29. 30139	92. 0	1. 50882	1. 78638	-0. 00556
4. 0	0. 00709	-21. 49354	-23. 28548	94. 0	1. 50309	1. 76985	-0. 02208
6. 0	0. 01592	-17. 97953	-19. 77147	96. 0	1. 49359	1. 74232	-0. 04961
8. 0	0. 02824	-15. 49172	-17. 28365	98. 0	1. 48038	1. 70372	-0. 08821
10. 0	0. 04398	-13. 56763	-15. 35957	100. 0	1. 46351	1. 65397	-0. 13797
12. 0	0. 06308	-12. 00129	-13. 79323	102. 0	1. 44310	1. 59295	-0. 19898
14. 0	0. 08545	-10. 68284	-12. 47477	104. 0	1. 41923	1. 52053	-0. 27141
16. 0	0. 11100	-9. 54670	-11. 33863	106. 0	1. 39204	1. 43652	-0. 35541
18. 0	0. 13962	-8. 55059	-10. 34253	108. 0	1. 36168	1. 34074	-0. 45119
20. 0	0. 17117	-7. 66567	-9. 45760	110. 0	1. 32830	1. 23295	-0. 55899
22. 0	0. 20552	-6. 87135	-8. 66329	112. 0	1. 29207	1. 11287	-0. 67906
24. 0	0. 24252	-6. 15248	-7. 94442	114. 0	1. 25320	0. 98020	-0. 81173
26. 0	0. 28200	-5. 49753	-7. 28947	116. 0	1. 21188	0. 83459	-0. 95735
28. 0	0. 32377	-4. 89758	-6. 68951	118. 0	1. 16832	0. 67562	-1. 11631
30. 0	0. 36766	-4. 34554	-6. 13748	120. 0	1. 12276	0. 50286	-1. 28908
32. 0	0. 41345	-3. 83575	-5. 62768	122. 0	1. 07542	0. 31577	-1. 47617
34. 0	0. 46094	-3. 36354	-5. 15548	124. 0	1. 02654	0. 11378	-1. 67816
36. 0	0. 50990	-2. 92511	-4. 71705	126. 0	0. 97639	-0. 10377	-1. 89571
38. 0	0. 56011	-2. 51724	-4. 30918	128. 0	0. 92520	-0. 33764	-2. 12957
40. 0	0. 61133	-2. 13726	-3. 92920	130. 0	0. 87324	-0. 58866	-2. 38060
42. 0	0. 66331	-1. 78287	-3. 57480	132. 0	0. 82076	-0. 85781	-2. 64975
44. 0	0. 71580	-1. 45211	-3. 24405	134. 0	0. 76803	-1. 14620	-2. 93814
46. 0	0. 76854	-1. 14331	-2. 93525	136. 0	0. 71531	-1. 45508	-3. 24702
48. 0	0. 82129	-0. 85502	-2. 64695	138. 0	0. 66284	-1. 78592	-3. 57785
50. 0	0. 87378	-0. 58596	-2. 37790	140. 0	0. 61089	-2. 14039	-3. 93232
52. 0	0. 92576	-0. 33503	-2. 12697	142. 0	0. 55970	-2. 52044	-4. 31238
54. 0	0. 97695	-0. 10128	-1. 89321	144. 0	0. 50952	-2. 92838	-4. 72031
56. 0	1. 02711	0. 11617	-1. 67577	146. 0	0. 46059	-3. 36687	-5. 15881
58. 0	1. 07598	0. 31804	-1. 47389	148. 0	0. 41313	-3. 83913	-5. 63107
60. 0	1. 12331	0. 50501	-1. 28692	150. 0	0. 36737	-4. 34898	-6. 14092
62. 0	1. 16887	0. 67766	-1. 11428	152. 0	0. 32351	-4. 90107	-6. 69300
64. 0	1. 21241	0. 83650	-0. 95544	154. 0	0. 28177	-5. 50107	-7. 29300
66. 0	1. 25371	0. 98198	-0. 80995	156. 0	0. 24232	-6. 15606	-7. 94799
68. 0	1. 29256	1. 11452	-0. 67742	158. 0	0. 20535	-6. 87497	-8. 66691
70. 0	1. 32876	1. 23446	-0. 55748	160. 0	0. 17103	-7. 66932	-9. 46125
72. 0	1. 36211	1. 34211	-0. 44983	162. 0	0. 13950	-8. 55427	-10. 34621
74. 0	1. 39243	1. 43775	-0. 35419	164. 0	0. 11091	-9. 55040	-11. 34234
76. 0	1. 41958	1. 52160	-0. 27033	166. 0	0. 08538	-10. 68657	-12. 47850
78. 0	1. 44340	1. 59388	-0. 19806	168. 0	0. 06302	-12. 00504	-13. 79698
80. 0	1. 46378	1. 65475	-0. 13719	170. 0	0. 04394	-13. 57140	-15. 36334
82. 0	1. 48059	1. 70435	-0. 08759	172. 0	0. 02821	-15. 49550	-17. 28744
84. 0	1. 49375	1. 74279	-0. 04915	174. 0	0. 01591	-17. 98333	-19. 77526
86. 0	1. 50320	1. 77017	-0. 02177	176. 0	0. 00708	-21. 49734	-23. 28928
88. 0	1. 50888	1. 78654	-0. 00540	178. 0	0. 00177	-27. 51326	-29. 30519
90. 0	1. 51075	1. 79194	0. 00000	180.	0. 00000	-INFINITY	-INFINI

TABLE A2-4

EPSLN= 1.0000(NORMALIZED GROUNDPLANE RADIUS)
MONOPOLE LENGTH= 0.250000(WAVELENGTH)
MONOPOLE RADIUS= 0.1000E-05(WAVELENGTH)
NUMBER OF MONOPOLE SEGMENTS= 1
NUMBER OF DISK SEGMENTS= 16
RADIATION RESISTANCE DETERMINED BY PATTERN= 0.202095E+02 OHMS

ELEVATION ANGLE (DEG)	DIRECTIVITY (NUMERIC)	DIRECTIVITY (DBI)	RELATIVE POWER (DB)	ELEVATION ANGLE (DEG)	DIRECTIVITY (NUMERIC)	DIRECTIVITY (DBI)	RELATIVE POWER (DB)
0.	0.00000	-INFINITY	-INFINITY				
2.0	0.00190	-27.20604	-28.92170	92.0	1.48262	1.71031	-0.00534
4.0	0.00760	-21.19146	-22.90712	94.0	1.47739	1.69494	-0.02072
6.0	0.01706	-17.67967	-19.39533	96.0	1.46875	1.66949	-0.04616
8.0	0.03023	-15.19495	-16.91060	98.0	1.45677	1.63390	-0.08176
10.0	0.04705	-13.27481	-14.99046	100.0	1.44147	1.58805	-0.12760
12.0	0.06740	-11.71326	-13.42891	102.0	1.42293	1.53182	-0.18383
14.0	0.09119	-10.40041	-12.11607	104.0	1.40121	1.46504	-0.25061
16.0	0.11829	-9.27068	-10.98633	106.0	1.37642	1.38751	-0.32814
18.0	0.14853	-8.28174	-9.99739	108.0	1.34865	1.29899	-0.41666
20.0	0.18177	-7.40472	-9.12037	110.0	1.31802	1.19921	-0.51644
22.0	0.21782	-6.61901	-8.33466	112.0	1.28465	1.08785	-0.62780
24.0	0.25648	-5.90939	-7.62505	114.0	1.24870	0.96457	-0.75109
26.0	0.29756	-5.26432	-6.97997	116.0	1.21030	0.82895	-0.88671
28.0	0.34082	-4.67481	-6.39047	118.0	1.16964	0.68054	-1.03512
30.0	0.38603	-4.13375	-5.84941	120.0	1.12689	0.51883	-1.19682
32.0	0.43297	-3.63540	-5.35105	122.0	1.08224	0.34325	-1.37240
34.0	0.48139	-3.17506	-4.89072	124.0	1.03590	0.15316	-1.56249
36.0	0.53102	-2.74886	-4.46451	126.0	0.98806	-0.05216	-1.76781
38.0	0.58163	-2.35353	-4.06919	128.0	0.93896	-0.27353	-1.98918
40.0	0.63295	-1.98634	-3.70199	130.0	0.88882	-0.51185	-2.22751
42.0	0.68471	-1.64492	-3.36058	132.0	0.83788	-0.76817	-2.48382
44.0	0.73667	-1.32728	-3.04293	134.0	0.78639	-1.04364	-2.75929
46.0	0.78856	-1.03166	-2.74731	136.0	0.73458	-1.33959	-3.05524
48.0	0.84013	-0.75654	-2.47219	138.0	0.68272	-1.65754	-3.37320
50.0	0.89113	-0.50060	-2.21625	140.0	0.63107	-1.99925	-3.71490
52.0	0.94131	-0.26267	-1.97832	142.0	0.57987	-2.36671	-4.08237
54.0	0.99044	-0.04171	-1.75737	144.0	0.52938	-2.76229	-4.47794
56.0	1.03829	0.16318	-1.55248	146.0	0.47987	-3.18872	-4.90438
58.0	1.08463	0.35281	-1.36285	148.0	0.43159	-3.64927	-5.36493
60.0	1.12925	0.52791	-1.18775	150.0	0.38478	-4.14782	-5.86348
62.0	1.17196	0.68911	-1.02654	152.0	0.33970	-4.68906	-6.40472
64.0	1.21255	0.83700	-0.87865	154.0	0.29657	-5.27874	-6.99439
66.0	1.25086	0.97209	-0.74357	156.0	0.25563	-5.92396	-7.63961
68.0	1.28671	1.09482	-0.62084	158.0	0.21708	-6.63371	-8.34936
70.0	1.31996	1.20560	-0.51005	160.0	0.18115	-7.41954	-9.13519
72.0	1.35045	1.30479	-0.41086	162.0	0.14802	-8.29667	-10.01232
74.0	1.37807	1.39271	-0.32295	164.0	0.11788	-9.28569	-11.00135
76.0	1.40269	1.46962	-0.24603	166.0	0.09088	-10.41551	-12.13117
78.0	1.42422	1.53577	-0.17988	168.0	0.06717	-11.72843	-13.44408
80.0	1.44257	1.59136	-0.12429	170.0	0.04688	-13.29004	-15.00569
82.0	1.45766	1.63655	-0.07910	172.0	0.03013	-15.21022	-16.92588
84.0	1.46943	1.67149	-0.04416	174.0	0.01700	-17.69499	-19.41064
86.0	1.47784	1.69627	-0.01938	176.0	0.00757	-21.20680	-22.92246
88.0	1.48285	1.71098	-0.00467	178.0	0.00190	-27.22140	-28.93705
90.0	1.48445	1.71565	0.00000	180.	0.00000	-INFINITY	-INFINITY

TABLE A2-5

EPSLN= 1.2500(NORMALIZED GROUNDPLANE RADIUS)
MONOPOLE LENGTH= 0.250000(WAVELENGTH)
MONOPOLE RADIUS= 0.1000E-05(WAVELENGTH)
NUMBER OF MONOPOLE SEGMENTS= 1
NUMBER OF DISK SEGMENTS= 16
RADIATION RESISTANCE DETERMINED BY PATTERN= 0.207151E+02 OHMS

ELEVATION ANGLE (DEG)	DIRECTIVITY (NUMERIC)	DIRECTIVITY (DBI)	RELATIVE POWER (DB)	ELEVATION ANGLE (DEG)	DIRECTIVITY (NUMERIC)	DIRECTIVITY (DBI)	RELATIVE POWER (DB)
0.	0.00000	-INFINITY	-INFINITY				
2.0	0.00209	-26.79196	-28.40031	92.0	1.44640	1.60287	-0.00548
4.0	0.00836	-20.77914	-22.38749	94.0	1.44168	1.58867	-0.01967
6.0	0.01875	-17.27027	-18.87862	96.0	1.43407	1.56571	-0.04264
8.0	0.03319	-14.78961	-16.39796	98.0	1.42360	1.53387	-0.07447
10.0	0.05159	-12.87468	-14.48302	100.0	1.41028	1.49305	-0.11529
12.0	0.07380	-11.31944	-12.92779	102.0	1.39414	1.44308	-0.16527
14.0	0.09968	-10.01400	-11.62235	104.0	1.37523	1.38374	-0.22460
16.0	0.12904	-8.89272	-10.50107	106.0	1.35357	1.31480	-0.29354
18.0	0.16169	-7.91327	-9.52161	108.0	1.32922	1.23597	-0.37238
20.0	0.19739	-7.04671	-8.65506	110.0	1.30224	1.14691	-0.46143
22.0	0.23592	-6.27240	-7.88075	112.0	1.27270	1.04726	-0.56109
24.0	0.27701	-5.57508	-7.18343	114.0	1.24068	0.93658	-0.67177
26.0	0.32039	-4.94314	-6.55149	116.0	1.20626	0.81441	-0.79394
28.0	0.36580	-4.36755	-5.97590	118.0	1.16956	0.68021	-0.92814
30.0	0.41294	-3.84114	-5.44948	120.0	1.13068	0.53339	-1.07495
32.0	0.46152	-3.35809	-4.96644	122.0	1.08976	0.37331	-1.23504
34.0	0.51125	-2.91366	-4.52201	124.0	1.04694	0.19923	-1.40912
36.0	0.56184	-2.50388	-4.11223	126.0	1.00239	0.01035	-1.59800
38.0	0.61299	-2.12544	-3.73379	128.0	0.95626	-0.19423	-1.80258
40.0	0.66443	-1.77551	-3.38385	130.0	0.90876	-0.41552	-2.02386
42.0	0.71587	-1.45166	-3.06000	132.0	0.86008	-0.65461	-2.26296
44.0	0.76704	-1.15179	-2.76014	134.0	0.81044	-0.91278	-2.52113
46.0	0.81769	-0.87409	-2.48244	136.0	0.76007	-1.19144	-2.79979
48.0	0.86757	-0.61696	-2.22530	138.0	0.70922	-1.49219	-3.10054
50.0	0.91644	-0.37896	-1.98731	140.0	0.65813	-1.81687	-3.42521
52.0	0.96408	-0.15887	-1.76721	142.0	0.60708	-2.16755	-3.77590
54.0	1.01029	0.04445	-1.56389	144.0	0.55633	-2.54669	-4.15503
56.0	1.05487	0.23200	-1.37635	146.0	0.50616	-2.95709	-4.56544
58.0	1.09766	0.40466	-1.20369	148.0	0.45687	-3.40210	-5.01045
60.0	1.13848	0.56325	-1.04510	150.0	0.40873	-3.88567	-5.49402
62.0	1.17720	0.70849	-0.89985	152.0	0.36203	-4.41256	-6.02091
64.0	1.21368	0.84104	-0.76730	154.0	0.31706	-4.98857	-6.59692
66.0	1.24781	0.96149	-0.64685	156.0	0.27410	-5.62089	-7.22923
68.0	1.27949	1.07038	-0.53797	158.0	0.23342	-6.31854	-7.92689
70.0	1.30863	1.16817	-0.44018	160.0	0.19529	-7.09315	-8.70150
72.0	1.33515	1.25530	-0.35305	162.0	0.15996	-7.95997	-9.56832
74.0	1.35898	1.33215	-0.27620	164.0	0.12765	-8.93965	-10.54800
76.0	1.38008	1.39905	-0.20930	166.0	0.09860	-10.06112	-11.66947
78.0	1.39839	1.45629	-0.15206	168.0	0.07300	-11.36673	-12.97508
80.0	1.41388	1.50413	-0.10422	170.0	0.05103	-12.92210	-14.53045
82.0	1.42652	1.54278	-0.06557	172.0	0.03283	-14.83715	-16.44549
84.0	1.43629	1.57241	-0.03593	174.0	0.01854	-17.31789	-18.92623
86.0	1.44316	1.59316	-0.01519	176.0	0.00827	-20.82682	-22.43517
88.0	1.44714	1.60512	-0.00323	178.0	0.00207	-26.83968	-28.44802
90.0	1.44822	1.60835	0.00000	180.	0.00000	-INFINITY	-INFINITY

TABLE A2-6

EPSLN= 1. 5000(NORMALIZED GROUNDPLANE RADIUS)
MONOPOLE LENGTH= 0 250000(WAVELENGTH)
MONOPOLE RADIUS= 0. 1000E-05(WAVELENGTH)
NUMBER OF MONOPOLE SEGMENTS= 1
NUMBER OF DISK SEGMENTS= 16
RADIATION RESISTANCE DETERMINED BY PATTERN= 0. 212521E+02 OHMS

ELEVATION ANGLE (DEG)	DIRECTIVITY (NUMERIC)	DIRECTIVITY (DBI)	RELATIVE POWER (DB)	ELEVATION ANGLE (DEG)	DIRECTIVITY (NUMERIC)	DIRECTIVITY (DBI)	RELATIVE POWER (DB)
0.	0. 00000	-INFINITY	-INFINITY				
2. 0	0. 00235	-26. 29046	-27. 79070	92. 0	1. 40824	1. 48675	-0. 01349
4. 0	0. 00938	-20. 27942	-21. 77966	94. 0	1. 40247	1. 46893	-0. 03131
6. 0	0. 02102	-16. 77352	-18. 27376	96. 0	1. 39434	1. 44368	-0. 05656
8. 0	0. 03718	-14. 29701	-15. 79725	98. 0	1. 38386	1. 41093	-0. 08931
10. 0	0. 05771	-12. 38738	-13. 88761	100. 0	1. 37105	1. 37053	-0. 12971
12. 0	0. 08244	-10. 83859	-12. 33882	102. 0	1. 35591	1. 32232	-0. 17792
14. 0	0. 11116	-9. 54071	-11. 04094	104. 0	1. 33846	1. 26605	-0. 23419
16. 0	0. 14361	-8. 42808	-9. 92832	106. 0	1. 31870	1. 20147	-0. 29877
18. 0	0. 17954	-7. 45835	-8. 95859	108. 0	1. 29665	1. 12824	-0. 37200
20. 0	0. 21865	-6. 60254	-8. 10278	110. 0	1. 27233	1. 04599	-0. 45425
22. 0	0. 26062	-5. 83997	-7. 34021	112. 0	1. 24574	0. 95428	-0. 54595
24. 0	0. 30512	-5. 15534	-6. 65557	114. 0	1. 21693	0. 85265	-0. 64759
26. 0	0. 35180	-4. 53699	-6. 03723	116. 0	1. 18592	0. 74054	-0. 75970
28. 0	0. 40033	-3. 97586	-5. 47610	118. 0	1. 15275	0. 61735	-0. 88289
30. 0	0. 45033	-3. 46470	-4. 96494	120. 0	1. 11749	0. 48243	-1. 01781
32. 0	0. 50146	-2. 99767	-4. 49791	122. 0	1. 08020	0. 33503	-1. 16521
34. 0	0. 55336	-2. 56995	-4. 07018	124. 0	1. 04096	0. 17434	-1. 32590
36. 0	0. 60569	-2. 17752	-3. 67776	126. 0	0. 99988	-0. 00052	-1. 50076
38. 0	0. 65811	-1. 81699	-3. 31723	128. 0	0. 95707	-0. 19056	-1. 69080
40. 0	0. 71032	-1. 48548	-2. 98572	130. 0	0. 91267	-0. 39685	-1. 89709
42. 0	0. 76199	-1. 18049	-2. 68073	132. 0	0. 86684	-0. 62062	-2. 12085
44. 0	0. 81286	-0. 89986	-2. 40010	134. 0	0. 81975	-0. 86320	-2. 36344
46. 0	0. 86264	-0. 64169	-2. 14193	136. 0	0. 77159	-1. 12612	-2. 62635
48. 0	0. 91111	-0. 40429	-1. 90453	138. 0	0. 72259	-1. 41106	-2. 91130
50. 0	0. 95803	-0. 18619	-1. 68643	140. 0	0. 67298	-1. 71995	-3. 22019
52. 0	1. 00322	0. 01396	-1. 48628	142. 0	0. 62302	-2. 05497	-3. 55521
54. 0	1. 04649	0. 19735	-1. 30289	144. 0	0. 57297	-2. 41866	-3. 91889
56. 0	1. 08769	0. 36505	-1. 13518	146. 0	0. 52313	-2. 81390	-4. 31414
58. 0	1. 12670	0. 51806	-0. 98217	148. 0	0. 47379	-3. 24413	-4. 74437
60. 0	1. 16339	0. 65726	-0. 84298	150. 0	0. 42527	-3. 71338	-5. 21362
62. 0	1. 19770	0. 78346	-0. 71677	152. 0	0. 37788	-4. 22649	-5. 72673
64. 0	1. 22953	0. 89741	-0. 60283	154. 0	0. 33195	-4. 78933	-6. 28957
66. 0	1. 25886	0. 99976	-0. 50048	156. 0	0. 28780	-5. 40916	-6. 90940
68. 0	1. 28563	1. 09115	-0. 40909	158. 0	0. 24575	-6. 09508	-7. 59532
70. 0	1. 30982	1. 17213	-0. 32811	160. 0	0. 20612	-6. 85876	-8. 35899
72. 0	1. 33143	1. 24320	-0. 25704	162. 0	0. 16922	-7. 71551	-9. 21575
74. 0	1. 35046	1. 30482	-0. 19542	164. 0	0. 13533	-8. 68604	-10. 18628
76. 0	1. 36691	1. 35740	-0. 14284	166. 0	0. 10473	-9. 79933	-11. 29957
78. 0	1. 38080	1. 40131	-0. 09893	168. 0	0. 07766	-11. 09776	-12. 59800
80. 0	1. 39215	1. 43686	-0. 06338	170. 0	0. 05436	-12. 64700	-14. 14724
82. 0	1. 40098	1. 46432	-0. 03592	172. 0	0. 03502	-14. 55698	-16. 05722
84. 0	1. 40732	1. 48392	-0. 01632	174. 0	0. 01980	-17. 03376	-18. 53400
86. 0	1. 41119	1. 49585	-0. 00439	176. 0	0. 00883	-20. 53985	-22. 04008
88. 0	1. 41262	1. 50024	0. 00000	178. 0	0. 00221	-26. 55099	-28. 05123
90. 0	1. 41162	1. 49719	-0. 00305	180.	0. 00000	-INFINITY	-INFINITY

TABLE A2-7

EPSLN= 1.7500(NORMALIZED GROUNDPLANE RADIUS)
MONOPOLE LENGTH= 0.250000(WAVELENGTH)
MONOPOLE RADIUS= 0.1000E-05(WAVELENGTH)
NUMBER OF MONOPOLE SEGMENTS= 1
NUMBER OF DISK SEGMENTS= 16
RADIATION RESISTANCE DETERMINED BY PATTERN= 0.224457E+02 OHMS

ELEVATION ANGLE (DEG)	DIRECTIVITY (NUMERIC)	DIRECTIVITY (DBI)	RELATIVE POWER (DB)	ELEVATION ANGLE (DEG)	DIRECTIVITY (NUMERIC)	DIRECTIVITY (DBI)	RELATIVE POWER (DB)
0.	0.00000	-INFINITY	-INFINITY				
2.0	0.00279	-25.55117	-26.81661	92.0	1.33361	1.25028	-0.01516
4.0	0.01111	-19.54315	-20.80860	94.0	1.32928	1.23617	-0.02928
6.0	0.02488	-16.04227	-17.30771	96.0	1.32357	1.21747	-0.04797
8.0	0.04393	-13.57276	-14.83821	98.0	1.31646	1.19408	-0.07137
10.0	0.06804	-11.67211	-12.93755	100.0	1.30791	1.16577	-0.09967
12.0	0.09696	-10.13423	-11.39967	102.0	1.29786	1.13229	-0.13316
14.0	0.13034	-8.84916	-10.11460	104.0	1.28626	1.09330	-0.17215
16.0	0.16783	-7.75121	-9.01665	106.0	1.27303	1.04839	-0.21705
18.0	0.20903	-6.79797	-8.06341	108.0	1.25809	0.99710	-0.26834
20.0	0.25349	-5.96041	-7.22586	110.0	1.24134	0.93890	-0.32654
22.0	0.30076	-5.21780	-6.48325	112.0	1.22270	0.87318	-0.39226
24.0	0.35037	-4.55477	-5.82022	114.0	1.20207	0.79929	-0.46615
26.0	0.40183	-3.95960	-5.22504	116.0	1.17937	0.71650	-0.54895
28.0	0.45466	-3.42312	-4.68856	118.0	1.15452	0.62401	-0.64144
30.0	0.50839	-2.93802	-4.20346	120.0	1.12745	0.52096	-0.74448
32.0	0.56255	-2.49836	-3.76380	122.0	1.09810	0.40643	-0.85901
34.0	0.61671	-2.09922	-3.36466	124.0	1.06645	0.27941	-0.98604
36.0	0.67043	-1.73648	-3.00192	126.0	1.03248	0.13880	-1.12665
38.0	0.72333	-1.40665	-2.67209	128.0	0.99619	-0.01657	-1.28201
40.0	0.77505	-1.10670	-2.37214	130.0	0.95764	-0.18798	-1.45342
42.0	0.82527	-0.83403	-2.09947	132.0	0.91689	-0.37681	-1.64225
44.0	0.87371	-0.58633	-1.85177	134.0	0.87406	-0.58460	-1.85004
46.0	0.92012	-0.36157	-1.62701	136.0	0.82928	-0.81301	-2.07845
48.0	0.96429	-0.15791	-1.42336	138.0	0.78272	-1.06391	-2.32935
50.0	1.00607	0.02627	-1.23917	140.0	0.73462	-1.33937	-2.60481
52.0	1.04532	0.19249	-1.07296	142.0	0.68522	-1.64172	-2.90716
54.0	1.08195	0.34209	-0.92335	144.0	0.63480	-1.97362	-3.23906
56.0	1.11592	0.47633	-0.78911	146.0	0.58370	-2.33810	-3.60354
58.0	1.14720	0.59638	-0.66907	148.0	0.53227	-2.73870	-4.00414
60.0	1.17579	0.70329	-0.56215	150.0	0.48089	-3.17957	-4.44502
62.0	1.20173	0.79807	-0.46738	152.0	0.42997	-3.66566	-4.93110
64.0	1.22508	0.88163	-0.38381	154.0	0.37993	-4.20294	-5.46838
66.0	1.24591	0.95485	-0.31059	156.0	0.33123	-4.79874	-6.06419
68.0	1.26431	1.01852	-0.24692	158.0	0.28430	-5.46226	-6.72771
70.0	1.28038	1.07338	-0.19206	160.0	0.23959	-6.20524	-7.47069
72.0	1.29423	1.12012	-0.14532	162.0	0.19756	-7.04307	-8.30851
74.0	1.30598	1.15937	-0.10608	164.0	0.15862	-7.99650	-9.26195
76.0	1.31574	1.19169	-0.07375	166.0	0.12318	-9.09458	-10.36002
78.0	1.32361	1.21761	-0.04783	168.0	0.09163	-10.37973	-11.64517
80.0	1.32971	1.23758	-0.02787	170.0	0.06430	-11.91766	-13.18310
82.0	1.33413	1.25199	-0.01345	172.0	0.04151	-13.81834	-15.08378
84.0	1.33696	1.26119	-0.00425	174.0	0.02351	-16.28786	-17.55330
86.0	1.33827	1.26544	0.00000	176.0	0.01050	-19.78874	-21.05418
88.0	1.33812	1.26496	-0.00048	178.0	0.00263	-25.79676	-27.06220
90.0	1.33656	1.25988	-0.00556	180.	0.00000	-INFINITY	-INFINITY

TABLE A2-8

EPSLN= 2.0000(NORMALIZED GROUNDPLANE RADIUS)
MONOPOLE LENGTH= 0.250000(WAVELENGTH)
MONOPOLE RADIUS= 0.1000E-05(WAVELENGTH)
NUMBER OF MONOPOLE SEGMENTS= 1
NUMBER OF DISK SEGMENTS= 16
RADIATION RESISTANCE DETERMINED BY PATTERN= 0.238990E+02 OHMS

ELEVATION ANGLE (DEG)	DIRECTIVITY (NUMERIC)	DIRECTIVITY (DBI)	RELATIVE POWER (DB)	ELEVATION ANGLE (DEG)	DIRECTIVITY (NUMERIC)	DIRECTIVITY (DBI)	RELATIVE POWER (DB)
0.	0.00000	-INFINITY	-INFINITY				
2.0	0.00340	-24.69016	-25.74217	92.0	1.25046	0.97070	-0.08131
4.0	0.01354	-18.68510	-19.73712	94.0	1.24530	0.95273	-0.09929
6.0	0.03028	-15.18915	-16.24116	96.0	1.23979	0.93347	-0.11854
8.0	0.05338	-12.72653	-13.77854	98.0	1.23390	0.91280	-0.13921
10.0	0.08251	-10.83469	-11.88670	100.0	1.22757	0.89048	-0.16153
12.0	0.11729	-9.30755	-10.35956	102.0	1.22072	0.86616	-0.18585
14.0	0.15721	-8.03512	-9.08713	104.0	1.21322	0.83941	-0.21260
16.0	0.20176	-6.95167	-8.00368	106.0	1.20495	0.80970	-0.24231
18.0	0.25034	-6.01475	-7.06676	108.0	1.19575	0.77640	-0.27561
20.0	0.30232	-5.19531	-6.24732	110.0	1.18544	0.73880	-0.31321
22.0	0.35706	-4.47257	-5.52458	112.0	1.17385	0.69612	-0.35589
24.0	0.41390	-3.83110	-4.88311	114.0	1.16077	0.64748	-0.40454
26.0	0.47216	-3.25913	-4.31114	116.0	1.14602	0.59192	-0.46009
28.0	0.53120	-2.74744	-3.79946	118.0	1.12939	0.52843	-0.52358
30.0	0.59038	-2.28867	-3.34068	120.0	1.11069	0.45591	-0.59610
32.0	0.64911	-1.87679	-2.92880	122.0	1.08973	0.37320	-0.67881
34.0	0.70683	-1.50683	-2.55884	124.0	1.06637	0.27906	-0.77295
36.0	0.76303	-1.17457	-2.22658	126.0	1.04044	0.17217	-0.87984
38.0	0.81725	-0.87645	-1.92846	128.0	1.01184	0.05113	-1.00088
40.0	0.86909	-0.60936	-1.66137	130.0	0.98049	-0.08556	-1.13757
42.0	0.91821	-0.37058	-1.42259	132.0	0.94635	-0.23950	-1.29151
44.0	0.96434	-0.15771	-1.20972	134.0	0.90941	-0.41240	-1.46441
46.0	1.00725	0.03139	-1.02063	136.0	0.86973	-0.60614	-1.65815
48.0	1.04681	0.19866	-0.85335	138.0	0.82742	-0.82275	-1.87476
50.0	1.08290	0.34588	-0.70613	140.0	0.78262	-1.06447	-2.11648
52.0	1.11550	0.47468	-0.57733	142.0	0.73556	-1.33380	-2.38581
54.0	1.14461	0.58656	-0.46545	144.0	0.68651	-1.63353	-2.68554
56.0	1.17029	0.68293	-0.36908	146.0	0.63579	-1.96685	-3.01886
58.0	1.19265	0.76512	-0.28690	148.0	0.58379	-2.33742	-3.38943
60.0	1.21181	0.83436	-0.21766	150.0	0.53094	-2.74951	-3.80152
62.0	1.22796	0.89183	-0.16018	152.0	0.47773	-3.20818	-4.26019
64.0	1.24127	0.93866	-0.11335	154.0	0.42467	-3.71950	-4.77151
66.0	1.25196	0.97591	-0.07610	156.0	0.37231	-4.29091	-5.34292
68.0	1.26026	1.00460	-0.04742	158.0	0.32124	-4.93168	-5.98369
70.0	1.26639	1.02567	-0.02634	160.0	0.27204	-5.65363	-6.70564
72.0	1.27058	1.04003	-0.01198	162.0	0.22531	-6.47222	-7.52423
74.0	1.27308	1.04855	-0.00346	164.0	0.18162	-7.40828	-8.46029
76.0	1.27409	1.05201	0.00000	166.0	0.14155	-8.49089	-9.54290
78.0	1.27384	1.05115	-0.00086	168.0	0.10562	-9.76254	-10.81455
80.0	1.27252	1.04664	-0.00537	170.0	0.07432	-11.28897	-12.34099
82.0	1.27030	1.03907	-0.01294	172.0	0.04808	-13.18020	-14.23222
84.0	1.26735	1.02897	-0.02304	174.0	0.02728	-15.64234	-16.69435
86.0	1.26380	1.01677	-0.03524	176.0	0.01220	-19.13793	-20.18994
88.0	1.25975	1.00284	-0.04917	178.0	0.00306	-25.14277	-26.19478
90.0	1.25529	0.98742	-0.06459	180.	0.00000	-INFINITY	-INFINITY

TABLE A2-9

EPSLN= 2.2500(NORMALIZED GROUNDPLANE RADIUS)
MONOPOLE LENGTH= 0.250000(WAVELENGTH)
MONOPOLE RADIUS= 0.1000E-05(WAVELENGTH)
NUMBER OF MONOPOLE SEGMENTS= 1
NUMBER OF DISK SEGMENTS= 16
RADIATION RESISTANCE DETERMINED BY PATTERN= 0.259969E+02 OHMS

ELEVATION ANGLE (DEG)	DIRECTIVITY (NUMERIC)	DIRECTIVITY (DBI)	RELATIVE POWER (DB)	ELEVATION ANGLE (DEG)	DIRECTIVITY (NUMERIC)	DIRECTIVITY (DBI)	RELATIVE POWER (DB)
0.	0.00000	-INFINITY	-INFINITY				
2.0	0.00429	-23.67209	-24.71056	92.0	1.14569	0.59069	-0.44778
4.0	0.01710	-17.67022	-18.70869	94.0	1.13828	0.56249	-0.47598
6.0	0.03820	-14.17957	-15.21804	96.0	1.13175	0.53749	-0.50098
8.0	0.06723	-11.72436	-12.76283	98.0	1.12606	0.51560	-0.52286
10.0	0.10370	-9.84203	-10.88050	100.0	1.12114	0.49660	-0.54186
12.0	0.14701	-8.32648	-9.36494	102.0	1.11689	0.48009	-0.55838
14.0	0.19644	-7.06768	-8.10614	104.0	1.11314	0.46550	-0.57297
16.0	0.25120	-5.99989	-7.03835	106.0	1.10972	0.45214	-0.58633
18.0	0.31041	-5.08063	-6.11910	108.0	1.10641	0.43916	-0.59930
20.0	0.37318	-4.28082	-5.31928	110.0	1.10296	0.42560	-0.61287
22.0	0.43857	-3.57962	-4.61809	112.0	1.09910	0.41036	-0.62811
24.0	0.50564	-2.96160	-4.00006	114.0	1.09452	0.39225	-0.64622
26.0	0.57347	-2.41492	-3.45339	116.0	1.08892	0.36998	-0.66849
28.0	0.64116	-1.93033	-2.96879	118.0	1.08198	0.34218	-0.69629
30.0	0.70788	-1.50040	-2.53886	120.0	1.07335	0.30740	-0.73107
32.0	0.77285	-1.11906	-2.15753	122.0	1.06271	0.26413	-0.77433
34.0	0.83536	-0.78128	-1.81975	124.0	1.04973	0.21079	-0.82768
36.0	0.89479	-0.48280	-1.52127	126.0	1.03412	0.14572	-0.89274
38.0	0.95061	-0.21997	-1.25844	128.0	1.01560	0.06722	-0.97125
40.0	1.00239	0.01038	-1.02809	130.0	0.99391	-0.02652	-1.06499
42.0	1.04979	0.21103	-0.82744	132.0	0.96886	-0.13738	-1.17585
44.0	1.09256	0.38446	-0.65400	134.0	0.94030	-0.26734	-1.30581
46.0	1.13055	0.53291	-0.50555	136.0	0.90813	-0.41851	-1.45698
48.0	1.16370	0.65841	-0.38005	138.0	0.87234	-0.59315	-1.63162
50.0	1.19202	0.76284	-0.27563	140.0	0.83297	-0.79371	-1.83218
52.0	1.21561	0.84793	-0.19054	142.0	0.79015	-1.02288	-2.06135
54.0	1.23461	0.91531	-0.12316	144.0	0.74411	-1.28363	-2.32209
56.0	1.24926	0.96653	-0.07193	146.0	0.69514	-1.57929	-2.61776
58.0	1.25981	1.00307	-0.03540	148.0	0.64362	-1.91368	-2.95215
60.0	1.26658	1.02633	-0.01214	150.0	0.59004	-2.29120	-3.32967
62.0	1.26990	1.03769	-0.00078	152.0	0.53493	-2.71702	-3.75549
64.0	1.27013	1.03847	0.00000	154.0	0.47892	-3.19734	-4.23580
66.0	1.26764	1.02996	-0.00851	156.0	0.42270	-3.73967	-4.77813
68.0	1.26282	1.01342	-0.02505	158.0	0.36700	-4.35338	-5.39185
70.0	1.25605	0.99008	-0.04839	160.0	0.31258	-5.05038	-6.08885
72.0	1.24771	0.96114	-0.07733	162.0	0.26024	-5.84619	-6.88465
74.0	1.23816	0.92776	-0.11071	164.0	0.21078	-6.76170	-7.80016
76.0	1.22774	0.89106	-0.14741	166.0	0.16497	-7.82605	-8.86452
78.0	1.21677	0.85210	-0.18636	168.0	0.12354	-9.08179	-10.12025
80.0	1.20557	0.81191	-0.22656	170.0	0.08720	-10.59468	-11.63315
82.0	1.19438	0.77141	-0.26705	172.0	0.05656	-12.47479	-13.51325
84.0	1.18344	0.73147	-0.30699	174.0	0.03215	-14.92824	-15.96671
86.0	1.17296	0.69284	-0.34562	176.0	0.01440	-18.41762	-19.45608
88.0	1.16310	0.65618	-0.38229	178.0	0.00362	-24.41872	-25.45719
90.0	1.15398	0.62200	-0.41647	180.	0.00000	-INFINITY	-INFINITY

TABLE A2-10

EPSLN= 2.5000(NORMALIZED GROUNDPLANE RADIUS)
MONOPOLE LENGTH= 0.250000(WAVELENGTH)
MONOPOLE RADIUS= 0.1000E-05(WAVELENGTH)
NUMBER OF MONOPOLE SEGMENTS= 1
NUMBER OF DISK SEGMENTS= 16
RADIATION RESISTANCE DETERMINED BY PATTERN= 0.290198E+02 OHMS

ELEVATION ANGLE (DEG)	DIRECTIVITY (NUMERIC)	DIRECTIVITY (DBI)	RELATIVE POWER (DB)
0.	0.00000	-INFINITY	-INFINITY
2.0	0.00558	-22.53282	-23.86563
4.0	0.02221	-16.53422	-17.86704
6.0	0.04956	-13.04902	-14.38183
8.0	0.08707	-10.60142	-11.93423
10.0	0.13400	-8.72885	-10.06166
12.0	0.18944	-7.22518	-8.55800
14.0	0.25233	-5.98038	-7.31320
16.0	0.32146	-4.92868	-6.26150
18.0	0.39559	-4.02757	-5.36039
20.0	0.47338	-3.24793	-4.58074
22.0	0.55349	-2.56890	-3.90172
24.0	0.63460	-1.97499	-3.30781
26.0	0.71542	-1.45436	-2.78718
28.0	0.79475	-0.99770	-2.33051
30.0	0.87146	-0.59754	-1.93036
32.0	0.94454	-0.24779	-1.58060
34.0	1.01313	0.05665	-1.27617
36.0	1.07648	0.32005	-1.01276
38.0	1.13400	0.54612	-0.78669
40.0	1.18524	0.73806	-0.59476
42.0	1.22990	0.89869	-0.43412
44.0	1.26781	1.03054	-0.30228
46.0	1.29894	1.13588	-0.19694
48.0	1.32336	1.21679	-0.11603
50.0	1.34128	1.27520	-0.05762
52.0	1.35297	1.31288	-0.01993
54.0	1.35880	1.33155	-0.00127
56.0	1.35920	1.33282	0.00000
58.0	1.35464	1.31824	-0.01458
60.0	1.34565	1.28933	-0.04348
62.0	1.33278	1.24759	-0.08522
64.0	1.31659	1.19449	-0.13832
66.0	1.29763	1.13151	-0.20131
68.0	1.27647	1.06011	-0.27271
70.0	1.25365	0.98178	-0.35104
72.0	1.22971	0.89802	-0.43480
74.0	1.20513	0.81033	-0.52249
76.0	1.18038	0.72022	-0.61259
78.0	1.15590	0.62922	-0.70360
80.0	1.13209	0.53880	-0.79402
82.0	1.10929	0.45044	-0.88238
84.0	1.08781	0.36554	-0.96728
86.0	1.06793	0.28542	-1.04740
88.0	1.04986	0.21131	-1.12151
90.0	1.03378	0.14426	-1.18855
92.0	1.01981	0.08519	-1.24763
94.0	1.00804	0.03476	-1.29805
96.0	0.99849	-0.00655	-1.33937
98.0	0.99116	-0.03856	-1.37138
100.0	0.98597	-0.06135	-1.39417
102.0	0.98282	-0.07525	-1.40807
104.0	0.98154	-0.08090	-1.41372
106.0	0.98193	-0.07919	-1.41200
108.0	0.98373	-0.07123	-1.40405
110.0	0.98665	-0.05838	-1.39120
112.0	0.99033	-0.04219	-1.37500
114.0	0.99441	-0.02433	-1.35715
116.0	0.99847	-0.00665	-1.33946
118.0	1.00206	0.00894	-1.32388
120.0	1.00471	0.02043	-1.31239
122.0	1.00595	0.02575	-1.30707
124.0	1.00526	0.02280	-1.31002
126.0	1.00217	0.00942	-1.32340
128.0	0.99620	-0.01656	-1.34937
130.0	0.98688	-0.05736	-1.39017
132.0	0.97381	-0.11525	-1.44807
134.0	0.95663	-0.19258	-1.52540
136.0	0.93502	-0.29177	-1.62459
138.0	0.90879	-0.41536	-1.74818
140.0	0.87780	-0.56605	-1.89887
142.0	0.84203	-0.74675	-2.07957
144.0	0.80156	-0.96062	-2.29344
146.0	0.75662	-1.21120	-2.54401
148.0	0.70755	-1.50242	-2.83524
150.0	0.65481	-1.83884	-3.17166
152.0	0.59900	-2.22575	-3.55857
154.0	0.54082	-2.66944	-4.00226
156.0	0.48111	-3.17754	-4.51036
158.0	0.42078	-3.75948	-5.09230
160.0	0.36081	-4.42725	-5.76007
162.0	0.30224	-5.19643	-6.52925
164.0	0.24615	-6.08798	-7.42080
166.0	0.19359	-7.13108	-8.46390
168.0	0.14560	-8.36831	-9.70113
170.0	0.10315	-9.86549	-11.19831
172.0	0.06710	-11.73270	-13.06552
174.0	0.03823	-14.17610	-15.50891
176.0	0.01715	-17.65828	-18.99110
178.0	0.00431	-23.65506	-24.98788
180.	0.00000	-INFINITY	-INFINITY

TABLE A2-11

EPSLN= 2.7500(NORMALIZED GROUNDPLANE RADIUS)
MONOPOLE LENGTH= 0.250000(WAVELENGTH)
MONOPOLE RADIUS= 0.1000E-05(WAVELENGTH)
NUMBER OF MONOPOLE SEGMENTS= 1
NUMBER OF DISK SEGMENTS= 16
RADIATION RESISTANCE DETERMINED BY PATTERN= 0.332407E+02 OHMS

ELEVATION ANGLE (DEG)	DIRECTIVITY (NUMERIC)	DIRECTIVITY (DBI)	RELATIVE POWER (DB)	ELEVATION ANGLE (DEG)	DIRECTIVITY (NUMERIC)	DIRECTIVITY (DBI)	RELATIVE POWER (DB)
0.	0.00000	-INFINITY	-INFINITY				
2.0	0.00734	-21.34301	-23.19298	92.0	0.88037	-0.55334	-2.40331
4.0	0.02919	-15.34767	-17.19764	94.0	0.86170	-0.64644	-2.49641
6.0	0.06504	-11.86788	-13.71785	96.0	0.84657	-0.72340	-2.57337
8.0	0.11408	-9.42784	-11.27781	98.0	0.83499	-0.78318	-2.63315
10.0	0.17519	-7.56495	-9.41492	100.0	0.82695	-0.82521	-2.67518
12.0	0.24700	-6.07308	-7.92305	102.0	0.82236	-0.84936	-2.69932
14.0	0.32793	-4.84215	-6.69212	104.0	0.82110	-0.85603	-2.70600
16.0	0.41626	-3.80637	-5.65634	106.0	0.82298	-0.84613	-2.69610
18.0	0.51013	-2.92319	-4.77316	108.0	0.82774	-0.82105	-2.67102
20.0	0.60765	-2.16346	-4.01343	110.0	0.83510	-0.78260	-2.63257
22.0	0.70692	-1.50628	-3.35625	112.0	0.84470	-0.73298	-2.58295
24.0	0.80610	-0.93611	-2.78608	114.0	0.85612	-0.67467	-2.52464
26.0	0.90343	-0.44107	-2.29104	116.0	0.86888	-0.61038	-2.46035
28.0	0.99728	-0.01181	-1.86178	118.0	0.88248	-0.54296	-2.39293
30.0	1.08622	0.35917	-1.49080	120.0	0.89632	-0.47536	-2.32532
32.0	1.16897	0.67802	-1.17194	122.0	0.90980	-0.41053	-2.26050
34.0	1.24448	0.94987	-0.90010	124.0	0.92226	-0.35146	-2.20143
36.0	1.31191	1.17903	-0.67094	126.0	0.93302	-0.30108	-2.15105
38.0	1.37064	1.36923	-0.48074	128.0	0.94139	-0.26229	-2.11226
40.0	1.42027	1.52370	-0.32627	130.0	0.94668	-0.23795	-2.08792
42.0	1.46059	1.64529	-0.20468	132.0	0.94823	-0.23088	-2.08085
44.0	1.49160	1.73652	-0.11345	134.0	0.94539	-0.24391	-2.09387
46.0	1.51345	1.79969	-0.05028	136.0	0.93759	-0.27987	-2.12984
48.0	1.52646	1.83687	-0.01310	138.0	0.92434	-0.34167	-2.19164
50.0	1.53108	1.84997	0.00000	140.0	0.90525	-0.43230	-2.28227
52.0	1.52784	1.84078	-0.00919	142.0	0.88005	-0.55493	-2.40490
54.0	1.51739	1.81097	-0.03900	144.0	0.84861	-0.71294	-2.56290
56.0	1.50042	1.76214	-0.08783	146.0	0.81096	-0.91001	-2.75998
58.0	1.47769	1.69583	-0.15414	148.0	0.76731	-1.15027	-3.00024
60.0	1.44995	1.61352	-0.23644	150.0	0.71806	-1.43837	-3.28834
62.0	1.41799	1.51672	-0.33325	152.0	0.66379	-1.77972	-3.62969
64.0	1.38257	1.40689	-0.44308	154.0	0.60524	-2.18070	-4.03067
66.0	1.34447	1.28552	-0.56445	156.0	0.54337	-2.64901	-4.49898
68.0	1.30442	1.15416	-0.69581	158.0	0.47927	-3.19415	-5.04412
70.0	1.26310	1.01438	-0.83559	160.0	0.41417	-3.82816	-5.67813
72.0	1.22119	0.86782	-0.98215	162.0	0.34941	-4.56669	-6.41665
74.0	1.17929	0.71620	-1.13376	164.0	0.28637	-5.43070	-7.28067
76.0	1.13798	0.56134	-1.28863	166.0	0.22649	-6.44944	-8.29940
78.0	1.09777	0.40512	-1.44484	168.0	0.17118	-7.66549	-9.51546
80.0	1.05914	0.24954	-1.60043	170.0	0.12177	-9.14472	-10.99469
82.0	1.02250	0.09665	-1.75332	172.0	0.07948	-10.99721	-12.84718
84.0	0.98823	-0.05143	-1.90140	174.0	0.04540	-13.42915	-15.27912
86.0	0.95663	-0.19256	-2.04253	176.0	0.02040	-16.90314	-18.75311
88.0	0.92798	-0.32459	-2.17456	178.0	0.00513	-22.89500	-24.74496
90.0	0.90251	-0.44549	-2.29546	180.	0.00000	-INFINITY	-INFINITY

TABLE A2-12

EPSLN= 3.0000(NORMALIZED GROUNDPLANE RADIUS)
MONOPOLE LENGTH= 0.250000(WAVELENGTH)
MONOPOLE RADIUS= 0.1000E-05(WAVELENGTH)
NUMBER OF MONOPOLE SEGMENTS= 1
NUMBER OF DISK SEGMENTS= 16
RADIATION RESISTANCE DETERMINED BY PATTERN= 0.386185E+02 OHMS

ELEVATION ANGLE (DEG)	DIRECTIVITY (NUMERIC)	DIRECTIVITY (DBI)	RELATIVE POWER (DB)	ELEVATION ANGLE (DEG)	DIRECTIVITY (NUMERIC)	DIRECTIVITY (DBI)	RELATIVE POWER (DB)
0.	0.00000	-INFINITY	-INFINITY				
2.0	0.00957	-20.18869	-22.64786	92.0	0.74435	-1.28224	-3.74142
4.0	0.03805	-14.19654	-16.65571	94.0	0.71630	-1.44904	-3.90822
6.0	0.08468	-10.72206	-13.18123	96.0	0.69281	-1.59384	-4.05301
8.0	0.14827	-8.28943	-10.74861	98.0	0.67397	-1.71362	-4.17279
10.0	0.22719	-6.43604	-8.89521	100.0	0.65980	-1.80588	-4.26506
12.0	0.31947	-4.95570	-7.41487	102.0	0.65030	-1.86884	-4.32801
14.0	0.42283	-3.73831	-6.19748	104.0	0.64543	-1.90154	-4.36071
16.0	0.53481	-2.71803	-5.17720	106.0	0.64505	-1.90404	-4.36321
18.0	0.65279	-1.85225	-4.31143	108.0	0.64902	-1.87740	-4.33657
20.0	0.77414	-1.11179	-3.57096	110.0	0.65710	-1.82367	-4.28284
22.0	0.89626	-0.47567	-2.93484	112.0	0.66900	-1.74576	-4.20493
24.0	1.01665	0.07170	-2.38748	114.0	0.68433	-1.64732	-4.10649
26.0	1.13299	0.54226	-1.91691	116.0	0.70267	-1.53248	-3.99165
28.0	1.24321	0.94544	-1.51373	118.0	0.72348	-1.40571	-3.86488
30.0	1.34549	1.28880	-1.17038	120.0	0.74617	-1.27161	-3.73078
32.0	1.43832	1.57854	-0.88063	122.0	0.77005	-1.13479	-3.59396
34.0	1.52050	1.81985	-0.63932	124.0	0.79438	-0.99973	-3.45890
36.0	1.59115	2.01712	-0.44205	126.0	0.81832	-0.87075	-3.32993
38.0	1.64972	2.17410	-0.28507	128.0	0.84102	-0.75196	-3.21113
40.0	1.69593	2.29408	-0.16510	130.0	0.86155	-0.64722	-3.10639
42.0	1.72979	2.37992	-0.07925	132.0	0.87898	-0.56022	-3.01940
44.0	1.75154	2.43419	-0.02498	134.0	0.89238	-0.49450	-2.95367
46.0	1.76164	2.45917	0.00000	136.0	0.90085	-0.45346	-2.91263
48.0	1.76073	2.45694	-0.00224	138.0	0.90356	-0.44044	-2.89961
50.0	1.74959	2.42937	-0.02980	140.0	0.89974	-0.45881	-2.91798
52.0	1.72910	2.37821	-0.08096	142.0	0.88879	-0.51199	-2.97116
54.0	1.70022	2.30506	-0.15411	144.0	0.87024	-0.60359	-3.06277
56.0	1.66396	2.21144	-0.24774	146.0	0.84383	-0.73747	-3.19664
58.0	1.62135	2.09876	-0.36041	148.0	0.80950	-0.91786	-3.37703
60.0	1.57340	1.96840	-0.49077	150.0	0.76745	-1.14951	-3.60868
62.0	1.52114	1.82169	-0.63748	152.0	0.71814	-1.43790	-3.89707
64.0	1.46552	1.65992	-0.79925	154.0	0.66230	-1.78946	-4.24863
66.0	1.40748	1.48442	-0.97475	156.0	0.60091	-2.21193	-4.67110
68.0	1.34788	1.29652	-1.16265	158.0	0.53520	-2.71485	-5.17402
70.0	1.28754	1.09760	-1.36157	160.0	0.46663	-3.31026	-5.76943
72.0	1.22719	0.88913	-1.57004	162.0	0.39684	-4.01383	-6.47300
74.0	1.16753	0.67269	-1.78648	164.0	0.32760	-4.84655	-7.30573
76.0	1.10918	0.45000	-2.00917	166.0	0.26076	-5.83766	-8.29683
78.0	1.05268	0.22295	-2.23622	168.0	0.19816	-7.02977	-9.48894
80.0	0.99853	-0.00637	-2.46554	170.0	0.14162	-8.48872	-10.94789
82.0	0.94719	-0.23563	-2.69480	172.0	0.09280	-10.32462	-12.78379
84.0	0.89903	-0.46225	-2.92142	174.0	0.05317	-12.74365	-15.20282
86.0	0.85440	-0.68340	-3.14257	176.0	0.02394	-16.20841	-18.66758
88.0	0.81358	-0.89599	-3.35517	178.0	0.00603	-22.19474	-24.65391
90.0	0.77683	-1.09675	-3.55592	180.	0.00000	-INFINITY	-INFINITY

TABLE A2-13

EPSLN= 3.2500(NORMALIZED GROUNDPLANE RADIUS)
MONOPOLE LENGTH= 0.250000(WAVELENGTH)
MONOPOLE RADIUS= 0.1000E-05(WAVELENGTH)
NUMBER OF MONOPOLE SEGMENTS= 1
NUMBER OF DISK SEGMENTS= 16
RADIATION RESISTANCE DETERMINED BY PATTERN= 0.441233E+02 OHMS

ELEVATION ANGLE (DEG)	DIRECTIVITY (NUMERIC)	DIRECTIVITY (DBI)	RELATIVE POWER (DB)	ELEVATION ANGLE (DEG)	DIRECTIVITY (NUMERIC)	DIRECTIVITY (DBI)	RELATIVE POWER (DB)
0.	0.00000	-INFINITY	-INFINITY				
2.0	0.01213	-19.16045	-22.19734	92.0	0.63650	-1.96205	-4.99894
4.0	0.04818	-13.17147	-16.20837	94.0	0.59793	-2.23349	-5.27039
6.0	0.10710	-9.70228	-12.73918	96.0	0.56439	-2.48423	-5.52113
8.0	0.18720	-7.27701	-10.31391	98.0	0.53601	-2.70830	-5.74520
10.0	0.28622	-5.43301	-8.46991	100.0	0.51291	-2.89962	-5.93651
12.0	0.40142	-3.96406	-7.00096	102.0	0.49518	-3.05238	-6.08927
14.0	0.52966	-2.76001	-5.79691	104.0	0.48288	-3.16160	-6.19850
16.0	0.66758	-1.75494	-4.79184	106.0	0.47603	-3.22368	-6.26057
18.0	0.81167	-0.90619	-3.94308	108.0	0.47459	-3.23679	-6.27369
20.0	0.95842	-0.18446	-3.22136	110.0	0.47849	-3.20128	-6.23818
22.0	1.10440	0.43127	-2.60563	112.0	0.48757	-3.11966	-6.15656
24.0	1.24644	0.95670	-2.08020	114.0	0.50160	-2.99644	-6.03334
26.0	1.38162	1.40389	-1.63301	116.0	0.52027	-2.83771	-5.87460
28.0	1.50742	1.78233	-1.25456	118.0	0.54317	-2.65060	-5.68750
30.0	1.62170	2.09972	-0.93718	120.0	0.56980	-2.44280	-5.47970
32.0	1.72280	2.36235	-0.67455	122.0	0.59951	-2.22203	-5.25893
34.0	1.80947	2.57551	-0.46138	124.0	0.63158	-1.99573	-5.03263
36.0	1.88092	2.74370	-0.29320	126.0	0.66515	-1.77083	-4.80773
38.0	1.93676	2.87076	-0.16614	128.0	0.69925	-1.55366	-4.59056
40.0	1.97700	2.96007	-0.07683	130.0	0.73284	-1.34993	-4.38682
42.0	2.00197	3.01457	-0.02233	132.0	0.76476	-1.16474	-4.20164
44.0	2.01229	3.03690	0.00000	134.0	0.79383	-1.00273	-4.03963
46.0	2.00881	3.02939	-0.00751	136.0	0.81882	-0.86812	-3.90502
48.0	1.99257	2.99414	-0.04276	138.0	0.83852	-0.76486	-3.80176
50.0	1.96474	2.93305	-0.10385	140.0	0.85178	-0.69673	-3.73363
52.0	1.92657	2.84784	-0.18905	142.0	0.85754	-0.66744	-3.70434
54.0	1.87936	2.74009	-0.29681	144.0	0.85491	-0.68078	-3.71768
56.0	1.82441	2.61122	-0.42567	146.0	0.84320	-0.74071	-3.77761
58.0	1.76301	2.46256	-0.57434	148.0	0.82195	-0.85153	-3.88843
60.0	1.69641	2.29530	-0.74160	150.0	0.79104	-1.01802	-4.05492
62.0	1.62577	2.11059	-0.92630	152.0	0.75065	-1.24564	-4.28254
64.0	1.55220	1.90949	-1.12741	154.0	0.70132	-1.54081	-4.57771
66.0	1.47673	1.69300	-1.34389	156.0	0.64398	-1.91124	-4.94814
68.0	1.40027	1.46211	-1.57478	158.0	0.57990	-2.36644	-5.40334
70.0	1.32367	1.21780	-1.81910	160.0	0.51069	-2.91842	-5.95532
72.0	1.24769	0.96106	-2.07584	162.0	0.43825	-3.58280	-6.61970
74.0	1.17299	0.69294	-2.34396	164.0	0.36471	-4.38055	-7.41744
76.0	1.10017	0.41460	-2.62229	166.0	0.29236	-5.34087	-8.37776
78.0	1.02976	0.12735	-2.90955	168.0	0.22355	-6.50634	-9.54324
80.0	0.96221	-0.16732	-3.20422	170.0	0.16059	-7.94280	-10.97970
82.0	0.89792	-0.46762	-3.50451	172.0	0.10567	-9.76032	-12.79722
84.0	0.83726	-0.77141	-3.80831	174.0	0.06074	-12.16507	-15.20197
86.0	0.78052	-1.07614	-4.11304	176.0	0.02742	-15.61965	-18.65655
88.0	0.72799	-1.37872	-4.41562	178.0	0.00692	-21.59986	-24.63676
90.0	0.67991	-1.67546	-4.71236	180.	0.00000	-INFINITY	-INFINITY

TABLE A2-14

EPSLN= 3.5000(NORMALIZED GROUNDPLANE RADIUS)
MONOPOLE LENGTH= 0.250000(WAVELENGTH)
MONOPOLE RADIUS= 0.1000E-05(WAVELENGTH)
NUMBER OF MONOPOLE SEGMENTS= 1
NUMBER OF DISK SEGMENTS= 16
RADIATION RESISTANCE DETERMINED BY PATTERN= 0.475769E+02 OHMS

ELEVATION ANGLE (DEG)	DIRECTIVITY (NUMERIC)	DIRECTIVITY (DBI)	RELATIVE POWER (DB)	ELEVATION ANGLE (DEG)	DIRECTIVITY (NUMERIC)	DIRECTIVITY (DBI)	RELATIVE POWER (DB)
0.	0.00000	-INFINITY	-INFINITY				
2.0	0.01475	-18.31222	-21.80048	92.0	0.57769	-2.38302	-5.87128
4.0	0.05853	-12.32650	-15.81476	94.0	0.52944	-2.76186	-6.25013
6.0	0.12994	-8.86269	-12.35095	96.0	0.48597	-3.13392	-6.62218
8.0	0.22673	-6.44492	-9.93318	98.0	0.44748	-3.49228	-6.98055
10.0	0.34590	-4.61048	-8.09874	100.0	0.41415	-3.82842	-7.31668
12.0	0.48383	-3.15309	-6.64135	102.0	0.38617	-4.13222	-7.62048
14.0	0.63643	-1.96251	-5.45078	104.0	0.36371	-4.39240	-7.88066
16.0	0.79933	-0.97276	-4.46103	106.0	0.34695	-4.59737	-8.08563
18.0	0.96804	-0.14107	-3.62933	108.0	0.33601	-4.73649	-8.22475
20.0	1.13814	0.56196	-2.92630	110.0	0.33101	-4.80163	-8.28989
22.0	1.30542	1.15750	-2.33076	112.0	0.33200	-4.78863	-8.27690
24.0	1.46601	1.66136	-1.82690	114.0	0.33898	-4.69826	-8.18652
26.0	1.61649	2.08574	-1.40252	116.0	0.35186	-4.53624	-8.02451
28.0	1.75400	2.44030	-1.04797	118.0	0.37047	-4.31248	-7.80074
30.0	1.87623	2.73286	-0.75540	120.0	0.39449	-4.03960	-7.52787
32.0	1.98148	2.96990	-0.51836	122.0	0.42351	-3.73141	-7.21967
34.0	2.06865	3.15687	-0.33139	124.0	0.45692	-3.40155	-6.88981
36.0	2.13718	3.29841	-0.18986	126.0	0.49401	-3.06260	-6.55087
38.0	2.18703	3.39854	-0.08972	128.0	0.53387	-2.72563	-6.21389
40.0	2.21860	3.46079	-0.02747	130.0	0.57543	-2.40006	-5.88832
42.0	2.23268	3.48826	0.00000	132.0	0.61748	-2.09379	-5.58205
44.0	2.23034	3.48371	-0.00455	134.0	0.65865	-1.81343	-5.30170
46.0	2.21289	3.44960	-0.03867	136.0	0.69750	-1.56457	-5.05283
48.0	2.18179	3.38814	-0.10013	138.0	0.73248	-1.35201	-4.84028
50.0	2.13861	3.30131	-0.18696	140.0	0.76207	-1.18006	-4.66833
52.0	2.08492	3.19090	-0.29737	142.0	0.78475	-1.05271	-4.54098
54.0	2.02232	3.05850	-0.42977	144.0	0.79913	-0.97385	-4.46212
56.0	1.95234	2.90555	-0.58272	146.0	0.80400	-0.94745	-4.43571
58.0	1.87643	2.73332	-0.75494	148.0	0.79841	-0.97772	-4.46599
60.0	1.79595	2.54295	-0.94532	150.0	0.78175	-1.06935	-4.55761
62.0	1.71216	2.33543	-1.15283	152.0	0.75376	-1.22766	-4.71592
64.0	1.62617	2.11165	-1.37661	154.0	0.71467	-1.45893	-4.94719
66.0	1.53899	1.87236	-1.61590	156.0	0.66516	-1.77074	-5.25901
68.0	1.45152	1.61822	-1.87004	158.0	0.60639	-2.17247	-5.66073
70.0	1.36452	1.34981	-2.13845	160.0	0.54001	-2.67601	-6.16427
72.0	1.27868	1.06762	-2.42064	162.0	0.46807	-3.29687	-6.78513
74.0	1.19457	0.77210	-2.71616	164.0	0.39301	-4.05592	-7.54418
76.0	1.11268	0.46368	-3.02458	166.0	0.31752	-4.98228	-8.47055
78.0	1.03343	0.14280	-3.34547	168.0	0.24443	-6.11847	-9.60673
80.0	0.95718	-0.19006	-3.67832	170.0	0.17659	-7.53025	-11.01851
82.0	0.88425	-0.53426	-4.02253	172.0	0.11674	-9.32764	-12.81591
84.0	0.81489	-0.88900	-4.37727	174.0	0.06735	-11.71678	-15.20504
86.0	0.74935	-1.25315	-4.74141	176.0	0.03048	-15.16022	-18.64849
88.0	0.68784	-1.62513	-5.11339	178.0	0.00770	-21.13377	-24.62203
90.0	0.63056	-2.00275	-5.49102	180.	0.00000	-INFINITY	-INFINITY

TABLE A2-15

EPSLN= 3.7500(NORMALIZED GROUNDPLANE RADIUS)
MONOPOLE LENGTH= 0.250000(WAVELENGTH)
MONOPOLE RADIUS= 0.1000E-05(WAVELENGTH)
NUMBER OF MONOPOLE SEGMENTS= 1
NUMBER OF DISK SEGMENTS= 11
RADIATION RESISTANCE DETERMINED BY PATTERN= 0.473479E+02 OHMS

ELEVATION ANGLE (DEG)	DIRECTIVITY (NUMERIC)	DIRECTIVITY (DBI)	RELATIVE POWER (DB)
0.	0.00000	-INFINITY	-INFINITY
2.0	0.01708	-17.67612	-21.44083
4.0	0.06771	-11.69375	-15.45846
6.0	0.15012	-8.23551	-12.00022
8.0	0.26149	-5.82547	-9.59018
10.0	0.39803	-4.00087	-7.76558
12.0	0.55522	-2.55535	-6.32006
14.0	0.72802	-1.37858	-5.14329
16.0	0.91108	-0.40445	-4.16916
18.0	1.09899	0.40992	-3.35479
20.0	1.28649	1.09408	-2.67063
22.0	1.46871	1.66936	-2.09535
24.0	1.64125	2.15174	-1.61297
26.0	1.80037	2.55361	-1.21110
28.0	1.94305	2.88484	-0.87987
30.0	2.06702	3.15345	-0.61126
32.0	2.17077	3.36615	-0.39856
34.0	2.25351	3.52859	-0.23612
36.0	2.31508	3.64567	-0.11904
38.0	2.35593	3.72162	-0.04309
40.0	2.37695	3.76019	-0.00452
42.0	2.37942	3.76471	0.00000
44.0	2.36490	3.73813	-0.02657
46.0	2.33514	3.68312	-0.08159
48.0	2.29195	3.60206	-0.16265
50.0	2.23722	3.49708	-0.26763
52.0	2.17275	3.37009	-0.39462
54.0	2.10030	3.22280	-0.54191
56.0	2.02149	3.05671	-0.70800
58.0	1.93782	2.87314	-0.89157
60.0	1.85064	2.67322	-1.09149
62.0	1.76113	2.45791	-1.30680
64.0	1.67033	2.22802	-1.53669
66.0	1.57913	1.98418	-1.78053
68.0	1.48829	1.72688	-2.03783
70.0	1.39845	1.45646	-2.30825
72.0	1.31012	1.17311	-2.59160
74.0	1.22374	0.87688	-2.88783
76.0	1.13965	0.56771	-3.19700
78.0	1.05814	0.24541	-3.51930
80.0	0.97943	-0.09028	-3.85499
82.0	0.90371	-0.43973	-4.20444
84.0	0.83113	-0.80331	-4.56802
86.0	0.76184	-1.18138	-4.94609
88.0	0.69595	-1.57420	-5.33891
90.0	0.63361	-1.98180	-5.74651
92.0	0.57493	-2.40383	-6.16854
94.0	0.52008	-2.83930	-6.60401
96.0	0.46921	-3.28629	-7.05100
98.0	0.42253	-3.74144	-7.50615
100.0	0.38024	-4.19944	-7.96415
102.0	0.34258	-4.65232	-8.41703
104.0	0.30983	-5.08876	-8.85347
106.0	0.28226	-5.49355	-9.25826
108.0	0.26016	-5.84767	-9.61238
110.0	0.24381	-6.12940	-9.89411
112.0	0.23351	-6.31700	-10.08171
114.0	0.22947	-6.39274	-10.15744
116.0	0.23189	-6.34721	-10.11192
118.0	0.24086	-6.18229	-9.94700
120.0	0.25640	-5.91090	-9.67561
122.0	0.27835	-5.55412	-9.31883
124.0	0.30643	-5.13672	-8.90143
126.0	0.34016	-4.68321	-8.44792
128.0	0.37885	-4.21532	-7.98003
130.0	0.42161	-3.75091	-7.51561
132.0	0.46730	-3.30402	-7.06873
134.0	0.51458	-2.88543	-6.65014
136.0	0.56191	-2.50335	-6.26806
138.0	0.60756	-2.16411	-5.92882
140.0	0.64971	-1.87281	-5.63752
142.0	0.68647	-1.63376	-5.39847
144.0	0.71600	-1.45087	-5.21558
146.0	0.73656	-1.32794	-5.09265
148.0	0.74663	-1.26894	-5.03365
150.0	0.74503	-1.27826	-5.04297
152.0	0.73098	-1.36093	-5.12564
154.0	0.70422	-1.52294	-5.28765
156.0	0.66503	-1.77159	-5.53630
158.0	0.61433	-2.11600	-5.88071
160.0	0.55362	-2.56786	-6.33257
162.0	0.48501	-3.14248	-6.90719
164.0	0.41110	-3.86057	-7.62528
166.0	0.33488	-4.75110	-8.51581
168.0	0.25963	-5.85645	-9.62116
170.0	0.18870	-7.24230	-11.00701
172.0	0.12535	-9.01860	-12.78331
174.0	0.07259	-11.39138	-15.15609
176.0	0.03294	-14.82319	-18.58790
178.0	0.00834	-20.78976	-24.55447
180.	0.00000	-INFINITY	-INFINITY

TABLE A2-16

EPSLN= 4.0000(NORMALIZED GROUNDPLANE RADIUS)
MONOPOLE LENGTH= 0.250000(WAVELENGTH)
MONOPOLE RADIUS= 0.1000E-05(WAVELENGTH)
NUMBER OF MONOPOLE SEGMENTS= 1
NUMBER OF DISK SEGMENTS= 16
RADIATION RESISTANCE DETERMINED BY PATTERN= 0.444339E+02 OHMS

ELEVATION ANGLE (DEG)	DIRECTIVITY (NUMERIC)	DIRECTIVITY (DBI)	RELATIVE POWER (DB)
0.	0. 00000	-INFINITY	-INFINITY
2. 0	0. 01922	-17. 16234	-21. 07366
4. 0	0. 07614	-11. 18380	-15. 09512
6. 0	0. 16858	-7. 73189	-11. 64321
8. 0	0. 29305	-5. 33064	-9. 24196
10. 0	0. 44492	-3. 51720	-7. 42852
12. 0	0. 61871	-2. 08513	-5. 99645
14. 0	0. 80836	-0. 92396	-4. 83528
16. 0	1. 00753	0. 03256	-3. 87876
18. 0	1. 20990	0. 82748	-3. 08384
20. 0	1. 40947	1. 49055	-2. 42077
22. 0	1. 60077	2. 04330	-1. 86802
24. 0	1. 77908	2. 50195	-1. 40937
26. 0	1. 94051	2. 87916	-1. 03216
28. 0	2. 08212	3. 18505	-0. 72627
30. 0	2. 20189	3. 42796	-0. 48336
32. 0	2. 29873	3. 61487	-0. 29645
34. 0	2. 37234	3. 75178	-0. 15954
36. 0	2. 42318	3. 84386	-0. 06746
38. 0	2. 45228	3. 89570	-0. 01563
40. 0	2. 46112	3. 91132	0. 00000
42. 0	2. 45152	3. 89435	-0. 01697
44. 0	2. 42551	3. 84803	-0. 06329
46. 0	2. 38521	3. 77527	-0. 13606
48. 0	2. 33273	3. 67865	-0. 23267
50. 0	2. 27014	3. 56052	-0. 35080
52. 0	2. 19935	3. 42294	-0. 48838
54. 0	2. 12213	3. 26771	-0. 64361
56. 0	2. 04006	3. 09643	-0. 81489
58. 0	1. 95454	2. 91044	-1. 00088
60. 0	1. 86677	2. 71090	-1. 20042
62. 0	1. 77776	2. 49874	-1. 41258
64. 0	1. 68838	2. 27471	-1. 63661
66. 0	1. 59932	2. 03935	-1. 87197
68. 0	1. 51114	1. 79305	-2. 11827
70. 0	1. 42429	1. 53600	-2. 37533
72. 0	1. 33913	1. 26822	-2. 64310
74. 0	1. 25591	0. 98958	-2. 92174
76. 0	1. 17483	0. 69975	-3. 21157
78. 0	1. 09604	0. 39825	-3. 51307
80. 0	1. 01962	0. 08440	-3. 82692
82. 0	0. 94566	-0. 24263	-4. 15395
84. 0	0. 87420	-0. 58389	-4. 49521
86. 0	0. 80527	-0. 94058	-4. 85190
88. 0	0. 73891	-1. 31408	-5. 22540
90. 0	0. 67516	-1. 70594	-5. 61726

ELEVATION ANGLE (DEG)	DIRECTIVITY (NUMERIC)	DIRECTIVITY (DBI)	RELATIVE POWER (DB)
92. 0	0. 61408	-2. 11778	-6. 02910
94. 0	0. 55574	-2. 55127	-6. 46259
96. 0	0. 50027	-3. 00795	-6. 91927
98. 0	0. 44781	-3. 48905	-7. 40037
100. 0	0. 39856	-3. 99505	-7. 90637
102. 0	0. 35276	-4. 52520	-8. 43652
104. 0	0. 31070	-5. 07654	-8. 98786
106. 0	0. 27273	-5. 64265	-9. 55397
108. 0	0. 23923	-6. 21179	-10. 12311
110. 0	0. 21063	-6. 76471	-10. 67603
112. 0	0. 18739	-7. 27253	-11. 18385
114. 0	0. 16996	-7. 69642	-11. 60774
116. 0	0. 15881	-7. 99120	-11. 90252
118. 0	0. 15434	-8. 11509	-12. 02641
120. 0	0. 15691	-8. 04344	-11. 95477
122. 0	0. 16675	-7. 77927	-11. 69059
124. 0	0. 18396	-7. 35270	-11. 26402
126. 0	0. 20845	-6. 80992	-10. 72124
128. 0	0. 23991	-6. 19952	-10. 11084
130. 0	0. 27776	-5. 56326	-9. 47458
132. 0	0. 32115	-4. 93287	-8. 84420
134. 0	0. 36892	-4. 33063	-8. 24196
136. 0	0. 41961	-3. 77151	-7. 68284
138. 0	0. 47148	-3. 26541	-7. 17673
140. 0	0. 52252	-2. 81894	-6. 73026
142. 0	0. 57059	-2. 43679	-6. 34811
144. 0	0. 61340	-2. 12260	-6. 03392
146. 0	0. 64870	-1. 87958	-5. 79090
148. 0	0. 67437	-1. 71102	-5. 62234
150. 0	0. 68856	-1. 62057	-5. 53189
152. 0	0. 68983	-1. 61256	-5. 52389
154. 0	0. 67727	-1. 69236	-5. 60369
156. 0	0. 65062	-1. 86671	-5. 77803
158. 0	0. 61035	-2. 14424	-6. 05557
160. 0	0. 55767	-2. 53621	-6. 44753
162. 0	0. 49459	-3. 05757	-6. 96889
164. 0	0. 42377	-3. 72872	-7. 64004
166. 0	0. 34847	-4. 57837	-8. 48969
168. 0	0. 27235	-5. 64869	-9. 56001
170. 0	0. 19929	-7. 00517	-10. 91649
172. 0	0. 13312	-8. 75762	-12. 66894
174. 0	0. 07741	-11. 11197	-15. 02329
176. 0	0. 03523	-14. 53067	-18. 44199
178. 0	0. 00893	-20. 48940	-24. 40072
180.	0. 00000	-INFINITY	-INFINITY

TABLE A2-17

EPSLN= 4.2500(NORMALIZED GROUNDPLANE RADIUS)
MONOPOLE LENGTH= 0.250000(WAVELENGTH)
MONOPOLE RADIUS= 0.1000E-05(WAVELENGTH)
NUMBER OF MONOPOLE SEGMENTS= 1
NUMBER OF DISK SEGMENTS= 16
RADIATION RESISTANCE DETERMINED BY PATTERN= 0.405777E+02 OHMS

ELEVATION ANGLE (DEG)	DIRECTIVITY (NUMERIC)	DIRECTIVITY (DBI)	RELATIVE POWER (DB)	ELEVATION ANGLE (DEG)	DIRECTIVITY (NUMERIC)	DIRECTIVITY (DBI)	RELATIVE POWER (DB)
0.	0.00000	-INFINITY	-INFINITY				
2.0	0.02096	-16.78670	-20.70236	92.0	0.67953	-1.67794	-5.59359
4.0	0.08294	-10.81236	-14.72802	94.0	0.62146	-2.06586	-5.98152
6.0	0.18334	-7.36741	-11.28307	96.0	0.56516	-2.47829	-6.39395
8.0	0.31799	-4.97582	-8.89148	98.0	0.51069	-2.91842	-6.83407
10.0	0.48143	-3.17464	-7.09030	100.0	0.45818	-3.38965	-7.30531
12.0	0.66722	-1.75729	-5.67295	102.0	0.40780	-3.89551	-7.81117
14.0	0.86833	-0.61317	-4.52883	104.0	0.35981	-4.43932	-8.35498
16.0	1.07749	0.32415	-3.59151	106.0	0.31451	-5.02366	-8.93932
18.0	1.28764	1.09794	-2.81771	108.0	0.27231	-5.64937	-9.56503
20.0	1.49218	1.73820	-2.17746	110.0	0.23368	-6.31385	-10.22951
22.0	1.68528	2.26672	-1.64894	112.0	0.19915	-7.00811	-10.92377
24.0	1.86210	2.70004	-1.21562	114.0	0.16935	-7.71225	-11.62791
26.0	2.01889	3.05113	-0.86452	116.0	0.14490	-8.38925	-12.30491
28.0	2.15302	3.33049	-0.58517	118.0	0.12649	-8.97939	-12.89505
30.0	2.26298	3.54681	-0.36885	120.0	0.11477	-9.40169	-13.31735
32.0	2.34827	3.70749	-0.20817	122.0	0.11034	-9.57249	-13.48815
34.0	2.40929	3.81890	-0.09676	124.0	0.11372	-9.44168	-13.35734
36.0	2.44717	3.88664	-0.02902	126.0	0.12525	-9.02234	-12.93800
38.0	2.46357	3.91566	0.00000	128.0	0.14507	-8.38413	-12.29979
40.0	2.46058	3.91038	-0.00528	130.0	0.17308	-7.61759	-11.53324
42.0	2.44049	3.87476	-0.04089	132.0	0.20882	-6.80221	-10.71787
44.0	2.40566	3.81235	-0.10331	134.0	0.25151	-5.99443	-9.91008
46.0	2.35847	3.72631	-0.18935	136.0	0.29995	-5.22945	-9.14510
48.0	2.30115	3.61944	-0.29621	138.0	0.35257	-4.52757	-8.44323
50.0	2.23575	3.49423	-0.42142	140.0	0.40739	-3.89991	-7.81557
52.0	2.16413	3.35283	-0.56282	142.0	0.46212	-3.35244	-7.26810
54.0	2.08790	3.19710	-0.71856	144.0	0.51422	-2.88853	-6.80419
56.0	2.00845	3.02860	-0.88706	146.0	0.56099	-2.51041	-6.42607
58.0	1.92692	2.84864	-1.06702	148.0	0.59977	-2.22015	-6.13581
60.0	1.84428	2.65827	-1.25739	150.0	0.62804	-2.02016	-5.93581
62.0	1.76129	2.45832	-1.45734	152.0	0.64363	-1.91365	-5.82931
64.0	1.67857	2.24941	-1.66625	154.0	0.64492	-1.90497	-5.82063
66.0	1.59660	2.03197	-1.88369	156.0	0.63096	-2.00000	-5.91566
68.0	1.51574	1.80625	-2.10941	158.0	0.60164	-2.20664	-6.12230
70.0	1.43627	1.57236	-2.34330	160.0	0.55776	-2.53551	-6.45116
72.0	1.35838	1.33022	-2.58544	162.0	0.50107	-3.00104	-6.91670
74.0	1.28222	1.07963	-2.83603	164.0	0.43419	-3.62321	-7.53887
76.0	1.20787	0.82022	-3.09544	166.0	0.36055	-4.43036	-8.34602
78.0	1.13539	0.55145	-3.36421	168.0	0.28416	-5.46438	-9.38003
80.0	1.06479	0.27263	-3.64303	170.0	0.20939	-6.79050	-10.70616
82.0	0.99606	-0.01715	-3.93281	172.0	0.14066	-8.51836	-12.43402
84.0	0.92918	-0.31899	-4.23465	174.0	0.08215	-10.85373	-14.76939
86.0	0.86412	-0.63424	-4.54989	176.0	0.03751	-14.25894	-18.17460
88.0	0.80085	-0.96450	-4.88015	178.0	0.00953	-20.20962	-24.12527
90.0	0.73932	-1.31166	-5.22732	180.	0.00000	-INFINITY	-INFINITY

TABLE A2-18

EPSLN= 4.5000(NORMALIZED GROUNDPLANE RADIUS)
MONOPOLE LENGTH= 0.250000(WAVELENGTH)
MONOPOLE RADIUS= 0.1000E-05(WAVELENGTH)
NUMBER OF MONOPOLE SEGMENTS= 1
NUMBER OF DISK SEGMENTS= 16
RADIATION RESISTANCE DETERMINED BY PATTERN= 0.371323E+02 OHMS

ELEVATION ANGLE (DEG)	DIRECTIVITY (NUMERIC)	DIRECTIVITY (DBI)	RELATIVE POWER (DB)	ELEVATION ANGLE (DEG)	DIRECTIVITY (NUMERIC)	DIRECTIVITY (DBI)	RELATIVE POWER (DB)
0.	0.00000	-INFINITY	-INFINITY				
2.0	0.02248	-16.48123	-20.31094	92.0	0.75173	-1.23939	-5.06911
4.0	0.08888	-10.51173	-14.34144	94.0	0.69652	-1.57066	-5.40037
6.0	0.19612	-7.07480	-10.90451	96.0	0.64224	-1.92304	-5.75275
8.0	0.33929	-4.69432	-8.52403	98.0	0.58885	-2.29999	-6.12970
10.0	0.51201	-2.90722	-6.73694	100.0	0.53635	-2.70554	-6.53525
12.0	0.70684	-1.50679	-5.33651	102.0	0.48481	-3.14433	-6.97404
14.0	0.91575	-0.38223	-4.21194	104.0	0.43435	-3.62162	-7.45133
16.0	1.13061	0.53313	-3.29659	106.0	0.38519	-4.14325	-7.97296
18.0	1.34364	1.28282	-2.54689	108.0	0.33764	-4.71546	-8.54518
20.0	1.54780	1.89715	-1.93256	110.0	0.29211	-5.34449	-9.17420
22.0	1.73711	2.39827	-1.43144	112.0	0.24914	-6.03564	-9.86535
24.0	1.90683	2.80313	-1.02659	114.0	0.20935	-6.79136	-10.62107
26.0	2.05358	3.12512	-0.70459	116.0	0.17349	-7.60736	-11.43707
28.0	2.17531	3.37522	-0.45450	118.0	0.14239	-8.46530	-12.29501
30.0	2.27124	3.56263	-0.26708	120.0	0.11694	-9.32028	-13.14999
32.0	2.34167	3.69527	-0.13445	122.0	0.09807	-10.08465	-13.91436
34.0	2.38783	3.78003	-0.04968	124.0	0.08666	-10.62175	-14.45146
36.0	2.41159	3.82303	-0.00668	126.0	0.08352	-10.78194	-14.61165
38.0	2.41530	3.82971	0.00000	128.0	0.08930	-10.49129	-14.32100
40.0	2.40156	3.80494	-0.02477	130.0	0.10442	-9.81220	-13.64191
42.0	2.37304	3.75306	-0.07666	132.0	0.12897	-8.89514	-12.72485
44.0	2.33233	3.67790	-0.15181	134.0	0.16267	-7.88704	-11.71675
46.0	2.28184	3.58285	-0.24686	136.0	0.20477	-6.88743	-10.71715
48.0	2.22373	3.47083	-0.35889	138.0	0.25403	-5.95120	-9.78091
50.0	2.15990	3.34433	-0.48539	140.0	0.30869	-5.10473	-8.93444
52.0	2.09191	3.20543	-0.62429	142.0	0.36651	-4.35913	-8.18885
54.0	2.02107	3.05582	-0.77389	144.0	0.42479	-3.71825	-7.54797
56.0	1.94844	2.89686	-0.93285	146.0	0.48052	-3.18292	-7.01263
58.0	1.87482	2.72959	-1.10013	148.0	0.53050	-2.75318	-6.58289
60.0	1.80085	2.55476	-1.27495	150.0	0.57156	-2.42942	-6.25913
62.0	1.72701	2.37294	-1.45677	152.0	0.60076	-2.21301	-6.04272
64.0	1.65366	2.18447	-1.64524	154.0	0.61564	-2.10670	-5.93641
66.0	1.58109	1.98956	-1.84015	156.0	0.61446	-2.11504	-5.94475
68.0	1.50949	1.78829	-2.04142	158.0	0.59637	-2.24482	-6.07453
70.0	1.43901	1.58065	-2.24907	160.0	0.56160	-2.50576	-6.33548
72.0	1.36978	1.36651	-2.46320	162.0	0.51150	-2.91157	-6.74128
74.0	1.30187	1.14568	-2.68403	164.0	0.44858	-3.48158	-7.31130
76.0	1.23534	0.91786	-2.91185	166.0	0.37638	-4.24368	-8.07339
78.0	1.17022	0.68266	-3.14705	168.0	0.29927	-5.23934	-9.06905
80.0	1.10651	0.43954	-3.39017	170.0	0.22215	-6.53348	-10.36320
82.0	1.04420	0.18782	-3.64189	172.0	0.15012	-8.23549	-12.06520
84.0	0.98325	-0.07338	-3.90310	174.0	0.08809	-10.55094	-14.38066
86.0	0.92360	-0.34517	-4.17488	176.0	0.04035	-13.94203	-17.77174
88.0	0.86518	-0.62891	-4.45862	178.0	0.01027	-19.88426	-23.71398
90.0	0.80792	-0.92630	-4.75602	180.	0.00000	-INFINITY	-INFINITY

TABLE A2-19

EPSLN= 4. 7500(NORMALIZED GROUNDPLANE RADIUS)
MONOPOLE LENGTH= 0. 250000(WAVELENGTH)
MONOPOLE RADIUS= 0. 1000E-05(WAVELENGTH)
NUMBER OF MONOPOLE SEGMENTS= 1
NUMBER OF DISK SEGMENTS= 14
RADIATION RESISTANCE DETERMINED BY PATTERN= 0. 344626E+02 OHMS

ELEVATION ANGLE (DEG)	DIRECTIVITY (NUMERIC)	DIRECTIVITY (DBI)	RELATIVE POWER (DB)	ELEVATION ANGLE (DEG)	DIRECTIVITY (NUMERIC)	DIRECTIVITY (DBI)	RELATIVE POWER (DB)
0.	0. 00000	-INFINITY	-INFINITY				
2. 0	0. 02388	-16. 21896	-19. 89249	92. 0	0. 81904	-0. 86694	-4. 54047
4. 0	0. 09430	-10. 25509	-13. 92862	94. 0	0. 76822	-1. 14513	-4. 81867
6. 0	0. 20761	-6. 82747	-10. 50101	96. 0	0. 71789	-1. 43943	-5. 11296
8. 0	0. 35811	-4. 45989	-8. 13343	98. 0	0. 66789	-1. 75296	-5. 42649
10. 0	0. 53837	-2. 68916	-6. 36270	100. 0	0. 61808	-2. 08952	-5. 76306
12. 0	0. 73988	-1. 30836	-4. 98189	102. 0	0. 56838	-2. 45364	-6. 12717
14. 0	0. 95357	-0. 20646	-3. 87999	104. 0	0. 51872	-2. 85065	-6. 52418
16. 0	1. 17044	0. 68349	-2. 99005	106. 0	0. 46916	-3. 28680	-6. 96033
18. 0	1. 38209	1. 40537	-2. 26816	108. 0	0. 41982	-3. 76934	-7. 44287
20. 0	1. 58121	1. 98989	-1. 68365	110. 0	0. 37097	-4. 30657	-7. 98011
22. 0	1. 76185	2. 45968	-1. 21385	112. 0	0. 32301	-4. 90788	-8. 58141
24. 0	1. 91964	2. 83221	-0. 84132	114. 0	0. 27648	-5. 58341	-9. 25694
26. 0	2. 05186	3. 12149	-0. 55205	116. 0	0. 23210	-6. 34332	-10. 01686
28. 0	2. 15731	3. 33912	-0. 33441	118. 0	0. 19074	-7. 19564	-10. 86917
30. 0	2. 23615	3. 49502	-0. 17852	120. 0	0. 15342	-8. 14117	-11. 81470
32. 0	2. 28970	3. 59778	-0. 07575	122. 0	0. 12128	-9. 16204	-12. 83558
34. 0	2. 32008	3. 65503	-0. 01851	124. 0	0. 09553	-10. 19849	-13. 87202
36. 0	2. 32999	3. 67353	0. 00000	126. 0	0. 07740	-11. 11287	-14. 78640
38. 0	2. 32239	3. 65935	-0. 01418	128. 0	0. 06803	-11. 67327	-15. 34680
40. 0	2. 30030	3. 61784	-0. 05569	130. 0	0. 06842	-11. 64825	-15. 32179
42. 0	2. 26659	3. 55373	-0. 11981	132. 0	0. 07930	-11. 00718	-14. 68071
44. 0	2. 22386	3. 47107	-0. 20247	134. 0	0. 10102	-9. 95574	-13. 62927
46. 0	2. 17435	3. 37329	-0. 30024	136. 0	0. 13345	-8. 74674	-12. 42028
48. 0	2. 11993	3. 26321	-0. 41032	138. 0	0. 17587	-7. 54799	-11. 22152
50. 0	2. 06207	3. 14304	-0. 53049	140. 0	0. 22694	-6. 44096	-10. 11450
52. 0	2. 00192	3. 01446	-0. 65907	142. 0	0. 28462	-5. 45737	-9. 13090
54. 0	1. 94030	2. 87868	-0. 79485	144. 0	0. 34626	-4. 60601	-8. 27954
56. 0	1. 87780	2. 73649	-0. 93704	146. 0	0. 40863	-3. 88666	-7. 56019
58. 0	1. 81484	2. 58838	-1. 08516	148. 0	0. 46812	-3. 29645	-6. 96998
60. 0	1. 75169	2. 43457	-1. 23896	150. 0	0. 52089	-2. 83257	-6. 50610
62. 0	1. 68855	2. 27514	-1. 39840	152. 0	0. 56318	-2. 49352	-6. 16706
64. 0	1. 62557	2. 11005	-1. 56349	154. 0	0. 59161	-2. 27964	-5. 95317
66. 0	1. 56287	1. 93922	-1. 73431	156. 0	0. 60346	-2. 19349	-5. 86702
68. 0	1. 50057	1. 76256	-1. 91097	158. 0	0. 59700	-2. 24026	-5. 91380
70. 0	1. 43879	1. 57999	-2. 09355	160. 0	0. 57169	-2. 42841	-6. 10195
72. 0	1. 37766	1. 39142	-2. 28211	162. 0	0. 52837	-2. 77062	-6. 44415
74. 0	1. 31729	1. 19682	-2. 47672	164. 0	0. 46931	-3. 28541	-6. 95894
76. 0	1. 25780	0. 99610	-2. 67743	166. 0	0. 39811	-4. 00000	-7. 67353
78. 0	1. 19928	0. 78919	-2. 88434	168. 0	0. 31949	-4. 95537	-8. 62890
80. 0	1. 14180	0. 57591	-3. 09762	170. 0	0. 23900	-6. 21603	-9. 88956
82. 0	1. 08542	0. 35598	-3. 31755	172. 0	0. 16252	-7. 89102	-11. 56456
84. 0	1. 03015	0. 12899	-3. 54455	174. 0	0. 09581	-10. 18571	-13. 85924
86. 0	0. 97595	-0. 10573	-3. 77926	176. 0	0. 04403	-13. 56208	-17. 23561
88. 0	0. 92277	-0. 34907	-4. 02260	178. 0	0. 01123	-19. 49553	-23. 16907
90. 0	0. 87051	-0. 60227	-4. 27580	180.	0. 00000	-INFINITY	-INFINITY

TABLE A2-20

EPSLN= 5.0000(NORMALIZED GROUNDPLANE RADIUS)
MONOPOLE LENGTH= 0.250000(WAVELENGTH)
MONOPOLE RADIUS= 0.1000E-05(WAVELENGTH)
NUMBER OF MONOPOLE SEGMENTS= 1
NUMBER OF DISK SEGMENTS= 16
RADIATION RESISTANCE DETERMINED BY PATTERN= 0.326839E+02 OHMS

ELEVATION ANGLE (DEG)	DIRECTIVITY (NUMERIC)	DIRECTIVITY (DBI)	RELATIVE POWER (DB)	ELEVATION ANGLE (DEG)	DIRECTIVITY (NUMERIC)	DIRECTIVITY (DBI)	RELATIVE POWER (DB)
0.	0.00000	-INFINITY	-INFINITY				
2.0	0.02533	-15.96361	-19.41636	92.0	0.87123	-0.59866	-4.05141
4.0	0.09985	-10.00660	-13.45935	94.0	0.82541	-0.83330	-4.28606
6.0	0.21926	-6.59037	-10.04312	96.0	0.78017	-1.07812	-4.53087
8.0	0.37683	-4.23855	-7.69130	98.0	0.73523	-1.33579	-4.78854
10.0	0.56392	-2.48781	-5.94056	100.0	0.69029	-1.60966	-5.06241
12.0	0.77073	-1.13099	-4.58375	102.0	0.64508	-1.90390	-5.35665
14.0	0.98701	-0.05678	-3.50954	104.0	0.59931	-2.22349	-5.67624
16.0	1.20286	0.80214	-2.65061	106.0	0.55279	-2.57440	-6.02715
18.0	1.40933	1.49014	-1.96262	108.0	0.50540	-2.96367	-6.41642
20.0	1.59899	2.03845	-1.41430	110.0	0.45713	-3.39959	-6.85234
22.0	1.76620	2.47039	-0.98236	112.0	0.40814	-3.89187	-7.34462
24.0	1.90730	2.80418	-0.64857	114.0	0.35877	-4.45188	-7.90463
26.0	2.02055	3.05469	-0.39806	116.0	0.30954	-5.09277	-8.54552
28.0	2.10596	3.23450	-0.21825	118.0	0.26125	-5.82948	-9.28223
30.0	2.16497	3.35452	-0.09823	120.0	0.21488	-6.67805	-10.13080
32.0	2.20008	3.42439	-0.02836	122.0	0.17167	-7.65299	-11.10574
34.0	2.21450	3.45275	0.00000	124.0	0.13305	-8.75972	-12.21247
36.0	2.21172	3.44730	-0.00546	126.0	0.10060	-9.97418	-13.42693
38.0	2.19525	3.41483	-0.03792	128.0	0.07594	-11.19539	-14.64814
40.0	2.16834	3.36128	-0.09147	130.0	0.06068	-12.16948	-15.62223
42.0	2.13384	3.29161	-0.16114	132.0	0.05626	-12.49812	-15.95087
44.0	2.09404	3.20986	-0.24289	134.0	0.06379	-11.95233	-15.40508
46.0	2.05074	3.11911	-0.33364	136.0	0.08393	-10.76061	-14.21336
48.0	2.00519	3.02156	-0.43120	138.0	0.11672	-9.32869	-12.78144
50.0	1.95820	2.91858	-0.53417	140.0	0.16142	-7.92048	-11.37323
52.0	1.91024	2.81089	-0.64187	142.0	0.21648	-6.64590	-10.09865
54.0	1.86150	2.69863	-0.75412	144.0	0.27945	-5.53698	-8.98974
56.0	1.81200	2.58158	-0.87117	146.0	0.34705	-4.59602	-8.04877
58.0	1.76169	2.45930	-0.99345	148.0	0.41531	-3.81633	-7.26908
60.0	1.71049	2.33121	-1.12154	150.0	0.47972	-3.19016	-6.64291
62.0	1.65835	2.19677	-1.25598	152.0	0.53561	-2.71148	-6.16423
64.0	1.60529	2.05553	-1.39722	154.0	0.57851	-2.37690	-5.82965
66.0	1.55137	1.90715	-1.54560	156.0	0.60450	-2.18603	-5.63878
68.0	1.49675	1.75150	-1.70126	158.0	0.61069	-2.14177	-5.59452
70.0	1.44164	1.58858	-1.86417	160.0	0.59556	-2.25077	-5.70352
72.0	1.38630	1.41858	-2.03417	162.0	0.55920	-2.52430	-5.97705
74.0	1.33101	1.24182	-2.21093	164.0	0.50352	-2.97981	-6.43257
76.0	1.27606	1.05870	-2.39405	166.0	0.43215	-3.64367	-7.09642
78.0	1.22171	0.86968	-2.58307	168.0	0.35025	-4.55617	-8.00892
80.0	1.16821	0.67520	-2.77755	170.0	0.26416	-5.78130	-9.23405
82.0	1.11574	0.47564	-2.97711	172.0	0.18081	-7.42772	-10.88047
84.0	1.06445	0.27123	-3.18152	174.0	0.10714	-9.70047	-13.15322
86.0	1.01438	0.06202	-3.39073	176.0	0.04942	-13.06132	-16.51407
88.0	0.96555	-0.15224	-3.60499	178.0	0.01263	-18.98553	-22.43828
90.0	0.91788	-0.37213	-3.82488	180.	0.00000	-INFINITY	-INFINITY

TABLE A2-21

EPSLN= 5. 2500(NORMALIZED GROUNDPLANE RADIUS)
MONOPOLE LENGTH= 0. 250000(WAVELENGTH)
MONOPOLE RADIUS= 0. 1000E-05(WAVELENGTH)
NUMBER OF MONOPOLE SEGMENTS= 1
NUMBER OF DISK SEGMENTS= 16
RADIATION RESISTANCE DETERMINED BY PATTERN= 0. 317052E+02 OHMS

ELEVATION ANGLE (DEG)	DIRECTIVITY (NUMERIC)	DIRECTIVITY (DBI)	RELATIVE POWER (DB)	ELEVATION ANGLE (DEG)	DIRECTIVITY (NUMERIC)	DIRECTIVITY (DBI)	RELATIVE POWER (DB)
0.	0. 00000	-INFINITY	-INFINITY				
2. 0	0. 02678	-15. 72116	-18. 88373	92. 0	0. 90345	-0. 44093	-3. 60351
4. 0	0. 10538	-9. 77247	-12. 93504	94. 0	0. 86224	-0. 64370	-3. 80628
6. 0	0. 23067	-6. 37000	-9. 53258	96. 0	0. 82227	-0. 84984	-4. 01241
8. 0	0. 39470	-4. 03730	-7. 19987	98. 0	0. 78316	-1. 06150	-4. 22408
10. 0	0. 58738	-2. 31081	-5. 47338	100. 0	0. 74445	-1. 28163	-4. 44421
12. 0	0. 79742	-0. 98312	-4. 14569	102. 0	0. 70566	-1. 51402	-4. 67659
14. 0	1. 01332	0. 05745	-3. 10513	104. 0	0. 66629	-1. 76335	-4. 92592
16. 0	1. 22425	0. 87869	-2. 28388	106. 0	0. 62585	-2. 03530	-5. 19788
18. 0	1. 42088	1. 52557	-1. 63701	108. 0	0. 58390	-2. 33662	-5. 49920
20. 0	1. 59592	2. 03010	-1. 13247	110. 0	0. 54010	-2. 67526	-5. 83783
22. 0	1. 74443	2. 41652	-0. 74605	112. 0	0. 49425	-3. 06053	-6. 22311
24. 0	1. 86388	2. 70419	-0. 45838	114. 0	0. 44633	-3. 50341	-6. 66598
26. 0	1. 95401	2. 90927	-0. 25330	116. 0	0. 39657	-4. 01681	-7. 17938
28. 0	2. 01641	3. 04580	-0. 11678	118. 0	0. 34546	-4. 61606	-7. 77864
30. 0	2. 05410	3. 12622	-0. 03635	120. 0	0. 29381	-5. 31935	-8. 48192
32. 0	2. 07099	3. 16178	-0. 00079	122. 0	0. 24277	-6. 14797	-9. 31054
34. 0	2. 07137	3. 16257	0. 00000	124. 0	0. 19384	-7. 12566	-10. 28823
36. 0	2. 05947	3. 13755	-0. 02503	126. 0	0. 14878	-8. 27454	-11. 43711
38. 0	2. 03910	3. 09439	-0. 06818	128. 0	0. 10964	-9. 60034	-12. 76292
40. 0	2. 01346	3. 03942	-0. 12315	130. 0	0. 07858	-11. 04672	-14. 20929
42. 0	1. 98493	2. 97746	-0. 18511	132. 0	0. 05778	-12. 38202	-15. 54459
44. 0	1. 95516	2. 91182	-0. 25076	134. 0	0. 04923	-13. 07751	-16. 24009
46. 0	1. 92503	2. 84438	-0. 31819	136. 0	0. 05454	-12. 63283	-15. 79540
48. 0	1. 89486	2. 77577	-0. 38681	138. 0	0. 07471	-11. 26611	-14. 42868
50. 0	1. 86448	2. 70559	-0. 45698	140. 0	0. 10993	-9. 58874	-12. 75132
52. 0	1. 83347	2. 63274	-0. 52984	142. 0	0. 15939	-7. 97552	-11. 13809
54. 0	1. 80123	2. 55569	-0. 60688	144. 0	0. 22113	-6. 55359	-9. 71616
56. 0	1. 76716	2. 47276	-0. 68981	146. 0	0. 29205	-5. 34543	-8. 50800
58. 0	1. 73075	2. 38234	-0. 78024	148. 0	0. 36796	-4. 34201	-7. 50458
60. 0	1. 69162	2. 28303	-0. 87954	150. 0	0. 44377	-3. 52843	-6. 69100
62. 0	1. 64960	2. 17379	-0. 98879	152. 0	0. 51384	-2. 89171	-6. 05429
64. 0	1. 60469	2. 05391	-1. 10866	154. 0	0. 57242	-2. 42286	-5. 58544
66. 0	1. 55708	1. 92311	-1. 23946	156. 0	0. 61416	-2. 11716	-5. 27974
68. 0	1. 50712	1. 78148	-1. 38109	158. 0	0. 63471	-1. 97423	-5. 13681
70. 0	1. 45527	1. 62944	-1. 53314	160. 0	0. 63121	-1. 99826	-5. 16084
72. 0	1. 40207	1. 46771	-1. 69487	162. 0	0. 60275	-2. 19865	-5. 36123
74. 0	1. 34811	1. 29725	-1. 86533	164. 0	0. 55063	-2. 59143	-5. 75400
76. 0	1. 29396	1. 11920	-2. 04338	166. 0	0. 47843	-3. 20185	-6. 36443
78. 0	1. 24017	0. 93481	-2. 22777	168. 0	0. 39180	-4. 06939	-7. 23197
80. 0	1. 18723	0. 74536	-2. 41722	170. 0	0. 29803	-5. 25740	-8. 41997
82. 0	1. 13556	0. 55210	-2. 61048	172. 0	0. 20540	-6. 87402	-10. 03660
84. 0	1. 08546	0. 35616	-2. 80642	174. 0	0. 12235	-9. 12395	-12. 28652
86. 0	1. 03716	0. 15846	-3. 00411	176. 0	0. 05664	-12. 46868	-15. 63125
88. 0	0. 99075	-0. 04037	-3. 20294	178. 0	0. 01451	-18. 38327	-21. 54585
90. 0	0. 94622	-0. 24009	-3. 40266	180.	0. 00000	-INFINITY	-INFINITY

TABLE A2-22

EPSLN= 5.5000(NORMALIZED GROUNDPLANE RADIUS)
MONOPOLE LENGTH= 0.250000(WAVELENGTH)
MONOPOLE RADIUS= 0.1000E-05(WAVELENGTH)
NUMBER OF MONOPOLE SEGMENTS= 1
NUMBER OF DISK SEGMENTS= 16
RADIATION RESISTANCE DETERMINED BY PATTERN= 0.315339E+02 OHMS

ELEVATION ANGLE (DEG)	DIRECTIVITY (NUMERIC)	DIRECTIVITY (DBI)	RELATIVE POWER (DB)	ELEVATION ANGLE (DEG)	DIRECTIVITY (NUMERIC)	DIRECTIVITY (DBI)	RELATIVE POWER (DB)
0.	0.00000	-INFINITY	-INFINITY				
2.0	0.02821	-15.49576	-18.27460	92.0	0.91052	-0.40711	-3.18596
4.0	0.11073	-9.55737	-12.33622	94.0	0.87253	-0.59218	-3.37103
6.0	0.24143	-6.17201	-8.95085	96.0	0.83706	-0.77243	-3.55128
8.0	0.41086	-3.86307	-6.64191	98.0	0.80364	-0.94938	-3.72823
10.0	0.60718	-2.16681	-4.94566	100.0	0.77170	-1.12552	-3.90437
12.0	0.81743	-0.87552	-3.65437	102.0	0.74057	-1.30433	-4.08318
14.0	1.02871	0.12291	-2.65593	104.0	0.70952	-1.49036	-4.26920
16.0	1.22939	0.89689	-1.88196	106.0	0.67777	-1.68920	-4.46804
18.0	1.41000	1.49219	-1.28665	108.0	0.64453	-1.90757	-4.68642
20.0	1.56385	1.94194	-0.83691	110.0	0.60906	-2.15337	-4.93221
22.0	1.68724	2.27176	-0.50708	112.0	0.57072	-2.43574	-5.21458
24.0	1.77942	2.50278	-0.27607	114.0	0.52902	-2.76528	-5.54413
26.0	1.84215	2.65325	-0.12559	116.0	0.48369	-3.15436	-5.93321
28.0	1.87915	2.73961	-0.03924	118.0	0.43476	-3.61749	-6.39634
30.0	1.89534	2.77688	-0.00197	120.0	0.38265	-4.17197	-6.95082
32.0	1.89620	2.77885	0.00000	122.0	0.32819	-4.83873	-7.61757
34.0	1.88706	2.75785	-0.02100	124.0	0.27269	-5.64327	-8.42211
36.0	1.87261	2.72448	-0.05437	126.0	0.21795	-6.61646	-9.39531
38.0	1.85659	2.68716	-0.09169	128.0	0.16621	-7.79338	-10.57222
40.0	1.84157	2.65187	-0.12698	130.0	0.12012	-9.20401	-11.98286
42.0	1.82895	2.62201	-0.15684	132.0	0.08254	-10.83354	-13.61238
44.0	1.81908	2.59852	-0.18032	134.0	0.05640	-12.48701	-15.26586
46.0	1.81147	2.58032	-0.19853	136.0	0.04444	-13.52234	-16.30119
48.0	1.80501	2.56479	-0.21405	138.0	0.04888	-13.10906	-15.88790
50.0	1.79823	2.54846	-0.23039	140.0	0.07113	-11.47939	-14.25824
52.0	1.78958	2.52751	-0.25134	142.0	0.11151	-9.52691	-12.30575
54.0	1.77757	2.49828	-0.28057	144.0	0.16894	-7.72257	-10.50142
56.0	1.76099	2.45758	-0.32127	146.0	0.24084	-6.18264	-8.96148
58.0	1.73894	2.40285	-0.37599	148.0	0.32307	-4.90708	-7.68593
60.0	1.71092	2.33229	-0.44656	150.0	0.41006	-3.87153	-6.65037
62.0	1.67679	2.24479	-0.53405	152.0	0.49520	-3.05220	-5.83104
64.0	1.63680	2.13995	-0.63889	154.0	0.57130	-2.43138	-5.21023
66.0	1.59147	2.01798	-0.76087	156.0	0.63126	-1.99791	-4.77675
68.0	1.54156	1.87960	-0.89925	158.0	0.66885	-1.74669	-4.52554
70.0	1.48799	1.72600	-1.05285	160.0	0.67944	-1.67849	-4.45734
72.0	1.43177	1.55872	-1.22013	162.0	0.66066	-1.80020	-4.57904
74.0	1.37392	1.37962	-1.39923	164.0	0.61292	-2.12595	-4.90479
76.0	1.31546	1.19077	-1.58808	166.0	0.53956	-2.67962	-5.45847
78.0	1.25730	0.99441	-1.78444	168.0	0.44672	-3.49962	-6.27847
80.0	1.20029	0.79285	-1.98600	170.0	0.34289	-4.64851	-7.42736
82.0	1.14510	0.58842	-2.19043	172.0	0.23802	-6.23383	-9.01268
84.0	1.09229	0.38337	-2.39548	174.0	0.14257	-8.45982	-11.23867
86.0	1.04226	0.17975	-2.59910	176.0	0.06626	-11.78767	-14.56651
88.0	0.99525	-0.02068	-2.79953	178.0	0.01701	-17.69222	-20.47107
90.0	0.95136	-0.21657	-2.99541	180.	0.00000	-INFINITY	-INFINITY

TABLE A2-23

EPSLN= 5.7500(NORMALIZED GROUNDPLANE RADIUS)
MONOPOLE LENGTH= 0.250000(WAVELENGTH)
MONOPOLE RADIUS= 0.1000E-05(WAVELENGTH)
NUMBER OF MONOPOLE SEGMENTS= 1
NUMBER OF DISK SEGMENTS= 16
RADIATION RESISTANCE DETERMINED BY PATTERN= 0.322632E+02 OHMS

ELEVATION ANGLE (DEG)	DIRECTIVITY (NUMERIC)	DIRECTIVITY (DBI)	RELATIVE POWER (DB)	ELEVATION ANGLE (DEG)	DIRECTIVITY (NUMERIC)	DIRECTIVITY (DBI)	RELATIVE POWER (DB)
0.	0.00000	-INFINITY	-INFINITY				
2.0	0.02945	-15.30901	-17.89665	92.0	0.88786	-0.51658	-3.10422
4.0	0.11525	-9.38361	-11.97125	94.0	0.85054	-0.70303	-3.29067
6.0	0.25004	-6.01983	-8.60747	96.0	0.81760	-0.87461	-3.46225
8.0	0.42257	-3.74099	-6.32863	98.0	0.78853	-1.03182	-3.61946
10.0	0.61899	-2.08319	-4.67083	100.0	0.76270	-1.17643	-3.76407
12.0	0.82444	-0.83842	-3.42606	102.0	0.73934	-1.31158	-3.89922
14.0	1.02469	0.10592	-2.48172	104.0	0.71751	-1.44174	-4.02938
16.0	1.20752	0.81895	-1.76869	106.0	0.69619	-1.57272	-4.16036
18.0	1.36382	1.34758	-1.24006	108.0	0.67428	-1.71163	-4.29927
20.0	1.48818	1.72654	-0.86110	110.0	0.65061	-1.86679	-4.45443
22.0	1.57896	1.98371	-0.60393	112.0	0.62406	-2.04776	-4.63540
24.0	1.63800	2.14314	-0.44450	114.0	0.59356	-2.26536	-4.85300
26.0	1.66985	2.22677	-0.36087	116.0	0.55823	-2.53187	-5.11951
28.0	1.68087	2.25534	-0.33230	118.0	0.51745	-2.86130	-5.44894
30.0	1.67822	2.24849	-0.33915	120.0	0.47098	-3.26998	-5.85762
32.0	1.66891	2.22433	-0.36331	122.0	0.41906	-3.77728	-6.36492
34.0	1.65903	2.19855	-0.38909	124.0	0.36251	-4.40681	-6.99445
36.0	1.65324	2.18336	-0.40428	126.0	0.30283	-5.18808	-7.77572
38.0	1.65445	2.18655	-0.40109	128.0	0.24219	-6.15845	-8.74609
40.0	1.66383	2.21109	-0.37655	130.0	0.18346	-7.36466	-9.95230
42.0	1.68095	2.25555	-0.33209	132.0	0.13006	-8.85865	-11.44629
44.0	1.70412	2.31501	-0.27263	134.0	0.08581	-10.66465	-13.25229
46.0	1.73079	2.38244	-0.20520	136.0	0.05464	-12.62461	-15.21225
48.0	1.75793	2.45001	-0.13763	138.0	0.04024	-13.95346	-16.54110
50.0	1.78246	2.51019	-0.07745	140.0	0.04560	-13.41010	-15.99774
52.0	1.80153	2.55642	-0.03122	142.0	0.07261	-11.39002	-13.97766
54.0	1.81280	2.58349	-0.00415	144.0	0.12158	-9.15121	-11.73885
56.0	1.81453	2.58764	0.00000	146.0	0.19095	-7.19078	-9.77842
58.0	1.80569	2.56644	-0.02120	148.0	0.27705	-5.57442	-8.16206
60.0	1.78593	2.51864	-0.06900	150.0	0.37416	-4.26942	-6.85706
62.0	1.75550	2.44400	-0.14364	152.0	0.47478	-3.23504	-5.82268
64.0	1.71518	2.34309	-0.24455	154.0	0.57019	-2.43977	-5.02741
66.0	1.66614	2.21712	-0.37052	156.0	0.65124	-1.86258	-4.45022
68.0	1.60984	2.06783	-0.51981	158.0	0.70933	-1.49153	-4.07917
70.0	1.54788	1.89738	-0.69026	160.0	0.73747	-1.32255	-3.91019
72.0	1.48193	1.70828	-0.87936	162.0	0.73129	-1.35912	-3.94676
74.0	1.41361	1.50331	-1.08433	164.0	0.68977	-1.61295	-4.20059
76.0	1.34447	1.28550	-1.30214	166.0	0.61573	-2.10609	-4.69373
78.0	1.27588	1.05810	-1.52954	168.0	0.51575	-2.87563	-5.46327
80.0	1.20906	0.82449	-1.76315	170.0	0.39965	-3.98315	-6.57079
82.0	1.14503	0.58816	-1.99948	172.0	0.27955	-5.53547	-8.12311
84.0	1.08458	0.35263	-2.23501	174.0	0.16841	-7.73628	-10.32392
86.0	1.02834	0.12135	-2.46629	176.0	0.07859	-11.04640	-13.63404
88.0	0.97669	-0.10244	-2.69008	178.0	0.02023	-16.94042	-19.52806
90.0	0.92985	-0.31586	-2.90350	180.	0.00000	-INFINITY	-INFINITY

TABLE A2-24

EPSLN= 6.0000(NORMALIZED GROUNDPLANE RADIUS)
MONOPOLE LENGTH= 0.250000(WAVELENGTH)
MONOPOLE RADIUS= 0.1000E-05(WAVELENGTH)
NUMBER OF MONOPOLE SEGMENTS= 1
NUMBER OF DISK SEGMENTS= 16
RADIATION RESISTANCE DETERMINED BY PATTERN= 0.340436E+02 OHMS

ELEVATION ANGLE (DEG)	DIRECTIVITY (NUMERIC)	DIRECTIVITY (DBI)	RELATIVE POWER (DB)	ELEVATION ANGLE (DEG)	DIRECTIVITY (NUMERIC)	DIRECTIVITY (DBI)	RELATIVE POWER (DB)
0.	0.00000	-INFINITY	-INFINITY	92.0	0.83394	-0.78863	-3.68598
2.0	0.03020	-15.19936	-18.09671	94.0	0.79352	-1.00440	-3.90175
4.0	0.11775	-9.29050	-12.18785	96.0	0.75971	-1.19353	-4.09088
6.0	0.25384	-5.95434	-8.85169	98.0	0.73207	-1.35449	-4.25185
8.0	0.42518	-3.71423	-6.61158	100.0	0.70999	-1.48750	-4.38486
10.0	0.61571	-2.10624	-5.00360	102.0	0.69267	-1.59473	-4.49208
12.0	0.80868	-0.92223	-3.81959	104.0	0.67914	-1.68039	-4.57774
14.0	0.98873	-0.04921	-2.94657	106.0	0.66824	-1.75067	-4.64803
16.0	1.14361	0.58278	-2.31458	108.0	0.65863	-1.81357	-4.71092
18.0	1.26535	1.02212	-1.87523	110.0	0.64884	-1.87861	-4.77596
20.0	1.35083	1.30600	-1.59136	112.0	0.63729	-1.95663	-4.85399
22.0	1.40154	1.46605	-1.43131	114.0	0.62235	-2.05966	-4.95702
24.0	1.42289	1.53171	-1.36564	116.0	0.60244	-2.20083	-5.09818
26.0	1.42301	1.53208	-1.36527	118.0	0.57616	-2.39455	-5.29191
28.0	1.41137	1.49642	-1.40094	120.0	0.54239	-2.65689	-5.55425
30.0	1.39742	1.45327	-1.44408	122.0	0.50047	-3.00621	-5.90356
32.0	1.38940	1.42826	-1.46910	124.0	0.45039	-3.46416	-6.36151
34.0	1.39348	1.44102	-1.45634	126.0	0.39289	-4.05726	-6.95462
36.0	1.41333	1.50244	-1.39492	128.0	0.32967	-4.81925	-7.71660
38.0	1.44996	1.61356	-1.28380	130.0	0.26337	-5.79441	-8.69176
40.0	1.50198	1.76666	-1.13070	132.0	0.19761	-7.04181	-9.93917
42.0	1.56610	1.94819	-0.94916	134.0	0.13688	-8.63672	-11.53407
44.0	1.63768	2.14228	-0.75507	136.0	0.08619	-10.64532	-13.54268
46.0	1.71142	2.33358	-0.56378	138.0	0.05078	-12.94283	-15.84019
48.0	1.78199	2.50906	-0.38830	140.0	0.03552	-14.49485	-17.39221
50.0	1.84449	2.65876	-0.23860	142.0	0.04433	-13.53310	-16.43045
52.0	1.89485	2.77574	-0.12162	144.0	0.07951	-10.99570	-13.89305
54.0	1.93005	2.85569	-0.04167	146.0	0.14119	-8.50207	-11.39942
56.0	1.94824	2.89642	-0.00094	148.0	0.22684	-6.44287	-9.34023
58.0	1.94866	2.89735	0.00000	150.0	0.33112	-4.80015	-7.69750
60.0	1.93157	2.85910	-0.03825	152.0	0.44602	-3.50647	-6.40382
62.0	1.89807	2.78313	-0.11423	154.0	0.56137	-2.50750	-5.40486
64.0	1.84990	2.67149	-0.22587	156.0	0.66578	-1.76669	-4.66404
66.0	1.78924	2.52668	-0.37067	158.0	0.74786	-1.26181	-4.15917
68.0	1.71851	2.35152	-0.54584	160.0	0.79763	-0.98197	-3.87933
70.0	1.64023	2.14906	-0.74830	162.0	0.80798	-0.92598	-3.82334
72.0	1.55688	1.92256	-0.97479	164.0	0.77584	-1.10228	-3.99964
74.0	1.47079	1.67551	-1.22185	166.0	0.70297	-1.53064	-4.42800
76.0	1.38406	1.41156	-1.48579	168.0	0.59614	-2.24648	-5.14384
78.0	1.29857	1.13465	-1.76271	170.0	0.46664	-3.31016	-6.20751
80.0	1.21589	0.84893	-2.04843	172.0	0.32903	-4.82761	-7.72497
82.0	1.13732	0.55883	-2.33853	174.0	0.19944	-7.00189	-9.89925
84.0	1.06391	0.26904	-2.62832	176.0	0.09347	-10.29336	-13.19071
86.0	0.99643	-0.01554	-2.91290	178.0	0.02412	-16.17629	-19.07365
88.0	0.93542	-0.28992	-3.18727	180.	0.00000	-INFINITY	-INFINITY
90.0	0.88122	-0.54915	-3.44650				

TABLE A2-25

EPSLN= 6.2500(NORMALIZED GROUNDPLANE RADIUS)
MONOPOLE LENGTH= 0.250000(WAVELENGTH)
MONOPOLE RADIUS= 0.1000E-05(WAVELENGTH)
NUMBER OF MONOPOLE SEGMENTS= 1
NUMBER OF DISK SEGMENTS= 16
RADIATION RESISTANCE DETERMINED BY PATTERN= 0.369423E+02 OHMS

ELEVATION ANGLE (DEG)	DIRECTIVITY (NUMERIC)	DIRECTIVITY (DBI)	RELATIVE POWER (DB)	ELEVATION ANGLE (DEG)	DIRECTIVITY (NUMERIC)	DIRECTIVITY (DBI)	RELATIVE POWER (DB)
0.	0.00000	-INFINITY	-INFINITY				
2.0	0.03004	-15.22289	-18.58903	92.0	0.75521	-1.21934	-4.58548
4.0	0.11654	-9.33518	-12.70132	94.0	0.70726	-1.50420	-4.87034
6.0	0.24920	-6.03452	-9.40066	96.0	0.66811	-1.75153	-5.11767
8.0	0.41261	-3.84460	-7.21074	98.0	0.63745	-1.95556	-5.32170
10.0	0.58858	-2.30192	-5.66806	100.0	0.61481	-2.11259	-5.47874
12.0	0.75882	-1.19863	-4.56477	102.0	0.59954	-2.22180	-5.58794
14.0	0.90744	-0.42183	-3.78797	104.0	0.59078	-2.28572	-5.65186
16.0	1.02306	0.09902	-3.26712	106.0	0.58744	-2.31035	-5.67649
18.0	1.10003	0.41406	-2.95208	108.0	0.58819	-2.30486	-5.67100
20.0	1.13872	0.56418	-2.80196	110.0	0.59143	-2.28100	-5.64714
22.0	1.14494	0.58782	-2.77832	112.0	0.59534	-2.25238	-5.61852
24.0	1.12860	0.52538	-2.84076	114.0	0.59788	-2.23384	-5.59998
26.0	1.10193	0.42154	-2.94460	116.0	0.59689	-2.24102	-5.60716
28.0	1.07759	0.32453	-3.04161	118.0	0.59017	-2.29020	-5.65634
30.0	1.06687	0.28112	-3.08502	120.0	0.57564	-2.39846	-5.76460
32.0	1.07838	0.32770	-3.03844	122.0	0.55154	-2.58421	-5.95035
34.0	1.11715	0.48113	-2.88501	124.0	0.51665	-2.86800	-6.23415
36.0	1.18445	0.73517	-2.63097	126.0	0.47055	-3.27392	-6.64006
38.0	1.27795	1.06513	-2.30101	128.0	0.41385	-3.83159	-7.19773
40.0	1.39240	1.43764	-1.92850	130.0	0.34838	-4.57941	-7.94555
42.0	1.52053	1.81996	-1.54618	132.0	0.27736	-5.56951	-8.93565
44.0	1.65402	2.18542	-1.18072	134.0	0.20533	-6.87539	-10.24153
46.0	1.78443	2.51499	-0.85115	136.0	0.13802	-8.60044	-11.96658
48.0	1.90399	2.79664	-0.56950	138.0	0.08197	-10.86353	-14.22967
50.0	2.00621	3.02376	-0.34238	140.0	0.04393	-13.57221	-16.93835
52.0	2.08624	3.19364	-0.17250	142.0	0.03016	-15.20591	-18.57206
54.0	2.14099	3.30614	-0.06000	144.0	0.04550	-13.42016	-16.78630
56.0	2.16910	3.36279	-0.00335	146.0	0.09251	-10.33817	-13.70431
58.0	2.17077	3.36614	0.00000	148.0	0.17069	-7.67799	-11.04413
60.0	2.14749	3.31932	-0.04682	150.0	0.27594	-5.59187	-8.95801
62.0	2.10173	3.22576	-0.14038	152.0	0.40047	-3.97428	-7.34043
64.0	2.03659	3.08903	-0.27711	154.0	0.53319	-2.73121	-6.09735
66.0	1.95557	2.91274	-0.45340	156.0	0.66063	-1.80042	-5.16656
68.0	1.86229	2.70048	-0.66566	158.0	0.76845	-1.14383	-4.50997
70.0	1.76028	2.45582	-0.91032	160.0	0.84323	-0.74054	-4.10668
72.0	1.65285	2.18233	-1.18381	162.0	0.87441	-0.58287	-3.94901
74.0	1.54298	1.88359	-1.48255	164.0	0.85607	-0.67490	-4.04104
76.0	1.43328	1.56333	-1.80282	166.0	0.78824	-1.03340	-4.39954
78.0	1.32599	1.22542	-2.14072	168.0	0.67738	-1.69168	-5.05782
80.0	1.22294	0.87405	-2.49209	170.0	0.53597	-2.70859	-6.07473
82.0	1.12559	0.51381	-2.85233	172.0	0.38115	-4.18899	-7.55513
84.0	1.03510	0.14982	-3.21632	174.0	0.23253	-6.33515	-9.70129
86.0	0.95232	-0.21219	-3.57833	176.0	0.10947	-9.60688	-12.97302
88.0	0.87785	-0.56582	-3.93196	178.0	0.02833	-15.47811	-18.84425
90.0	0.81208	-0.90403	-4.27017	180.	0.00000	-INFINITY	-INFINITY

TABLE A2-26

EPSLN= 6.5000(NORMALIZED GROUNDPLANE RADIUS)
MONOPOLE LENGTH= 0.250000(WAVELENGTH)
MONOPOLE RADIUS= 0.1000E-05(WAVELENGTH)
NUMBER OF MONOPOLE SEGMENTS= 1
NUMBER OF DISK SEGMENTS= 16
RADIATION RESISTANCE DETERMINED BY PATTERN= 0.405661E+02 OHMS

ELEVATION ANGLE (DEG)	DIRECTIVITY (NUMERIC)	DIRECTIVITY (DBI)	RELATIVE POWER (DB)	ELEVATION ANGLE (DEG)	DIRECTIVITY (NUMERIC)	DIRECTIVITY (DBI)	RELATIVE POWER (DB)
0.	0.00000	-INFINITY	-INFINITY				
2.0	0.02867	-15.42621	-19.32198	92.0	0.67013	-1.73840	-5.63417
4.0	0.11053	-9.56530	-13.46107	94.0	0.61098	-2.13974	-6.03551
6.0	0.23388	-6.31000	-10.20577	96.0	0.56207	-2.50209	-6.39786
8.0	0.38150	-4.18504	-8.08081	98.0	0.52330	-2.81251	-6.70828
10.0	0.53355	-2.72827	-6.62404	100.0	0.49444	-3.05886	-6.95463
12.0	0.67089	-1.73350	-5.62927	102.0	0.47514	-3.23181	-7.12758
14.0	0.77815	-1.08936	-4.98513	104.0	0.46485	-3.32690	-7.22267
16.0	0.84601	-0.72622	-4.62199	106.0	0.46281	-3.34600	-7.24177
18.0	0.87239	-0.59292	-4.48869	108.0	0.46798	-3.29775	-7.19352
20.0	0.86231	-0.64335	-4.53912	110.0	0.47900	-3.19663	-7.09240
22.0	0.82678	-0.82609	-4.72186	112.0	0.49416	-3.06129	-6.95705
24.0	0.78064	-1.07547	-4.97124	114.0	0.51138	-2.91255	-6.80832
26.0	0.74014	-1.30685	-5.20261	116.0	0.52822	-2.77184	-6.66761
28.0	0.72044	-1.42403	-5.31980	118.0	0.54197	-2.66024	-6.55601
30.0	0.73354	-1.34575	-5.24152	120.0	0.54976	-2.59825	-6.49402
32.0	0.78689	-1.04083	-4.93660	122.0	0.54876	-2.60615	-6.50192
34.0	0.88276	-0.54159	-4.43736	124.0	0.53644	-2.70482	-6.60059
36.0	1.01833	0.07889	-3.81688	126.0	0.51086	-2.91701	-6.81278
38.0	1.18655	0.74284	-3.15293	128.0	0.47106	-3.26923	-7.16500
40.0	1.37723	1.39006	-2.50571	130.0	0.41740	-3.79443	-7.69020
42.0	1.57851	1.98246	-1.91331	132.0	0.35186	-4.53634	-8.43211
44.0	1.77818	2.49975	-1.39602	134.0	0.27819	-5.55660	-9.45237
46.0	1.96489	2.93338	-0.96239	136.0	0.20198	-6.94697	-10.84274
48.0	2.12904	3.28184	-0.61393	138.0	0.13036	-8.84851	-12.74428
50.0	2.26336	3.54753	-0.34824	140.0	0.07152	-11.45572	-15.35149
52.0	2.36311	3.73485	-0.16092	142.0	0.03386	-14.70290	-18.59867
54.0	2.42609	3.84907	-0.04670	144.0	0.02497	-16.02609	-19.92186
56.0	2.45232	3.89577	0.00000	146.0	0.05040	-12.97553	-16.87130
58.0	2.44370	3.88047	-0.01529	148.0	0.11253	-9.48734	-13.38311
60.0	2.40352	3.80847	-0.08729	150.0	0.20957	-6.78671	-10.68248
62.0	2.33601	3.68475	-0.21102	152.0	0.33507	-4.74860	-8.64437
64.0	2.24591	3.51393	-0.38184	154.0	0.47802	-3.20555	-7.10132
66.0	2.13811	3.30030	-0.59547	156.0	0.62366	-2.05054	-5.94631
68.0	2.01735	3.04781	-0.84796	158.0	0.75508	-1.22008	-5.11585
70.0	1.88803	2.76010	-1.13567	160.0	0.85540	-0.67832	-4.57409
72.0	1.75411	2.44057	-1.45520	162.0	0.91025	-0.40839	-4.30416
74.0	1.61898	2.09242	-1.80335	164.0	0.91023	-0.40848	-4.30425
76.0	1.48550	1.71873	-2.17704	166.0	0.85281	-0.69146	-4.58723
78.0	1.35600	1.32258	-2.57319	168.0	0.74337	-1.28793	-5.18370
80.0	1.23231	0.90719	-2.98858	170.0	0.59499	-2.25489	-6.15066
82.0	1.11585	0.47607	-3.41970	172.0	0.42699	-3.69586	-7.59163
84.0	1.00769	0.03328	-3.86249	174.0	0.26229	-5.81218	-9.70795
86.0	0.90859	-0.41630	-4.31207	176.0	0.12408	-9.06301	-12.95878
88.0	0.81909	-0.86668	-4.76245	178.0	0.03220	-14.92186	-18.81763
90.0	0.73953	-1.31042	-5.20619	180.	0.00000	-INFINITY	-INFINITY

TABLE A2-27

EPSLN= 6.7500(NORMALIZED GROUNDPLANE RADIUS)
MONOPOLE LENGTH= 0.250000(WAVELENGTH)
MONOPOLE RADIUS= 0.1000E-05(WAVELENGTH)
NUMBER OF MONOPOLE SEGMENTS= 1
NUMBER OF DISK SEGMENTS= 16
RADIATION RESISTANCE DETERMINED BY PATTERN= 0.435313E+02 OHMS

ELEVATION ANGLE (DEG)	DIRECTIVITY (NUMERIC)	DIRECTIVITY (DBI)	RELATIVE POWER (DB)	ELEVATION ANGLE (DEG)	DIRECTIVITY (NUMERIC)	DIRECTIVITY (DBI)	RELATIVE POWER (DB)
0.	0.00000	-INFINITY	-INFINITY				
2.0	0.02612	-15.82993	-20.19410	92.0	0.60775	-2.16278	-6.52695
4.0	0.09996	-10.00186	-14.36603	94.0	0.53659	-2.70357	-7.06774
6.0	0.20880	-6.80277	-11.16694	96.0	0.47578	-3.22597	-7.59014
8.0	0.33422	-4.75965	-9.12382	98.0	0.42537	-3.71230	-8.07647
10.0	0.45564	-3.41375	-7.77792	100.0	0.38544	-4.14048	-8.50465
12.0	0.55419	-2.56341	-6.92758	102.0	0.35597	-4.48585	-8.85002
14.0	0.61618	-2.10293	-6.46710	104.0	0.33689	-4.72508	-9.08925
16.0	0.63551	-1.96877	-6.33294	106.0	0.32796	-4.84177	-9.20594
18.0	0.61460	-2.11405	-6.47822	108.0	0.32871	-4.83186	-9.19603
20.0	0.56375	-2.48913	-6.85330	110.0	0.33836	-4.70620	-9.07037
22.0	0.49916	-3.01758	-7.38175	112.0	0.35574	-4.48873	-8.85290
24.0	0.44010	-3.56447	-7.92864	114.0	0.37919	-4.21143	-8.57560
26.0	0.40572	-3.91773	-8.28190	116.0	0.40656	-3.90875	-8.27292
28.0	0.41215	-3.84950	-8.21367	118.0	0.43516	-3.61353	-7.97770
30.0	0.47027	-3.27649	-7.64066	120.0	0.46183	-3.35521	-7.71938
32.0	0.58454	-2.33184	-6.69601	122.0	0.48308	-3.15984	-7.52401
34.0	0.75275	-1.23352	-5.59769	124.0	0.49532	-3.05114	-7.41531
36.0	0.96677	-0.14675	-4.51092	126.0	0.49520	-3.05222	-7.41639
38.0	1.21404	0.84232	-3.52185	128.0	0.47999	-3.18764	-7.55181
40.0	1.47927	1.70049	-2.66368	130.0	0.44812	-3.48604	-7.85021
42.0	1.74640	2.42143	-1.94274	132.0	0.39960	-3.98380	-8.34797
44.0	2.00018	3.01068	-1.35349	134.0	0.33645	-4.73080	-9.09497
46.0	2.22754	3.47825	-0.88592	136.0	0.26300	-5.80048	-10.16465
48.0	2.41841	3.83530	-0.52887	138.0	0.18583	-7.30888	-11.67305
50.0	2.56610	4.09274	-0.27143	140.0	0.11347	-9.45115	-13.81532
52.0	2.66730	4.26072	-0.10345	142.0	0.05567	-12.54378	-16.90795
54.0	2.72172	4.34844	-0.01573	144.0	0.02228	-16.52122	-20.88539
56.0	2.73160	4.36417	0.00000	146.0	0.02185	-16.60538	-20.96955
58.0	2.70103	4.31530	-0.04887	148.0	0.06011	-12.21066	-16.57483
60.0	2.63538	4.20842	-0.15575	150.0	0.13851	-8.58529	-12.94946
62.0	2.54063	4.04941	-0.31476	152.0	0.25320	-5.96541	-10.32958
64.0	2.42296	3.84345	-0.52072	154.0	0.39466	-4.03774	-8.40191
66.0	2.28831	3.59516	-0.76901	156.0	0.54824	-2.61031	-6.97448
68.0	2.14217	3.30855	-1.05562	158.0	0.69561	-1.57635	-5.94052
70.0	1.98937	2.98715	-1.37702	160.0	0.81719	-0.87677	-5.24094
72.0	1.83401	2.63401	-1.73016	162.0	0.89509	-0.48133	-4.84550
74.0	1.67947	2.25173	-2.11244	164.0	0.91620	-0.38008	-4.74425
76.0	1.52846	1.84255	-2.52162	166.0	0.87485	-0.58067	-4.94484
78.0	1.38306	1.40840	-2.95577	168.0	0.77441	-1.11026	-5.47443
80.0	1.24480	0.95099	-3.41318	170.0	0.62755	-2.02349	-6.38766
82.0	1.11480	0.47196	-3.89221	172.0	0.45475	-3.42223	-7.78640
84.0	0.99382	-0.02694	-4.39111	174.0	0.28140	-5.50670	-9.87087
86.0	0.88236	-0.54356	-4.90773	176.0	0.13380	-8.73529	-13.09946
88.0	0.78074	-1.07494	-5.43911	178.0	0.03483	-14.58098	-18.94515
90.0	0.68916	-1.61681	-5.98098	180.	0.00000	-INFINITY	-INFINITY

TABLE A2-28

EPSLN= 7.0000(NORMALIZED GROUNDPLANE RADIUS)
MONOPOLE LENGTH= 0.250000(WAVELENGTH)
MONOPOLE RADIUS= 0.1000E-05(WAVELENGTH)
NUMBER OF MONOPOLE SEGMENTS= 1
NUMBER OF DISK SEGMENTS= 16
RADIATION RESISTANCE DETERMINED BY PATTERN= 0.441969E+02 OHMS

ELEVATION ANGLE (DEG)	DIRECTIVITY (NUMERIC)	DIRECTIVITY (DBI)	RELATIVE POWER (DB)	ELEVATION ANGLE (DEG)	DIRECTIVITY (NUMERIC)	DIRECTIVITY (DBI)	RELATIVE POWER (DB)
0.	0.00000	-INFINITY	-INFINITY				
2.0	0.02306	-16.37058	-21.06374	92.0	0.58973	-2.29345	-6.98661
4.0	0.08750	-10.58008	-15.27324	94.0	0.50961	-2.92761	-7.62077
6.0	0.18005	-7.44608	-12.13924	96.0	0.43849	-3.58037	-8.27353
8.0	0.28186	-5.49973	-10.19289	98.0	0.37654	-4.24189	-8.93505
10.0	0.37249	-4.28885	-8.98201	100.0	0.32400	-4.89461	-9.58777
12.0	0.43433	-3.62181	-8.31497	102.0	0.28117	-5.51025	-10.20341
14.0	0.45627	-3.40779	-8.10095	104.0	0.24842	-6.04816	-10.74132
16.0	0.43606	-3.60449	-8.29765	106.0	0.22604	-6.45810	-11.15126
18.0	0.38085	-4.19249	-8.88565	108.0	0.21425	-6.69079	-11.38395
20.0	0.30582	-5.14533	-9.83849	110.0	0.21303	-6.71551	-11.40867
22.0	0.23150	-6.35452	-11.04768	112.0	0.22206	-6.53535	-11.22851
24.0	0.18010	-7.44487	-12.13803	114.0	0.24052	-6.18841	-10.88157
26.0	0.17185	-7.64851	-12.34167	116.0	0.26707	-5.73382	-10.42698
28.0	0.22184	-6.53952	-11.23268	118.0	0.29964	-5.23402	-9.92718
30.0	0.33797	-4.71122	-9.40438	120.0	0.33548	-4.74327	-9.43643
32.0	0.52011	-2.83903	-7.53219	122.0	0.37116	-4.30437	-8.99753
34.0	0.76061	-1.18836	-5.88152	124.0	0.40269	-3.95034	-8.64350
36.0	1.04574	0.19423	-4.49893	126.0	0.42579	-3.70805	-8.40121
38.0	1.35779	1.32833	-3.36483	128.0	0.43634	-3.60179	-8.29495
40.0	1.67743	2.24644	-2.44672	130.0	0.43083	-3.65691	-8.35007
42.0	1.98587	2.97950	-1.71366	132.0	0.40705	-3.90352	-8.59668
44.0	2.26665	3.55385	-1.13931	134.0	0.36464	-4.38131	-9.07447
46.0	2.50686	3.99130	-0.70186	136.0	0.30568	-5.14726	-9.84042
48.0	2.69770	4.30994	-0.38322	138.0	0.23498	-6.28964	-10.98280
50.0	2.83459	4.52490	-0.16826	140.0	0.16006	-7.95722	-12.65038
52.0	2.91675	4.64899	-0.04417	142.0	0.09067	-10.42542	-15.11858
54.0	2.94656	4.69316	0.00000	144.0	0.03783	-14.22139	-18.91455
56.0	2.92879	4.66689	-0.02627	146.0	0.01238	-19.07226	-23.76542
58.0	2.86973	4.57841	-0.11475	148.0	0.02319	-16.34623	-21.03939
60.0	2.77646	4.43491	-0.25825	150.0	0.07534	-11.22955	-15.92271
62.0	2.65625	4.24268	-0.45048	152.0	0.16851	-7.73365	-12.42681
64.0	2.51605	4.00719	-0.68597	154.0	0.29605	-5.28633	-9.97949
66.0	2.36220	3.73317	-0.95999	156.0	0.44498	-3.51661	-8.20977
68.0	2.20022	3.42467	-1.26849	158.0	0.59718	-2.23897	-6.93213
70.0	2.03473	3.08507	-1.60809	160.0	0.73173	-1.35651	-6.04967
72.0	1.86946	2.71716	-1.97600	162.0	0.82814	-0.81896	-5.51212
74.0	1.70731	2.32312	-2.37004	164.0	0.86999	-0.60487	-5.29803
76.0	1.55044	1.90454	-2.78862	166.0	0.84823	-0.71485	-5.40801
78.0	1.40039	1.46248	-3.23068	168.0	0.76356	-1.17154	-5.86470
80.0	1.25819	0.99746	-3.69570	170.0	0.62710	-2.02660	-6.71976
82.0	1.12448	0.50950	-4.18366	172.0	0.45921	-3.37987	-8.07303
84.0	0.99959	-0.00178	-4.69494	174.0	0.28641	-5.43017	-10.12333
86.0	0.88367	-0.53708	-5.23024	176.0	0.13693	-8.63495	-13.32811
88.0	0.77675	-1.09719	-5.79035	178.0	0.03576	-14.46658	-19.15974
90.0	0.67878	-1.68271	-6.37587	180.	0.00000	-INFINITY	-INFINITY

TABLE A2-29

EPSLN= 7.2500(NORMALIZED GROUNDPLANE RADIUS)
MONOPOLE LENGTH= 0.250000(WAVELENGTH)
MONOPOLE RADIUS= 0.1000E-05(WAVELENGTH)
NUMBER OF MONOPOLE SEGMENTS= 1
NUMBER OF DISK SEGMENTS= 16
RADIATION RESISTANCE DETERMINED BY PATTERN= 0.422974E+02 OHMS

ELEVATION ANGLE (DEG)	DIRECTIVITY (NUMERIC)	DIRECTIVITY (DBI)	RELATIVE POWER (DB)	ELEVATION ANGLE (DEG)	DIRECTIVITY (NUMERIC)	DIRECTIVITY (DBI)	RELATIVE POWER (DB)
0.	0.00000	-INFINITY	-INFINITY				
2.0	0.02033	-16.91763	-21.76379	92.0	0.61882	-2.08439	-6.93054
4.0	0.07647	-11.16526	-16.01142	94.0	0.53518	-2.71502	-7.56117
6.0	0.15495	-8.09817	-12.94432	96.0	0.45832	-3.38830	-8.23445
8.0	0.23696	-6.25324	-11.09940	98.0	0.38836	-4.10762	-8.95378
10.0	0.30282	-5.18821	-10.03436	100.0	0.32558	-4.87345	-9.71961
12.0	0.33665	-4.72824	-9.57440	102.0	0.27041	-5.67981	-10.52596
14.0	0.33024	-4.81168	-9.65784	104.0	0.22344	-6.50834	-11.35449
16.0	0.28516	-5.44911	-10.29526	106.0	0.18539	-7.31924	-12.16540
18.0	0.21274	-6.72144	-11.56759	108.0	0.15699	-8.04135	-12.88750
20.0	0.13209	-8.79132	-13.63748	110.0	0.13896	-8.57121	-13.41737
22.0	0.06654	-11.76938	-16.61554	112.0	0.13183	-8.79984	-13.64600
24.0	0.03946	-14.03870	-18.88485	114.0	0.13583	-8.67013	-13.51629
26.0	0.07023	-11.53452	-16.38068	116.0	0.15067	-8.21966	-13.06582
28.0	0.17116	-7.66586	-12.51202	118.0	0.17543	-7.55905	-12.40521
30.0	0.34577	-4.61213	-9.45828	120.0	0.20833	-6.81239	-11.65855
32.0	0.58864	-2.30148	-7.14764	122.0	0.24673	-6.07776	-10.92391
34.0	0.88667	-0.52237	-5.36853	124.0	0.28704	-5.42056	-10.26672
36.0	1.22124	0.86802	-3.97813	126.0	0.32491	-4.88240	-9.72855
38.0	1.57094	1.96160	-2.88456	128.0	0.35550	-4.49165	-9.33781
40.0	1.91422	2.81993	-2.02623	130.0	0.37396	-4.27170	-9.11785
42.0	2.23172	3.48640	-1.35976	132.0	0.37611	-4.24687	-9.09303
44.0	2.50786	3.99303	-0.85313	134.0	0.35911	-4.44767	-9.29382
46.0	2.73176	4.36443	-0.48173	136.0	0.32233	-4.91699	-9.76314
48.0	2.89746	4.62018	-0.22598	138.0	0.26791	-5.72013	-10.56628
50.0	3.00351	4.77629	-0.06987	140.0	0.20117	-6.96438	-11.81054
52.0	3.05222	4.84616	0.00000	142.0	0.13050	-8.84381	-13.68996
54.0	3.04873	4.84119	-0.00497	144.0	0.06671	-11.75831	-16.60447
56.0	2.99999	4.77120	-0.07495	146.0	0.02170	-16.63487	-21.48103
58.0	2.91390	4.64474	-0.20141	148.0	0.00672	-21.72495	-26.57110
60.0	2.79848	4.46923	-0.37693	150.0	0.03019	-15.20206	-20.04821
62.0	2.66140	4.25111	-0.59505	152.0	0.09563	-10.19390	-15.04006
64.0	2.50954	3.99595	-0.85021	154.0	0.20016	-6.98623	-11.83238
66.0	2.34882	3.70851	-1.13765	156.0	0.33379	-4.76532	-9.61148
68.0	2.18413	3.39278	-1.45338	158.0	0.48011	-3.18656	-8.03271
70.0	2.01932	3.05205	-1.79411	160.0	0.61838	-2.08742	-6.93357
72.0	1.85733	2.68888	-2.15728	162.0	0.72673	-1.38625	-6.23241
74.0	1.70028	2.30520	-2.54096	164.0	0.78617	-1.04483	-5.89099
76.0	1.54962	1.90226	-2.94390	166.0	0.78451	-1.05404	-5.90020
78.0	1.40626	1.48066	-3.36550	168.0	0.71936	-1.43055	-6.27671
80.0	1.27068	1.04035	-3.80581	170.0	0.59950	-2.22213	-7.06829
82.0	1.14303	0.58059	-4.26556	172.0	0.44401	-3.52607	-8.37223
84.0	1.02329	0.09999	-4.74617	174.0	0.27929	-5.53946	-10.38562
86.0	0.91125	-0.40361	-5.24977	176.0	0.13432	-8.71859	-13.56475
88.0	0.80667	-0.93306	-5.77922	178.0	0.03520	-14.53512	-19.38127
90.0	0.70926	-1.49193	-6.33808	180.	0.00000	-INFINITY	-INFINITY

TABLE A2-30

EPSLN= 7.5000(NORMALIZED GROUNDPLANE RADIUS)
MONOPOLE LENGTH= 0.250000(WAVELENGTH)
MONOPOLE RADIUS= 0.1000E-05(WAVELENGTH)
NUMBER OF MONOPOLE SEGMENTS= 1
NUMBER OF DISK SEGMENTS= 16
RADIATION RESISTANCE DETERMINED BY PATTERN= 0.391003E+02 OHMS

ELEVATION ANGLE (DEG)	DIRECTIVITY (NUMERIC)	DIRECTIVITY (DBI)	RELATIVE POWER (DB)	ELEVATION ANGLE (DEG)	DIRECTIVITY (NUMERIC)	DIRECTIVITY (DBI)	RELATIVE POWER (DB)
0.	0.00000	-INFINITY	-INFINITY				
2.0	0.01853	-17.32209	-22.18205	92.0	0.68087	-1.66937	-6.52933
4.0	0.06915	-11.60200	-16.46196	94.0	0.59916	-2.22460	-7.08457
6.0	0.13830	-8.59176	-13.45173	96.0	0.52191	-2.82400	-7.68397
8.0	0.20731	-6.83376	-11.69372	98.0	0.44909	-3.47663	-8.33659
10.0	0.25730	-5.89555	-10.75551	100.0	0.38083	-4.19272	-9.05269
12.0	0.27414	-5.62029	-10.48025	102.0	0.31746	-4.98308	-9.84304
14.0	0.25228	-5.98118	-10.84115	104.0	0.25958	-5.85728	-10.71724
16.0	0.19660	-7.06426	-11.92423	106.0	0.20799	-6.81953	-11.67950
18.0	0.12180	-9.14346	-14.00342	108.0	0.16371	-7.85920	-12.71917
20.0	0.04971	-13.03524	-17.89521	110.0	0.12790	-8.93134	-13.79131
22.0	0.00506	-22.96253	-27.82249	112.0	0.10176	-9.92433	-14.78430
24.0	0.01080	-19.66616	-24.52612	114.0	0.08640	-10.63504	-15.49500
26.0	0.08403	-10.75546	-15.61543	116.0	0.08264	-10.82807	-15.68803
28.0	0.23314	-6.32382	-11.18378	118.0	0.09081	-10.41853	-15.27849
30.0	0.45669	-3.40379	-8.26375	120.0	0.11050	-9.56631	-14.42627
32.0	0.74406	-1.28393	-6.14389	122.0	0.14034	-8.52807	-13.38804
34.0	1.07745	0.32399	-4.53597	124.0	0.17787	-7.49904	-12.35900
36.0	1.43478	1.56785	-3.29211	126.0	0.21945	-6.58665	-11.44661
38.0	1.79274	2.53516	-2.32480	128.0	0.26044	-5.84297	-10.70293
40.0	2.12965	3.28307	-1.57689	130.0	0.29548	-5.29476	-10.15473
42.0	2.42758	3.85173	-1.00823	132.0	0.31908	-4.96102	-9.82099
44.0	2.67363	4.27101	-0.58895	134.0	0.32639	-4.86268	-9.72264
46.0	2.86035	4.56419	-0.29577	136.0	0.31409	-5.02942	-9.88938
48.0	2.98543	4.75007	-0.10990	138.0	0.28136	-5.50741	-10.36737
50.0	3.05087	4.84424	-0.01572	140.0	0.23056	-6.37225	-11.23221
52.0	3.06194	4.85996	0.00000	142.0	0.16764	-7.75635	-12.61631
54.0	3.02596	4.80863	-0.05134	144.0	0.10190	-9.91820	-14.77816
56.0	2.95128	4.70011	-0.15986	146.0	0.04508	-13.46042	-18.32038
58.0	2.84642	4.54298	-0.31698	148.0	0.00965	-20.15405	-25.01401
60.0	2.71938	4.34469	-0.51527	150.0	0.00667	-21.76071	-26.62067
62.0	2.57728	4.11162	-0.74834	152.0	0.04330	-13.63556	-18.49553
64.0	2.42616	3.84920	-1.01076	154.0	0.12069	-9.18339	-14.04335
66.0	2.27090	3.56198	-1.29798	156.0	0.23264	-6.33310	-11.19307
68.0	2.11528	3.25367	-1.60629	158.0	0.36559	-4.37001	-9.22997
70.0	1.96211	2.92724	-1.93272	160.0	0.50016	-3.00893	-7.86889
72.0	1.81340	2.58494	-2.27503	162.0	0.61421	-2.11686	-6.97682
74.0	1.67046	2.22837	-2.63159	164.0	0.68700	-1.63045	-6.49041
76.0	1.53409	1.85852	-3.00144	166.0	0.70358	-1.52688	-6.38685
78.0	1.40468	1.47578	-3.38418	168.0	0.65845	-1.81475	-6.67472
80.0	1.28231	1.07993	-3.78003	170.0	0.55760	-2.53680	-7.39677
82.0	1.16685	0.67014	-4.18982	172.0	0.41811	-3.78707	-8.64704
84.0	1.05801	0.24490	-4.61507	174.0	0.26543	-5.76050	-10.62047
86.0	0.95542	-0.19807	-5.05804	176.0	0.12847	-8.91196	-13.77193
88.0	0.85864	-0.66188	-5.52184	178.0	0.03379	-14.71221	-19.57217
90.0	0.76726	-1.15059	-6.01055	180.	0.00000	-INFINITY	-INFINITY

TABLE A2-31

EPSLN= 7.7500(NORMALIZED GROUNDPLANE RADIUS)
MONOPOLE LENGTH= 0.250000(WAVELENGTH)
MONOPOLE RADIUS= 0.1000E-05(WAVELENGTH)
NUMBER OF MONOPOLE SEGMENTS= 1
NUMBER OF DISK SEGMENTS= 16
RADIATION RESISTANCE DETERMINED BY PATTERN= 0.359610E+02 OHMS

ELEVATION ANGLE (DEG)	DIRECTIVITY (NUMERIC)	DIRECTIVITY (DBI)	RELATIVE POWER (DB)
0.	0.00000	-INFINITY	-INFINITY
2.0	0.01784	-17.48706	-22.24659
4.0	0.06625	-11.78837	-16.54789
6.0	0.13136	-8.81524	-13.57477
8.0	0.19442	-7.11251	-11.87204
10.0	0.23703	-6.25193	-11.01146
12.0	0.24640	-6.08351	-10.84304
14.0	0.21920	-6.59151	-11.35104
16.0	0.16301	-7.87779	-12.63732
18.0	0.09515	-10.21591	-14.97544
20.0	0.03921	-14.06578	-18.82531
22.0	0.02026	-16.93328	-21.69280
24.0	0.05983	-12.23063	-16.99016
26.0	0.17186	-7.64820	-12.40773
28.0	0.36032	-4.43317	-9.19270
30.0	0.61881	-2.08440	-6.84393
32.0	0.93210	-0.30536	-5.06489
34.0	1.27884	1.06817	-3.69136
36.0	1.63500	2.13518	-2.62434
38.0	1.97721	2.96052	-1.79901
40.0	2.28542	3.58967	-1.16986
42.0	2.54474	4.05644	-0.70309
44.0	2.74611	4.38718	-0.37235
46.0	2.88616	4.60320	-0.15633
48.0	2.96638	4.72227	-0.03725
50.0	2.99194	4.75953	0.00000
52.0	2.97030	4.72800	-0.03153
54.0	2.91005	4.63900	-0.12053
56.0	2.81987	4.50230	-0.25723
58.0	2.70787	4.32628	-0.43325
60.0	2.58110	4.11806	-0.64147
62.0	2.44542	3.88354	-0.87599
64.0	2.30545	3.62756	-1.13197
66.0	2.16469	3.35395	-1.40558
68.0	2.02568	3.06571	-1.69382
70.0	1.89021	2.76509	-1.99443
72.0	1.75945	2.45376	-2.30577
74.0	1.63412	2.13284	-2.62669
76.0	1.51462	1.80303	-2.95649
78.0	1.40108	1.46464	-3.29489
80.0	1.29347	1.11755	-3.64198
82.0	1.19158	0.76123	-3.99830
84.0	1.09513	0.39465	-4.36488
86.0	1.00372	0.01614	-4.74339
88.0	0.91692	-0.37668	-5.13621
90.0	0.83424	-0.78711	-5.54664
92.0	0.75517	-1.21953	-5.97906
94.0	0.67926	-1.67962	-6.43915
96.0	0.60609	-2.17460	-6.93413
98.0	0.53537	-2.71350	-7.47303
100.0	0.46693	-3.30745	-8.06697
102.0	0.40086	-3.97003	-8.72955
104.0	0.33748	-4.71754	-9.47707
106.0	0.27740	-5.56900	-10.32853
108.0	0.22155	-6.54521	-11.30474
110.0	0.17120	-7.66496	-12.42449
112.0	0.12785	-8.93307	-13.69260
114.0	0.09317	-10.30714	-15.06667
116.0	0.06885	-11.62068	-16.38021
118.0	0.05636	-12.49043	-17.24996
120.0	0.05665	-12.46764	-17.22717
122.0	0.06992	-11.55413	-16.31366
124.0	0.09523	-10.21247	-14.97200
126.0	0.13033	-8.84953	-13.60906
128.0	0.17156	-7.65572	-12.41525
130.0	0.21395	-6.69692	-11.45645
132.0	0.25157	-5.99340	-10.75293
134.0	0.27826	-5.55542	-10.31495
136.0	0.28854	-5.39799	-10.15752
138.0	0.27867	-5.54905	-10.30858
140.0	0.24782	-6.05866	-10.81819
142.0	0.19881	-7.01560	-11.77513
144.0	0.13848	-8.58600	-13.34553
146.0	0.07723	-11.12196	-15.88149
148.0	0.02770	-15.57486	-20.33439
150.0	0.00267	-25.72946	-30.48899
152.0	0.01243	-19.05387	-23.81340
154.0	0.06212	-12.06758	-16.82711
156.0	0.14968	-8.24840	-13.00793
158.0	0.26505	-5.76666	-10.52619
160.0	0.39109	-4.07721	-8.83674
162.0	0.50623	-2.95655	-7.71608
164.0	0.58861	-2.30175	-7.06128
166.0	0.62088	-2.06995	-6.82948
168.0	0.59449	-2.25855	-7.01808
170.0	0.51244	-2.90359	-7.66311
172.0	0.38950	-4.09488	-8.85441
174.0	0.24977	-6.02465	-10.78418
176.0	0.12173	-9.14600	-13.90553
178.0	0.03215	-14.92858	-19.68811
180.	0.00000	-INFINITY	-INFINITY

TABLE A2-32

EPSLN= 8.0000(NORMALIZED GROUNDPLANE RADIUS)
MONOPOLE LENGTH= 0.250000(WAVELENGTH)
MONOPOLE RADIUS= 0.1000E-05(WAVELENGTH)
NUMBER OF MONOPOLE SEGMENTS= 1
NUMBER OF DISK SEGMENTS= 16
RADIATION RESISTANCE DETERMINED BY PATTERN= 0.334954E+02 OHMS

ELEVATION ANGLE (DEG)	DIRECTIVITY (NUMERIC)	DIRECTIVITY (DBI)	RELATIVE POWER (DB)	ELEVATION ANGLE (DEG)	DIRECTIVITY (NUMERIC)	DIRECTIVITY (DBI)	RELATIVE POWER (DB)
0.	0.00000	-INFINITY	-INFINITY				
2.0	0.01822	-17.39420	-21.95972	92.0	0.82468	-0.83717	-5.40269
4.0	0.06752	-11.70576	-16.27128	94.0	0.75628	-1.21315	-5.77867
6.0	0.13338	-8.74916	-13.31468	96.0	0.68987	-1.61231	-6.17783
8.0	0.19643	-7.06787	-11.63338	98.0	0.62486	-2.04215	-6.60767
10.0	0.23822	-6.23020	-10.79572	100.0	0.56076	-2.51222	-7.07774
12.0	0.24677	-6.07715	-10.64267	102.0	0.49722	-3.03454	-7.60006
14.0	0.22042	-6.56744	-11.13296	104.0	0.43411	-3.62403	-8.18955
16.0	0.16899	-7.72144	-12.28695	106.0	0.37160	-4.29919	-8.86470
18.0	0.11186	-9.51318	-14.07870	108.0	0.31025	-5.08284	-9.64835
20.0	0.07383	-11.31738	-15.88290	110.0	0.25102	-6.00297	-10.56849
22.0	0.07968	-10.98636	-15.55188	112.0	0.19530	-7.09293	-11.65845
24.0	0.14892	-8.27038	-12.83590	114.0	0.14493	-8.38846	-12.95398
26.0	0.29194	-5.34707	-9.91259	116.0	0.10203	-9.91258	-14.47810
28.0	0.50820	-2.93963	-7.50515	118.0	0.06889	-11.61826	-16.18378
30.0	0.78675	-1.04166	-5.60718	120.0	0.04766	-13.21859	-17.78411
32.0	1.10851	0.44741	-4.11811	122.0	0.04001	-13.97816	-18.54368
34.0	1.44987	1.61328	-2.95224	124.0	0.04676	-13.30092	-17.86643
36.0	1.78637	2.51972	-2.04580	126.0	0.06746	-11.70953	-16.27505
38.0	2.09616	3.21424	-1.35128	128.0	0.10005	-9.99765	-14.56317
40.0	2.36227	3.73329	-0.83223	130.0	0.14075	-8.51544	-13.08096
42.0	2.57388	4.10589	-0.45963	132.0	0.18413	-7.34885	-11.91437
44.0	2.72638	4.35586	-0.20966	134.0	0.22356	-6.50598	-11.07149
46.0	2.82053	4.50331	-0.06220	136.0	0.25211	-5.98417	-10.54968
48.0	2.86122	4.56552	0.00000	138.0	0.26359	-5.79068	-10.35620
50.0	2.85590	4.55743	-0.00809	140.0	0.25400	-5.95170	-10.51722
52.0	2.81318	4.49198	-0.07354	142.0	0.22270	-6.52277	-11.08829
54.0	2.74174	4.38026	-0.18526	144.0	0.17336	-7.61049	-12.17600
56.0	2.64947	4.23159	-0.33393	146.0	0.11409	-9.42763	-13.99315
58.0	2.54309	4.05361	-0.51191	148.0	0.05665	-12.46805	-17.03357
60.0	2.42795	3.85239	-0.71312	150.0	0.01465	-18.34132	-22.90684
62.0	2.30813	3.63261	-0.93291	152.0	0.00086	-30.67217	-35.23769
64.0	2.18660	3.39769	-1.16783	154.0	0.02413	-16.17406	-20.73958
66.0	2.06544	3.15012	-1.41540	156.0	0.08669	-10.62006	-15.18558
68.0	1.94608	2.89160	-1.67392	158.0	0.18245	-7.38849	-11.95401
70.0	1.82951	2.62336	-1.94216	160.0	0.29709	-5.27114	-9.83666
72.0	1.71643	2.34626	-2.21926	162.0	0.41020	-3.87004	-8.43556
74.0	1.60731	2.06099	-2.50452	164.0	0.49929	-3.01644	-7.58196
76.0	1.50250	1.76814	-2.79738	166.0	0.54486	-2.63714	-7.20266
78.0	1.40223	1.46818	-3.09733	168.0	0.53535	-2.71361	-7.27913
80.0	1.30663	1.16153	-3.40399	170.0	0.47068	-3.27278	-7.83830
82.0	1.21573	0.84837	-3.71714	172.0	0.36317	-4.39893	-8.96445
84.0	1.12944	0.52864	-4.03688	174.0	0.23546	-6.28080	-10.84631
86.0	1.04756	0.20179	-4.36372	176.0	0.11563	-9.36926	-13.93478
88.0	0.96977	-0.13331	-4.69883	178.0	0.03067	-15.13261	-19.69813
90.0	0.89565	-0.47863	-5.04415	180.	0.00000	-INFINITY	-INFINITY

TABLE A2-33

EPSLN= 8.2500(NORMALIZED GROUNDPLANE RADIUS)
MONOPOLE LENGTH= 0.250000(WAVELENGTH)
MONOPOLE RADIUS= 0.1000E-05(WAVELENGTH)
NUMBER OF MONOPOLE SEGMENTS= 1
NUMBER OF DISK SEGMENTS= 16
RADIATION RESISTANCE DETERMINED BY PATTERN= 0.318841E+02 OHMS

ELEVATION ANGLE (DEG)	DIRECTIVITY (NUMERIC)	DIRECTIVITY (DBI)	RELATIVE POWER (DB)
0.	0.00000	-INFINITY	-INFINITY
2.0	0.01951	-17.09849	-21.40430
4.0	0.07222	-11.41326	-15.71906
6.0	0.14255	-8.46026	-12.76606
8.0	0.20992	-6.77947	-11.08528
10.0	0.25516	-5.93182	-10.23763
12.0	0.26656	-5.74202	-10.04783
14.0	0.24374	-6.13081	-10.43662
16.0	0.19837	-7.02533	-11.33114
18.0	0.15163	-8.19216	-12.49797
20.0	0.12919	-8.88770	-13.19351
22.0	0.15519	-8.09141	-12.39722
24.0	0.24682	-6.07612	-10.38193
26.0	0.41080	-3.86369	-8.16950
28.0	0.64228	-1.92276	-6.22856
30.0	0.92630	-0.33246	-4.63827
32.0	1.24104	0.93786	-3.36795
34.0	1.56190	1.93654	-2.36926
36.0	1.86556	2.70809	-1.59772
38.0	2.13304	3.29000	-1.01581
40.0	2.35160	3.71364	-0.59217
42.0	2.51511	4.00558	-0.30023
44.0	2.62344	4.18872	-0.11709
46.0	2.68102	4.28301	-0.02280
48.0	2.69514	4.30581	0.00000
50.0	2.67430	4.27209	-0.03372
52.0	2.62691	4.19445	-0.11136
54.0	2.56041	4.08309	-0.22272
56.0	2.48079	3.94590	-0.35991
58.0	2.39255	3.78861	-0.51720
60.0	2.29882	3.61505	-0.69076
62.0	2.20167	3.42753	-0.87828
64.0	2.10246	3.22727	-1.07854
66.0	2.00207	3.01480	-1.29101
68.0	1.90123	2.79035	-1.51546
70.0	1.80058	2.55413	-1.75168
72.0	1.70079	2.30651	-1.99930
74.0	1.60255	2.04812	-2.25769
76.0	1.50659	1.77994	-2.52587
78.0	1.41357	1.50318	-2.80262
80.0	1.32412	1.21927	-3.08654
82.0	1.23870	0.92966	-3.37615
84.0	1.15764	0.63572	-3.67009
86.0	1.08106	0.33850	-3.96731
88.0	1.00890	0.03850	-4.26731
90.0	0.94091	-0.26453	-4.57034
92.0	0.87662	-0.57187	-4.87768
94.0	0.81544	-0.88608	-5.19189
96.0	0.75661	-1.21126	-5.51707
98.0	0.69932	-1.55326	-5.85907
100.0	0.64269	-1.91996	-6.22577
102.0	0.58594	-2.32150	-6.62731
104.0	0.52836	-2.77069	-7.07650
106.0	0.46953	-3.28339	-7.58920
108.0	0.40933	-3.87925	-8.18506
110.0	0.34812	-4.58267	-8.88848
112.0	0.28680	-5.42425	-9.73006
114.0	0.22685	-6.44259	-10.74840
116.0	0.17038	-7.68571	-11.99152
118.0	0.12001	-9.20786	-13.51367
120.0	0.07865	-11.04288	-15.34869
122.0	0.04923	-13.07732	-17.38313
124.0	0.03423	-14.65581	-18.96162
126.0	0.03516	-14.54011	-18.84592
128.0	0.05203	-12.83783	-17.14364
130.0	0.08293	-10.81305	-15.11886
132.0	0.12379	-9.07317	-13.37898
134.0	0.16852	-7.73348	-12.03929
136.0	0.20958	-6.78642	-11.09223
138.0	0.23906	-6.21490	-10.52071
140.0	0.25010	-6.01885	-10.32466
142.0	0.23854	-6.22438	-10.53019
144.0	0.20436	-6.89597	-11.20178
146.0	0.15256	-8.16554	-12.47135
148.0	0.09303	-10.31400	-14.61981
150.0	0.03920	-14.06663	-18.37244
152.0	0.00560	-22.52162	-26.82743
154.0	0.00441	-23.55649	-27.86230
156.0	0.04216	-13.75131	-18.05712
158.0	0.11706	-9.31583	-13.62164
160.0	0.21819	-6.61167	-10.91748
162.0	0.32686	-4.85642	-9.16222
164.0	0.42035	-3.76394	-8.06975
166.0	0.47719	-3.21305	-7.51886
168.0	0.48284	-3.16194	-7.46775
170.0	0.43402	-3.62492	-7.93073
172.0	0.34049	-4.67891	-8.98471
174.0	0.22345	-6.50812	-10.81392
176.0	0.11065	-9.56063	-13.86643
178.0	0.02949	-15.30301	-19.60882
180.	0.00000	-INFINITY	-INFINITY

TABLE A2-34

EPSLN= 8.5000(NORMALIZED GROUNDPLANE RADIUS)
MONOPOLE LENGTH= 0.250000(WAVELENGTH)
MONOPOLE RADIUS= 0.1000E-05(WAVELENGTH)
NUMBER OF MONOPOLE SEGMENTS= 1
NUMBER OF DISK SEGMENTS= 16
RADIATION RESISTANCE DETERMINED BY PATTERN= 0.311644E+02 OHMS

ELEVATION ANGLE (DEG)	DIRECTIVITY (NUMERIC)	DIRECTIVITY (DBI)	RELATIVE POWER (DB)	ELEVATION ANGLE (DEG)	DIRECTIVITY (NUMERIC)	DIRECTIVITY (DBI)	RELATIVE POWER (DB)
0.	0.00000	-INFINITY	-INFINITY				
2.0	0.02145	-16.68529	-20.65873	92.0	0.90179	-0.44893	-4.42236
4.0	0.07941	-11.00120	-14.97465	94.0	0.84562	-0.72826	-4.70170
6.0	0.15674	-8.04830	-12.02175	96.0	0.79330	-1.00560	-4.97905
8.0	0.23104	-6.36307	-10.33652	98.0	0.74385	-1.28517	-5.25862
10.0	0.28187	-5.49946	-9.47291	100.0	0.69609	-1.57334	-5.54679
12.0	0.29729	-5.26823	-9.24168	102.0	0.64881	-1.87883	-5.85228
14.0	0.27786	-5.56175	-9.53519	104.0	0.60077	-2.21295	-6.18640
16.0	0.23696	-6.25330	-10.22675	106.0	0.55084	-2.58978	-6.56322
18.0	0.19737	-7.04715	-11.02060	108.0	0.49813	-3.02654	-6.99999
20.0	0.18545	-7.31784	-11.29128	110.0	0.44216	-3.54423	-7.51768
22.0	0.22447	-6.48844	-10.46189	112.0	0.38295	-4.16860	-8.14205
24.0	0.32918	-4.82571	-8.79916	114.0	0.32124	-4.93174	-8.90519
26.0	0.50266	-2.98728	-6.96073	116.0	0.25856	-5.87443	-9.84788
28.0	0.73619	-1.33009	-5.30354	118.0	0.19728	-7.04927	-11.02272
30.0	1.01169	0.05048	-3.92297	120.0	0.14052	-8.52265	-12.49610
32.0	1.30581	1.15880	-2.81464	122.0	0.09195	-10.36455	-14.33800
34.0	1.59453	2.02634	-1.94711	124.0	0.05536	-12.56775	-16.54120
36.0	1.85714	2.68845	-1.28500	126.0	0.03414	-14.66738	-18.64083
38.0	2.07886	3.17826	-0.79519	128.0	0.03055	-15.14986	-19.12331
40.0	2.25189	3.52547	-0.44798	130.0	0.04506	-13.46181	-17.43526
42.0	2.37495	3.75655	-0.21690	132.0	0.07576	-11.20554	-15.17899
44.0	2.45189	3.89500	-0.07845	134.0	0.11806	-9.27891	-13.25236
46.0	2.48968	3.96144	-0.01201	136.0	0.16491	-7.82760	-11.80105
48.0	2.49658	3.97345	0.00000	138.0	0.20755	-6.82870	-10.80215
50.0	2.48050	3.94540	-0.02805	140.0	0.23696	-6.25320	-10.22665
52.0	2.44807	3.88824	-0.08521	142.0	0.24565	-6.09688	-10.07033
54.0	2.40414	3.80960	-0.16385	144.0	0.22964	-6.38952	-10.36297
56.0	2.35180	3.71400	-0.25945	146.0	0.19010	-7.21019	-11.18364
58.0	2.29270	3.60347	-0.36998	148.0	0.13402	-8.72815	-12.70160
60.0	2.22754	3.47825	-0.49520	150.0	0.07364	-11.32915	-15.30260
62.0	2.15650	3.33749	-0.63596	152.0	0.02424	-16.15425	-20.12770
64.0	2.07966	3.17992	-0.79353	154.0	0.00085	-30.70211	-34.67556
66.0	1.99728	3.00438	-0.96907	156.0	0.01416	-18.48817	-22.46162
68.0	1.90993	2.81017	-1.16328	158.0	0.06703	-11.73701	-15.71046
70.0	1.81853	2.59719	-1.37626	160.0	0.15249	-8.16760	-12.14105
72.0	1.72428	2.36608	-1.60737	162.0	0.25421	-5.94805	-9.92149
74.0	1.62861	2.11816	-1.85529	164.0	0.34976	-4.56231	-8.53576
76.0	1.53298	1.85538	-2.11807	166.0	0.41599	-3.80922	-7.78267
78.0	1.43887	1.58022	-2.39323	168.0	0.43533	-3.61181	-7.58526
80.0	1.34758	1.29554	-2.67791	170.0	0.40119	-3.96654	-7.93999
82.0	1.26022	1.00445	-2.96900	172.0	0.32060	-4.94036	-8.91381
84.0	1.17763	0.71009	-3.26336	174.0	0.21323	-6.71161	-10.68506
86.0	1.10038	0.41543	-3.55802	176.0	0.10654	-9.72478	-13.69823
88.0	1.02873	0.12300	-3.85045	178.0	0.02855	-15.44433	-19.41778
90.0	0.96264	-0.16537	-4.13882	180.	0.00000	-INFINITY	-INFINITY

TABLE A2-35

EPSLN= 8. 5000(NORMALIZED GROUNDPLANE RADIUS)
MONOPOLE LENGTH= 0. 250000(WAVELENGTH)
MONOPOLE RADIUS= 0. 1000E-05(WAVELENGTH)
NUMBER OF MONOPOLE SEGMENTS= 1
NUMBER OF DISK SEGMENTS= 25
NUMBER OF ANNULAR ZONES PER DISK SEGEMENT= 1
RADIATION RESISTANCE DETERMINED BY PATTERN= 0. 315105E+02 OHMS

ELEVATION ANGLE (DEG)	DIRECTIVITY (NUMERIC)	DIRECTIVITY (dBi)	RELATIVE POWER (dB)	ELEVATION ANGLE (DEG)	DIRECTIVITY (NUMERIC)	DIRECTIVITY (dBi)	RELATIVE POWER (dB)
0.	0. 00000	-INFINITY	-INFINITY	92. 0	0. 89041	-0. 50412	-4. 48282
2. 0	0. 02167	-16. 64156	-20. 62027	94. 0	0. 83372	-0. 78980	-4. 76850
4. 0	0. 08021	-10. 95788	-14. 93659	96. 0	0. 78119	-1. 07242	-5. 05112
6. 0	0. 15828	-8. 00572	-11. 98442	98. 0	0. 73181	-1. 35600	-5. 33470
8. 0	0. 23326	-6. 32167	-10. 30037	100. 0	0. 68441	-1. 64684	-5. 62555
10. 0	0. 28445	-5. 45988	-9. 43859	102. 0	0. 63772	-1. 95371	-5. 93241
12. 0	0. 29981	-5. 23156	-9. 21027	104. 0	0. 59047	-2. 28799	-6. 26669
14. 0	0. 27991	-5. 52989	-9. 50859	106. 0	0. 54151	-2. 66394	-6. 64265
16. 0	0. 23825	-6. 22962	-10. 20833	108. 0	0. 48989	-3. 09905	-7. 07775
18. 0	0. 19786	-7. 03632	-11. 01502	110. 0	0. 43505	-3. 61456	-7. 59326
20. 0	0. 18532	-7. 32088	-11. 29958	112. 0	0. 37701	-4. 23651	-8. 21521
22. 0	0. 22407	-6. 49615	-10. 47486	114. 0	0. 31643	-4. 99718	-8. 97589
24. 0	0. 32893	-4. 82895	-8. 80766	116. 0	0. 25483	-5. 93742	-9. 91613
26. 0	0. 50293	-2. 98489	-6. 96359	118. 0	0. 19457	-7. 10933	-11. 08803
28. 0	0. 73721	-1. 32407	-5. 30277	120. 0	0. 13876	-8. 57722	-12. 55593
30. 0	1. 01349	0. 05819	-3. 92051	122. 0	0. 09112	-10. 40385	-14. 38255
32. 0	1. 30823	1. 16685	-2. 81185	124. 0	0. 05548	-12. 55835	-16. 53705
34. 0	1. 59731	2. 03389	-1. 94481	126. 0	0. 03529	-14. 52323	-18. 50193
36. 0	1. 85998	2. 69507	-1. 28363	128. 0	0. 03288	-14. 83076	-18. 80946
38. 0	2. 08151	3. 18379	-0. 79491	130. 0	0. 04876	-13. 11967	-17. 09838
40. 0	2. 25424	3. 53001	-0. 44869	132. 0	0. 08101	-10. 91450	-14. 89321
42. 0	2. 37707	3. 76042	-0. 21828	134. 0	0. 12501	-9. 03038	-13. 00909
44. 0	2. 45397	3. 89870	-0. 08001	136. 0	0. 17360	-7. 60458	-11. 58329
46. 0	2. 49205	3. 96557	-0. 01313	138. 0	0. 21784	-6. 61857	-10. 59727
48. 0	2. 49960	3. 97870	0. 00000	140. 0	0. 24850	-6. 04674	-10. 02544
50. 0	2. 48451	3. 95241	-0. 02629	142. 0	0. 25786	-5. 88621	-9. 86492
52. 0	2. 45332	3. 89755	-0. 08115	144. 0	0. 24176	-6. 16624	-10. 14494
54. 0	2. 41075	3. 82152	-0. 15719	146. 0	0. 20124	-6. 96275	-10. 94146
56. 0	2. 35973	3. 72862	-0. 25009	148. 0	0. 14333	-8. 43650	-12. 41521
58. 0	2. 30177	3. 62062	-0. 35809	150. 0	0. 08039	-10. 94786	-14. 92656
60. 0	2. 23743	3. 49749	-0. 48121	152. 0	0. 02801	-15. 52684	-19. 50554
62. 0	2. 16680	3. 35818	-0. 62052	154. 0	0. 00156	-28. 07083	-32. 04954
64. 0	2. 08990	3. 20126	-0. 77744	156. 0	0. 01213	-19. 16061	-23. 13931
66. 0	2. 00700	3. 02546	-0. 95324	158. 0	0. 06291	-12. 01246	-15. 99116
68. 0	1. 91867	2. 82999	-1. 14871	160. 0	0. 14714	-8. 32262	-12. 30132
70. 0	1. 82589	2. 61474	-1. 36397	162. 0	0. 24854	-6. 04605	-10. 02476
72. 0	1. 72994	2. 38032	-1. 59838	164. 0	0. 34454	-4. 62763	-8. 60634
74. 0	1. 63233	2. 12809	-1. 85061	166. 0	0. 41174	-3. 85376	-7. 83246
76. 0	1. 53464	1. 86005	-2. 11865	168. 0	0. 43228	-3. 64232	-7. 62102
78. 0	1. 43839	1. 57876	-2. 39994	170. 0	0. 39928	-3. 98724	-7. 96595
80. 0	1. 34499	1. 28719	-2. 69151	172. 0	0. 31958	-4. 95415	-8. 93285
82. 0	1. 25562	0. 98858	-2. 99012	174. 0	0. 21279	-6. 72059	-10. 69929
84. 0	1. 17119	0. 68627	-3. 29244	176. 0	0. 10640	-9. 73060	-13. 70930
86. 0	1. 09230	0. 38343	-3. 59527	178. 0	0. 02852	-15. 44833	-19. 42704
88. 0	1. 01926	0. 08287	-3. 89584	180	0. 00000	-INFINITY	-INFINITY
90. 0	0. 95206	-0. 21333	-4. 19204				

TABLE A2-36

EPSLN= 8.7500(NORMALIZED GROUNDPLANE RADIUS)
MONOPOLE LENGTH= 0.250000(WAVELENGTH)
MONOPOLE RADIUS= 0.1000E-05(WAVELENGTH)
NUMBER OF MONOPOLE SEGMENTS= 1
NUMBER OF DISK SEGMENTS= 26
NUMBER OF ANNULAR ZONES PER DISK SEGEMENT= 1
RADIATION RESISTANCE DETERMINED BY PATTERN= 0.317823E+02 OHMS

ELEVATION ANGLE (DEG)	DIRECTIVITY (NUMERIC)	DIRECTIVITY (dBi)	RELATIVE POWER (dB)	ELEVATION ANGLE (DEG)	DIRECTIVITY (NUMERIC)	DIRECTIVITY (dBi)	RELATIVE POWER (dB)
0.	0.00000	-INFINITY	-INFINITY				
2.0	0.02410	-16.18012	-19.84221	92.0	0.88059	-0.55228	-4.21437
4.0	0.08915	-10.49883	-14.16092	94.0	0.82469	-0.83711	-4.49920
6.0	0.17582	-7.54944	-11.21153	96.0	0.77544	-1.10453	-4.76661
8.0	0.25904	-5.86625	-9.52834	98.0	0.73177	-1.35623	-5.01832
10.0	0.31629	-4.99921	-8.66130	100.0	0.69237	-1.59663	-5.25872
12.0	0.33483	-4.75177	-8.41386	102.0	0.65568	-1.83309	-5.49518
14.0	0.31596	-5.00364	-8.66573	104.0	0.62001	-2.07601	-5.73810
16.0	0.27474	-5.61073	-9.27282	106.0	0.58360	-2.33886	-6.00095
18.0	0.23569	-6.27666	-9.93875	108.0	0.54473	-2.63819	-6.30028
20.0	0.22590	-6.46081	-10.12290	110.0	0.50191	-2.99374	-6.65583
22.0	0.26787	-5.72070	-9.38279	112.0	0.45406	-3.42890	-7.09099
24.0	0.37395	-4.27185	-7.93394	114.0	0.40072	-3.97156	-7.63365
26.0	0.54389	-2.64488	-6.30697	116.0	0.34232	-4.65566	-8.31775
28.0	0.76574	-1.15919	-4.82128	118.0	0.28031	-5.52354	-9.18563
30.0	1.01938	0.08335	-3.57874	120.0	0.21730	-6.62931	-10.29140
32.0	1.28143	1.07696	-2.58513	122.0	0.15698	-8.04150	-11.70359
34.0	1.53013	1.84729	-1.81480	124.0	0.10387	-9.83524	-13.49733
36.0	1.74894	2.42776	-1.23433	126.0	0.06281	-12.01988	-15.68197
38.0	1.92841	2.85199	-0.81010	128.0	0.03825	-14.17390	-17.83599
40.0	2.06612	3.15156	-0.51053	130.0	0.03332	-14.77316	-18.43525
42.0	2.16533	3.35523	-0.30686	132.0	0.04890	-13.10707	-16.76916
44.0	2.23270	3.48830	-0.17379	134.0	0.08286	-10.81678	-14.47887
46.0	2.27608	3.57187	-0.09022	136.0	0.12971	-8.87031	-12.53240
48.0	2.30261	3.62221	-0.03988	138.0	0.18095	-7.42450	-11.08659
50.0	2.31755	3.65030	-0.01179	140.0	0.22615	-6.45607	-10.11816
52.0	2.32385	3.66209	0.00000	142.0	0.25487	-5.93681	-9.59890
54.0	2.32231	3.65921	-0.00288	144.0	0.25903	-5.86650	-9.52859
56.0	2.31217	3.64020	-0.02189	146.0	0.23528	-6.28408	-9.94616
58.0	2.29184	3.60185	-0.06024	148.0	0.18672	-7.28801	-10.95010
60.0	2.25963	3.54036	-0.12172	150.0	0.12322	-9.09304	-12.75513
62.0	2.21425	3.45227	-0.20982	152.0	0.06000	-12.21820	-15.88029
64.0	2.15524	3.33497	-0.32712	154.0	0.01438	-18.42110	-22.08319
66.0	2.08303	3.18695	-0.47514	156.0	0.00133	-28.75558	-32.41767
68.0	1.99890	3.00791	-0.65418	158.0	0.02889	-15.39267	-19.05476
70.0	1.90485	2.79862	-0.86347	160.0	0.09489	-10.22778	-13.88987
72.0	1.80336	2.56081	-1.10128	162.0	0.18623	-7.29953	-10.96161
74.0	1.69711	2.29709	-1.36500	164.0	0.28134	-5.50776	-9.16985
76.0	1.58882	2.01075	-1.65134	166.0	0.35556	-4.49081	-8.15290
78.0	1.48105	1.70570	-1.95639	168.0	0.38817	-4.10980	-7.77189
80.0	1.37605	1.38635	-2.27574	170.0	0.36878	-4.33228	-7.99437
82.0	1.27571	1.05750	-2.60459	172.0	0.30130	-5.21002	-8.87210
84.0	1.18148	0.72427	-2.93782	174.0	0.20358	-6.91272	-10.57481
86.0	1.09442	0.39185	-3.27024	176.0	0.10281	-9.87981	-13.54190
88.0	1.01516	0.06533	-3.59676	178.0	0.02772	-15.57274	-19.23483
90.0	0.94392	-0.25064	-3.91273	180.	0.00000	-INFINITY	-INFINITY

TABLE A2-37

EPSLN= 9.0000(NORMALIZED GROUNDPLANE RADIUS)
MONOPOLE LENGTH= 0.250000(WAVELENGTH)
MONOPOLE RADIUS= 0.1000E-05(WAVELENGTH)
NUMBER OF MONOPOLE SEGMENTS= 1
NUMBER OF DISK SEGMENTS= 27
NUMBER OF ANNULAR ZONES PER DISK SEGEMENT= 1
RADIATION RESISTANCE DETERMINED BY PATTERN= 0.328995E+02 OHMS

ELEVATION ANGLE (DEG)	DIRECTIVITY (NUMERIC)	DIRECTIVITY (dBi)	RELATIVE POWER (dB)	ELEVATION ANGLE (DEG)	DIRECTIVITY (NUMERIC)	DIRECTIVITY (dBi)	RELATIVE POWER (dB)
0.	0.00000	-INFINITY	-INFINITY				
2.0	0.02659	-15.75210	-19.52164	92.0	0.84199	-0.74694	-4.51648
4.0	0.09828	-10.07542	-13.84495	94.0	0.78223	-1.06666	-4.83620
6.0	0.19350	-7.13315	-10.90269	96.0	0.73193	-1.35533	-5.12487
8.0	0.28453	-5.45871	-9.22825	98.0	0.69010	-1.61088	-5.38042
10.0	0.34669	-4.60055	-8.37009	100.0	0.65547	-1.83449	-5.60402
12.0	0.36643	-4.36007	-8.12961	102.0	0.62645	-2.03116	-5.80070
14.0	0.34555	-4.61494	-8.38448	104.0	0.60117	-2.21000	-5.97954
16.0	0.30041	-5.22288	-8.99242	106.0	0.57755	-2.38409	-6.15363
18.0	0.25673	-5.90518	-9.67472	108.0	0.55333	-2.57018	-6.33972
20.0	0.24185	-6.16453	-9.93406	110.0	0.52622	-2.78830	-6.55783
22.0	0.27708	-5.57391	-9.34345	112.0	0.49413	-3.06157	-6.83111
24.0	0.37248	-4.28897	-8.05851	114.0	0.45536	-3.41642	-7.18595
26.0	0.52519	-2.79683	-6.56637	116.0	0.40895	-3.88332	-7.65286
28.0	0.72143	-1.41808	-5.18762	118.0	0.35495	-4.49839	-8.26792
30.0	0.94096	-0.26427	-4.03381	120.0	0.29472	-5.30586	-9.07540
32.0	1.16254	0.65406	-3.11548	122.0	0.23112	-6.36164	-10.13117
34.0	1.36852	1.36252	-2.40701	124.0	0.16841	-7.73633	-11.50586
36.0	1.54786	1.89731	-1.87222	126.0	0.11200	-9.50783	-13.27737
38.0	1.69675	2.29618	-1.47335	128.0	0.06779	-11.68835	-15.45788
40.0	1.81755	2.59486	-1.17468	130.0	0.04123	-13.84816	-17.61770
42.0	1.91640	2.82486	-0.94468	132.0	0.03614	-14.42059	-18.19012
44.0	2.00063	3.01166	-0.75788	134.0	0.05353	-12.71362	-16.48316
46.0	2.07646	3.17323	-0.59631	136.0	0.09072	-10.42276	-14.19229
48.0	2.14758	3.31950	-0.45004	138.0	0.14098	-8.50838	-12.27792
50.0	2.21468	3.45311	-0.31643	140.0	0.19415	-7.11873	-10.88827
52.0	2.27575	3.57125	-0.19829	142.0	0.23823	-6.23012	-9.99966
54.0	2.32698	3.66793	-0.10161	144.0	0.26190	-5.81867	-9.58820
56.0	2.36383	3.73616	-0.03337	146.0	0.25744	-5.89322	-9.66276
58.0	2.38207	3.76954	0.00000	148.0	0.22336	-6.51001	-10.27955
60.0	2.37853	3.76309	-0.00645	150.0	0.16583	-7.80331	-11.57284
62.0	2.35161	3.71366	-0.05588	152.0	0.09828	-10.07518	-13.84471
64.0	2.30136	3.61984	-0.14970	154.0	0.03867	-14.12583	-17.89537
66.0	2.22935	3.48178	-0.28776	156.0	0.00497	-23.03919	-26.80872
68.0	2.13839	3.30087	-0.46866	158.0	0.00980	-20.08930	-23.85884
70.0	2.03214	3.07953	-0.69001	160.0	0.05593	-12.52319	-16.29273
72.0	1.91466	2.82091	-0.94862	162.0	0.13421	-8.72219	-12.49173
74.0	1.79012	2.52882	-1.24072	164.0	0.22496	-6.47890	-10.24844
76.0	1.66247	2.20753	-1.56201	166.0	0.30310	-5.18417	-8.95371
78.0	1.53526	1.86182	-1.90772	168.0	0.34547	-4.61595	-8.38549
80.0	1.41153	1.49690	-2.27264	170.0	0.33836	-4.70621	-8.47574
82.0	1.29374	1.11846	-2.65108	172.0	0.28255	-5.48906	-9.25860
84.0	1.18377	0.73267	-3.03687	174.0	0.19388	-7.12461	-10.89415
86.0	1.08297	0.34618	-3.42336	176.0	0.09893	-10.04676	-13.81630
88.0	0.99220	-0.03400	-3.80354	178.0	0.02683	-15.71382	-19.48336
90.0	0.91187	-0.40068	-4.17022	180	0.00000	-INFINITY	-INFINITY

TABLE A2-38

EPSLN= 9. 2500(NORMALIZED GROUNDPLANE RADIUS)
MONOPOLE LENGTH= 0. 250000(WAVELENGTH)
MONOPOLE RADIUS= 0. 1000E-05(WAVELENGTH)
NUMBER OF MONOPOLE SEGMENTS= 1
NUMBER OF DISK SEGMENTS= 27
NUMBER OF ANNULAR ZONES PER DISK SEGEMENT= 1
RADIATION RESISTANCE DETERMINED BY PATTERN= 0. 346397E+02 OHMS

ELEVATION ANGLE (DEG)	DIRECTIVITY (NUMERIC)	DIRECTIVITY (dBi)	RELATIVE POWER (dB)
0.	0. 00000	-INFINITY	-INFINITY
2. 0	0. 02890	-15. 39035	-19. 49540
4. 0	0. 10666	-9. 72003	-13. 82509
6. 0	0. 20949	-6. 78837	-10. 89343
8. 0	0. 30698	-5. 12893	-9. 23399
10. 0	0. 37232	-4. 29080	-8. 39586
12. 0	0. 39107	-4. 07748	-8. 18254
14. 0	0. 36536	-4. 37280	-8. 47786
16. 0	0. 31258	-5. 05034	-9. 15540
18. 0	0. 25925	-5. 86278	-9. 96784
20. 0	0. 23255	-6. 33475	-10. 43981
22. 0	0. 25255	-5. 97653	-10. 08159
24. 0	0. 32737	-4. 84957	-8. 95463
26. 0	0. 45255	-3. 44336	-7. 54842
28. 0	0. 61400	-2. 11831	-6. 22337
30. 0	0. 79328	-1. 00571	-5. 11077
32. 0	0. 97306	-0. 11861	-4. 22367
34. 0	1. 14120	0. 57362	-3. 53144
36. 0	1. 29261	1. 11468	-2. 99038
38. 0	1. 42868	1. 54934	-2. 55572
40. 0	1. 55503	1. 91738	-2. 18768
42. 0	1. 67857	2. 24940	-1. 85566
44. 0	1. 80478	2. 56424	-1. 54082
46. 0	1. 93584	2. 86870	-1. 23636
48. 0	2. 07006	3. 15983	-0. 94523
50. 0	2. 20225	3. 42867	-0. 67639
52. 0	2. 32497	3. 66417	-0. 44088
54. 0	2. 43001	3. 85608	-0. 24898
56. 0	2. 50984	3. 99646	-0. 10860
58. 0	2. 55874	4. 08026	-0. 02480
60. 0	2. 57339	4. 10506	0. 00000
62. 0	2. 55310	4. 07068	-0. 03438
64. 0	2. 49953	3. 97859	-0. 12647
66. 0	2. 41628	3. 83147	-0. 27359
68. 0	2. 30822	3. 63278	-0. 47228
70. 0	2. 18098	3. 38651	-0. 71855
72. 0	2. 04031	3. 09696	-1. 00809
74. 0	1. 89176	2. 76866	-1. 33640
76. 0	1. 74032	2. 40628	-1. 69878
78. 0	1. 59027	2. 01470	-2. 09036
80. 0	1. 44512	1. 59904	-2. 50602
82. 0	1. 30762	1. 16482	-2. 94024
84. 0	1. 17981	0. 71811	-3. 38695
86. 0	1. 06309	0. 26569	-3. 83937
88. 0	0. 95835	-0. 18474	-4. 28980
90. 0	0. 86606	-0. 62453	-4. 72959
92. 0	0. 78629	-1. 04419	-5. 14925
94. 0	0. 71882	-1. 43380	-5. 53886
96. 0	0. 66316	-1. 78381	-5. 88887
98. 0	0. 61854	-2. 08632	-6. 19138
100. 0	0. 58390	-2. 33658	-6. 44164
102. 0	0. 55789	-2. 53451	-6. 63957
104. 0	0. 53880	-2. 68572	-6. 79078
106. 0	0. 52458	-2. 80186	-6. 90692
108. 0	0. 51285	-2. 90010	-7. 00516
110. 0	0. 50094	-3. 00214	-7. 10719
112. 0	0. 48606	-3. 13308	-7. 23814
114. 0	0. 46551	-3. 32074	-7. 42580
116. 0	0. 43696	-3. 59558	-7. 70063
118. 0	0. 39890	-3. 99137	-8. 09643
120. 0	0. 35100	-4. 54697	-8. 65203
122. 0	0. 29451	-5. 30901	-9. 41407
124. 0	0. 23251	-6. 33553	-10. 44059
126. 0	0. 16988	-7. 69846	-11. 80351
128. 0	0. 11292	-9. 47229	-13. 57735
130. 0	0. 06852	-11. 64182	-15. 74688
132. 0	0. 04298	-13. 66719	-17. 77225
134. 0	0. 04053	-13. 92190	-18. 02696
136. 0	0. 06191	-12. 08227	-16. 18733
138. 0	0. 10337	-9. 85607	-13. 96113
140. 0	0. 15654	-8. 05377	-12. 15882
142. 0	0. 20947	-6. 78882	-10. 89388
144. 0	0. 24889	-6. 03996	-10. 14502
146. 0	0. 26340	-5. 79378	-9. 89884
148. 0	0. 24688	-6. 07520	-10. 18025
150. 0	0. 20100	-6. 96794	-11. 07300
152. 0	0. 13606	-8. 66283	-12. 76789
154. 0	0. 06910	-11. 60501	-15. 71007
156. 0	0. 01979	-17. 03657	-21. 14163
158. 0	0. 00452	-23. 44414	-27. 54920
160. 0	0. 03090	-15. 10029	-19. 20535
162. 0	0. 09416	-10. 26148	-14. 36654
164. 0	0. 17742	-7. 51009	-11. 61515
166. 0	0. 25607	-5. 91640	-10. 02145
168. 0	0. 30533	-5. 15227	-9. 25733
170. 0	0. 30856	-5. 10661	-9. 21167
172. 0	0. 26345	-5. 79306	-9. 89812
174. 0	0. 18362	-7. 36089	-11. 46595
176. 0	0. 09467	-10. 23795	-14. 34301
178. 0	0. 02583	-15. 87912	-19. 98418
180	0. 00000	-INFINITY	-INFINITY

TABLE A2-39

EPSLN= 9.5000(NORMALIZED GROUNDPLANE RADIUS)
MONOPOLE LENGTH= 0.250000(WAVELENGTH)
MONOPOLE RADIUS= 0.1000E-05(WAVELENGTH)
NUMBER OF MONOPOLE SEGMENTS= 1
NUMBER OF DISK SEGMENTS= 28
NUMBER OF ANNULAR ZONES PER DISK SEGEMENT= 1
RADIATION RESISTANCE DETERMINED BY PATTERN= 0.365792E+02 OHMS

ELEVATION ANGLE (DEG)	DIRECTIVITY (NUMERIC)	DIRECTIVITY (dBi)	RELATIVE POWER (dB)	ELEVATION ANGLE (DEG)	DIRECTIVITY (NUMERIC)	DIRECTIVITY (dBi)	RELATIVE POWER (dB)
0.	0.00000	-INFINITY	-INFINITY				
2.0	0.03092	-15.09731	-19.56780	92.0	0.72979	-1.36804	-5.83853
4.0	0.11392	-9.43396	-13.90444	94.0	0.65287	-1.85173	-6.32221
6.0	0.22314	-6.51417	-10.98466	96.0	0.58905	-2.29850	-6.76899
8.0	0.32568	-4.87215	-9.34264	98.0	0.53780	-2.69379	-7.16427
10.0	0.39276	-4.05873	-8.52922	100.0	0.49843	-3.02399	-7.49448
12.0	0.40909	-3.88178	-8.35226	102.0	0.46997	-3.27926	-7.74974
14.0	0.37705	-4.23606	-8.70655	104.0	0.45119	-3.45640	-7.92689
16.0	0.31470	-5.02109	-9.49157	106.0	0.44044	-3.56114	-8.03162
18.0	0.24894	-6.03913	-10.50962	108.0	0.43565	-3.60862	-8.07911
20.0	0.20650	-6.85087	-11.32136	110.0	0.43429	-3.62217	-8.09265
22.0	0.20619	-6.85724	-11.32772	112.0	0.43343	-3.63086	-8.10135
24.0	0.25479	-5.93820	-10.40868	114.0	0.42982	-3.66709	-8.13758
26.0	0.34729	-4.59308	-9.06357	116.0	0.42024	-3.76499	-8.23548
28.0	0.47086	-3.27108	-7.74156	118.0	0.40179	-3.96001	-8.43050
30.0	0.61042	-2.14368	-6.61417	120.0	0.37241	-4.28984	-8.76033
32.0	0.75382	-1.22730	-5.69778	122.0	0.33140	-4.79643	-9.26691
34.0	0.89496	-0.48196	-4.95245	124.0	0.27996	-5.52904	-9.99953
36.0	1.03427	0.14636	-4.32413	126.0	0.22142	-6.54786	-11.01834
38.0	1.17695	0.70758	-3.76291	128.0	0.16127	-7.92451	-12.39500
40.0	1.32980	1.23787	-3.23261	130.0	0.10664	-9.72085	-14.19133
42.0	1.49807	1.75532	-2.71516	132.0	0.06526	-11.85324	-16.32372
44.0	1.68305	2.26097	-2.20952	134.0	0.04397	-13.56878	-18.03927
46.0	1.88111	2.74415	-1.72634	136.0	0.04689	-13.28906	-17.75955
48.0	2.08412	3.18923	-1.28126	138.0	0.07391	-11.31311	-15.78359
50.0	2.28087	3.58100	-0.88948	140.0	0.11968	-9.21992	-13.69041
52.0	2.45907	3.90771	-0.56278	142.0	0.17389	-7.59717	-12.06765
54.0	2.60730	4.16192	-0.30857	144.0	0.22299	-6.51706	-10.98755
56.0	2.71658	4.34022	-0.13027	146.0	0.25322	-5.96505	-10.43554
58.0	2.78124	4.44239	-0.02810	148.0	0.25444	-5.94422	-10.41471
60.0	2.79929	4.47049	0.00000	150.0	0.22369	-6.50357	-10.97406
62.0	2.77210	4.42808	-0.04240	152.0	0.16718	-7.76819	-12.23868
64.0	2.70375	4.31966	-0.15083	154.0	0.09968	-10.01403	-14.48452
66.0	2.60027	4.15018	-0.32031	156.0	0.04098	-13.87390	-18.34439
68.0	2.46876	3.92478	-0.54570	158.0	0.01010	-19.95706	-24.42755
70.0	2.31664	3.64859	-0.82190	160.0	0.01879	-17.26167	-21.73216
72.0	2.15109	3.32658	-1.14390	162.0	0.06675	-11.75580	-16.22629
74.0	1.97858	2.96354	-1.50694	164.0	0.14047	-8.52417	-12.99466
76.0	1.80470	2.56406	-1.90643	166.0	0.21672	-6.64099	-11.11148
78.0	1.63401	2.13256	-2.33793	168.0	0.26992	-5.68762	-10.15811
80.0	1.47008	1.67342	-2.79707	170.0	0.28111	-5.51124	-9.98173
82.0	1.31557	1.19113	-3.27936	172.0	0.24517	-6.10527	-10.57576
84.0	1.17233	0.69050	-3.77999	174.0	0.17345	-7.60832	-12.07881
86.0	1.04160	0.17701	-4.29347	176.0	0.09032	-10.44231	-14.91279
88.0	0.92409	-0.34285	-4.81333	178.0	0.02478	-16.05877	-20.52925
90.0	0.82014	-0.86113	-5.33162	180	0.00000	-INFINITY	-INFINITY

TABLE A2-40

EPSLN= 9. 7500(NORMALIZED GROUNDPLANE RADIUS)
MONOPOLE LENGTH= 0. 250000(WAVELENGTH)
MONOPOLE RADIUS= 0. 1000E-05(WAVELENGTH)
NUMBER OF MONOPOLE SEGMENTS= 1
NUMBER OF DISK SEGMENTS= 29
NUMBER OF ANNULAR ZONES PER DISK SEGEMENT= 1
RADIATION RESISTANCE DETERMINED BY PATTERN= 0. 380863E+02 OHMS

ELEVATION ANGLE (DEG)	DIRECTIVITY (NUMERIC)	DIRECTIVITY (dBi)	RELATIVE POWER (dB)
0.	0. 00000	-INFINITY	-INFINITY
2. 0	0. 03276	-14. 84643	-19. 61948
4. 0	0. 12052	-9. 18932	-13. 96236
6. 0	0. 23550	-6. 28011	-11. 05316
8. 0	0. 34249	-4. 65357	-9. 42662
10. 0	0. 41095	-3. 86209	-8. 63514
12. 0	0. 42483	-3. 71783	-8. 49087
14. 0	0. 38664	-4. 12692	-8. 89997
16. 0	0. 31492	-5. 01795	-9. 79099
18. 0	0. 23667	-6. 25860	-11. 03165
20. 0	0. 17799	-7. 49607	-12. 26912
22. 0	0. 15661	-8. 05193	-12. 82498
24. 0	0. 17847	-7. 48427	-12. 25731
26. 0	0. 23903	-6. 21544	-10. 98849
28. 0	0. 32779	-4. 84408	-9. 61713
30. 0	0. 43396	-3. 62552	-8. 39857
32. 0	0. 55096	-2. 58882	-7. 36187
34. 0	0. 67831	-1. 68573	-6. 45877
36. 0	0. 82077	-0. 85778	-5. 63082
38. 0	0. 98544	-0. 06368	-4. 83673
40. 0	1. 17815	0. 71201	-4. 06103
42. 0	1. 40043	1. 46260	-3. 31045
44. 0	1. 64790	2. 16932	-2. 60372
46. 0	1. 91045	2. 81136	-1. 96168
48. 0	2. 17370	3. 37199	-1. 40106
50. 0	2. 42139	3. 84066	-0. 93239
52. 0	2. 63797	4. 21270	-0. 56035
54. 0	2. 81061	4. 48801	-0. 28504
56. 0	2. 93065	4. 66964	-0. 10341
58. 0	2. 99406	4. 76260	-0. 01044
60. 0	3. 00127	4. 77305	0. 00000
62. 0	2. 95638	4. 70761	-0. 06544
64. 0	2. 86618	4. 57303	-0. 20002
66. 0	2. 73899	4. 37590	-0. 39715
68. 0	2. 58373	4. 12247	-0. 65058
70. 0	2. 40911	3. 81856	-0. 95449
72. 0	2. 22305	3. 46950	-1. 30355
74. 0	2. 03241	3. 08011	-1. 69294
76. 0	1. 84276	2. 65469	-2. 11836
78. 0	1. 65847	2. 19708	-2. 57596
80. 0	1. 48277	1. 71074	-3. 06230
82. 0	1. 31791	1. 19885	-3. 57420
84. 0	1. 16533	0. 66451	-4. 10854
86. 0	1. 02589	0. 11102	-4. 66203
88. 0	0. 89998	-0. 45768	-5. 23073
90. 0	0. 78768	-1. 03648	-5. 80952

ELEVATION ANGLE (DEG)	DIRECTIVITY (NUMERIC)	DIRECTIVITY (dBi)	RELATIVE POWER (dB)
92. 0	0. 68892	-1. 61833	-6. 39138
94. 0	0. 60348	-2. 19341	-6. 96645
96. 0	0. 53110	-2. 74820	-7. 52124
98. 0	0. 47151	-3. 26511	-8. 03815
100. 0	0. 42433	-3. 72301	-8. 49605
102. 0	0. 38908	-4. 09960	-8. 87264
104. 0	0. 36510	-4. 37593	-9. 14897
106. 0	0. 35138	-4. 54219	-9. 31524
108. 0	0. 34653	-4. 60262	-9. 37567
110. 0	0. 34859	-4. 57691	-9. 34995
112. 0	0. 35502	-4. 49748	-9. 27053
114. 0	0. 36272	-4. 40431	-9. 17735
116. 0	0. 36813	-4. 33993	-9. 11298
118. 0	0. 36757	-4. 34657	-9. 11962
120. 0	0. 35764	-4. 46559	-9. 23864
122. 0	0. 33582	-4. 73888	-9. 51192
124. 0	0. 30119	-5. 21159	-9. 98464
126. 0	0. 25493	-5. 93573	-10. 70878
128. 0	0. 20076	-6. 97321	-11. 74626
130. 0	0. 14480	-8. 39225	-13. 16530
132. 0	0. 09493	-10. 22583	-14. 99888
134. 0	0. 05945	-12. 25883	-17. 03187
136. 0	0. 04520	-13. 44885	-18. 22190
138. 0	0. 05559	-12. 54980	-17. 32285
140. 0	0. 08893	-10. 50938	-15. 28243
142. 0	0. 13778	-8. 60818	-13. 38123
144. 0	0. 18984	-7. 21609	-11. 98913
146. 0	0. 23057	-6. 37206	-11. 14511
148. 0	0. 24701	-6. 07283	-10. 84588
150. 0	0. 23207	-6. 34380	-11. 11685
152. 0	0. 18760	-7. 26773	-12. 04078
154. 0	0. 12510	-9. 02734	-13. 80039
156. 0	0. 06321	-11. 99197	-16. 76501
158. 0	0. 02219	-16. 53925	-21. 31229
160. 0	0. 01700	-17. 69659	-22. 46963
162. 0	0. 05139	-12. 89147	-17. 66451
164. 0	0. 11536	-9. 37934	-14. 15239
166. 0	0. 18759	-7. 26796	-12. 04101
168. 0	0. 24239	-6. 15482	-10. 92786
170. 0	0. 25911	-5. 86509	-10. 63813
172. 0	0. 23026	-6. 37787	-11. 15092
174. 0	0. 16506	-7. 82351	-12. 59656
176. 0	0. 08671	-10. 61912	-15. 39217
178. 0	0. 02391	-16. 21350	-20. 98655
180.	0. 00000	-INFINITY	-INFINITY

TABLE A2-41

EPSLN= 10. 0000(NORMALIZED GROUNDPLANE RADIUS)
MONOPOLE LENGTH= 0. 250000(WAVELENGTH)
MONOPOLE RADIUS= 0. 1000E-05(WAVELENGTH)
NUMBER OF MONOPOLE SEGMENTS= 1
NUMBER OF DISK SEGMENTS= 30
NUMBER OF ANNULAR ZONES PER DISK SEGEMENT= 1
RADIATION RESISTANCE DETERMINED BY PATTERN= 0. 387759E+02 OHMS

ELEVATION ANGLE (DEG)	DIRECTIVITY (NUMERIC)	DIRECTIVITY (dBi)	RELATIVE POWER (dB)	ELEVATION ANGLE (DEG)	DIRECTIVITY (NUMERIC)	DIRECTIVITY (dBi)	RELATIVE POWER (dB)
0.	0. 00000	-INFINITY	-INFINITY				
2. 0	0. 03473	-14. 59347	-19. 58491	92. 0	0. 67036	-1. 73695	-6. 72838
4. 0	0. 12761	-8. 94129	-13. 93272	94. 0	0. 57930	-2. 37097	-7. 36241
6. 0	0. 24887	-6. 04019	-11. 03163	96. 0	0. 50025	-3. 00816	-7. 99960
8. 0	0. 36100	-4. 42494	-9. 41638	98. 0	0. 43301	-3. 63500	-8. 62643
10. 0	0. 43167	-3. 64846	-8. 63989	100. 0	0. 37749	-4. 23100	-9. 22243
12. 0	0. 44410	-3. 52517	-8. 51661	102. 0	0. 33358	-4. 76801	-9. 75944
14. 0	0. 40098	-3. 96879	-8. 96023	104. 0	0. 30116	-5. 21207	-10. 20351
16. 0	0. 32132	-4. 93065	-9. 92208	106. 0	0. 27991	-5. 52982	-10. 52125
18. 0	0. 23221	-6. 34119	-11. 33262	108. 0	0. 26921	-5. 69903	-10. 69047
20. 0	0. 15922	-7. 98002	-12. 97146	110. 0	0. 26796	-5. 71927	-10. 71071
22. 0	0. 11920	-9. 23742	-14. 22885	112. 0	0. 27442	-5. 61589	-10. 60733
24. 0	0. 11770	-9. 29233	-14. 28377	114. 0	0. 28610	-5. 43478	-10. 42621
26. 0	0. 15113	-8. 20640	-13. 19784	116. 0	0. 29980	-5. 23165	-10. 22309
28. 0	0. 21183	-6. 74021	-11. 73165	118. 0	0. 31169	-5. 06272	-10. 05415
30. 0	0. 29346	-5. 32453	-10. 31596	120. 0	0. 31769	-4. 97995	-9. 97139
32. 0	0. 39466	-4. 03780	-9. 02923	122. 0	0. 31398	-5. 03102	-10. 02245
34. 0	0. 51959	-2. 84338	-7. 83482	124. 0	0. 29772	-5. 26185	-10. 25328
36. 0	0. 67582	-1. 70169	-6. 69312	126. 0	0. 26788	-5. 72057	-10. 71200
38. 0	0. 87058	-0. 60193	-5. 59337	128. 0	0. 22586	-6. 46156	-11. 45299
40. 0	1. 10702	0. 44154	-4. 54989	130. 0	0. 17591	-7. 54711	-12. 53854
42. 0	1. 38176	1. 40432	-3. 58712	132. 0	0. 12489	-9. 03487	-14. 02631
44. 0	1. 68431	2. 26422	-2. 72722	134. 0	0. 08133	-10. 89756	-15. 88899
46. 0	1. 99839	3. 00681	-1. 98463	136. 0	0. 05376	-12. 69535	-17. 68678
48. 0	2. 30453	3. 62582	-1. 36562	138. 0	0. 04849	-13. 14338	-18. 13481
50. 0	2. 58310	4. 12142	-0. 87002	140. 0	0. 06743	-11. 71162	-16. 70305
52. 0	2. 81715	4. 49810	-0. 49334	142. 0	0. 10660	-9. 72240	-14. 71383
54. 0	2. 99429	4. 76294	-0. 22849	144. 0	0. 15612	-8. 06543	-13. 05687
56. 0	3. 10770	4. 92439	-0. 06704	146. 0	0. 20200	-6. 94639	-11. 93782
58. 0	3. 15605	4. 99143	0. 00000	148. 0	0. 22979	-6. 38664	-11. 37807
60. 0	3. 14274	4. 97308	-0. 01835	150. 0	0. 22909	-6. 39993	-11. 39137
62. 0	3. 07472	4. 87806	-0. 11338	152. 0	0. 19765	-7. 04093	-12. 03236
64. 0	2. 96113	4. 71458	-0. 27685	154. 0	0. 14335	-8. 43591	-13. 42734
66. 0	2. 81208	4. 49028	-0. 50116	156. 0	0. 08279	-10. 82008	-15. 81152
68. 0	2. 63760	4. 21209	-0. 77934	158. 0	0. 03645	-14. 38318	-19. 37461
70. 0	2. 44696	3. 88626	-1. 10517	160. 0	0. 02158	-16. 65960	-21. 65103
72. 0	2. 24818	3. 51830	-1. 47313	162. 0	0. 04536	-13. 43369	-18. 42513
74. 0	2. 04786	3. 11299	-1. 87844	164. 0	0. 10105	-9. 95448	-14. 94591
76. 0	1. 85114	2. 67439	-2. 31704	166. 0	0. 16933	-7. 71278	-12. 70421
78. 0	1. 66184	2. 20588	-2. 78555	168. 0	0. 22469	-6. 48416	-11. 47560
80. 0	1. 48258	1. 71017	-3. 28126	170. 0	0. 24515	-6. 10574	-11. 09717
82. 0	1. 31504	1. 18939	-3. 80204	172. 0	0. 22118	-6. 55246	-11. 54390
84. 0	1. 16016	0. 64519	-4. 34625	174. 0	0. 16031	-7. 95040	-12. 94184
86. 0	1. 01833	0. 07889	-4. 91254	176. 0	0. 08485	-10. 71372	-15. 70515
88. 0	0. 88957	-0. 50821	-5. 49964	178. 0	0. 02350	-16. 28937	-21. 28080
90. 0	0. 77368	-1. 11441	-6. 10584	180	0. 00000	-INFINITY	-INFINITY

TABLE A2-42

EPSLN= 10.2500(NORMALIZED GROUNDPLANE RADIUS)
MONOPOLE LENGTH= 0.250000(WAVELENGTH)
MONOPOLE RADIUS= 0.1000E-05(WAVELENGTH)
NUMBER OF MONOPOLE SEGMENTS= 1
NUMBER OF DISK SEGMENTS= 30
NUMBER OF ANNULAR ZONES PER DISK SEGEMENT= 1
RADIATION RESISTANCE DETERMINED BY PATTERN= 0.385880E+02 OHMS

ELEVATION ANGLE (DEG)	DIRECTIVITY (NUMERIC)	DIRECTIVITY (dBi)	RELATIVE POWER (dB)
0.	0.00000	-INFINITY	-INFINITY
2.0	0.03708	-14.30827	-19.42820
4.0	0.13615	-8.65979	-13.77972
6.0	0.26519	-5.76440	-10.88433
8.0	0.38404	-4.15618	-9.27611
10.0	0.45844	-3.38719	-8.50712
12.0	0.47083	-3.27139	-8.39132
14.0	0.42421	-3.72418	-8.84411
16.0	0.33829	-4.70707	-9.82700
18.0	0.24048	-6.18921	-11.30914
20.0	0.15599	-8.06902	-13.18894
22.0	0.10100	-9.95700	-15.07693
24.0	0.08086	-10.92256	-16.04249
26.0	0.09313	-10.30931	-15.42924
28.0	0.13294	-8.76360	-13.88353
30.0	0.19815	-7.03016	-12.15009
32.0	0.29188	-5.34791	-10.46784
34.0	0.42184	-3.74850	-8.86843
36.0	0.59700	-2.24028	-7.36020
38.0	0.82334	-0.84422	-5.96414
40.0	1.10031	0.41516	-4.70477
42.0	1.41918	1.52038	-3.59955
44.0	1.76363	2.46407	-2.65586
46.0	2.11225	3.24745	-1.87247
48.0	2.44201	3.87748	-1.24245
50.0	2.73172	4.36436	-0.75556
52.0	2.96477	4.71990	-0.40003
54.0	3.13066	4.95636	-0.16357
56.0	3.22540	5.08584	-0.03409
58.0	3.25082	5.11993	0.00000
60.0	3.21328	5.06949	-0.05044
62.0	3.12217	4.94457	-0.17536
64.0	2.98835	4.75431	-0.36562
66.0	2.82291	4.50697	-0.61296
68.0	2.63628	4.20991	-0.91002
70.0	2.43757	3.86958	-1.25035
72.0	2.23437	3.49155	-1.62838
74.0	2.03261	3.08053	-2.03940
76.0	1.83671	2.64042	-2.47951
78.0	1.64978	2.17426	-2.94566
80.0	1.47379	1.68435	-3.43558
82.0	1.30984	1.17218	-3.94775
84.0	1.15838	0.63851	-4.48142
86.0	1.01940	0.08343	-5.03650
88.0	0.89257	-0.49356	-5.61349
90.0	0.77744	-1.09332	-6.21324

ELEVATION ANGLE (DEG)	DIRECTIVITY (NUMERIC)	DIRECTIVITY (dBi)	RELATIVE POWER (dB)
92.0	0.67349	-1.71669	-6.83662
94.0	0.58025	-2.36385	-7.48378
96.0	0.49737	-3.03319	-8.15312
98.0	0.42466	-3.71956	-8.83949
100.0	0.36211	-4.41164	-9.53157
102.0	0.30984	-5.08857	-10.20850
104.0	0.26812	-5.71673	-10.83666
106.0	0.23718	-6.24926	-11.36918
108.0	0.21712	-6.63290	-11.75283
110.0	0.20773	-6.82496	-11.94489
112.0	0.20823	-6.81456	-11.93448
114.0	0.21710	-6.63331	-11.75324
116.0	0.23195	-6.34614	-11.46606
118.0	0.24942	-6.03062	-11.15054
120.0	0.26545	-5.76022	-10.88015
122.0	0.27556	-5.59788	-10.71780
124.0	0.27559	-5.59734	-10.71727
126.0	0.26254	-5.80812	-10.92804
128.0	0.23545	-6.28096	-11.40089
130.0	0.19624	-7.07218	-12.19211
132.0	0.14991	-8.24159	-13.36152
134.0	0.10419	-9.82154	-14.94147
136.0	0.06816	-11.66478	-16.78471
138.0	0.05012	-12.99990	-18.11983
140.0	0.05512	-12.58719	-17.70712
142.0	0.08273	-10.82345	-15.94337
144.0	0.12609	-8.99320	-14.11313
146.0	0.17281	-7.62432	-12.74424
148.0	0.20802	-6.81892	-11.93884
150.0	0.21899	-6.59582	-11.71575
152.0	0.19989	-6.99211	-12.11204
154.0	0.15498	-8.09736	-13.21729
156.0	0.09842	-10.06923	-15.18916
158.0	0.05025	-12.98836	-18.10829
160.0	0.02932	-15.32831	-20.44824
162.0	0.04564	-13.40691	-18.52684
164.0	0.09536	-10.20648	-15.32641
166.0	0.16090	-7.93433	-13.05426
168.0	0.21691	-6.63722	-11.75715
170.0	0.24009	-6.19624	-11.31617
172.0	0.21913	-6.59291	-11.71284
174.0	0.16020	-7.95331	-13.07324
176.0	0.08530	-10.69064	-15.81057
178.0	0.02371	-16.25104	-21.37097
180	0.00000	-INFINITY	-INFINITY

TABLE A2-43

EPSLN= 10. 5000(NORMALIZED GROUNDPLANE RADIUS)
MONOPOLE LENGTH= 0. 250000(WAVELENGTH)
MONOPOLE RADIUS= 0. 1000E-05(WAVELENGTH)
NUMBER OF MONOPOLE SEGMENTS= 1
NUMBER OF DISK SEGMENTS= 31
NUMBER OF ANNULAR ZONES PER DISK SEGEMENT= 1
RADIATION RESISTANCE DETERMINED BY PATTERN= 0. 378627E+02 OHMS

ELEVATION ANGLE (DEG)	DIRECTIVITY (NUMERIC)	DIRECTIVITY (dBi)	RELATIVE POWER (dB)
0.	0. 00000	-INFINITY	-INFINITY
2. 0	0. 04012	-13. 96648	-19. 13043
4. 0	0. 14719	-8. 32129	-13. 48524
6. 0	0. 28638	-5. 43065	-10. 59459
8. 0	0. 41425	-3. 82737	-8. 99132
10. 0	0. 49413	-3. 06157	-8. 22551
12. 0	0. 50766	-2. 94431	-8. 10825
14. 0	0. 45841	-3. 38748	-8. 55143
16. 0	0. 36723	-4. 35060	-9. 51454
18. 0	0. 26230	-5. 81201	-10. 97596
20. 0	0. 16877	-7. 72697	-12. 89092
22. 0	0. 10225	-9. 90342	-15. 06737
24. 0	0. 06786	-11. 68378	-16. 84773
26. 0	0. 06412	-11. 93006	-17. 09401
28. 0	0. 08872	-10. 51967	-15. 68362
30. 0	0. 14323	-8. 43964	-13. 60359
32. 0	0. 23457	-6. 29724	-11. 46119
34. 0	0. 37304	-4. 28244	-9. 44638
36. 0	0. 56802	-2. 45637	-7. 62032
38. 0	0. 82338	-0. 84402	-6. 00796
40. 0	1. 13435	0. 54745	-4. 61649
42. 0	1. 48689	1. 72278	-3. 44116
44. 0	1. 85954	2. 69406	-2. 46989
46. 0	2. 22697	3. 47714	-1. 68681
48. 0	2. 56408	4. 08931	-1. 07464
50. 0	2. 84964	4. 54791	-0. 61604
52. 0	3. 06874	4. 86960	-0. 29435
54. 0	3. 21369	5. 07004	-0. 09390
56. 0	3. 28377	5. 16373	-0. 00021
58. 0	3. 28394	5. 16395	0. 00000
60. 0	3. 22311	5. 08276	-0. 08119
62. 0	3. 11246	4. 93104	-0. 23291
64. 0	2. 96381	4. 71850	-0. 44545
66. 0	2. 78851	4. 45373	-0. 71022
68. 0	2. 59669	4. 14421	-1. 01974
70. 0	2. 39685	3. 79640	-1. 36755
72. 0	2. 19573	3. 41579	-1. 74816
74. 0	1. 99845	3. 00693	-2. 15702
76. 0	1. 80863	2. 57351	-2. 59044
78. 0	1. 62869	2. 11838	-3. 04557
80. 0	1. 46002	1. 64359	-3. 52035
82. 0	1. 30328	1. 15039	-4. 01356
84. 0	1. 15856	0. 63918	-4. 52476
86. 0	1. 02555	0. 10956	-5. 05438
88. 0	0. 90370	-0. 43976	-5. 60370
90. 0	0. 79234	-1. 01090	-6. 17485

ELEVATION ANGLE (DEG)	DIRECTIVITY (NUMERIC)	DIRECTIVITY (dBi)	RELATIVE POWER (dB)
92. 0	0. 69076	-1. 60675	-6. 77069
94. 0	0. 59832	-2. 23066	-7. 39460
96. 0	0. 51452	-2. 88594	-8. 04988
98. 0	0. 43906	-3. 57477	-8. 73872
100. 0	0. 37185	-4. 29633	-9. 46027
102. 0	0. 31307	-5. 04352	-10. 20746
104. 0	0. 26315	-5. 79798	-10. 96193
106. 0	0. 22266	-6. 52360	-11. 68754
108. 0	0. 19225	-7. 16134	-12. 32529
110. 0	0. 17245	-7. 63343	-12. 79738
112. 0	0. 16343	-7. 86674	-13. 03069
114. 0	0. 16475	-7. 83179	-12. 99574
116. 0	0. 17509	-7. 56751	-12. 73145
118. 0	0. 19204	-7. 16610	-12. 33004
120. 0	0. 21209	-6. 73471	-11. 89866
122. 0	0. 23081	-6. 36745	-11. 53139
124. 0	0. 24333	-6. 13812	-11. 30207
126. 0	0. 24516	-6. 10542	-11. 26937
128. 0	0. 23328	-6. 32126	-11. 48520
130. 0	0. 20711	-6. 83809	-12. 00204
132. 0	0. 16935	-7. 71218	-12. 87613
134. 0	0. 12611	-8. 99260	-14. 15655
136. 0	0. 08606	-10. 65202	-15. 81597
138. 0	0. 05862	-12. 31976	-17. 48371
140. 0	0. 05131	-12. 89804	-18. 06199
142. 0	0. 06705	-11. 73584	-16. 89979
144. 0	0. 10227	-9. 90271	-15. 06665
146. 0	0. 14679	-8. 33296	-13. 49690
148. 0	0. 18620	-7. 30027	-12. 46422
150. 0	0. 20618	-6. 85762	-12. 02156
152. 0	0. 19787	-7. 03615	-12. 20010
154. 0	0. 16211	-7. 90185	-13. 06579
156. 0	0. 11054	-9. 56498	-14. 72893
158. 0	0. 06246	-12. 04431	-17. 20825
160. 0	0. 03795	-14. 20798	-19. 37193
162. 0	0. 04953	-13. 05140	-18. 21535
164. 0	0. 09587	-10. 18318	-15. 34713
166. 0	0. 16076	-7. 93809	-13. 10204
168. 0	0. 21855	-6. 60453	-11. 76848
170. 0	0. 24434	-6. 12001	-11. 28395
172. 0	0. 22498	-6. 47859	-11. 64254
174. 0	0. 16561	-7. 80917	-12. 97312
176. 0	0. 08861	-10. 52536	-15. 68931
178. 0	0. 02470	-16. 07319	-21. 23713
180.	0. 00000	-INFINITY	-INFINITY

TABLE A2-44

EPSLN= 10. 7500(NORMALIZED GROUNDPLANE RADIUS)
MONOPOLE LENGTH= 0. 250000(WAVELENGTH)
MONOPOLE RADIUS= 0. 1000E-05(WAVELENGTH)
NUMBER OF MONOPOLE SEGMENTS= 1
NUMBER OF DISK SEGMENTS= 32
NUMBER OF ANNULAR ZONES PER DISK SEGEMENT= 1
RADIATION RESISTANCE DETERMINED BY PATTERN= 0. 368938E+02 OHMS

ELEVATION ANGLE (DEG)	DIRECTIVITY (NUMERIC)	DIRECTIVITY (dBi)	RELATIVE POWER (dB)	ELEVATION ANGLE (DEG)	DIRECTIVITY (NUMERIC)	DIRECTIVITY (dBi)	RELATIVE POWER (dB)
0.	0. 00000	-INFINITY	-INFINITY				
2. 0	0. 04402	-13. 56330	-18. 73479	92. 0	0. 71581	-1. 45200	-6. 62348
4. 0	0. 16136	-7. 92195	-13. 09344	94. 0	0. 62636	-2. 03174	-7. 20323
6. 0	0. 31356	-5. 03682	-10. 20831	96. 0	0. 54410	-2. 64322	-7. 81470
8. 0	0. 45298	-3. 43917	-8. 61065	98. 0	0. 46851	-3. 29277	-8. 46425
10. 0	0. 53996	-2. 67639	-7. 84788	100. 0	0. 39935	-3. 98649	-9. 15798
12. 0	0. 55525	-2. 55514	-7. 72662	102. 0	0. 33664	-4. 72840	-9. 89988
14. 0	0. 50347	-2. 98029	-8. 15177	104. 0	0. 28074	-5. 51691	-10. 68840
16. 0	0. 40728	-3. 90108	-9. 07256	106. 0	0. 23235	-6. 33858	-11. 51007
18. 0	0. 29619	-5. 28424	-10. 45572	108. 0	0. 19239	-7. 15823	-12. 32972
20. 0	0. 19559	-7. 08656	-12. 25805	110. 0	0. 16191	-7. 90733	-13. 07881
22. 0	0. 12043	-9. 19254	-14. 36402	112. 0	0. 14187	-8. 48120	-13. 65268
24. 0	0. 07534	-11. 22959	-16. 40108	114. 0	0. 13284	-8. 76675	-13. 93824
26. 0	0. 05942	-12. 26043	-17. 43192	116. 0	0. 13468	-8. 70705	-13. 87854
28. 0	0. 07252	-11. 39556	-16. 56705	118. 0	0. 14618	-8. 35107	-13. 52256
30. 0	0. 11945	-9. 22807	-14. 39956	120. 0	0. 16485	-7. 82903	-13. 00052
32. 0	0. 21042	-6. 76915	-11. 94063	122. 0	0. 18685	-7. 28501	-12. 45650
34. 0	0. 35768	-4. 46500	-9. 63648	124. 0	0. 20727	-6. 83469	-12. 00618
36. 0	0. 57039	-2. 43825	-7. 60973	126. 0	0. 22076	-6. 56071	-11. 73219
38. 0	0. 84981	-0. 70676	-5. 87825	128. 0	0. 22261	-6. 52447	-11. 69596
40. 0	1. 18683	0. 74390	-4. 42759	130. 0	0. 20994	-6. 77896	-11. 95044
42. 0	1. 56246	1. 93810	-3. 23339	132. 0	0. 18293	-7. 37711	-12. 54860
44. 0	1. 95092	2. 90240	-2. 26909	134. 0	0. 14551	-8. 37105	-13. 54254
46. 0	2. 32414	3. 66263	-1. 50886	136. 0	0. 10518	-9. 78084	-14. 95233
48. 0	2. 65632	4. 24280	-0. 92868	138. 0	0. 07159	-11. 45117	-16. 62266
50. 0	2. 92742	4. 66486	-0. 50663	140. 0	0. 05410	-12. 66816	-17. 83965
52. 0	3. 12512	4. 94867	-0. 22282	142. 0	0. 05858	-12. 32267	-17. 49416
54. 0	3. 24501	5. 11216	-0. 05932	144. 0	0. 08478	-10. 71689	-15. 88838
56. 0	3. 28964	5. 17149	0. 00000	146. 0	0. 12520	-9. 02397	-14. 19545
58. 0	3. 26669	5. 14108	-0. 03040	148. 0	0. 16643	-7. 78778	-12. 95927
60. 0	3. 18700	5. 03382	-0. 13767	150. 0	0. 19320	-7. 13995	-12. 31144
62. 0	3. 06270	4. 86105	-0. 31044	152. 0	0. 19406	-7. 12065	-12. 29214
64. 0	2. 90584	4. 63271	-0. 53878	154. 0	0. 16663	-7. 78258	-12. 95407
66. 0	2. 72734	4. 35739	-0. 81410	156. 0	0. 12004	-9. 20681	-14. 37830
68. 0	2. 53654	4. 04241	-1. 12908	158. 0	0. 07284	-11. 37630	-16. 54778
70. 0	2. 34098	3. 69398	-1. 47751	160. 0	0. 04629	-13. 34545	-18. 51694
72. 0	2. 14647	3. 31726	-1. 85423	162. 0	0. 05529	-12. 57349	-17. 74497
74. 0	1. 95726	2. 91649	-2. 25499	164. 0	0. 10082	-9. 96454	-15. 13602
76. 0	1. 77628	2. 49512	-2. 67636	166. 0	0. 16760	-7. 75730	-12. 92878
78. 0	1. 60540	2. 05583	-3. 11565	168. 0	0. 22904	-6. 40093	-11. 57242
80. 0	1. 44564	1. 60060	-3. 57088	170. 0	0. 25805	-5. 88295	-11. 05443
82. 0	1. 29739	1. 13070	-4. 04078	172. 0	0. 23930	-6. 21049	-11. 38198
84. 0	1. 16055	0. 64665	-4. 52484	174. 0	0. 17717	-7. 51605	-12. 68753
86. 0	1. 03470	0. 14814	-5. 02335	176. 0	0. 09519	-10. 21421	-15. 38569
88. 0	0. 91916	-0. 36609	-5. 53757	178. 0	0. 02660	-15. 75121	-20. 92270
90. 0	0. 81315	-0. 89832	-6. 06980	180.	0. 00000	-INFINITY	-INFINITY

TABLE A2-45

EPSLN= 11. 0000(NORMALIZED GROUNDPLANE RADIUS)
MONOPOLE LENGTH= 0. 250000(WAVELENGTH)
MONOPOLE RADIUS= 0. 1000E-05(WAVELENGTH)
NUMBER OF MONOPOLE SEGMENTS= 1
NUMBER OF DISK SEGMENTS= 33
NUMBER OF ANNULAR ZONES PER DISK SEGEMENT= 1
RADIATION RESISTANCE DETERMINED BY PATTERN= 0. 359200E+02 OHMS

ELEVATION ANGLE (DEG)	DIRECTIVITY (NUMERIC)	DIRECTIVITY (dBi)	RELATIVE POWER (dB)	ELEVATION ANGLE (DEG)	DIRECTIVITY (NUMERIC)	DIRECTIVITY (dBi)	RELATIVE POWER (dB)
0.	0. 00000	-INFINITY	-INFINITY				
2. 0	0. 04884	-13. 11211	-18. 23569	92. 0	0. 74289	-1. 29073	-6. 41431
4. 0	0. 17881	-7. 47614	-12. 59971	94. 0	0. 65773	-1. 81955	-6. 94312
6. 0	0. 34681	-4. 59903	-9. 72261	96. 0	0. 57880	-2. 37475	-7. 49832
8. 0	0. 49999	-3. 01038	-8. 13396	98. 0	0. 50534	-2. 96413	-8. 08770
10. 0	0. 59503	-2. 25458	-7. 37815	100. 0	0. 43682	-3. 59693	-8. 72050
12. 0	0. 61193	-2. 13301	-7. 25658	102. 0	0. 37299	-4. 28308	-9. 40665
14. 0	0. 55698	-2. 54161	-7. 66518	104. 0	0. 31394	-5. 03152	-10. 15509
16. 0	0. 45551	-3. 41498	-8. 53855	106. 0	0. 26022	-5. 84664	-10. 97022
18. 0	0. 33899	-4. 69814	-9. 82172	108. 0	0. 21275	-6. 72123	-11. 84480
20. 0	0. 23313	-6. 32406	-11. 44763	110. 0	0. 17284	-7. 62358	-12. 74715
22. 0	0. 15196	-8. 18275	-13. 30632	112. 0	0. 14195	-8. 47856	-13. 60214
24. 0	0. 09903	-10. 04214	-15. 16572	114. 0	0. 12150	-9. 15415	-14. 27772
26. 0	0. 07352	-11. 33577	-16. 45934	116. 0	0. 11246	-9. 48998	-14. 61356
28. 0	0. 07701	-11. 13468	-16. 25825	118. 0	0. 11493	-9. 39548	-14. 51906
30. 0	0. 11728	-9. 30761	-14. 43118	120. 0	0. 12775	-8. 93654	-14. 06011
32. 0	0. 20753	-6. 82913	-11. 95270	122. 0	0. 14814	-8. 29333	-13. 41691
34. 0	0. 36166	-4. 41700	-9. 54058	124. 0	0. 17176	-7. 65070	-12. 77428
36. 0	0. 58823	-2. 30454	-7. 42811	126. 0	0. 19307	-7. 14276	-12. 26633
38. 0	0. 88569	-0. 52720	-5. 65077	128. 0	0. 20621	-6. 85690	-11. 98047
40. 0	1. 24066	0. 93652	-4. 18706	130. 0	0. 20631	-6. 85483	-11. 97840
42. 0	1. 62969	2. 12105	-3. 00252	132. 0	0. 19101	-7. 18945	-12. 31302
44. 0	2. 02358	3. 06120	-2. 06238	134. 0	0. 16172	-7. 91236	-13. 03593
46. 0	2. 39264	3. 78878	-1. 33479	136. 0	0. 12410	-9. 06222	-14. 18580
48. 0	2. 71153	4. 33214	-0. 79143	138. 0	0. 08730	-10. 58965	-15. 71322
50. 0	2. 96235	4. 71636	-0. 40722	140. 0	0. 06177	-12. 09230	-17. 21587
52. 0	3. 13589	4. 96361	-0. 15996	142. 0	0. 05595	-12. 52222	-17. 64579
54. 0	3. 23114	5. 09356	-0. 03001	144. 0	0. 07288	-11. 37408	-16. 49765
56. 0	3. 25355	5. 12357	0. 00000	146. 0	0. 10798	-9. 66655	-14. 79012
58. 0	3. 21286	5. 06892	-0. 05465	148. 0	0. 14938	-8. 25714	-13. 38071
60. 0	3. 12098	4. 94291	-0. 18066	150. 0	0. 18132	-7. 41558	-12. 53915
62. 0	2. 99022	4. 75702	-0. 36655	152. 0	0. 19005	-7. 21131	-12. 33488
64. 0	2. 83205	4. 52100	-0. 60257	154. 0	0. 17009	-7. 69310	-12. 81667
66. 0	2. 65646	4. 24304	-0. 88054	156. 0	0. 12810	-8. 92462	-14. 04820
68. 0	2. 47167	3. 92991	-1. 19366	158. 0	0. 08186	-10. 86921	-15. 99278
70. 0	2. 28414	3. 58722	-1. 53635	160. 0	0. 05395	-12. 67992	-17. 80349
72. 0	2. 09873	3. 21956	-1. 90401	162. 0	0. 06184	-12. 08738	-17. 21095
74. 0	1. 91899	2. 83072	-2. 29286	164. 0	0. 10877	-9. 63473	-14. 75830
76. 0	1. 74736	2. 42381	-2. 69976	166. 0	0. 18006	-7. 44583	-12. 56941
78. 0	1. 58542	2. 00144	-3. 12214	168. 0	0. 24747	-6. 06477	-11. 18834
80. 0	1. 43406	1. 56566	-3. 55791	170. 0	0. 28088	-5. 51487	-10. 63844
82. 0	1. 29362	1. 11807	-4. 00550	172. 0	0. 26222	-5. 81333	-10. 93691
84. 0	1. 16403	0. 65966	-4. 46392	174. 0	0. 19519	-7. 09552	-12. 21910
86. 0	1. 04489	0. 19072	-4. 93286	176. 0	0. 10527	-9. 77685	-14. 90042
88. 0	0. 93555	-0. 28933	-5. 41290	178. 0	0. 02949	-15. 30374	-20. 42731
90. 0	0. 83519	-0. 78216	-5. 90573	180	0. 00000	-INFINITY	-INFINITY

TABLE A2-46

EPSLN= 11.2500(NORMALIZED GROUNDPLANE RADIUS)
MONOPOLE LENGTH= 0.250000(WAVELENGTH)
MONOPOLE RADIUS= 0.1000E-05(WAVELENGTH)
NUMBER OF MONOPOLE SEGMENTS= 1
NUMBER OF DISK SEGMENTS= 33
NUMBER OF ANNULAR ZONES PER DISK SEGEMENT= 1
RADIATION RESISTANCE DETERMINED BY PATTERN= 0.350770E+02 OHMS

ELEVATION ANGLE (DEG)	DIRECTIVITY (NUMERIC)	DIRECTIVITY (dBi)	RELATIVE POWER (dB)	ELEVATION ANGLE (DEG)	DIRECTIVITY (NUMERIC)	DIRECTIVITY (dBi)	RELATIVE POWER (dB)
0.	0.00000	-INFINITY	-INFINITY				
2.0	0.05454	-12.63293	-17.66518	92.0	0.76774	-1.14787	-6.18012
4.0	0.19932	-7.00456	-12.03681	94.0	0.68721	-1.62911	-6.66136
6.0	0.38554	-4.13927	-9.17151	96.0	0.61260	-2.12820	-7.16045
8.0	0.55399	-2.56499	-7.59724	98.0	0.54290	-2.65281	-7.68506
10.0	0.65720	-1.82302	-6.85527	100.0	0.47721	-3.21294	-8.24519
12.0	0.67467	-1.70906	-6.74131	102.0	0.41488	-3.82076	-8.85301
14.0	0.61533	-2.10891	-7.14116	104.0	0.35563	-4.49004	-9.52229
16.0	0.50815	-2.94012	-7.97237	106.0	0.29960	-5.23464	-10.26689
18.0	0.38709	-4.12192	-9.15417	108.0	0.24746	-6.06500	-11.09724
20.0	0.27813	-5.55750	-10.58975	110.0	0.20042	-6.98054	-12.01279
22.0	0.19374	-7.12781	-12.16006	112.0	0.16017	-7.95420	-12.98645
24.0	0.13559	-8.67770	-13.70995	114.0	0.12866	-8.90573	-13.93798
26.0	0.10225	-9.90340	-14.93565	116.0	0.10778	-9.67468	-14.70692
28.0	0.09667	-10.14694	-15.17919	118.0	0.09892	-10.04737	-15.07962
30.0	0.12950	-8.87719	-13.90944	120.0	0.10237	-9.89846	-14.93071
32.0	0.21686	-6.63830	-11.67055	122.0	0.11683	-9.32434	-14.35659
34.0	0.37418	-4.26922	-9.30147	124.0	0.13910	-8.56684	-13.59909
36.0	0.60933	-2.15148	-7.18372	126.0	0.16407	-7.84978	-12.88203
38.0	0.91792	-0.37196	-5.40421	128.0	0.18540	-7.31883	-12.35107
40.0	1.28256	1.08079	-3.95146	130.0	0.19673	-7.06133	-12.09358
42.0	1.67600	2.24273	-2.78952	132.0	0.19331	-7.13735	-12.16960
44.0	2.06654	3.15244	-1.87980	134.0	0.17383	-7.59870	-12.63095
46.0	2.42406	3.84543	-1.18682	136.0	0.14156	-8.49071	-13.52296
48.0	2.72465	4.35310	-0.67915	138.0	0.10436	-9.81449	-14.84674
50.0	2.95324	4.70299	-0.32926	140.0	0.07307	-11.36240	-16.39465
52.0	3.10400	4.91922	-0.11303	142.0	0.05820	-12.35071	-17.38296
54.0	3.17897	5.02287	-0.00938	144.0	0.06595	-11.80756	-16.83981
56.0	3.18585	5.03225	0.00000	146.0	0.09494	-10.22549	-15.25774
58.0	3.13552	4.96309	-0.06916	148.0	0.13526	-8.68834	-13.72059
60.0	3.03998	4.82870	-0.20355	150.0	0.17111	-7.66712	-12.69937
62.0	2.91082	4.64015	-0.39210	152.0	0.18671	-7.28826	-12.32051
64.0	2.75833	4.40646	-0.62579	154.0	0.17350	-7.60707	-12.63932
66.0	2.59112	4.13488	-0.89737	156.0	0.13554	-8.67922	-13.71147
68.0	2.41613	3.83121	-1.20104	158.0	0.08993	-10.46080	-15.49305
70.0	2.23877	3.50009	-1.53216	160.0	0.06076	-12.16363	-17.19588
72.0	2.06316	3.14533	-1.88692	162.0	0.06841	-11.64903	-16.68128
74.0	1.89242	2.77017	-2.26208	164.0	0.11861	-9.25891	-14.29116
76.0	1.72881	2.37746	-2.65478	166.0	0.19702	-7.05494	-12.08719
78.0	1.57392	1.96982	-3.06242	168.0	0.27304	-5.63771	-10.66996
80.0	1.42878	1.54966	-3.48259	170.0	0.31249	-5.05161	-10.08386
82.0	1.29394	1.11913	-3.91312	172.0	0.29381	-5.31933	-10.35158
84.0	1.16952	0.68008	-4.35217	174.0	0.21991	-6.57751	-11.60975
86.0	1.05532	0.23383	-4.79842	176.0	0.11908	-9.24176	-14.27401
88.0	0.95081	-0.21905	-5.25130	178.0	0.03343	-14.75846	-19.79071
90.0	0.85526	-0.67901	-5.71126	180	0.00000	-INFINITY	-INFINITY

TABLE A2-47

EPSLN= 11.5000(NORMALIZED GROUNDPLANE RADIUS)
MONOPOLE LENGTH= 0.250000(WAVELENGTH)
MONOPOLE RADIUS= 0.1000E-05(WAVELENGTH)
NUMBER OF MONOPOLE SEGMENTS= 1
NUMBER OF DISK SEGMENTS= 34
NUMBER OF ANNULAR ZONES PER DISK SEGEMENT= 1
RADIATION RESISTANCE DETERMINED BY PATTERN= 0.345396E+02 OHMS

ELEVATION ANGLE (DEG)	DIRECTIVITY (NUMERIC)	DIRECTIVITY (dBi)	RELATIVE POWER (dB)	ELEVATION ANGLE (DEG)	DIRECTIVITY (NUMERIC)	DIRECTIVITY (dBi)	RELATIVE POWER (dB)
0.	0.00000	-INFINITY	-INFINITY				
2.0	0.06109	-12.14065	-17.05348	92.0	0.78459	-1.05356	-5.96640
4.0	0.22270	-6.52286	-11.43570	94.0	0.70818	-1.49859	-6.41142
6.0	0.42909	-3.67450	-8.58733	96.0	0.63811	-1.95106	-6.86389
8.0	0.61348	-2.12201	-7.03484	98.0	0.57312	-2.41756	-7.33040
10.0	0.72383	-1.40363	-6.31646	100.0	0.51197	-2.90755	-7.82039
12.0	0.73976	-1.30909	-6.22192	102.0	0.45358	-3.43350	-8.34633
14.0	0.67407	-1.71297	-6.62580	104.0	0.39712	-4.01078	-8.92361
16.0	0.56061	-2.51343	-7.42626	106.0	0.34220	-4.65715	-9.56998
18.0	0.43639	-3.60124	-8.51408	108.0	0.28898	-5.39132	-10.30415
20.0	0.32728	-4.85084	-9.76368	110.0	0.23826	-6.22941	-11.14224
22.0	0.24312	-6.14173	-11.05457	112.0	0.19157	-7.17674	-12.08957
24.0	0.18257	-7.38576	-12.29859	114.0	0.15103	-8.20946	-13.12229
26.0	0.14273	-8.45473	-13.36757	116.0	0.11914	-9.23933	-14.15217
28.0	0.12762	-8.94081	-13.85365	118.0	0.09836	-10.07184	-14.98468
30.0	0.15077	-8.21686	-13.12969	120.0	0.09045	-10.43612	-15.34896
32.0	0.23144	-6.35569	-11.26852	122.0	0.09579	-10.18688	-15.09971
34.0	0.38677	-4.12548	-9.03831	124.0	0.11276	-9.47861	-14.39144
36.0	0.62402	-2.04799	-6.96082	126.0	0.13736	-8.62126	-13.53410
38.0	0.93616	-0.28652	-5.19935	128.0	0.16350	-7.86484	-12.77767
40.0	1.30223	1.14687	-3.76596	130.0	0.18388	-7.35467	-12.26750
42.0	1.69194	2.28385	-2.62899	132.0	0.19175	-7.17271	-12.08555
44.0	2.07220	3.16432	-1.74851	134.0	0.18299	-7.37574	-12.28857
46.0	2.41349	3.82646	-1.08638	136.0	0.15806	-8.01187	-12.92470
48.0	2.69422	4.30432	-0.60851	138.0	0.12284	-9.10649	-14.01932
50.0	2.90254	4.62778	-0.28506	140.0	0.08779	-10.56566	-15.47850
52.0	3.03580	4.82273	-0.09010	142.0	0.06493	-11.87573	-16.78856
54.0	3.09846	4.91146	-0.00137	144.0	0.06348	-11.97334	-16.88617
56.0	3.09944	4.91283	0.00000	146.0	0.08550	-10.68059	-15.59342
58.0	3.04963	4.84247	-0.07036	148.0	0.12352	-9.08250	-13.99533
60.0	2.96006	4.71300	-0.19983	150.0	0.16219	-7.89980	-12.81263
62.0	2.84075	4.53433	-0.37851	152.0	0.18386	-7.35506	-12.26790
64.0	2.70025	4.31405	-0.59879	154.0	0.17684	-7.52417	-12.43700
66.0	2.54560	4.05790	-0.85494	156.0	0.14243	-8.46402	-13.37686
68.0	2.38246	3.77026	-1.14258	158.0	0.09694	-10.13479	-15.04762
70.0	2.21544	3.45460	-1.45823	160.0	0.06630	-11.78511	-16.69795
72.0	2.04829	3.11391	-1.79893	162.0	0.07429	-11.29080	-16.20363
74.0	1.88405	2.75093	-2.16190	164.0	0.12957	-8.87510	-13.78794
76.0	1.72522	2.36845	-2.54438	166.0	0.21802	-6.61495	-11.52779
78.0	1.57374	1.96934	-2.94350	168.0	0.30588	-5.14453	-10.05737
80.0	1.43106	1.55659	-3.35625	170.0	0.35362	-4.51458	-9.42741
82.0	1.29815	1.13326	-3.77957	172.0	0.33514	-4.74775	-9.66059
84.0	1.17554	0.70236	-4.21048	174.0	0.25235	-5.97992	-10.89275
86.0	1.06330	0.26658	-4.64626	176.0	0.13721	-8.62603	-13.53887
88.0	0.96118	-0.17194	-5.08477	178.0	0.03862	-14.13202	-19.04485
90.0	0.86857	-0.61196	-5.52479	180.	0.00000	-INFINITY	-INFINITY

TABLE A2-48

EPSLN= 11.7500(NORMALIZED GROUNDPLANE RADIUS)
MONOPOLE LENGTH= 0.250000(WAVELENGTH)
MONOPOLE RADIUS= 0.1000E-05(WAVELENGTH)
NUMBER OF MONOPOLE SEGMENTS= 1
NUMBER OF DISK SEGMENTS= 35
NUMBER OF ANNULAR ZONES PER DISK SEGEMENT= 1
RADIATION RESISTANCE DETERMINED BY PATTERN= 0.350588E+02 OHMS

ELEVATION ANGLE (DEG)	DIRECTIVITY (NUMERIC)	DIRECTIVITY (dBi)	RELATIVE POWER (dB)	ELEVATION ANGLE (DEG)	DIRECTIVITY (NUMERIC)	DIRECTIVITY (dBi)	RELATIVE POWER (dB)
0.	0.00000	-INFINITY	-INFINITY				
2.0	0.06752	-11.70558	-16.50360	92.0	0.77281	-1.11929	-5.91731
4.0	0.24558	-6.09808	-10.89610	94.0	0.69896	-1.55549	-6.35351
6.0	0.47143	-3.26579	-8.06381	96.0	0.63290	-1.98663	-6.78465
8.0	0.67096	-1.73303	-6.53105	98.0	0.57323	-2.41674	-7.21476
10.0	0.78820	-1.03365	-5.83167	100.0	0.51843	-2.85307	-7.65110
12.0	0.80374	-0.94885	-5.74687	102.0	0.46704	-3.30642	-8.10445
14.0	0.73492	-1.33757	-6.13560	104.0	0.41772	-3.79115	-8.58917
16.0	0.62051	-2.07254	-6.87056	106.0	0.36942	-4.32483	-9.12285
18.0	0.49979	-3.01216	-7.81018	108.0	0.32156	-4.92735	-9.72537
20.0	0.39721	-4.00974	-8.80777	110.0	0.27423	-5.61883	-10.41685
22.0	0.31876	-4.96532	-9.76334	112.0	0.22830	-6.41497	-11.21299
24.0	0.25919	-5.86375	-10.66178	114.0	0.18549	-7.31679	-12.11481
26.0	0.21397	-6.69654	-11.49456	116.0	0.14829	-8.28899	-13.08701
28.0	0.18834	-7.25053	-12.04855	118.0	0.11960	-9.22259	-14.02061
30.0	0.19908	-7.00970	-11.80772	120.0	0.10223	-9.90439	-14.70242
32.0	0.26872	-5.70707	-10.50509	122.0	0.09803	-10.08644	-14.88446
34.0	0.41604	-3.80863	-8.60666	124.0	0.10711	-9.70170	-14.49972
36.0	0.64767	-1.88645	-6.68447	126.0	0.12708	-8.95939	-13.75742
38.0	0.95419	-0.20363	-5.00165	128.0	0.15280	-8.15868	-12.95670
40.0	1.31192	1.17909	-3.61894	130.0	0.17698	-7.52064	-12.31866
42.0	1.68880	2.27579	-2.52224	132.0	0.19160	-7.17593	-11.97395
44.0	2.05181	3.12138	-1.67665	134.0	0.19021	-7.20761	-12.00563
46.0	2.37333	3.75357	-1.04445	136.0	0.17042	-7.68479	-12.48281
48.0	2.63481	4.20750	-0.59052	138.0	0.13570	-8.67409	-13.47211
50.0	2.82765	4.51425	-0.28377	140.0	0.09547	-10.20151	-14.99954
52.0	2.95164	4.70063	-0.09739	142.0	0.06271	-12.02638	-16.82440
54.0	3.01240	4.78912	-0.00890	144.0	0.04955	-13.04961	-17.84763
56.0	3.01858	4.79802	0.00000	146.0	0.06187	-12.08495	-16.88297
58.0	2.97965	4.74166	-0.05637	148.0	0.09560	-10.19559	-14.99362
60.0	2.90455	4.63079	-0.16723	150.0	0.13672	-8.64168	-13.43970
62.0	2.80103	4.47318	-0.32484	152.0	0.16639	-7.78874	-12.58676
64.0	2.67563	4.27426	-0.52376	154.0	0.16966	-7.70412	-12.50214
66.0	2.53387	4.03784	-0.76018	156.0	0.14429	-8.40758	-13.20560
68.0	2.38054	3.76676	-1.03126	158.0	0.10466	-9.80216	-14.60018
70.0	2.21994	3.46341	-1.33462	160.0	0.07743	-11.11103	-15.90905
72.0	2.05596	3.13014	-1.66788	162.0	0.08924	-10.49463	-15.29265
74.0	1.89214	2.76953	-2.02849	164.0	0.15159	-8.19318	-12.99120
76.0	1.73161	2.38450	-2.41353	166.0	0.25092	-6.00465	-10.80267
78.0	1.57702	1.97838	-2.81964	168.0	0.35049	-4.55321	-9.35123
80.0	1.43051	1.55492	-3.24310	170.0	0.40555	-3.91956	-8.71758
82.0	1.29366	1.11819	-3.67983	172.0	0.38534	-4.14153	-8.93955
84.0	1.16748	0.67249	-4.12553	174.0	0.29096	-5.36165	-10.15967
86.0	1.05248	0.22213	-4.57589	176.0	0.15857	-7.99787	-12.79589
88.0	0.94868	-0.22879	-5.02681	178.0	0.04469	-13.49750	-18.29552
90.0	0.85571	-0.67676	-5.47478	180.	0.00000	-INFINITY	-INFINITY

TABLE A2-49

EPSLN= 12.0000(NORMALIZED GROUNDPLANE RADIUS)
MONOPOLE LENGTH= 0.250000(WAVELENGTH)
MONOPOLE RADIUS= 0.1000E-05(WAVELENGTH)
NUMBER OF MONOPOLE SEGMENTS= 1
NUMBER OF DISK SEGMENTS= 36
NUMBER OF ANNULAR ZONES PER DISK SEGEMENT= 1
RADIATION RESISTANCE DETERMINED BY PATTERN= 0.344984E+02 OHMS

ELEVATION ANGLE (DEG)	DIRECTIVITY (NUMERIC)	DIRECTIVITY (dBi)	RELATIVE POWER (dB)	ELEVATION ANGLE (DEG)	DIRECTIVITY (NUMERIC)	DIRECTIVITY (dBi)	RELATIVE POWER (dB)
0.	0.00000	-INFINITY	-INFINITY				
2.0	0.07468	-11.26769	-15.92317	92.0	0.78616	-1.04490	-5.70038
4.0	0.27028	-5.68194	-10.33742	94.0	0.71317	-1.46808	-6.12356
6.0	0.51449	-2.88621	-7.54169	96.0	0.64933	-1.87534	-6.53082
8.0	0.72363	-1.40482	-6.06031	98.0	0.59310	-2.26875	-6.92423
10.0	0.83747	-0.77030	-5.42578	100.0	0.54273	-2.65415	-7.30963
12.0	0.83969	-0.75879	-5.41428	102.0	0.49640	-3.04171	-7.69719
14.0	0.75578	-1.21607	-5.87155	104.0	0.45226	-3.44615	-8.10163
16.0	0.63273	-1.98778	-6.64327	106.0	0.40863	-3.88674	-8.54222
18.0	0.51373	-2.89268	-7.54816	108.0	0.36417	-4.38691	-9.04239
20.0	0.42103	-3.75690	-8.41238	110.0	0.31817	-4.97346	-9.62894
22.0	0.35421	-4.50743	-9.16291	112.0	0.27072	-5.67484	-10.33032
24.0	0.30116	-5.21205	-9.86753	114.0	0.22300	-6.51698	-11.17246
26.0	0.25339	-5.96210	-10.61758	116.0	0.17732	-7.51231	-12.16779
28.0	0.21643	-6.64682	-11.30230	118.0	0.13701	-8.63261	-13.28809
30.0	0.21046	-6.76835	-11.42383	120.0	0.10590	-9.75091	-14.40639
32.0	0.26218	-5.81401	-10.46949	122.0	0.08763	-10.57336	-15.22884
34.0	0.39312	-4.05480	-8.71028	124.0	0.08451	-10.73116	-15.38664
36.0	0.61023	-2.14508	-6.80057	126.0	0.09642	-10.15843	-14.81391
38.0	0.90265	-0.44483	-5.10031	128.0	0.12005	-9.20653	-13.86201
40.0	1.24486	0.95121	-3.70427	130.0	0.14879	-8.27420	-12.92968
42.0	1.60408	2.05225	-2.60323	132.0	0.17384	-7.59853	-12.25402
44.0	1.94833	2.89663	-1.75885	134.0	0.18638	-7.29595	-11.95143
46.0	2.25265	3.52693	-1.12855	136.0	0.18063	-7.43221	-12.08769
48.0	2.50181	3.98255	-0.67293	138.0	0.15656	-8.05317	-12.70865
50.0	2.69009	4.29766	-0.35782	140.0	0.12123	-9.16373	-13.81921
52.0	2.81884	4.50070	-0.15478	142.0	0.08735	-10.58722	-15.24270
54.0	2.89353	4.61428	-0.04120	144.0	0.06895	-11.61450	-16.26998
56.0	2.92111	4.65548	0.00000	146.0	0.07527	-11.23360	-15.88909
58.0	2.90833	4.63643	-0.01905	148.0	0.10537	-9.77280	-14.42828
60.0	2.86099	4.56517	-0.09031	150.0	0.14654	-8.34037	-12.99586
62.0	2.78401	4.44670	-0.20878	152.0	0.17859	-7.48133	-12.13681
64.0	2.68168	4.28407	-0.37141	154.0	0.18334	-7.36733	-12.02281
66.0	2.55817	4.07930	-0.57618	156.0	0.15547	-8.08340	-12.73888
68.0	2.41774	3.83409	-0.82139	158.0	0.10883	-9.63248	-14.28796
70.0	2.26483	3.55036	-1.10512	160.0	0.07319	-11.35556	-16.01104
72.0	2.10400	3.23046	-1.42502	162.0	0.08083	-10.92438	-15.57986
74.0	1.93971	2.87736	-1.77812	164.0	0.14826	-8.28983	-12.94531
76.0	1.77608	2.49462	-2.16086	166.0	0.26275	-5.80460	-10.46008
78.0	1.61675	2.08644	-2.56904	168.0	0.38272	-4.17123	-8.82671
80.0	1.46471	1.65753	-2.99795	170.0	0.45459	-3.42380	-8.07928
82.0	1.32223	1.21308	-3.44240	172.0	0.43943	-3.57107	-8.22655
84.0	1.19087	0.75865	-3.89683	174.0	0.33564	-4.74126	-9.39674
86.0	1.07153	0.30006	-4.35542	176.0	0.18428	-7.34523	-12.00071
88.0	0.96451	-0.15691	-4.81239	178.0	0.05216	-12.82667	-17.48215
90.0	0.86960	-0.60678	-5.26226	180.	0.00000	-INFINITY	-INFINITY

TABLE A2-50

EPSLN= 12.2500(NORMALIZED GROUNDPLANE RADIUS)
MONOPOLE LENGTH= 0.250000(WAVELENGTH)
MONOPOLE RADIUS= 0.1000E-05(WAVELENGTH)
NUMBER OF MONOPOLE SEGMENTS= 1
NUMBER OF DISK SEGMENTS= 36
NUMBER OF ANNULAR ZONES PER DISK SEGEMENT= 1
RADIATION RESISTANCE DETERMINED BY PATTERN= 0.350058E+02 OHMS

ELEVATION ANGLE (DEG)	DIRECTIVITY (NUMERIC)	DIRECTIVITY (dBi)	RELATIVE POWER (dB)	ELEVATION ANGLE (DEG)	DIRECTIVITY (NUMERIC)	DIRECTIVITY (dBi)	RELATIVE POWER (dB)
0.	0.00000	-INFINITY	-INFINITY				
2.0	0.08047	-10.94356	-15.54015	92.0	0.76992	-1.13553	-5.73212
4.0	0.28973	-5.38000	-9.97658	94.0	0.69516	-1.57916	-6.17575
6.0	0.54683	-2.62147	-7.21805	96.0	0.63147	-1.99649	-6.59307
8.0	0.75999	-1.19192	-5.78850	98.0	0.57733	-2.38578	-6.98237
10.0	0.86667	-0.62144	-5.21802	100.0	0.53096	-2.74939	-7.34597
12.0	0.85544	-0.67811	-5.27469	102.0	0.49038	-3.09468	-7.69126
14.0	0.76077	-1.18745	-5.78404	104.0	0.45345	-3.43469	-8.03127
16.0	0.63743	-1.95568	-6.55227	106.0	0.41801	-3.78818	-8.38476
18.0	0.53063	-2.75206	-7.34865	108.0	0.38201	-4.17924	-8.77582
20.0	0.45815	-3.38996	-7.98655	110.0	0.34383	-4.63651	-9.23309
22.0	0.41116	-3.85994	-8.45652	112.0	0.30255	-5.19202	-9.78860
24.0	0.36966	-4.32193	-8.91851	114.0	0.25828	-5.87904	-10.47562
26.0	0.32120	-4.93221	-9.52879	116.0	0.21247	-6.72693	-11.32351
28.0	0.27208	-5.65297	-10.24956	118.0	0.16798	-7.74733	-12.34391
30.0	0.24641	-6.08344	-10.68002	120.0	0.12886	-8.89867	-13.49526
32.0	0.27526	-5.60259	-10.19917	122.0	0.09973	-10.01181	-14.60840
34.0	0.38290	-4.16916	-8.76574	124.0	0.08467	-10.72259	-15.31917
36.0	0.57688	-2.38912	-6.98570	126.0	0.08591	-10.65932	-15.25591
38.0	0.84570	-0.72783	-5.32441	128.0	0.10249	-9.89317	-14.48976
40.0	1.16349	0.65764	-3.93894	130.0	0.12952	-8.87668	-13.47327
42.0	1.49855	1.75670	-2.83988	132.0	0.15860	-7.99700	-12.59358
44.0	1.82159	2.60452	-1.99207	134.0	0.17967	-7.45525	-12.05184
46.0	2.11123	3.24536	-1.35122	136.0	0.18416	-7.34810	-11.94468
48.0	2.35551	3.72085	-0.87573	138.0	0.16856	-7.73257	-12.32915
50.0	2.55049	4.06623	-0.53035	140.0	0.13694	-8.63466	-13.23124
52.0	2.69732	4.30932	-0.28726	142.0	0.10085	-9.96338	-14.55996
54.0	2.79931	4.47050	-0.12608	144.0	0.07559	-11.21538	-15.81196
56.0	2.85989	4.56349	-0.03310	146.0	0.07373	-11.32335	-15.91993
58.0	2.88176	4.59658	0.00000	148.0	0.09821	-10.07859	-14.67517
60.0	2.86700	4.57428	-0.02230	150.0	0.13876	-8.57724	-13.17382
62.0	2.81760	4.49879	-0.09779	152.0	0.17487	-7.57291	-12.16949
64.0	2.73611	4.37134	-0.22524	154.0	0.18531	-7.32096	-11.91755
66.0	2.62606	4.19305	-0.40353	156.0	0.16090	-7.93432	-12.53090
68.0	2.49202	3.96551	-0.63108	158.0	0.11341	-9.45354	-14.05012
70.0	2.33937	3.69098	-0.90560	160.0	0.07414	-11.29959	-15.89617
72.0	2.17396	3.37251	-1.22407	162.0	0.08011	-10.96340	-15.55999
74.0	2.00163	3.01384	-1.58274	164.0	0.15291	-8.15577	-12.75236
76.0	1.82783	2.61935	-1.97723	166.0	0.28151	-5.50513	-10.10171
78.0	1.65729	2.19399	-2.40259	168.0	0.42044	-3.76297	-8.35955
80.0	1.49389	1.74318	-2.85340	170.0	0.50790	-2.94218	-7.53877
82.0	1.34055	1.27284	-3.32374	172.0	0.49681	-3.03808	-7.63466
84.0	1.19932	0.78935	-3.80723	174.0	0.38263	-4.17224	-8.76882
86.0	1.07140	0.29953	-4.29705	176.0	0.21125	-6.75213	-11.34871
88.0	0.95732	-0.18944	-4.78602	178.0	0.05998	-12.21969	-16.81627
90.0	0.85700	-0.67018	-5.26677	180	0.00000	-INFINITY	-INFINITY

TABLE A2-51

EPSLN= 12.5000(NORMALIZED GROUNDPLANE RADIUS)
MONOPOLE LENGTH= 0.250000(WAVELENGTH)
MONOPOLE RADIUS= 0.1000E-05(WAVELENGTH)
NUMBER OF MONOPOLE SEGMENTS= 1
NUMBER OF DISK SEGMENTS= 37
NUMBER OF ANNULAR ZONES PER DISK SEGEMENT= 1
RADIATION RESISTANCE DETERMINED BY PATTERN= 0.357880E+02 OHMS

ELEVATION ANGLE (DEG)	DIRECTIVITY (NUMERIC)	DIRECTIVITY (dBi)	RELATIVE POWER (dB)	ELEVATION ANGLE (DEG)	DIRECTIVITY (NUMERIC)	DIRECTIVITY (dBi)	RELATIVE POWER (dB)
0.	0.00000	-INFINITY	-INFINITY				
2.0	0.08453	-10.73005	-15.37846	92.0	0.74565	-1.27467	-5.92308
4.0	0.30246	-5.19325	-9.84166	94.0	0.66700	-1.75877	-6.40718
6.0	0.56492	-2.48016	-7.12857	96.0	0.60119	-2.20988	-6.85829
8.0	0.77354	-1.11518	-5.76359	98.0	0.54683	-2.62145	-7.26986
10.0	0.86566	-0.62655	-5.27496	100.0	0.50227	-2.99061	-7.63902
12.0	0.83693	-0.77309	-5.42150	102.0	0.46558	-3.32001	-7.96842
14.0	0.73229	-1.35317	-6.00158	104.0	0.43459	-3.61919	-8.26760
16.0	0.61413	-2.11737	-6.76578	106.0	0.40690	-3.90515	-8.55356
18.0	0.52818	-2.77221	-7.42062	108.0	0.38001	-4.20201	-8.85042
20.0	0.48529	-3.13994	-7.78835	110.0	0.35157	-4.53992	-9.18833
22.0	0.46612	-3.31504	-7.96345	112.0	0.31962	-4.95360	-9.60201
24.0	0.44149	-3.55084	-8.19925	114.0	0.28310	-5.48060	-10.12901
26.0	0.39469	-4.03745	-8.68586	116.0	0.24219	-6.15837	-10.80678
28.0	0.33318	-4.77321	-9.42162	118.0	0.19872	-7.01750	-11.66591
30.0	0.28543	-5.44496	-10.09337	120.0	0.15621	-8.06288	-12.71129
32.0	0.28719	-5.41824	-10.06665	122.0	0.11952	-9.22560	-13.87401
34.0	0.36572	-4.36846	-9.01687	124.0	0.09395	-10.27108	-14.91949
36.0	0.52970	-2.75969	-7.40810	126.0	0.08380	-10.76742	-15.41583
38.0	0.76804	-1.14615	-5.79456	128.0	0.09071	-10.42333	-15.07174
40.0	1.05606	0.23690	-4.41151	130.0	0.11224	-9.49850	-14.14691
42.0	1.36473	1.35045	-3.29796	132.0	0.14147	-8.49343	-13.14184
44.0	1.66861	2.22356	-2.42485	134.0	0.16817	-7.74243	-12.39084
46.0	1.95017	2.90073	-1.74768	136.0	0.18178	-7.40453	-12.05294
48.0	2.19997	3.42417	-1.22424	138.0	0.17543	-7.55903	-12.20744
50.0	2.41434	3.82798	-0.82043	140.0	0.14967	-8.24860	-12.89701
52.0	2.59229	4.13683	-0.51158	142.0	0.11384	-9.43708	-14.08549
54.0	2.73312	4.36659	-0.28182	144.0	0.08347	-10.78493	-15.43334
56.0	2.83539	4.52613	-0.12228	146.0	0.07383	-11.31743	-15.96584
58.0	2.89704	4.61955	-0.02886	148.0	0.09187	-10.36830	-15.01671
60.0	2.91636	4.64841	0.00000	150.0	0.13052	-8.84337	-13.49178
62.0	2.89293	4.61338	-0.03503	152.0	0.16976	-7.70161	-12.35002
64.0	2.82836	4.51534	-0.13307	154.0	0.18583	-7.30891	-11.95732
66.0	2.72644	4.35596	-0.29245	156.0	0.16538	-7.81513	-12.46354
68.0	2.59288	4.13782	-0.51059	158.0	0.11731	-9.30654	-13.95495
70.0	2.43468	3.86443	-0.78398	160.0	0.07372	-11.32421	-15.97262
72.0	2.25943	3.53999	-1.10842	162.0	0.07621	-11.17975	-15.82816
74.0	2.07458	3.16930	-1.47911	164.0	0.15214	-8.17754	-12.82595
76.0	1.88693	2.75755	-1.89086	166.0	0.29325	-5.32759	-9.97600
78.0	1.70226	2.31027	-2.33814	168.0	0.45101	-3.45811	-8.10652
80.0	1.52520	1.83327	-2.81514	170.0	0.55542	-2.55375	-7.20216
82.0	1.35915	1.33268	-3.31573	172.0	0.55052	-2.59226	-7.24067
84.0	1.20644	0.81505	-3.83336	174.0	0.42789	-3.68670	-8.33511
86.0	1.06842	0.28743	-4.36098	176.0	0.23767	-6.24029	-10.88870
88.0	0.94570	-0.24245	-4.89086	178.0	0.06772	-11.69277	-16.34118
90.0	0.83827	-0.76616	-5.41457	180	0.00000	-INFINITY	-INFINITY

TABLE A2-52

EPSLN= 12.7500(NORMALIZED GROUNDPLANE RADIUS)
MONOPOLE LENGTH= 0.250000(WAVELENGTH)
MONOPOLE RADIUS= 0.1000E-05(WAVELENGTH)
NUMBER OF MONOPOLE SEGMENTS= 1
NUMBER OF DISK SEGMENTS= 38
NUMBER OF ANNULAR ZONES PER DISK SEGEMENT= 1
RADIATION RESISTANCE DETERMINED BY PATTERN= 0.366734E+02 OHMS

ELEVATION ANGLE (DEG)	DIRECTIVITY (NUMERIC)	DIRECTIVITY (dBi)	RELATIVE POWER (dB)	ELEVATION ANGLE (DEG)	DIRECTIVITY (NUMERIC)	DIRECTIVITY (dBi)	RELATIVE POWER (dB)
0.	0.00000	-INFINITY	-INFINITY				
2.0	0.08672	-10.61894	-15.38977	92.0	0.71920	-1.43153	-6.20237
4.0	0.30807	-5.11352	-9.88436	94.0	0.63544	-1.96924	-6.74007
6.0	0.56827	-2.45444	-7.22527	96.0	0.56577	-2.47361	-7.24445
8.0	0.76421	-1.16785	-5.93868	98.0	0.50898	-2.93303	-7.70387
10.0	0.83532	-0.78145	-5.55228	100.0	0.46366	-3.33804	-8.10888
12.0	0.78636	-1.04377	-5.81460	102.0	0.42815	-3.68406	-8.45489
14.0	0.67358	-1.71613	-6.48697	104.0	0.40049	-3.97411	-8.74495
16.0	0.56641	-2.46872	-7.23956	106.0	0.37837	-4.22082	-8.99165
18.0	0.50921	-2.93106	-7.70190	108.0	0.35919	-4.44670	-9.21754
20.0	0.50376	-2.97777	-7.74861	110.0	0.34018	-4.68296	-9.45380
22.0	0.51849	-2.85260	-7.62344	112.0	0.31863	-4.96719	-9.73803
24.0	0.51447	-2.88639	-7.65722	114.0	0.29235	-5.34093	-10.11176
26.0	0.47102	-3.26958	-8.04042	116.0	0.26019	-5.84706	-10.61790
28.0	0.39730	-4.00883	-8.77967	118.0	0.22254	-6.52588	-11.29672
30.0	0.32652	-4.86089	-9.63173	120.0	0.18175	-7.40520	-12.17604
32.0	0.29917	-5.24084	-10.01168	122.0	0.14213	-8.47316	-13.24400
34.0	0.34554	-4.61499	-9.38582	124.0	0.10935	-9.61178	-14.38261
36.0	0.47588	-3.22505	-7.99589	126.0	0.08918	-10.49717	-15.26800
38.0	0.68050	-1.67172	-6.44256	128.0	0.08564	-10.67307	-15.44391
40.0	0.93725	-0.28144	-5.05227	130.0	0.09907	-10.04065	-14.81149
42.0	1.22099	0.86713	-3.90371	132.0	0.12487	-9.03553	-13.80636
44.0	1.51072	1.79185	-2.97899	134.0	0.15380	-8.13031	-12.90114
46.0	1.79238	2.53429	-2.23654	136.0	0.17437	-7.58536	-12.35620
48.0	2.05779	3.13401	-1.63682	138.0	0.17693	-7.52196	-12.29280
50.0	2.30180	3.62068	-1.15016	140.0	0.15840	-8.00233	-12.77316
52.0	2.51951	4.01315	-0.75768	142.0	0.12510	-9.02730	-13.79813
54.0	2.70485	4.32144	-0.44940	144.0	0.09168	-10.37743	-15.14826
56.0	2.85084	4.54972	-0.22111	146.0	0.07525	-11.23510	-16.00593
58.0	2.95075	4.69933	-0.07151	148.0	0.08652	-10.62870	-15.39954
60.0	2.99974	4.77084	0.00000	150.0	0.12216	-9.13079	-13.90162
62.0	2.99599	4.76540	-0.00544	152.0	0.16356	-7.86329	-12.63413
64.0	2.94122	4.68528	-0.08556	154.0	0.18503	-7.32760	-12.09844
66.0	2.84046	4.53389	-0.23694	156.0	0.16906	-7.71949	-12.49033
68.0	2.70126	4.31567	-0.45517	158.0	0.12094	-9.17433	-13.94517
70.0	2.53264	4.03573	-0.73510	160.0	0.07260	-11.39064	-16.16147
72.0	2.34405	3.69967	-1.07116	162.0	0.06981	-11.56094	-16.33178
74.0	2.14452	3.31331	-1.45753	164.0	0.14610	-8.35363	-13.12447
76.0	1.94201	2.88252	-1.88832	166.0	0.29711	-5.27080	-10.04164
78.0	1.74308	2.41317	-2.35767	168.0	0.47241	-3.25678	-8.02762
80.0	1.55278	1.91110	-2.85973	170.0	0.59431	-2.25986	-7.03069
82.0	1.37474	1.38221	-3.38863	172.0	0.59760	-2.23586	-7.00670
84.0	1.21131	0.83255	-3.93829	174.0	0.46908	-3.28753	-8.05837
86.0	1.06380	0.26861	-4.50223	176.0	0.26224	-5.81301	-10.58384
88.0	0.93274	-0.30242	-5.07325	178.0	0.07500	-11.24944	-16.02028
90.0	0.81803	-0.87230	-5.64314	180.	0.00000	-INFINITY	-INFINITY

TABLE A2-53

EPSLN= 13. 0000(NORMALIZED GROUNDPLANE RADIUS)
MONOPOLE LENGTH= 0. 250000(WAVELENGTH)
MONOPOLE RADIUS= 0. 1000E-05(WAVELENGTH)
NUMBER OF MONOPOLE SEGMENTS= 1
NUMBER OF DISK SEGMENTS= 39
NUMBER OF ANNULAR ZONES PER DISK SEGEMENT= 1
RADIATION RESISTANCE DETERMINED BY PATTERN= 0. 374754E+02 OHMS

ELEVATION ANGLE (DEG)	DIRECTIVITY (NUMERIC)	DIRECTIVITY (dBi)	RELATIVE POWER (dB)	ELEVATION ANGLE (DEG)	DIRECTIVITY (NUMERIC)	DIRECTIVITY (dBi)	RELATIVE POWER (dB)
0.	0. 00000	-INFINITY	-INFINITY				
2. 0	0. 08752	-10. 57877	-15. 50015	92. 0	0. 69633	-1. 57184	-6. 49322
4. 0	0. 30839	-5. 10901	-10. 03039	94. 0	0. 60762	-2. 16371	-7. 08509
6. 0	0. 56076	-2. 51220	-7. 43358	96. 0	0. 53347	-2. 72889	-7. 65027
8. 0	0. 73821	-1. 31818	-6. 23956	98. 0	0. 47290	-3. 25229	-8. 17367
10. 0	0. 78407	-1. 05646	-5. 97784	100. 0	0. 42480	-3. 71815	-8. 63953
12. 0	0. 71367	-1. 46500	-6. 38638	102. 0	0. 38788	-4. 11304	-9. 03442
14. 0	0. 59507	-2. 25433	-7. 17571	104. 0	0. 36057	-4. 43008	-9. 35146
16. 0	0. 50388	-2. 97674	-7. 89812	106. 0	0. 34094	-4. 67319	-9. 59457
18. 0	0. 48133	-3. 17556	-8. 09694	108. 0	0. 32662	-4. 85963	-9. 78101
20. 0	0. 51832	-2. 85398	-7. 77536	110. 0	0. 31480	-5. 01967	-9. 94105
22. 0	0. 57009	-2. 44060	-7. 36198	112. 0	0. 30244	-5. 19360	-10. 11498
24. 0	0. 58800	-2. 30625	-7. 22763	114. 0	0. 28657	-5. 42770	-10. 34908
26. 0	0. 54817	-2. 61085	-7. 53223	116. 0	0. 26482	-5. 77051	-10. 69189
28. 0	0. 46226	-3. 35114	-8. 27253	118. 0	0. 23608	-6. 26932	-11. 19070
30. 0	0. 36859	-4. 33454	-9. 25592	120. 0	0. 20117	-6. 96441	-11. 88579
32. 0	0. 31230	-5. 05431	-9. 97569	122. 0	0. 16316	-7. 87396	-12. 79534
34. 0	0. 32657	-4. 86027	-9. 78165	124. 0	0. 12726	-8. 95292	-13. 87430
36. 0	0. 42338	-3. 73269	-8. 65407	126. 0	0. 09989	-10. 00489	-14. 92627
38. 0	0. 59513	-2. 25392	-7. 17530	128. 0	0. 08684	-10. 61262	-15. 53400
40. 0	0. 82302	-0. 84592	-5. 76730	130. 0	0. 09113	-10. 40342	-15. 32480
42. 0	1. 08638	0. 35982	-4. 56156	132. 0	0. 11093	-9. 54937	-14. 47075
44. 0	1. 36854	1. 36257	-3. 55881	134. 0	0. 13894	-8. 57177	-13. 49315
46. 0	1. 65801	2. 19586	-2. 72552	136. 0	0. 16380	-7. 85679	-12. 77817
48. 0	1. 94642	2. 89236	-2. 02902	138. 0	0. 17402	-7. 59399	-12. 51537
50. 0	2. 22554	3. 47435	-1. 44703	140. 0	0. 16314	-7. 87452	-12. 79590
52. 0	2. 48531	3. 95380	-0. 96758	142. 0	0. 13402	-8. 72816	-13. 64954
54. 0	2. 71378	4. 33574	-0. 58564	144. 0	0. 09952	-10. 02093	-14. 94231
56. 0	2. 89860	4. 62188	-0. 29950	146. 0	0. 07764	-11. 09891	-16. 02029
58. 0	3. 02923	4. 81332	-0. 10806	148. 0	0. 08243	-10. 83923	-15. 76061
60. 0	3. 09887	4. 91203	-0. 00935	150. 0	0. 11449	-9. 41246	-14. 33384
62. 0	3. 10555	4. 92138	0. 00000	152. 0	0. 15738	-8. 03061	-12. 95199
64. 0	3. 05218	4. 84611	-0. 07527	154. 0	0. 18417	-7. 34791	-12. 26929
66. 0	2. 94578	4. 69201	-0. 22938	156. 0	0. 17326	-7. 61310	-12. 53448
68. 0	2. 79611	4. 46554	-0. 45584	158. 0	0. 12568	-9. 00726	-13. 92864
70. 0	2. 61428	4. 17353	-0. 74785	160. 0	0. 07230	-11. 40887	-16. 33025
72. 0	2. 41146	3. 82280	-1. 09858	162. 0	0. 06245	-12. 04469	-16. 96607
74. 0	2. 19789	3. 42006	-1. 50132	164. 0	0. 13617	-8. 65916	-13. 58054
76. 0	1. 98230	2. 97170	-1. 94968	166. 0	0. 29411	-5. 31497	-10. 23635
78. 0	1. 77163	2. 48374	-2. 43764	168. 0	0. 48515	-3. 14123	-8. 06261
80. 0	1. 57104	1. 96186	-2. 95952	170. 0	0. 62461	-2. 04388	-6. 96526
82. 0	1. 38403	1. 41145	-3. 50993	172. 0	0. 63783	-1. 95295	-6. 87433
84. 0	1. 21276	0. 83776	-4. 08362	174. 0	0. 50592	-2. 95922	-7. 88060
86. 0	1. 05832	0. 24617	-4. 67521	176. 0	0. 28477	-5. 45503	-10. 37641
88. 0	0. 92099	-0. 35744	-5. 27882	178. 0	0. 08176	-10. 87459	-15. 79597
90. 0	0. 80053	-0. 96625	-5. 88763	180.	0. 00000	-INFINITY	-INFINITY

TABLE A2-54

EPSLN= 13.2500(NORMALIZED GROUNDPLANE RADIUS)
MONOPOLE LENGTH= 0.250000(WAVELENGTH)
MONOPOLE RADIUS= 0.1000E-05(WAVELENGTH)
NUMBER OF MONOPOLE SEGMENTS= 1
NUMBER OF DISK SEGMENTS= 39
NUMBER OF ANNULAR ZONES PER DISK SEGEMENT= 1
RADIATION RESISTANCE DETERMINED BY PATTERN= 0.380403E+02 OHMS

ELEVATION ANGLE (DEG)	DIRECTIVITY (NUMERIC)	DIRECTIVITY (dBi)	RELATIVE POWER (dB)	ELEVATION ANGLE (DEG)	DIRECTIVITY (NUMERIC)	DIRECTIVITY (dBi)	RELATIVE POWER (dB)
0.	0.00000	-INFINITY	-INFINITY				
2.0	0.08791	-10.55976	-15.61588	92.0	0.68078	-1.66996	-6.72608
4.0	0.30696	-5.12914	-10.18526	94.0	0.58814	-2.30520	-7.36132
6.0	0.54930	-2.60190	-7.65802	96.0	0.50986	-2.92548	-7.98160
8.0	0.70564	-1.51419	-6.57031	98.0	0.44511	-3.51537	-8.57150
10.0	0.72419	-1.40147	-6.45759	100.0	0.39304	-4.05567	-9.11179
12.0	0.63196	-1.99311	-7.04923	102.0	0.35276	-4.52522	-9.58134
14.0	0.50920	-2.93115	-7.98727	104.0	0.32319	-4.90541	-9.96153
16.0	0.43713	-3.59394	-8.65006	106.0	0.30292	-5.18670	-10.24282
18.0	0.45251	-3.44368	-8.49980	108.0	0.29006	-5.37505	-10.43117
20.0	0.53407	-2.72405	-7.78017	110.0	0.28217	-5.49488	-10.55100
22.0	0.62329	-2.05313	-7.10925	112.0	0.27625	-5.58699	-10.64311
24.0	0.66230	-1.78942	-6.84554	114.0	0.26898	-5.70277	-10.75889
26.0	0.62503	-2.04097	-7.09709	116.0	0.25717	-5.89787	-10.95399
28.0	0.52654	-2.78572	-7.84184	118.0	0.23839	-6.22712	-11.28324
30.0	0.41061	-3.86575	-8.92187	120.0	0.21184	-6.73999	-11.79611
32.0	0.32688	-4.85613	-9.91225	122.0	0.17899	-7.47164	-12.52776
34.0	0.31116	-5.07011	-10.12623	124.0	0.14394	-8.41822	-13.47434
36.0	0.37714	-4.23497	-9.29110	126.0	0.11287	-9.47435	-14.53047
38.0	0.51952	-2.84396	-7.90009	128.0	0.09262	-10.33309	-15.38921
40.0	0.72321	-1.40734	-6.46346	130.0	0.08837	-10.53690	-15.59302
42.0	0.97200	-0.12332	-5.17944	132.0	0.10112	-9.95157	-15.00770
44.0	1.25292	0.97924	-4.07689	134.0	0.12602	-8.99553	-14.05165
46.0	1.55592	1.91987	-3.13625	136.0	0.15282	-8.15812	-13.21424
48.0	1.87108	2.72092	-2.33520	138.0	0.16906	-7.71950	-12.77562
50.0	2.18595	3.39640	-1.65973	140.0	0.16547	-7.81283	-12.86895
52.0	2.48478	3.95288	-1.10325	142.0	0.14141	-8.49522	-13.55134
54.0	2.74991	4.39319	-0.66293	144.0	0.10728	-9.69496	-14.75109
56.0	2.96453	4.71956	-0.33656	146.0	0.08118	-10.90548	-15.96160
58.0	3.11551	4.93530	-0.12083	148.0	0.07995	-10.97205	-16.02817
60.0	3.19546	5.04533	-0.01079	150.0	0.10819	-9.65804	-14.71417
62.0	3.20341	5.05612	0.00000	152.0	0.15216	-8.17693	-13.23305
64.0	3.14427	4.97520	-0.08092	154.0	0.18426	-7.34577	-12.40189
66.0	3.02741	4.81072	-0.24540	156.0	0.17890	-7.47380	-12.52992
68.0	2.86487	4.57105	-0.48507	158.0	0.13240	-8.78102	-13.83714
70.0	2.66958	4.26443	-0.79169	160.0	0.07374	-11.32277	-16.37889
72.0	2.45401	3.89876	-1.15737	162.0	0.05543	-12.56242	-17.61854
74.0	2.22914	3.48138	-1.57474	164.0	0.12432	-9.05467	-14.11079
76.0	2.00404	3.01906	-2.03706	166.0	0.28702	-5.42088	-10.47701
78.0	1.78562	2.51789	-2.53823	168.0	0.49277	-3.07351	-8.12964
80.0	1.57882	1.98332	-3.07280	170.0	0.65025	-1.86921	-6.92533
82.0	1.38682	1.42020	-3.63592	172.0	0.67485	-1.70794	-6.76406
84.0	1.21142	0.83294	-4.22318	174.0	0.54114	-2.66691	-7.72303
86.0	1.05334	0.22569	-4.83043	176.0	0.30676	-5.13197	-10.18810
88.0	0.91258	-0.39731	-5.45343	178.0	0.08843	-10.53411	-15.59023
90.0	0.78864	-1.03123	-6.08735	180.	0.00000	-INFINITY	-INFINITY

TABLE A2-55

EPSLN= 13.5000(NORMALIZED GROUNDPLANE RADIUS)
MONOPOLE LENGTH= 0.250000(WAVELENGTH)
MONOPOLE RADIUS= 0.1000E-05(WAVELENGTH)
NUMBER OF MONOPOLE SEGMENTS= 1
NUMBER OF DISK SEGMENTS= 40
NUMBER OF ANNULAR ZONES PER DISK SEGEMENT= 1
RADIATION RESISTANCE DETERMINED BY PATTERN= 0.383927E+02 OHMS

ELEVATION ANGLE (DEG)	DIRECTIVITY (NUMERIC)	DIRECTIVITY (dBi)	RELATIVE POWER (dB)	ELEVATION ANGLE (DEG)	DIRECTIVITY (NUMERIC)	DIRECTIVITY (dBi)	RELATIVE POWER (dB)
0.	0.00000	-INFINITY	-INFINITY				
2.0	0.08912	-10.50009	-15.65753	92.0	0.67197	-1.72652	-6.88397
4.0	0.30827	-5.11062	-10.26807	94.0	0.57709	-2.38759	-7.54504
6.0	0.54226	-2.65789	-7.81533	96.0	0.49587	-3.04632	-8.20377
8.0	0.67801	-1.68766	-6.84510	98.0	0.42756	-3.69003	-8.84748
10.0	0.66871	-1.74762	-6.90506	100.0	0.37152	-4.30017	-9.45762
12.0	0.55398	-2.56506	-7.72251	102.0	0.32718	-4.85211	-10.00956
14.0	0.42723	-3.69336	-8.85081	104.0	0.29392	-5.31776	-10.47520
16.0	0.37539	-4.25512	-9.41257	106.0	0.27088	-5.67227	-10.82971
18.0	0.42995	-3.66583	-8.82327	108.0	0.25679	-5.90415	-11.06159
20.0	0.55632	-2.54676	-7.70421	110.0	0.24979	-6.02428	-11.18172
22.0	0.68175	-1.66375	-6.82120	112.0	0.24727	-6.06834	-11.22579
24.0	0.73953	-1.31046	-6.46790	114.0	0.24598	-6.09108	-11.24852
26.0	0.70241	-1.53408	-6.69152	116.0	0.24230	-6.15638	-11.31383
28.0	0.58992	-2.29204	-7.44949	118.0	0.23289	-6.32856	-11.48601
30.0	0.45186	-3.44991	-8.60736	120.0	0.21546	-6.66639	-11.82383
32.0	0.34236	-4.65521	-9.81266	122.0	0.18980	-7.21693	-12.37438
34.0	0.29952	-5.23575	-10.39319	124.0	0.15847	-8.00050	-13.15795
36.0	0.33848	-4.70470	-9.86215	126.0	0.12679	-8.96912	-14.12657
38.0	0.45616	-3.40885	-8.56630	128.0	0.10188	-9.91917	-15.07662
40.0	0.64101	-1.93134	-7.08879	130.0	0.09048	-10.43448	-15.59192
42.0	0.88086	-0.55094	-5.70838	132.0	0.09613	-10.17152	-15.32897
44.0	1.16559	0.66547	-4.49197	134.0	0.11666	-9.33076	-14.48821
46.0	1.48551	1.71875	-3.43870	136.0	0.14355	-8.43008	-13.58753
48.0	1.82806	2.61992	-2.53753	138.0	0.16416	-7.84729	-13.00473
50.0	2.17598	3.37655	-1.78090	140.0	0.16703	-7.77218	-12.92963
52.0	2.50783	3.99299	-1.16446	142.0	0.14816	-8.29268	-13.45013
54.0	2.80096	4.47307	-0.68437	144.0	0.11520	-9.38541	-14.54286
56.0	3.03521	4.82189	-0.33555	146.0	0.08579	-10.66581	-15.82326
58.0	3.19622	5.04636	-0.11108	148.0	0.07911	-11.01764	-16.17509
60.0	3.27718	5.15501	-0.00244	150.0	0.10378	-9.83895	-14.99640
62.0	3.27902	5.15745	0.00000	152.0	0.14903	-8.26719	-13.42464
64.0	3.20910	5.06383	-0.09362	154.0	0.18692	-7.28354	-12.44098
66.0	3.07921	4.88439	-0.27306	156.0	0.18782	-7.26263	-12.42008
68.0	2.90339	4.62905	-0.52840	158.0	0.14280	-8.45262	-13.61006
70.0	2.69600	4.30720	-0.85025	160.0	0.07844	-11.05442	-16.21187
72.0	2.47031	3.92752	-1.22993	162.0	0.05034	-12.98121	-18.13866
74.0	2.23762	3.49785	-1.65959	164.0	0.11282	-9.47618	-14.63363
76.0	2.00688	3.02521	-2.13224	166.0	0.27951	-5.53608	-10.69353
78.0	1.78472	2.51569	-2.64175	168.0	0.50059	-3.00521	-8.16265
80.0	1.57564	1.97457	-3.18288	170.0	0.67771	-1.68955	-6.84699
82.0	1.38238	1.40628	-3.75117	172.0	0.71519	-1.45578	-6.61322
84.0	1.20629	0.81453	-4.34291	174.0	0.57992	-2.36632	-7.52376
86.0	1.04773	0.20248	-4.95497	176.0	0.33113	-4.80004	-9.95748
88.0	0.90634	-0.42709	-5.58453	178.0	0.09584	-10.18445	-15.34189
90.0	0.78140	-1.07128	-6.22872	180.	0.00000	-INFINITY	-INFINITY

TABLE A2-56

EPSLN= 13.7500(NORMALIZED GROUNDPLANE RADIUS)
MONOPOLE LENGTH= 0.250000(WAVELENGTH)
MONOPOLE RADIUS= 0.1000E-05(WAVELENGTH)
NUMBER OF MONOPOLE SEGMENTS= 1
NUMBER OF DISK SEGMENTS= 41
NUMBER OF ANNULAR ZONES PER DISK SEGEMENT= 1
RADIATION RESISTANCE DETERMINED BY PATTERN= 0.385409E+02 OHMS

ELEVATION ANGLE (DEG)	DIRECTIVITY (NUMERIC)	DIRECTIVITY (dBi)	RELATIVE POWER (dB)	ELEVATION ANGLE (DEG)	DIRECTIVITY (NUMERIC)	DIRECTIVITY (dBi)	RELATIVE POWER (dB)
0.	0.00000	-INFINITY	-INFINITY				
2.0	0.09237	-10.34473	-15.57915	92.0	0.66921	-1.74436	-6.97878
4.0	0.31648	-4.99654	-10.23096	94.0	0.57371	-2.41309	-7.64751
6.0	0.54702	-2.62000	-7.85442	96.0	0.49092	-3.08988	-8.32430
8.0	0.66464	-1.77414	-7.00856	98.0	0.42009	-3.76654	-9.00096
10.0	0.62696	-2.02761	-7.26203	100.0	0.36069	-4.42866	-9.66308
12.0	0.48751	-3.12021	-8.35463	102.0	0.31236	-5.05346	-10.28788
14.0	0.35479	-4.50027	-9.73469	104.0	0.27484	-5.60925	-10.84367
16.0	0.32251	-4.91459	-10.14901	106.0	0.24778	-6.05930	-11.29372
18.0	0.41647	-3.80414	-9.03856	108.0	0.23055	-6.37233	-11.60675
20.0	0.58759	-2.30923	-7.54365	110.0	0.22195	-6.53753	-11.77195
22.0	0.74789	-1.26164	-6.49606	112.0	0.21999	-6.57597	-11.81039
24.0	0.82181	-0.85226	-6.08668	114.0	0.22182	-6.53990	-11.77432
26.0	0.78190	-1.06850	-6.30292	116.0	0.22384	-6.50069	-11.73511
28.0	0.65327	-1.84908	-7.08350	118.0	0.22211	-6.53429	-11.76871
30.0	0.49250	-3.07597	-8.31039	120.0	0.21326	-6.71095	-11.94537
32.0	0.35831	-4.45744	-9.69186	122.0	0.19548	-7.08896	-12.32338
34.0	0.29084	-5.36349	-10.59791	124.0	0.16963	-7.70492	-12.93934
36.0	0.30626	-5.13913	-10.37355	126.0	0.13976	-8.54630	-13.78072
38.0	0.40336	-3.94307	-9.17749	128.0	0.11262	-9.48381	-14.71823
40.0	0.57378	-2.41252	-7.64694	130.0	0.09591	-10.18126	-15.41568
42.0	0.80882	-0.92149	-6.15590	132.0	0.09528	-10.20982	-15.44424
44.0	1.10045	0.41569	-4.81873	134.0	0.11121	-9.53858	-14.77300
46.0	1.43841	1.57881	-3.65561	136.0	0.13721	-8.62603	-13.86045
48.0	1.80687	2.56927	-2.66515	138.0	0.16108	-7.92970	-13.16412
50.0	2.18349	3.39152	-1.84290	140.0	0.16963	-7.70499	-12.93941
52.0	2.54157	4.05102	-1.18340	142.0	0.15578	-8.07479	-13.30921
54.0	2.85429	4.55498	-0.67944	144.0	0.12429	-9.05569	-14.29011
56.0	3.09934	4.91269	-0.32173	146.0	0.09200	-10.36235	-15.59677
58.0	3.26223	5.13514	-0.09928	148.0	0.08025	-10.95560	-16.19002
60.0	3.33766	5.23442	0.00000	150.0	0.10169	-9.92732	-15.16174
62.0	3.32893	5.22304	-0.01138	152.0	0.14878	-8.27446	-13.50888
64.0	3.24596	5.11343	-0.12099	154.0	0.19330	-7.13767	-12.37209
66.0	3.10279	4.91752	-0.31690	156.0	0.20127	-6.96229	-12.19671
68.0	2.91506	4.64647	-0.58794	158.0	0.15790	-8.01618	-13.25060
70.0	2.69808	4.31055	-0.92387	160.0	0.08698	-10.60600	-15.84042
72.0	2.46546	3.91898	-1.31544	162.0	0.04758	-13.22551	-18.45993
74.0	2.22841	3.47995	-1.75446	164.0	0.10274	-9.88260	-15.11702
76.0	1.99554	3.00061	-2.23381	166.0	0.27430	-5.61774	-10.85216
78.0	1.77299	2.48707	-2.74735	168.0	0.51361	-2.89362	-8.12804
80.0	1.56475	1.94445	-3.28997	170.0	0.71399	-1.46310	-6.69752
82.0	1.37307	1.37693	-3.85749	172.0	0.76639	-1.15550	-6.38992
84.0	1.19888	0.78777	-4.44665	174.0	0.62848	-2.01711	-7.25153
86.0	1.04218	0.17945	-5.05497	176.0	0.36147	-4.41930	-9.65372
88.0	0.90235	-0.44627	-5.68069	178.0	0.10505	-9.78594	-15.02036
90.0	0.77839	-1.08801	-6.32243	180.	0.00000	-INFINITY	-INFINITY

TABLE A2-57

EPSLN= 14.0000(NORMALIZED GROUNDPLANE RADIUS)
MONOPOLE LENGTH= 0.250000(WAVELENGTH)
MONOPOLE RADIUS= 0.1000E-05(WAVELENGTH)
NUMBER OF MONOPOLE SEGMENTS= 1
NUMBER OF DISK SEGMENTS= 42
NUMBER OF ANNULAR ZONES PER DISK SEGEMENT= 1
RADIATION RESISTANCE DETERMINED BY PATTERN= 0.386276E+02 OHMS

ELEVATION ANGLE (DEG)	DIRECTIVITY (NUMERIC)	DIRECTIVITY (dBi)	RELATIVE POWER (dB)	ELEVATION ANGLE (DEG)	DIRECTIVITY (NUMERIC)	DIRECTIVITY (dBi)	RELATIVE (dB)
0.	0.00000	-INFINITY	-INFINITY				
2.0	0.09904	-10.04177	-15.32676	92.0	0.66917	-1.74466	-7.02965
4.0	0.33635	-4.73209	-10.01707	94.0	0.57449	-2.40715	-7.69214
6.0	0.57175	-2.42794	-7.71292	96.0	0.49156	-3.08425	-8.36924
8.0	0.67539	-1.70443	-6.98942	98.0	0.41954	-3.77230	-9.05729
10.0	0.60808	-2.16043	-7.44541	100.0	0.35790	-4.46243	-9.74742
12.0	0.43931	-3.57227	-8.85726	102.0	0.30639	-5.13724	-10.42223
14.0	0.29609	-5.28578	-10.57077	104.0	0.26500	-5.76761	-11.05260
16.0	0.28110	-5.51143	-10.79642	106.0	0.23377	-6.31220	-11.59719
18.0	0.41442	-3.82560	-9.11059	108.0	0.21261	-6.72413	-12.00912
20.0	0.63072	-2.00166	-7.28665	110.0	0.20102	-6.96769	-12.25268
22.0	0.82501	-0.83540	-6.12039	112.0	0.19772	-7.03958	-12.32457
24.0	0.91240	-0.39813	-5.68312	114.0	0.20046	-6.97981	-12.26480
26.0	0.86601	-0.62476	-5.90975	116.0	0.20593	-6.86287	-12.14786
28.0	0.71800	-1.43873	-6.72371	118.0	0.21003	-6.77709	-12.06208
30.0	0.53280	-2.73433	-8.01932	120.0	0.20860	-6.80685	-12.09183
32.0	0.37406	-4.27055	-9.55554	122.0	0.19847	-7.02297	-12.30796
34.0	0.28361	-5.47273	-10.75772	124.0	0.17883	-7.47556	-12.76054
36.0	0.27811	-5.55785	-10.84283	126.0	0.15220	-8.17581	-13.46080
38.0	0.35770	-4.46485	-9.74983	128.0	0.12457	-9.04572	-14.33071
40.0	0.51675	-2.86721	-8.15219	130.0	0.10407	-9.82684	-15.11182
42.0	0.74954	-1.25202	-6.53701	132.0	0.09806	-10.08522	-15.37020
44.0	1.04955	0.21003	-5.07495	134.0	0.10951	-9.60542	-14.89041
46.0	1.40535	1.47785	-3.80714	136.0	0.13417	-8.72333	-14.00832
48.0	1.79743	2.54653	-2.73846	138.0	0.16059	-7.94274	-13.22772
50.0	2.19841	3.42108	-1.86390	140.0	0.17428	-7.58764	-12.87262
52.0	2.57674	4.11070	-1.17428	142.0	0.16518	-7.82049	-13.10547
54.0	2.90231	4.62744	-0.65755	144.0	0.13510	-8.69335	-13.97834
56.0	3.15158	4.98529	-0.29970	146.0	0.09998	-10.00096	-15.28595
58.0	3.31081	5.19934	-0.08565	148.0	0.08332	-10.79262	-16.07760
60.0	3.37675	5.28499	0.00000	150.0	0.10204	-9.91244	-15.19743
62.0	3.35530	5.25731	-0.02768	152.0	0.15208	-8.17923	-13.46422
64.0	3.25886	5.13065	-0.15433	154.0	0.20480	-6.88661	-12.17160
66.0	3.10337	4.91834	-0.36665	156.0	0.22117	-6.55282	-11.83780
68.0	2.90569	4.63250	-0.65249	158.0	0.17962	-7.45648	-12.74146
70.0	2.68164	4.28401	-1.00098	160.0	0.10081	-9.96499	-15.24997
72.0	2.44482	3.88247	-1.40251	162.0	0.04824	-13.16597	-18.45096
74.0	2.20611	3.43626	-1.84872	164.0	0.09558	-10.19639	-15.48137
76.0	1.97360	2.95258	-2.33240	166.0	0.27464	-5.61229	-10.89728
78.0	1.75288	2.43752	-2.84746	168.0	0.53785	-2.69341	-7.97839
80.0	1.54744	1.89613	-3.38886	170.0	0.76761	-1.14857	-6.43356
82.0	1.35908	1.33244	-3.95255	172.0	0.83789	-0.76812	-6.05311
84.0	1.18837	0.74952	-4.53546	174.0	0.69476	-1.58167	-6.86665
86.0	1.03502	0.14951	-5.13548	176.0	0.40244	-3.95300	-9.23799
88.0	0.89817	-0.46640	-5.75139	178.0	0.11743	-9.30226	-14.58725
90.0	0.77665	-1.09776	-6.38275	180.	0.00000	-INFINITY	-INFINITY

Tables A2-58–A2-66; Experimental Monopole Antennas: $h/\lambda \approx 0.25$, $b = 0.25$ inch, $N = 4$

TABLE A2-58

EPSLN= 0.7660(NORMALIZED GROUNDPLANE RADIUS)
MONOPOLE LENGTH= 0.239600(WAVELENGTH)
MONOPOLE RADIUS= 0.6350E-03(WAVELENGTH)
NUMBER OF MONOPOLE SEGMENTS= 4
NUMBER OF DISK SEGMENTS= 12
RADIATION RESISTANCE DETERMINED BY PATTERN= 0.177438E+02 OHMS

ELEVATION ANGLE (DEG)	DIRECTIVITY (NUMERIC)	DIRECTIVITY (DBI)	RELATIVE POWER (DB)	ELEVATION ANGLE (DEG)	DIRECTIVITY (NUMERIC)	DIRECTIVITY (DBI)	RELATIVE POWER (DB)
0.	0.00000	-INFINITY	-INFINITY				
2.0	0.00181	-27.41896	-29.19799	92.0	1.50391	1.77221	-0.00683
4.0	0.00724	-21.40335	-23.18238	94.0	1.49781	1.75457	-0.02446
6.0	0.01626	-17.88985	-19.66889	96.0	1.48802	1.72609	-0.05295
8.0	0.02882	-15.40275	-17.18178	98.0	1.47458	1.68670	-0.09234
10.0	0.04488	-13.47957	-15.25860	100.0	1.45758	1.63631	-0.14272
12.0	0.06435	-11.91433	-13.69337	102.0	1.43708	1.57481	-0.20422
14.0	0.08715	-10.59717	-12.37621	104.0	1.41321	1.50206	-0.27697
16.0	0.11317	-9.46251	-11.24154	106.0	1.38608	1.41789	-0.36115
18.0	0.14230	-8.46806	-10.24710	108.0	1.35584	1.32208	-0.45696
20.0	0.17438	-7.58497	-9.36400	110.0	1.32263	1.21440	-0.56464
22.0	0.20928	-6.79265	-8.57168	112.0	1.28664	1.09457	-0.68446
24.0	0.24684	-6.07593	-7.85496	114.0	1.24804	0.96228	-0.81675
26.0	0.28686	-5.42328	-7.20232	116.0	1.20703	0.81717	-0.96186
28.0	0.32917	-4.82577	-6.60480	118.0	1.16381	0.65883	-1.12021
30.0	0.37357	-4.27630	-6.05534	120.0	1.11861	0.48679	-1.29225
32.0	0.41984	-3.76920	-5.54823	122.0	1.07165	0.30053	-1.47850
34.0	0.46776	-3.29980	-5.07883	124.0	1.02317	0.09947	-1.67957
36.0	0.51710	-2.86426	-4.64330	126.0	0.97340	-0.11708	-1.89611
38.0	0.56763	-2.45938	-4.23842	128.0	0.92260	-0.34986	-2.12889
40.0	0.61909	-2.08246	-3.86149	130.0	0.87102	-0.59973	-2.37877
42.0	0.67124	-1.73119	-3.51023	132.0	0.81890	-0.86768	-2.64672
44.0	0.72383	-1.40361	-3.18265	134.0	0.76651	-1.15481	-2.93385
46.0	0.77660	-1.09803	-2.87707	136.0	0.71410	-1.46241	-3.24144
48.0	0.82928	-0.81298	-2.59202	138.0	0.66192	-1.79193	-3.57096
50.0	0.88162	-0.54719	-2.32622	140.0	0.61023	-2.14506	-3.92409
52.0	0.93336	-0.29953	-2.07856	142.0	0.55927	-2.52377	-4.30281
54.0	0.98423	-0.06902	-1.84806	144.0	0.50929	-2.93036	-4.70940
56.0	1.03400	0.14519	-1.63385	146.0	0.46052	-3.36753	-5.14657
58.0	1.08240	0.34386	-1.43517	148.0	0.41319	-3.83848	-5.61752
60.0	1.12919	0.52768	-1.25136	150.0	0.36753	-4.34705	-6.12609
62.0	1.17415	0.69722	-1.08182	152.0	0.32375	-4.89789	-6.67693
64.0	1.21703	0.85302	-0.92601	154.0	0.28205	-5.49670	-7.27573
66.0	1.25764	0.99555	-0.78349	156.0	0.24263	-6.15054	-7.92958
68.0	1.29575	1.12521	-0.65382	158.0	0.20567	-6.86837	-8.64741
70.0	1.33118	1.24237	-0.53666	160.0	0.17133	-7.66171	-9.44074
72.0	1.36375	1.34736	-0.43168	162.0	0.13977	-8.54573	-10.32476
74.0	1.39330	1.44043	-0.33860	164.0	0.11115	-9.54101	-11.32004
76.0	1.41966	1.52185	-0.25718	166.0	0.08558	-10.67640	-12.45544
78.0	1.44272	1.59182	-0.18721	168.0	0.06318	-11.99420	-13.77324
80.0	1.46235	1.65052	-0.12852	170.0	0.04406	-13.55998	-15.33901
82.0	1.47846	1.69808	-0.08095	172.0	0.02829	-15.48360	-17.26264
84.0	1.49095	1.73464	-0.04440	174.0	0.01595	-17.97105	-19.75009
86.0	1.49978	1.76028	-0.01876	176.0	0.00710	-21.48480	-23.26383
88.0	1.50490	1.77506	-0.00397	178.0	0.00178	-27.50055	-29.27959
90.0	1.50627	1.77903	0.00000	180.	0.00000	-INFINITY	-INFINITY

TABLE A2-59

EPSLN= 0.9190(NORMALIZED GROUNDPLANE RADIUS)
MONOPOLE LENGTH= 0.239600(WAVELENGTH)
MONOPOLE RADIUS= 0.7630E-03(WAVELENGTH)
NUMBER OF MONOPOLE SEGMENTS= 4
NUMBER OF DISK SEGMENTS= 11
RADIATION RESISTANCE DETERMINED BY PATTERN= 0.183449E+02 OHMS

ELEVATION ANGLE (DEG)	DIRECTIVITY (NUMERIC)	DIRECTIVITY (DBI)	RELATIVE POWER (DB)	ELEVATION ANGLE (DEG)	DIRECTIVITY (NUMERIC)	DIRECTIVITY (DBI)	RELATIVE POWER (DB)
0.	0.00000	-INFINITY	-INFINITY				
2.0	0.00189	-27.24413	-28.97891	92.0	1.48872	1.72812	-0.00665
4.0	0.00753	-21.22930	-22.96407	94.0	1.48293	1.71121	-0.02357
6.0	0.01692	-17.71708	-19.45186	96.0	1.47366	1.68398	-0.05079
8.0	0.02998	-15.23177	-16.96655	98.0	1.46096	1.64639	-0.08839
10.0	0.04666	-13.31088	-15.04566	100.0	1.44488	1.59832	-0.13646
12.0	0.06686	-11.74842	-13.48320	102.0	1.42549	1.53965	-0.19512
14.0	0.09048	-10.43451	-12.16928	104.0	1.40289	1.47023	-0.26455
16.0	0.11739	-9.30356	-11.03833	106.0	1.37716	1.38985	-0.34492
18.0	0.14746	-8.31326	-10.04804	108.0	1.34844	1.29831	-0.43647
20.0	0.18052	-7.43475	-9.16952	110.0	1.31684	1.19533	-0.53945
22.0	0.21640	-6.64741	-8.38219	112.0	1.28251	1.08061	-0.65416
24.0	0.25491	-5.93605	-7.67083	114.0	1.24561	0.95382	-0.78096
26.0	0.29586	-5.28913	-7.02390	116.0	1.20630	0.81456	-0.92022
28.0	0.33903	-4.69766	-6.43244	118.0	1.16477	0.66239	-1.07239
30.0	0.38419	-4.15456	-5.88933	120.0	1.12120	0.49682	-1.23795
32.0	0.43111	-3.65408	-5.38885	122.0	1.07580	0.31730	-1.41748
34.0	0.47956	-3.19154	-4.92632	124.0	1.02877	0.12319	-1.61159
36.0	0.52929	-2.76309	-4.49786	126.0	0.98035	-0.08620	-1.82098
38.0	0.58003	-2.36547	-4.10024	128.0	0.93075	-0.31168	-2.04645
40.0	0.63155	-1.99594	-3.73072	130.0	0.88021	-0.55413	-2.28890
42.0	0.68357	-1.65218	-3.38696	132.0	0.82898	-0.81457	-2.54935
44.0	0.73584	-1.33219	-3.06697	134.0	0.77729	-1.09415	-2.82893
46.0	0.78809	-1.03423	-2.76900	136.0	0.72541	-1.39419	-3.12896
48.0	0.84008	-0.75679	-2.49157	138.0	0.67357	-1.71617	-3.45095
50.0	0.89154	-0.49858	-2.23335	140.0	0.62204	-2.06184	-3.79662
52.0	0.94223	-0.25842	-1.99320	142.0	0.57106	-2.43320	-4.16797
54.0	0.99190	-0.03531	-1.77009	144.0	0.52089	-2.83256	-4.56734
56.0	1.04031	0.17165	-1.56313	146.0	0.47177	-3.26268	-4.99745
58.0	1.08724	0.36326	-1.37152	148.0	0.42396	-3.72678	-5.46155
60.0	1.13246	0.54022	-1.19455	150.0	0.37768	-4.22874	-5.96351
62.0	1.17576	0.70317	-1.03160	152.0	0.33318	-4.77324	-6.50801
64.0	1.21693	0.85267	-0.88211	154.0	0.29067	-5.36600	-7.10078
66.0	1.25580	0.98920	-0.74557	156.0	0.25037	-6.01413	-7.74891
68.0	1.29218	1.11322	-0.62155	158.0	0.21249	-6.72660	-8.46138
70.0	1.32590	1.22512	-0.50965	160.0	0.17722	-7.51496	-9.24973
72.0	1.35683	1.32524	-0.40953	162.0	0.14473	-8.39440	-10.12917
74.0	1.38481	1.41389	-0.32088	164.0	0.11520	-9.38552	-11.12030
76.0	1.40972	1.49133	-0.24344	166.0	0.08877	-10.51720	-12.25198
78.0	1.43146	1.55780	-0.17698	168.0	0.06559	-11.83175	-13.56653
80.0	1.44993	1.61348	-0.12129	170.0	0.04576	-13.39475	-15.12953
82.0	1.46506	1.65854	-0.07623	172.0	0.02940	-15.31608	-17.05086
84.0	1.47677	1.69312	-0.04166	174.0	0.01659	-17.80174	-19.53652
86.0	1.48501	1.71730	-0.01747	176.0	0.00739	-21.31420	-23.04897
88.0	1.48976	1.73117	-0.00360	178.0	0.00185	-27.32918	-29.06396
90.0	1.49100	1.73478	0.00000	180.	0.00000	-INFINITY	-INFINITY

TABLE A2-60

EPSLN= 1.0970(NORMALIZED GROUNDPLANE RADIUS)
MONOPOLE LENGTH= 0.238500(WAVELENGTH)
MONOPOLE RADIUS= 0.9110E-03(WAVELENGTH)
NUMBER OF MONOPOLE SEGMENTS= 4
NUMBER OF DISK SEGMENTS= 13
RADIATION RESISTANCE DETERMINED BY PATTERN= 0.187752E+02 OHMS

ELEVATION ANGLE (DEG)	DIRECTIVITY (NUMERIC)	DIRECTIVITY (DBI)	RELATIVE POWER (DB)	ELEVATION ANGLE (DEG)	DIRECTIVITY (NUMERIC)	DIRECTIVITY (DBI)	RELATIVE POWER (DB)
0.	0.00000	-INFINITY	-INFINITY				
2.0	0.00200	-26.98412	-28.65160	92.0	1.46588	1.66098	-0.00650
4.0	0.00800	-20.97041	-22.63790	94.0	1.46051	1.64504	-0.02245
6.0	0.01795	-17.46008	-19.12756	96.0	1.45198	1.61960	-0.04788
8.0	0.03179	-14.97738	-16.64486	98.0	1.44032	1.58460	-0.08288
10.0	0.04943	-13.05983	-14.72732	100.0	1.42558	1.53992	-0.12756
12.0	0.07077	-11.50143	-13.16892	102.0	1.40780	1.48541	-0.18207
14.0	0.09567	-10.19227	-11.85976	104.0	1.38704	1.42090	-0.24659
16.0	0.12397	-9.06675	-10.73424	106.0	1.36338	1.34615	-0.32133
18.0	0.15551	-8.08254	-9.75003	108.0	1.33688	1.26093	-0.40656
20.0	0.19008	-7.21073	-8.87822	110.0	1.30765	1.16492	-0.50257
22.0	0.22747	-6.43071	-8.09819	112.0	1.27579	1.05779	-0.60969
24.0	0.26747	-5.72722	-7.39471	114.0	1.24141	0.93916	-0.72832
26.0	0.30983	-5.08870	-6.75619	116.0	1.20465	0.80860	-0.85889
28.0	0.35431	-4.50613	-6.17362	118.0	1.16563	0.66561	-1.00188
30.0	0.40065	-3.97238	-5.63987	120.0	1.12452	0.50966	-1.15783
32.0	0.44857	-3.48167	-5.14915	122.0	1.08147	0.34013	-1.32736
34.0	0.49782	-3.02927	-4.69676	124.0	1.03666	0.15635	-1.51114
36.0	0.54812	-2.61127	-4.27876	126.0	0.99028	-0.04244	-1.70992
38.0	0.59919	-2.22439	-3.89187	128.0	0.94252	-0.25708	-1.92457
40.0	0.65075	-1.86583	-3.53331	130.0	0.89361	-0.48853	-2.15602
42.0	0.70255	-1.53321	-3.20069	132.0	0.84375	-0.73786	-2.40535
44.0	0.75431	-1.22447	-2.89196	134.0	0.79318	-1.00628	-2.67376
46.0	0.80578	-0.93785	-2.60533	136.0	0.74214	-1.29514	-2.96263
48.0	0.85669	-0.67176	-2.33925	138.0	0.69087	-1.60601	-3.27349
50.0	0.90680	-0.42486	-2.09235	140.0	0.63964	-1.94067	-3.60815
52.0	0.95589	-0.19593	-1.86341	142.0	0.58869	-2.30117	-3.96865
54.0	1.00372	0.01610	-1.65138	144.0	0.53828	-2.68989	-4.35738
56.0	1.05007	0.21219	-1.45529	146.0	0.48869	-3.10963	-4.77711
58.0	1.09476	0.39318	-1.27430	148.0	0.44018	-3.56366	-5.23115
60.0	1.13759	0.55984	-1.10764	150.0	0.39301	-4.05591	-5.72340
62.0	1.17838	0.71285	-0.95464	152.0	0.34745	-4.59110	-6.25859
64.0	1.21697	0.85281	-0.81468	154.0	0.30374	-5.17500	-6.84249
66.0	1.25322	0.98027	-0.68721	156.0	0.26213	-5.81476	-7.48225
68.0	1.28698	1.09573	-0.57175	158.0	0.22288	-6.51938	-8.18687
70.0	1.31814	1.19963	-0.46786	160.0	0.18619	-7.30044	-8.96793
72.0	1.34659	1.29234	-0.37515	162.0	0.15229	-8.17318	-9.84067
74.0	1.37221	1.37422	-0.29327	164.0	0.12139	-9.15823	-10.82571
76.0	1.39494	1.44557	-0.22192	166.0	0.09366	-10.28448	-11.95197
78.0	1.41470	1.50664	-0.16085	168.0	0.06927	-11.59428	-13.26177
80.0	1.43142	1.55767	-0.10982	170.0	0.04838	-13.15322	-14.82071
82.0	1.44505	1.59884	-0.06865	172.0	0.03111	-15.07121	-16.73869
84.0	1.45556	1.63030	-0.03718	174.0	0.01756	-17.55425	-19.22173
86.0	1.46291	1.65218	-0.01530	176.0	0.00783	-21.06483	-22.73231
88.0	1.46709	1.66456	-0.00293	178.0	0.00196	-27.07868	-28.74617
90.0	1.46808	1.66749	0.00000	180.	0.00000	-INFINITY	-INFINITY

TABLE A2-61

EPSLN= 1.3790(NORMALIZED GROUNDPLANE RADIUS)
MONOPOLE LENGTH= 0.238200(WAVELENGTH)
MONOPOLE RADIUS= 0.1143E-02(WAVELENGTH)
NUMBER OF MONOPOLE SEGMENTS= 4
NUMBER OF DISK SEGMENTS= 20
RADIATION RESISTANCE DETERMINED BY PATTERN= 0.198933E+02 OHMS

ELEVATION ANGLE (DEG)	DIRECTIVITY (NUMERIC)	DIRECTIVITY (DBI)	RELATIVE POWER (DB)	ELEVATION ANGLE (DEG)	DIRECTIVITY (NUMERIC)	DIRECTIVITY (DBI)	RELATIVE POWER (DB)
0.	0.00000	-INFINITY	-INFINITY				
2.0	0.00227	-26.44481	-27.96848	92.0	1.41797	1.51666	-0.00700
4.0	0.00905	-20.43334	-21.95701	94.0	1.41318	1.50196	-0.02170
6.0	0.02029	-16.92672	-18.45039	96.0	1.40589	1.47951	-0.04416
8.0	0.03590	-14.44920	-15.97287	98.0	1.39611	1.44920	-0.07447
10.0	0.05574	-12.53828	-14.06195	100.0	1.38385	1.41090	-0.11277
12.0	0.07965	-10.98793	-12.51159	102.0	1.36912	1.36441	-0.15926
14.0	0.10744	-9.68820	-11.21187	104.0	1.35192	1.30950	-0.21417
16.0	0.13888	-8.57347	-10.09714	106.0	1.33226	1.24588	-0.27778
18.0	0.17373	-7.60137	-9.12503	108.0	1.31016	1.17325	-0.35042
20.0	0.21169	-6.74293	-8.26660	110.0	1.28564	1.09120	-0.43246
22.0	0.25249	-5.97748	-7.50115	112.0	1.25873	0.99933	-0.52433
24.0	0.29582	-5.28974	-6.81340	114.0	1.22946	0.89716	-0.62651
26.0	0.34135	-4.66805	-6.19171	116.0	1.19789	0.78415	-0.73951
28.0	0.38875	-4.10334	-5.62701	118.0	1.16405	0.65972	-0.86395
30.0	0.43768	-3.58840	-5.11207	120.0	1.12803	0.52321	-1.00045
32.0	0.48782	-3.11738	-4.64104	122.0	1.08991	0.37392	-1.14975
34.0	0.53883	-2.68547	-4.20913	124.0	1.04979	0.21103	-1.31263
36.0	0.59038	-2.28867	-3.81234	126.0	1.00779	0.03369	-1.48997
38.0	0.64215	-1.92360	-3.44727	128.0	0.96404	-0.15907	-1.68273
40.0	0.69384	-1.58739	-3.11106	130.0	0.91869	-0.36831	-1.89198
42.0	0.74515	-1.27755	-2.80121	132.0	0.87192	-0.59523	-2.11890
44.0	0.79581	-0.99192	-2.51559	134.0	0.82392	-0.84114	-2.36481
46.0	0.84555	-0.72862	-2.25229	136.0	0.77490	-1.10753	-2.63120
48.0	0.89413	-0.48599	-2.00966	138.0	0.72509	-1.39606	-2.91972
50.0	0.94134	-0.26255	-1.78622	140.0	0.67474	-1.70862	-3.23229
52.0	0.98697	-0.05698	-1.58065	142.0	0.62411	-2.04737	-3.57104
54.0	1.03084	0.13191	-1.39176	144.0	0.57348	-2.41480	-3.93847
56.0	1.07280	0.30517	-1.21849	146.0	0.52314	-2.81380	-4.33747
58.0	1.11270	0.46379	-1.05988	148.0	0.47340	-3.24774	-4.77141
60.0	1.15043	0.60862	-0.91505	150.0	0.42456	-3.72064	-5.24430
62.0	1.18590	0.74046	-0.78320	152.0	0.37694	-4.23730	-5.76097
64.0	1.21900	0.86005	-0.66362	154.0	0.33086	-4.80358	-6.32725
66.0	1.24969	0.96802	-0.55565	156.0	0.28663	-5.42671	-6.95038
68.0	1.27790	1.06498	-0.45868	158.0	0.24458	-6.11576	-7.63942
70.0	1.30361	1.15148	-0.37219	160.0	0.20500	-6.88237	-8.40604
72.0	1.32678	1.22798	-0.29569	162.0	0.16820	-7.74185	-9.26552
74.0	1.34739	1.29493	-0.22873	164.0	0.13444	-8.71487	-10.23854
76.0	1.36544	1.35273	-0.17094	166.0	0.10398	-9.83041	-11.35407
78.0	1.38092	1.40170	-0.12197	168.0	0.07708	-11.13081	-12.65448
80.0	1.39384	1.44214	-0.08153	170.0	0.05393	-12.68175	-14.20541
82.0	1.40421	1.47431	-0.04936	172.0	0.03473	-14.59313	-16.11680
84.0	1.41202	1.49840	-0.02526	174.0	0.01963	-17.07101	-18.59468
86.0	1.41729	1.51459	-0.00908	176.0	0.00875	-20.57789	-22.10156
88.0	1.42003	1.52299	-0.00068	178.0	0.00219	-26.58952	-28.11319
90.0	1.42026	1.52367	0.00000	180.	0.00000	-INFINITY	-INFINITY

TABLE A2-62

EPSLN= 1.5930(NORMALIZED GROUNDPLANE RADIUS)
MONOPOLE LENGTH= 0.237900(WAVELENGTH)
MONOPOLE RADIUS= 0.1321E-02(WAVELENGTH)
NUMBER OF MONOPOLE SEGMENTS= 4
NUMBER OF DISK SEGMENTS= 26
RADIATION RESISTANCE DETERMINED BY PATTERN= 0.209239E+02 OHMS

ELEVATION ANGLE (DEG)	DIRECTIVITY (NUMERIC)	DIRECTIVITY (DBI)	RELATIVE POWER (DB)	ELEVATION ANGLE (DEG)	DIRECTIVITY (NUMERIC)	DIRECTIVITY (DBI)	RELATIVE POWER (DB)
0.	0.00000	-INFINITY	-INFINITY				
2.0	0.00256	-25.91172	-27.28861	92.0	1.36949	1.36559	-0.01131
4.0	0.01023	-19.90232	-21.27922	94.0	1.36493	1.35109	-0.02580
6.0	0.02291	-16.39915	-17.77605	96.0	1.35852	1.33064	-0.04625
8.0	0.04049	-13.92645	-15.30334	98.0	1.35024	1.30412	-0.07277
10.0	0.06278	-12.02170	-13.39859	100.0	1.34009	1.27134	-0.10555
12.0	0.08956	-10.47884	-11.85573	102.0	1.32803	1.23208	-0.14481
14.0	0.12056	-9.18792	-10.56481	104.0	1.31403	1.18604	-0.19085
16.0	0.15548	-8.08326	-9.46016	106.0	1.29804	1.13289	-0.24401
18.0	0.19398	-7.12248	-8.49937	108.0	1.28003	1.07220	-0.30469
20.0	0.23569	-6.27657	-7.65346	110.0	1.25995	1.00353	-0.37336
22.0	0.28023	-5.52482	-6.90171	112.0	1.23776	0.92635	-0.45054
24.0	0.32720	-4.85188	-6.22877	114.0	1.21341	0.84008	-0.53681
26.0	0.37618	-4.24606	-5.62296	116.0	1.18689	0.74409	-0.63280
28.0	0.42675	-3.69825	-5.07514	118.0	1.15816	0.63768	-0.73921
30.0	0.47850	-3.20115	-4.57804	120.0	1.12722	0.52008	-0.85681
32.0	0.53102	-2.74886	-4.12575	122.0	1.09407	0.39045	-0.98644
34.0	0.58391	-2.33651	-3.71340	124.0	1.05874	0.24788	-1.12901
36.0	0.63679	-1.96001	-3.33690	126.0	1.02126	0.09138	-1.28551
38.0	0.68930	-1.61592	-2.99282	128.0	0.98172	-0.08014	-1.45703
40.0	0.74109	-1.30128	-2.67817	130.0	0.94019	-0.26785	-1.64474
42.0	0.79186	-1.01351	-2.39041	132.0	0.89679	-0.47307	-1.84996
44.0	0.84132	-0.75038	-2.12728	134.0	0.85168	-0.69723	-2.07412
46.0	0.88922	-0.50991	-1.88680	136.0	0.80502	-0.94193	-2.31882
48.0	0.93533	-0.29033	-1.66722	138.0	0.75702	-1.20893	-2.58582
50.0	0.97948	-0.09006	-1.46696	140.0	0.70791	-1.50024	-2.87713
52.0	1.02148	0.09231	-1.28458	142.0	0.65794	-1.81812	-3.19501
54.0	1.06123	0.25808	-1.11881	144.0	0.60742	-2.16514	-3.54203
56.0	1.09861	0.40843	-0.96846	146.0	0.55663	-2.54430	-3.92119
58.0	1.13356	0.54443	-0.83246	148.0	0.50593	-2.95906	-4.33595
60.0	1.16602	0.66707	-0.70982	150.0	0.45567	-3.41351	-4.79040
62.0	1.19599	0.77727	-0.59963	152.0	0.40621	-3.91255	-5.28944
64.0	1.22345	0.87585	-0.50104	154.0	0.35792	-4.46209	-5.83898
66.0	1.24842	0.96360	-0.41330	156.0	0.31121	-5.06944	-6.44633
68.0	1.27093	1.04122	-0.33568	158.0	0.26646	-5.74373	-7.12062
70.0	1.29103	1.10936	-0.26753	160.0	0.22404	-6.49668	-7.87357
72.0	1.30877	1.16863	-0.20826	162.0	0.18435	-7.34362	-8.72052
74.0	1.32421	1.21956	-0.15733	164.0	0.14773	-8.30531	-9.68220
76.0	1.33741	1.26264	-0.11425	166.0	0.11453	-9.41073	-10.78762
78.0	1.34844	1.29831	-0.07858	168.0	0.08507	-10.70229	-12.07918
80.0	1.35736	1.32695	-0.04994	170.0	0.05963	-12.24568	-13.62257
82.0	1.36423	1.34888	-0.02801	172.0	0.03845	-14.15086	-15.52775
84.0	1.36911	1.36437	-0.01252	174.0	0.02176	-16.62389	-18.00078
86.0	1.37204	1.37366	-0.00323	176.0	0.00971	-20.12729	-21.50418
88.0	1.37306	1.37689	0.00000	178.0	0.00243	-26.13682	-27.51371
90.0	1.37220	1.37418	-0.00271	180.	0.00000	-INFINITY	-INFINITY

TABLE A2-63

EPSLN= 1.9150(NORMALIZED GROUNDPLANE RADIUS)
MONOPOLE LENGTH= 0.237400(WAVELENGTH)
MONOPOLE RADIUS= 0.1589E-02(WAVELENGTH)
NUMBER OF MONOPOLE SEGMENTS= 4
NUMBER OF DISK SEGMENTS= 27
RADIATION RESISTANCE DETERMINED BY PATTERN= 0.229245E+02 OHMS

ELEVATION ANGLE (DEG)	DIRECTIVITY (NUMERIC)	DIRECTIVITY (DBI)	RELATIVE POWER (DB)	ELEVATION ANGLE (DEG)	DIRECTIVITY (NUMERIC)	DIRECTIVITY (DBI)	RELATIVE POWER (DB)
0.	0.00000	-INFINITY	-INFINITY				
2.0	0.00324	-24.88942	-25.99127	92.0	1.27250	1.04659	-0.05527
4.0	0.01293	-18.88366	-19.98551	94.0	1.26738	1.02906	-0.07279
6.0	0.02893	-15.38654	-16.48840	96.0	1.26164	1.00935	-0.09251
8.0	0.05102	-12.92230	-14.02415	98.0	1.25525	0.98729	-0.11456
10.0	0.07892	-11.02838	-12.13023	100.0	1.24815	0.96268	-0.13917
12.0	0.11224	-9.49871	-10.60056	102.0	1.24028	0.93519	-0.16667
14.0	0.15055	-8.22329	-9.32514	104.0	1.23152	0.90440	-0.19745
16.0	0.19336	-7.13642	-8.23828	106.0	1.22175	0.86984	-0.23201
18.0	0.24012	-6.19565	-7.29750	108.0	1.21085	0.83092	-0.27094
20.0	0.29027	-5.37193	-6.47379	110.0	1.19867	0.78698	-0.31487
22.0	0.34320	-4.64450	-5.74635	112.0	1.18503	0.73730	-0.36456
24.0	0.39830	-3.99793	-5.09979	114.0	1.16978	0.68105	-0.42080
26.0	0.45494	-3.42048	-4.52233	116.0	1.15275	0.61737	-0.48449
28.0	0.51252	-2.90293	-4.00478	118.0	1.13378	0.54529	-0.55657
30.0	0.57044	-2.43792	-3.53978	120.0	1.11270	0.46379	-0.63807
32.0	0.62814	-2.01946	-3.12132	122.0	1.08937	0.37177	-0.73009
34.0	0.68508	-1.64257	-2.74443	124.0	1.06367	0.26806	-0.83379
36.0	0.74079	-1.30308	-2.40493	126.0	1.03548	0.15141	-0.95044
38.0	0.79480	-0.99741	-2.09926	128.0	1.00473	0.02049	-1.08137
40.0	0.84674	-0.72248	-1.82434	130.0	0.97137	-0.12615	-1.22800
42.0	0.89627	-0.47561	-1.57746	132.0	0.93540	-0.29004	-1.39189
44.0	0.94310	-0.25441	-1.35626	134.0	0.89684	-0.47285	-1.57471
46.0	0.98702	-0.05675	-1.15860	136.0	0.85577	-0.67641	-1.77826
48.0	1.02784	0.11928	-0.98258	138.0	0.81233	-0.90268	-2.00453
50.0	1.06547	0.27542	-0.82644	140.0	0.76668	-1.15388	-2.25573
52.0	1.09983	0.41326	-0.68859	142.0	0.71904	-1.43245	-2.53430
54.0	1.13092	0.53430	-0.56755	144.0	0.66971	-1.74116	-2.84301
56.0	1.15875	0.63989	-0.46196	146.0	0.61899	-2.08315	-3.18500
58.0	1.18340	0.73132	-0.37053	148.0	0.56728	-2.46205	-3.56390
60.0	1.20498	0.80979	-0.29206	150.0	0.51498	-2.88211	-3.98396
62.0	1.22361	0.87644	-0.22541	152.0	0.46256	-3.34834	-4.45020
64.0	1.23946	0.93233	-0.16952	154.0	0.41051	-3.86681	-4.96867
66.0	1.25270	0.97848	-0.12338	156.0	0.35934	-4.44493	-5.54678
68.0	1.26352	1.01583	-0.08602	158.0	0.30960	-5.09194	-6.19379
70.0	1.27213	1.04530	-0.05655	160.0	0.26184	-5.81965	-6.92150
72.0	1.27871	1.06772	-0.03413	162.0	0.21660	-6.64350	-7.74535
74.0	1.28348	1.08389	-0.01796	164.0	0.17441	-7.58430	-8.68616
76.0	1.28663	1.09453	-0.00732	166.0	0.13580	-8.67113	-9.77298
78.0	1.28834	1.10032	-0.00154	168.0	0.10124	-9.94646	-11.04831
80.0	1.28880	1.10185	0.00000	170.0	0.07119	-11.47602	-12.57787
82.0	1.28815	1.09968	-0.00218	172.0	0.04603	-13.36982	-14.47167
84.0	1.28655	1.09425	-0.00760	174.0	0.02610	-15.83396	-16.93581
86.0	1.28409	1.08597	-0.01589	176.0	0.01167	-19.33099	-20.43284
88.0	1.28089	1.07513	-0.02673	178.0	0.00293	-25.33669	-26.43855
90.0	1.27701	1.06196	-0.03990	180.	0.00000	-INFINITY	-INFINITY

TABLE A2-64

EPSLN= 2. 1970(NORMALIZED GROUNDPLANE RADIUS)
MONOPOLE LENGTH= 0. 236600(WAVELENGTH)
MONOPOLE RADIUS= 0. 1822E-02(WAVELENGTH)
NUMBER OF MONOPOLE SEGMENTS= 4
NUMBER OF DISK SEGMENTS= 27
RADIATION RESISTANCE DETERMINED BY PATTERN= 0. 253505E+02 OHMS

ELEVATION ANGLE (DEG)	DIRECTIVITY (NUMERIC)	DIRECTIVITY (DBI)	RELATIVE POWER (DB)	ELEVATION ANGLE (DEG)	DIRECTIVITY (NUMERIC)	DIRECTIVITY (DBI)	RELATIVE POWER (DB)
0.	0. 00000	-INFINITY	-INFINITY				
2. 0	0. 00420	-23. 76991	-24. 80549	92. 0	1. 15871	0. 63976	-0. 39583
4. 0	0. 01672	-17. 76771	-18. 80329	94. 0	1. 15139	0. 61221	-0. 42337
6. 0	0. 03735	-14. 27651	-15. 31210	96. 0	1. 14479	0. 58725	-0. 44833
8. 0	0. 06576	-11. 82054	-12. 85612	98. 0	1. 13888	0. 56479	-0. 47079
10. 0	0. 10146	-9. 93723	-10. 97281	100. 0	1. 13359	0. 54458	-0. 49101
12. 0	0. 14386	-8. 42048	-9. 45607	102. 0	1. 12882	0. 52623	-0. 50935
14. 0	0. 19230	-7. 16029	-8. 19587	104. 0	1. 12440	0. 50921	-0. 52637
16. 0	0. 24599	-6. 09089	-7. 12648	106. 0	1. 12017	0. 49284	-0. 54274
18. 0	0. 30410	-5. 16983	-6. 20541	108. 0	1. 11592	0. 47632	-0. 55926
20. 0	0. 36576	-4. 36801	-5. 40359	110. 0	1. 11140	0. 45871	-0. 57688
22. 0	0. 43007	-3. 66462	-4. 70020	112. 0	1. 10636	0. 43897	-0. 59662
24. 0	0. 49611	-3. 04420	-4. 07978	114. 0	1. 10051	0. 41595	-0. 61963
26. 0	0. 56300	-2. 49495	-3. 53053	116. 0	1. 09356	0. 38843	-0. 64715
28. 0	0. 62985	-2. 00760	-3. 04318	118. 0	1. 08520	0. 35509	-0. 68049
30. 0	0. 69587	-1. 57474	-2. 61032	120. 0	1. 07511	0. 31453	-0. 72105
32. 0	0. 76027	-1. 19030	-2. 22588	122. 0	1. 06299	0. 26529	-0. 77030
34. 0	0. 82238	-0. 84925	-1. 88483	124. 0	1. 04854	0. 20584	-0. 82975
36. 0	0. 88159	-0. 54734	-1. 58292	126. 0	1. 03147	0. 13457	-0. 90101
38. 0	0. 93736	-0. 28092	-1. 31650	128. 0	1. 01154	0. 04982	-0. 98576
40. 0	0. 98928	-0. 04683	-1. 08241	130. 0	0. 98851	-0. 05017	-1. 08575
42. 0	1. 03698	0. 15771	-0. 87787	132. 0	0. 96222	-0. 16724	-1. 20282
44. 0	1. 08023	0. 33518	-0. 70041	134. 0	0. 93254	-0. 30334	-1. 33892
46. 0	1. 11887	0. 48779	-0. 54779	136. 0	0. 89938	-0. 46055	-1. 49614
48. 0	1. 15281	0. 61759	-0. 41799	138. 0	0. 86276	-0. 64111	-1. 67669
50. 0	1. 18207	0. 72644	-0. 30915	140. 0	0. 82273	-0. 84743	-1. 88302
52. 0	1. 20672	0. 81606	-0. 21952	142. 0	0. 77944	-1. 08219	-2. 11777
54. 0	1. 22690	0. 88809	-0. 14750	144. 0	0. 73311	-1. 34832	-2. 38391
56. 0	1. 24281	0. 94405	-0. 09154	146. 0	0. 68404	-1. 64916	-2. 68475
58. 0	1. 25470	0. 98540	-0. 05018	148. 0	0. 63263	-1. 98850	-3. 02409
60. 0	1. 26286	1. 01355	-0. 02203	150. 0	0. 57933	-2. 37073	-3. 40631
62. 0	1. 26761	1. 02984	-0. 00574	152. 0	0. 52469	-2. 80101	-3. 83659
64. 0	1. 26928	1. 03558	0. 00000	154. 0	0. 46930	-3. 28550	-4. 32109
66. 0	1. 26825	1. 03204	-0. 00355	156. 0	0. 41383	-3. 83175	-4. 86733
68. 0	1. 26486	1. 02044	-0. 01515	158. 0	0. 35900	-4. 44909	-5. 48467
70. 0	1. 25950	1. 00198	-0. 03361	160. 0	0. 30553	-5. 14942	-6. 18500
72. 0	1. 25251	0. 97782	-0. 05776	162. 0	0. 25420	-5. 94826	-6. 98385
74. 0	1. 24425	0. 94909	-0. 08650	164. 0	0. 20575	-6. 86650	-7. 90209
76. 0	1. 23505	0. 91685	-0. 11873	166. 0	0. 16094	-7. 93328	-8. 96886
78. 0	1. 22522	0. 88213	-0. 15345	168. 0	0. 12047	-9. 19112	-10. 22670
80. 0	1. 21504	0. 84590	-0. 18969	170. 0	0. 08500	-10. 70581	-11. 74139
82. 0	1. 20477	0. 80903	-0. 22655	172. 0	0. 05511	-12. 58738	-13. 62296
84. 0	1. 19463	0. 77233	-0. 26325	174. 0	0. 03132	-15. 04198	-16. 07756
86. 0	1. 18482	0. 73652	-0. 29906	176. 0	0. 01402	-18. 53218	-19. 56776
88. 0	1. 17549	0. 70219	-0. 33339	178. 0	0. 00352	-24. 53377	-25. 56935
90. 0	1. 16676	0. 66982	-0. 36576	180.	0. 00000	-INFINITY	-INFINITY

TABLE A2-65

EPSLN= 3.0000(NORMALIZED GROUNDPLANE RADIUS)
MONOPOLE LENGTH= 0.235500(WAVELENGTH)
MONOPOLE RADIUS= 0.2478E-02(WAVELENGTH)
NUMBER OF MONOPOLE SEGMENTS= 4
NUMBER OF DISK SEGMENTS= 27
RADIATION RESISTANCE DETERMINED BY PATTERN= 0.393427E+02 OHMS

ELEVATION ANGLE (DEG)	DIRECTIVITY (NUMERIC)	DIRECTIVITY (DBI)	RELATIVE POWER (DB)	ELEVATION ANGLE (DEG)	DIRECTIVITY (NUMERIC)	DIRECTIVITY (DBI)	RELATIVE POWER (DB)
0.	0.00000	-INFINITY	-INFINITY				
2.0	0.00987	-20.05844	-22.60144	92.0	0.73392	-1.34351	-3.88652
4.0	0.03921	-14.06658	-16.60958	94.0	0.70441	-1.52173	-4.06474
6.0	0.08725	-10.59258	-13.13559	96.0	0.67950	-1.67809	-4.22109
8.0	0.15273	-8.16063	-10.70364	98.0	0.65928	-1.80932	-4.35233
10.0	0.23399	-6.30811	-8.85111	100.0	0.64378	-1.91265	-4.45566
12.0	0.32894	-4.82882	-7.37183	102.0	0.63300	-1.98594	-4.52895
14.0	0.43524	-3.61267	-6.15568	104.0	0.62691	-2.02796	-4.57096
16.0	0.55032	-2.59381	-5.13682	106.0	0.62540	-2.03845	-4.58146
18.0	0.67148	-1.72965	-4.27265	108.0	0.62831	-2.01826	-4.56127
20.0	0.79598	-0.99096	-3.53397	110.0	0.63544	-1.96928	-4.51229
22.0	0.92113	-0.35679	-2.89980	112.0	0.64649	-1.89436	-4.43737
24.0	1.04435	0.18845	-2.35456	114.0	0.66113	-1.79715	-4.34016
26.0	1.16325	0.65673	-1.88627	116.0	0.67891	-1.68188	-4.22489
28.0	1.27570	1.05747	-1.48554	118.0	0.69933	-1.55316	-4.09617
30.0	1.37982	1.39823	-1.14477	120.0	0.72181	-1.41578	-3.95879
32.0	1.47409	1.68524	-0.85777	122.0	0.74567	-1.27455	-3.81756
34.0	1.55728	1.92367	-0.61934	124.0	0.77017	-1.13416	-3.67717
36.0	1.62851	2.11791	-0.42510	126.0	0.79449	-0.99912	-3.54213
38.0	1.68723	2.27175	-0.27126	128.0	0.81777	-0.87371	-3.41672
40.0	1.73318	2.38845	-0.15456	130.0	0.83908	-0.76198	-3.30499
42.0	1.76640	2.47090	-0.07211	132.0	0.85748	-0.66775	-3.21076
44.0	1.78717	2.52165	-0.02136	134.0	0.87203	-0.59467	-3.13768
46.0	1.79598	2.54301	0.00000	136.0	0.88181	-0.54625	-3.08926
48.0	1.79351	2.53703	-0.00597	138.0	0.88595	-0.52592	-3.06893
50.0	1.78058	2.50562	-0.03739	140.0	0.88367	-0.53710	-3.08011
52.0	1.75812	2.45049	-0.09251	142.0	0.87432	-0.58328	-3.12629
54.0	1.72714	2.37327	-0.16974	144.0	0.85741	-0.66811	-3.21112
56.0	1.68867	2.27545	-0.26756	146.0	0.83263	-0.79548	-3.33849
58.0	1.64378	2.15845	-0.38456	148.0	0.79990	-0.96965	-3.51265
60.0	1.59354	2.02362	-0.51939	150.0	0.75938	-1.19540	-3.73840
62.0	1.53896	1.87228	-0.67073	152.0	0.71151	-1.47822	-4.02122
64.0	1.48105	1.70571	-0.83730	154.0	0.65697	-1.82455	-4.36756
66.0	1.42076	1.52520	-1.01781	156.0	0.59674	-2.24216	-4.78517
68.0	1.35895	1.33205	-1.21096	158.0	0.53204	-2.74058	-5.28359
70.0	1.29647	1.12761	-1.41539	160.0	0.46431	-3.33188	-5.87488
72.0	1.23405	0.91332	-1.62969	162.0	0.39521	-4.03171	-6.57471
74.0	1.17238	0.69069	-1.85232	164.0	0.32651	-4.86107	-7.40408
76.0	1.11209	0.46139	-2.08162	166.0	0.26006	-5.84921	-8.39222
78.0	1.05372	0.22726	-2.31575	168.0	0.19775	-7.03874	-9.58175
80.0	0.99778	-0.00965	-2.55266	170.0	0.14140	-8.49551	-11.03852
82.0	0.94470	-0.24707	-2.79008	172.0	0.09269	-10.32963	-12.87263
84.0	0.89486	-0.48246	-3.02547	174.0	0.05312	-12.74726	-15.29027
86.0	0.84859	-0.71300	-3.25601	176.0	0.02393	-16.21103	-18.75404
88.0	0.80619	-0.93561	-3.47861	178.0	0.00603	-22.19676	-24.73977
90.0	0.76790	-1.14695	-3.68995	180.	0.00000	-INFINITY	-INFINITY

TABLE A2-66

EPSLN= 4.0000(NORMALIZED GROUNDPLANE RADIUS)
MONOPOLE LENGTH= 0.234600(WAVELENGTH)
MONOPOLE RADIUS= 0.3304E-02(WAVELENGTH)
NUMBER OF MONOPOLE SEGMENTS= 4
NUMBER OF DISK SEGMENTS= 27
RADIATION RESISTANCE DETERMINED BY PATTERN= 0.405539E+02 OHMS

ELEVATION ANGLE (DEG)	DIRECTIVITY (NUMERIC)	DIRECTIVITY (DBI)	RELATIVE POWER (DB)	ELEVATION ANGLE (DEG)	DIRECTIVITY (NUMERIC)	DIRECTIVITY (DBI)	RELATIVE POWER (DB)
0.	0.00000	-INFINITY	-INFINITY				
2.0	0.01927	-17.15085	-21.06158	92.0	0.62892	-2.01408	-5.92481
4.0	0.07634	-11.17250	-15.08323	94.0	0.57077	-2.43538	-6.34611
6.0	0.16901	-7.72092	-11.63165	96.0	0.51534	-2.87905	-6.78978
8.0	0.29376	-5.32011	-9.23085	98.0	0.46275	-3.34651	-7.25724
10.0	0.44594	-3.50724	-7.41798	100.0	0.41318	-3.83864	-7.74937
12.0	0.62004	-2.07583	-5.98656	102.0	0.36683	-4.35533	-8.26606
14.0	0.80995	-0.91541	-4.82614	104.0	0.32399	-4.89471	-8.80544
16.0	1.00932	0.04028	-3.87046	106.0	0.28497	-5.45205	-9.36279
18.0	1.21180	0.83432	-3.07642	108.0	0.25014	-6.01815	-9.92889
20.0	1.41139	1.49646	-2.41428	110.0	0.21992	-6.57726	-10.48799
22.0	1.60260	2.04826	-1.86247	112.0	0.19477	-7.10484	-11.01557
24.0	1.78073	2.50598	-1.40475	114.0	0.17514	-7.56626	-11.47700
26.0	1.94191	2.88228	-1.02845	116.0	0.16149	-7.91847	-11.82920
28.0	2.08320	3.18731	-0.72342	118.0	0.15428	-8.11702	-12.02775
30.0	2.20264	3.42943	-0.48130	120.0	0.15386	-8.12870	-12.03944
32.0	2.29914	3.61566	-0.29507	122.0	0.16053	-7.94440	-11.85513
34.0	2.37246	3.75200	-0.15874	124.0	0.17443	-7.58369	-11.49442
36.0	2.42306	3.84365	-0.06708	126.0	0.19554	-7.08758	-10.99831
38.0	2.45200	3.89521	-0.01553	128.0	0.22362	-6.50496	-10.41569
40.0	2.46078	3.91073	0.00000	130.0	0.25816	-5.88107	-9.79180
42.0	2.45124	3.89386	-0.01688	132.0	0.29841	-5.25192	-9.16266
44.0	2.42539	3.84782	-0.06291	134.0	0.34327	-4.64358	-8.55432
46.0	2.38537	3.77555	-0.13519	136.0	0.39139	-4.07391	-7.98464
48.0	2.33327	3.67965	-0.23109	138.0	0.44108	-3.55479	-7.46553
50.0	2.27114	3.56244	-0.34829	140.0	0.49044	-3.09418	-7.00491
52.0	2.20090	3.42600	-0.48473	142.0	0.53733	-2.69758	-6.60832
54.0	2.12430	3.27215	-0.63858	144.0	0.57954	-2.36917	-6.27991
56.0	2.04290	3.10247	-0.80826	146.0	0.61482	-2.11249	-6.02322
58.0	1.95809	2.91832	-0.99241	148.0	0.64107	-1.93098	-5.84171
60.0	1.87105	2.72086	-1.18987	150.0	0.65639	-1.82839	-5.73912
62.0	1.78280	2.51104	-1.39970	152.0	0.65931	-1.80911	-5.71985
64.0	1.69419	2.28961	-1.62112	154.0	0.64885	-1.87852	-5.78926
66.0	1.60589	2.05716	-1.85357	156.0	0.62469	-2.04336	-5.95409
68.0	1.51848	1.81409	-2.09664	158.0	0.58719	-2.31223	-6.22297
70.0	1.43240	1.56064	-2.35010	160.0	0.53748	-2.69638	-6.60712
72.0	1.34799	1.29686	-2.61387	162.0	0.47745	-3.21075	-7.12148
74.0	1.26552	1.02267	-2.88806	164.0	0.40967	-3.87570	-7.78644
76.0	1.18517	0.73781	-3.17292	166.0	0.33729	-4.71993	-8.63067
78.0	1.10709	0.44185	-3.46889	168.0	0.26390	-5.78560	-9.69633
80.0	1.03138	0.13417	-3.77656	170.0	0.19328	-7.13817	-11.04890
82.0	0.95808	-0.18599	-4.09673	172.0	0.12920	-8.88743	-12.79817
84.0	0.88724	-0.51960	-4.43034	174.0	0.07517	-11.23933	-15.15006
86.0	0.81888	-0.86781	-4.77855	176.0	0.03423	-14.65627	-18.56700
88.0	0.75302	-1.23195	-5.14268	178.0	0.00868	-20.61395	-24.52469
90.0	0.68968	-1.61349	-5.52423	180.	0.00000	-INFINITY	-INFINITY

A.3 OBLATE SPHEROIDAL WAVE FUNCTIONS

Tables A3-1–A3-4; $ka = 3, 4, 5, \sqrt{42}$, $h/\lambda = 0.25$

TABLE A3-1. DIRECTIVITY AND RADIATION RESISTANCE OF QUARTERWAVE MONOPOLE MOUNTED ON CIRCULAR GROUNDPLANE OF FINITE EXTENT, $2\pi a/\lambda = 3.0$

$2\pi a/\lambda$ = 3.00

radiation resistance = 37.4765 ohms

ELEVATION ANGLE, θ (DEG)	DIRECTIVITY d(θ) (NUMERIC)	DIRECTIVITY D(θ) (dBi)	RELATIVE POWER (dB)
0.00000	0.00000	-INFINITY	-INFINITY
2.00000	0.00993	-20.03252	-22.55502
4.00000	0.03943	-14.04121	-16.56371
6.00000	0.08774	-10.56811	-13.09061
8.00000	0.15355	-8.13740	-10.65990
10.00000	0.23516	-6.28645	-8.80895
12.00000	0.33044	-4.80905	-7.33155
14.00000	0.43701	-3.59507	-6.11757
16.00000	0.55225	-2.57866	-5.10116
18.00000	0.67342	-1.71716	-4.23966
20.00000	0.79775	-0.98135	-3.50385
22.00000	0.92253	-0.35021	-2.87271
24.00000	1.04517	0.19189	-2.33061
26.00000	1.16331	0.65696	-1.86554
28.00000	1.27482	1.05449	-1.46801
30.00000	1.37787	1.39209	-1.13041
32.00000	1.47098	1.67606	-0.84644
34.00000	1.55297	1.91164	-0.61086
36.00000	1.62305	2.10331	-0.41919
38.00000	1.68071	2.25492	-0.26758
40.00000	1.72577	2.36983	-0.15267
42.00000	1.75833	2.45100	-0.07150
44.00000	1.77873	2.50110	-0.02140
46.00000	1.78752	2.52250	0.00000
48.00000	1.78541	2.51738	-0.00512
50.00000	1.77325	2.48769	-0.03481
52.00000	1.75198	2.43528	-0.08722
54.00000	1.72259	2.36182	-0.16068
56.00000	1.68612	2.26889	-0.25361
58.00000	1.64360	2.15797	-0.36453
60.00000	1.59604	2.03045	-0.49205
62.00000	1.54442	1.88766	-0.63484
64.00000	1.48967	1.73089	-0.79161
66.00000	1.43264	1.56138	-0.96112
68.00000	1.37416	1.38037	-1.14213
70.00000	1.31494	1.18908	-1.33342
72.00000	1.25567	0.98876	-1.53374
74.00000	1.19694	0.78071	-1.74179
76.00000	1.13928	0.56630	-1.95620
78.00000	1.08318	0.34700	-2.17550
80.00000	1.02906	0.12440	-2.39810
82.00000	0.97729	-0.09975	-2.62225
84.00000	0.92822	-0.32348	-2.84598
86.00000	0.88214	-0.54464	-3.06714
88.00000	0.83930	-0.76082	-3.28332
90.00000	0.79995	-0.96938	-3.49188

ELEVATION ANGLE, θ (DEG)	DIRECTIVITY d(θ) (NUMERIC)	DIRECTIVITY D(θ) (dBi)	RELATIVE POWER (dB)
92.00000	0.76429	-1.16744	-3.68994
94.00000	0.73249	-1.35197	-3.87447
96.00000	0.70472	-1.51980	-4.04230
98.00000	0.68111	-1.66780	-4.19030
100.00000	0.66176	-1.79299	-4.31549
102.00000	0.64674	-1.89273	-4.41523
104.00000	0.63607	-1.96494	-4.48744
106.00000	0.62976	-2.00827	-4.53077
108.00000	0.62774	-2.02223	-4.54473
110.00000	0.62989	-2.00735	-4.52985
112.00000	0.63604	-1.96515	-4.48765
114.00000	0.64593	-1.89811	-4.42061
116.00000	0.65924	-1.80957	-4.33207
118.00000	0.67554	-1.70350	-4.22600
120.00000	0.69433	-1.58436	-4.10686
122.00000	0.71501	-1.45689	-3.97939
124.00000	0.73690	-1.32593	-3.84843
126.00000	0.75922	-1.19631	-3.71881
128.00000	0.78114	-1.07272	-3.59522
130.00000	0.80174	-0.95966	-3.48216
132.00000	0.82007	-0.86147	-3.38397
134.00000	0.83517	-0.78226	-3.30476
136.00000	0.84606	-0.72598	-3.24848
138.00000	0.85183	-0.69648	-3.21898
140.00000	0.85162	-0.69755	-3.22005
142.00000	0.84470	-0.73298	-3.25548
144.00000	0.83048	-0.80670	-3.32920
146.00000	0.80857	-0.92282	-3.44532
148.00000	0.77879	-1.08580	-3.60830
150.00000	0.74121	-1.30059	-3.82309
152.00000	0.69618	-1.57279	-4.09529
154.00000	0.64432	-1.90897	-4.43147
156.00000	0.58655	-2.31696	-4.83946
158.00000	0.52404	-2.80637	-5.32887
160.00000	0.45821	-3.38933	-5.91183
162.00000	0.39070	-4.08152	-6.60402
164.00000	0.32330	-4.90400	-7.42650
166.00000	0.25787	-5.88601	-8.40851
168.00000	0.19633	-7.07020	-9.59270
170.00000	0.14053	-8.52244	-11.04494
172.00000	0.09220	-10.35283	-12.87533
174.00000	0.05287	-12.76756	-15.29006
176.00000	0.02383	-16.22925	-18.75175
178.00000	0.00601	-22.21373	-24.73623
180.00000	0.00000	-INFINITY	-INFINITY

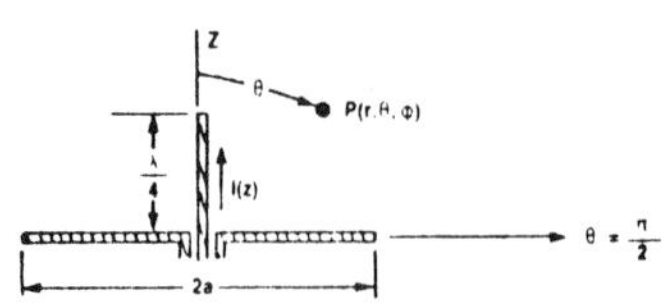

TABLE A3-2. DIRECTIVITY AND RADIATION RESISTANCE OF QUARTERWAVE MONOPOLE MOUNTED ON CIRCULAR GROUNDPLANE OF FINITE EXTENT, $2\pi a/\lambda = 4.0$

$2\pi a/\lambda$ = 4.00

radiation resistance = 42.6724 ohms

ELEVATION ANGLE, θ (DEG)	DIRECTIVITY d(θ) (NUMERIC)	DIRECTIVITY D(θ) (dBi)	RELATIVE POWER (dB)	ELEVATION ANGLE, θ (DEG)	DIRECTIVITY d(θ) (NUMERIC)	DIRECTIVITY D(θ) (dBi)	RELATIVE POWER (dB)
0.00000	0.00000	-INFINITY	-INFINITY	92.00000	0.64167	-1.92691	-5.82634
2.00000	0.01944	-17.11225	-21.01168	94.00000	0.58310	-2.34255	-6.24198
4.00000	0.07701	-11.13439	-15.03382	96.00000	0.52695	-2.78228	-6.68171
6.00000	0.17047	-7.68361	-11.58304	98.00000	0.47336	-3.24811	-7.14754
8.00000	0.29622	-5.28393	-9.18336	100.00000	0.42251	-3.74161	-7.64104
10.00000	0.44952	-3.47249	-7.37192	102.00000	0.37466	-4.26361	-8.16304
12.00000	0.62477	-2.04281	-5.94224	104.00000	0.33011	-4.81337	-8.71280
14.00000	0.81575	-0.88441	-4.78384	106.00000	0.28923	-5.38758	-9.28701
16.00000	1.01601	0.06897	-3.83046	108.00000	0.25243	-5.97863	-9.87806
18.00000	1.21911	0.86044	-3.03899	110.00000	0.22017	-6.57234	-10.47177
20.00000	1.41899	1.51978	-2.37965	112.00000	0.19297	-7.14509	-11.04452
22.00000	1.61011	2.06856	-1.83087	114.00000	0.17134	-7.66144	-11.56087
24.00000	1.78774	2.52305	-1.37638	116.00000	0.15580	-8.07432	-11.97375
26.00000	1.94803	2.89595	-1.00348	118.00000	0.14685	-8.33133	-12.23076
28.00000	2.08808	3.19746	-0.70197	120.00000	0.14491	-8.38898	-12.28841
30.00000	2.20595	3.43597	-0.46346	122.00000	0.15032	-8.22971	-12.12914
32.00000	2.30066	3.61852	-0.28091	124.00000	0.16328	-7.87069	-11.77012
34.00000	2.37202	3.75119	-0.14824	126.00000	0.18379	-7.35684	-11.25627
36.00000	2.42059	3.83922	-0.06021	128.00000	0.21164	-6.74413	-10.64356
38.00000	2.44752	3.88726	-0.01217	130.00000	0.24635	-6.08456	-9.98399
40.00000	2.45439	3.89943	0.00000	132.00000	0.28715	-5.41895	-9.31838
42.00000	2.44311	3.87943	-0.02000	134.00000	0.33296	-4.77613	-8.67556
44.00000	2.41578	3.83057	-0.06886	136.00000	0.38237	-4.17521	-8.07464
46.00000	2.37456	3.75583	-0.14360	138.00000	0.43367	-3.62844	-7.52787
48.00000	2.32161	3.65789	-0.24154	140.00000	0.48488	-3.14367	-7.04310
50.00000	2.25899	3.53915	-0.36028	142.00000	0.53381	-2.72610	-6.62553
52.00000	2.18863	3.40173	-0.49770	144.00000	0.57816	-2.37953	-6.27896
54.00000	2.11228	3.24752	-0.65191	146.00000	0.61559	-2.10711	-6.00654
56.00000	2.03149	3.07816	-0.82127	148.00000	0.64388	-1.91197	-5.81140
58.00000	1.94762	2.89504	-1.00439	150.00000	0.66107	-1.79754	-5.69697
60.00000	1.86182	2.69937	-1.20006	152.00000	0.66558	-1.76798	-5.66741
62.00000	1.77506	2.49212	-1.40731	154.00000	0.65638	-1.82843	-5.72786
64.00000	1.68813	2.27407	-1.62536	156.00000	0.63307	-1.98547	-5.88490
66.00000	1.60170	2.04580	-1.85363	158.00000	0.59600	-2.24757	-6.14700
68.00000	1.51625	1.80770	-2.09173	160.00000	0.54628	-2.62584	-6.52527
70.00000	1.43218	1.55998	-2.33945	162.00000	0.48583	-3.13511	-7.03454
72.00000	1.34979	1.30267	-2.59676	164.00000	0.41728	-3.79569	-7.69512
74.00000	1.26929	1.03562	-2.86381	166.00000	0.34386	-4.63619	-8.53562
76.00000	1.19082	0.75847	-3.14096	168.00000	0.26923	-5.69872	-9.59815
78.00000	1.11448	0.47071	-3.42872	170.00000	0.19730	-7.04870	-10.94813
80.00000	1.04031	0.17162	-3.72781	172.00000	0.13195	-8.79590	-12.69533
82.00000	0.96834	-0.13972	-4.03915	174.00000	0.07680	-11.14620	-15.04563
84.00000	0.89858	-0.46444	-4.36387	176.00000	0.03498	-14.56203	-18.46146
86.00000	0.83102	-0.80387	-4.70330	178.00000	0.00887	-20.51905	-24.41848
88.00000	0.76567	-1.15956	-5.05899	180.00000	0.00000	-INFINITY	-INFINITY
90.00000	0.70254	-1.53327	-5.43270				

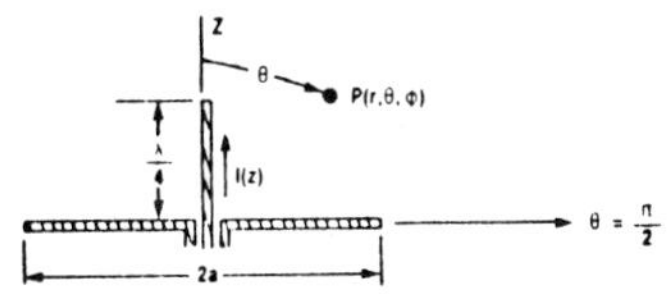

TABLE A3-3. DIRECTIVITY AND RADIATION RESISTANCE OF QUARTERWAVE MONOPOLE MOUNTED ON CIRCULAR GROUNDPLANE OF FINITE EXTENT, $2\pi a/\lambda = 5.0$

$2\pi a/\lambda$ = 5.00 radiation resistance = 32.5653 ohms

ELEVATION ANGLE, θ (DEG)	DIRECTIVITY d(θ) (NUMERIC)	DIRECTIVITY D(θ) (dBi)	RELATIVE POWER (dB)	ELEVATION ANGLE, θ (DEG)	DIRECTIVITY d(θ) (NUMERIC)	DIRECTIVITY D(θ) (dBi)	RELATIVE POWER (dB)
0.00000	0.00000	-INFINITY	-INFINITY	92.00000	0.87449	-0.58246	-3.95421
2.00000	0.02482	-16.05208	-19.42383	94.00000	0.82949	-0.81188	-4.18363
4.00000	0.09782	-10.09571	-13.46746	96.00000	0.78534	-1.04941	-4.42116
6.00000	0.21476	-6.68050	-10.05225	98.00000	0.74175	-1.29741	-4.66916
8.00000	0.36897	-4.33006	-7.70181	100.00000	0.69841	-1.55892	-4.93067
10.00000	0.55196	-2.58096	-5.95271	102.00000	0.65499	-1.83766	-5.20941
12.00000	0.75406	-1.22596	-4.59771	104.00000	0.61120	-2.13818	-5.50993
14.00000	0.96524	-0.15364	-3.52539	106.00000	0.56678	-2.46583	-5.83758
16.00000	1.17584	0.70348	-2.66827	108.00000	0.52156	-2.82695	-6.19870
18.00000	1.37717	1.38988	-1.98187	110.00000	0.47546	-3.22886	-6.60061
20.00000	1.56207	1.93701	-1.43474	112.00000	0.42854	-3.68005	-7.05180
22.00000	1.72518	2.36835	-1.00340	114.00000	0.38105	-4.19022	-7.56197
24.00000	1.86308	2.70232	-0.66943	116.00000	0.33340	-4.77031	-8.14206
26.00000	1.97423	2.95399	-0.41776	118.00000	0.28627	-5.43226	-8.80401
28.00000	2.05878	3.13610	-0.23565	120.00000	0.24054	-6.18819	-9.55994
30.00000	2.11823	3.25972	-0.11203	122.00000	0.19733	-7.04807	-10.41982
32.00000	2.15506	3.33460	-0.03715	124.00000	0.15798	-8.01386	-11.38561
34.00000	2.17240	3.36939	-0.00236	126.00000	0.12400	-9.06566	-12.43741
36.00000	2.17358	3.37175	0.00000	128.00000	0.09699	-10.13273	-13.50448
38.00000	2.16190	3.34835	-0.02340	130.00000	0.07855	-11.04864	-14.42039
40.00000	2.14037	3.30488	-0.06687	132.00000	0.07016	-11.53903	-14.91078
42.00000	2.11155	3.24602	-0.12573	134.00000	0.07305	-11.36388	-14.73563
44.00000	2.07752	3.17544	-0.19631	136.00000	0.08801	-10.55472	-13.92647
46.00000	2.03979	3.09585	-0.27590	138.00000	0.11527	-9.38302	-12.75477
48.00000	1.99943	3.00905	-0.36270	140.00000	0.15432	-8.11571	-11.48746
50.00000	1.95708	2.91608	-0.45567	142.00000	0.20386	-6.90657	-10.27832
52.00000	1.91308	2.81734	-0.55441	144.00000	0.26171	-5.82188	-9.19363
54.00000	1.86757	2.71277	-0.65898	146.00000	0.32480	-4.88378	-8.25553
56.00000	1.82054	2.60201	-0.76974	148.00000	0.38938	-4.09623	-7.46798
58.00000	1.77196	2.48454	-0.88721	150.00000	0.45112	-3.45707	-6.82882
60.00000	1.72180	2.35982	-1.01193	152.00000	0.50545	-2.96325	-6.33500
62.00000	1.67008	2.22739	-1.14436	154.00000	0.54790	-2.61299	-5.98474
64.00000	1.61692	2.08689	-1.28486	156.00000	0.57454	-2.40678	-5.77853
66.00000	1.56249	1.93818	-1.43357	158.00000	0.58237	-2.34797	-5.71972
68.00000	1.50705	1.78127	-1.59048	160.00000	0.56971	-2.44350	-5.81525
70.00000	1.45090	1.61639	-1.75536	162.00000	0.53644	-2.70476	-6.07651
72.00000	1.39441	1.44390	-1.92785	164.00000	0.48425	-3.14928	-6.52103
74.00000	1.33793	1.26434	-2.10741	166.00000	0.41654	-3.80345	-7.17520
76.00000	1.28183	1.07831	-2.29344	168.00000	0.33825	-4.70757	-8.07932
78.00000	1.22644	0.88647	-2.48528	170.00000	0.25553	-5.92564	-9.29739
80.00000	1.17204	0.68944	-2.68231	172.00000	0.17513	-7.56629	-10.93804
82.00000	1.11887	0.48782	-2.88393	174.00000	0.10388	-9.83457	-13.20632
84.00000	1.06709	0.28202	-3.08973	176.00000	0.04795	-13.19224	-16.56399
86.00000	1.01679	0.07231	-3.29944	178.00000	0.01226	-19.11454	-22.48629
88.00000	0.96797	-0.14137	-3.51312	180.00000	0.00000	-INFINITY	-INFINITY
90.00000	0.92058	-0.35937	-3.73112				

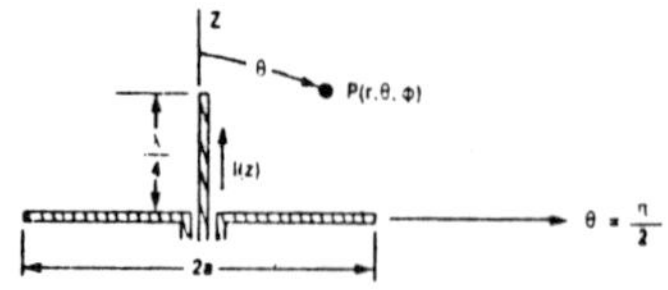

TABLE A3-4. DIRECTIVITY AND RADIATION RESISTANCE OF QUARTERWAVE MONOPOLE MOUNTED ON CIRCULAR GROUNDPLANE OF FINITE EXTENT, $2\pi a/\lambda = \sqrt{42}$

$2\pi a/\lambda$ = 6.48 radiation resistance = 40.1666 ohms

ELEVATION ANGLE, θ (DEG)	DIRECTIVITY d(θ) (NUMERIC)	DIRECTIVITY D(θ) (dBi)	RELATIVE POWER (dB)	ELEVATION ANGLE, θ (DEG)	DIRECTIVITY d(θ) (NUMERIC)	DIRECTIVITY D(θ) (dBi)	RELATIVE POWER (dB)
0.00000	0.00000	-INFINITY	-INFINITY	92.00000	0.67231	-1.72428	-5.79331
2.00000	0.02090	-16.79764	-20.86667	94.00000	0.60756	-2.16409	-6.23312
4.00000	0.08056	-10.93877	-15.00780	96.00000	0.55213	-2.57960	-6.64863
6.00000	0.17036	-7.68624	-11.75527	98.00000	0.50596	-2.95882	-7.02785
8.00000	0.27774	-5.56365	-9.63268	100.00000	0.46895	-3.28869	-7.35772
10.00000	0.38842	-4.10697	-8.17600	102.00000	0.44091	-3.55647	-7.62550
12.00000	0.48900	-3.10694	-7.17597	104.00000	0.42153	-3.75175	-7.82078
14.00000	0.56920	-2.44734	-6.51637	106.00000	0.41033	-3.86868	-7.93771
16.00000	0.62359	-2.05104	-6.12007	108.00000	0.40664	-3.90789	-7.97692
18.00000	0.65226	-1.85579	-5.92482	110.00000	0.40953	-3.87714	-7.94617
20.00000	0.66065	-1.80030	-5.86933	112.00000	0.41776	-3.79075	-7.85978
22.00000	0.65833	-1.81553	-5.88456	114.00000	0.42975	-3.66784	-7.73687
24.00000	0.65732	-1.82220	-5.89123	116.00000	0.44359	-3.53020	-7.59923
26.00000	0.66999	-1.73930	-5.80833	118.00000	0.45704	-3.40044	-7.46947
28.00000	0.70715	-1.50491	-5.57394	120.00000	0.46764	-3.30084	-7.36987
30.00000	0.77647	-1.09876	-5.16779	122.00000	0.47283	-3.25299	-7.32202
32.00000	0.88155	-0.54752	-4.61655	124.00000	0.47011	-3.27797	-7.34700
34.00000	1.02161	0.09285	-3.97618	126.00000	0.45739	-3.39717	-7.46620
36.00000	1.19179	0.76201	-3.30702	128.00000	0.43317	-3.63342	-7.70245
38.00000	1.38401	1.41140	-2.65763	130.00000	0.39696	-4.01254	-8.08157
40.00000	1.58806	2.00867	-2.06036	132.00000	0.34951	-4.56539	-8.63442
42.00000	1.79284	2.53541	-1.53362	134.00000	0.29306	-5.33046	-9.39949
44.00000	1.98751	2.98310	-1.08593	136.00000	0.23140	-6.35645	-10.42548
46.00000	2.16247	3.34950	-0.71953	138.00000	0.16977	-7.70145	-11.77048
48.00000	2.31001	3.63614	-0.43289	140.00000	0.11453	-9.41087	-13.47990
50.00000	2.42471	3.84660	-0.22243	142.00000	0.07254	-11.39394	-15.46297
52.00000	2.50351	3.98550	-0.08353	144.00000	0.05038	-12.97752	-17.04655
54.00000	2.54561	4.05792	-0.01111	146.00000	0.05332	-12.73134	-16.80037
56.00000	2.55213	4.06903	0.00000	148.00000	0.08437	-10.73804	-14.80707
58.00000	2.52571	4.02384	-0.04519	150.00000	0.14342	-8.43399	-12.50302
60.00000	2.47012	3.92718	-0.14185	152.00000	0.22664	-6.44658	-10.51561
62.00000	2.38976	3.78355	-0.28548	154.00000	0.32649	-4.86135	-8.93038
64.00000	2.28936	3.59714	-0.47189	156.00000	0.43216	-3.64353	-7.71256
66.00000	2.17362	3.37184	-0.69719	158.00000	0.53081	-2.75063	-6.81966
68.00000	2.04701	3.11119	-0.95784	160.00000	0.60915	-2.15273	-6.22176
70.00000	1.91357	2.81845	-1.25058	162.00000	0.65552	-1.83417	-5.90320
72.00000	1.77689	2.49660	-1.57243	164.00000	0.66177	-1.79291	-5.86194
74.00000	1.63998	2.14840	-1.92063	166.00000	0.62500	-2.04117	-6.11020
76.00000	1.50536	1.77639	-2.29264	168.00000	0.54842	-2.60887	-6.67790
78.00000	1.37500	1.38304	-2.68599	170.00000	0.44133	-3.55241	-7.62144
80.00000	1.25048	0.97077	-3.09826	172.00000	0.31807	-4.97483	-9.04386
82.00000	1.13296	0.54214	-3.52689	174.00000	0.19602	-7.07709	-11.14612
84.00000	1.02328	0.09996	-3.96907	176.00000	0.09294	-10.31807	-14.38710
86.00000	0.92205	-0.35244	-4.42147	178.00000	0.02415	-16.17109	-20.24012
88.00000	0.82967	-0.81097	-4.88000	180.00000	0.00000	-INFINITY	-INFINITY
90.00000	0.74637	-1.27046	-5.33949				

A.4 VARIATIONAL METHOD

Table A4-1; 28.0 < *ka* < ∞, h/λ = 0.25

TABLE A4-1 .

$2\pi a/\lambda$	$R - R_\infty$ (OHMS)	$X - X_\infty$ (OHMS)	R (OHMS)	X (OHMS)	$d(\pi/2)$ (NUMERIC)	$D(\pi/2)$ (dBi)
28.2743340	-0.0428709	-1.0999534	36.4966393	20.1576080	0.8214043	-0.8544304
29.0597324	1.0697033	0.0411037	37.6092110	21.2986641	0.7971051	-0.9848440
29.8451309	-0.0340973	0.9679080	36.5054131	22.2254696	0.8212068	-0.8554745
30.6305294	-0.9435477	0.0328219	35.5959625	21.2903824	0.8421881	-0.7459092
31.4159279	-0.0364609	-0.9881375	36.5030479	20.2694225	0.8212600	-0.8551931
32.2013245	0.9636334	0.0351067	37.5031433	21.2926674	0.7993595	-0.9725786
32.9867249	-0.0294509	0.8773347	36.5100594	22.1348953	0.8211023	-0.8560272
33.7721214	-0.8572898	0.0284486	35.6822205	21.2860088	0.8401521	-0.7564207
34.5575218	-0.0314970	-0.8968816	36.5080109	20.3606796	0.8211484	-0.8557835
35.3429184	0.8766313	0.0304365	37.4161415	21.2879963	0.8012182	-0.9624919
36.1283150	-0.0257698	0.8023182	36.5137405	22.0598793	0.8210195	-0.8564651
36.9137154	-0.7855343	0.0249676	35.7539749	21.2825279	0.8384660	-0.7651453
37.6991119	-0.0275676	-0.8210008	36.5119400	20.4365597	0.8210601	-0.8562508
38.4845123	0.8039876	0.0267142	37.3434982	21.2842751	0.8027768	-0.9540519
39.2699089	-0.0228035	0.7391611	36.5167046	21.9967213	0.8209529	-0.8568177
40.0553093	-0.7249014	0.0221447	35.8146057	21.2797050	0.8370466	-0.7725036
40.8407059	-0.0243846	-0.7569175	36.5151253	20.5006428	0.8209884	-0.8566298
41.6261024	0.7424243	0.0236855	37.2819328	21.2812462	0.8041025	-0.9468860
42.4115028	-0.0203678	0.6852531	36.5191422	21.9428139	0.8208981	-0.8571074
43.1968994	-0.6729873	0.0198227	35.8665199	21.2773838	0.8358350	-0.7787943
43.9822998	-0.0217679	-0.7020830	36.5177422	20.5554771	0.8209296	-0.8569410
44.7676964	0.6895890	0.0211916	37.2290993	21.2787514	0.8052436	-0.9407272
45.5530930	-0.0183338	0.6386985	36.5211754	21.8962593	0.8208524	-0.8573493
46.3384933	-0.6280351	0.0178772	35.9114723	21.2754383	0.8347888	-0.7842339
47.1238899	-0.0195912	-0.6546321	36.5199165	20.6029282	0.8208807	-0.8571995
47.9092903	0.6437514	0.0191055	37.1832619	21.2766666	0.8062363	-0.9353768
48.6946869	-0.0166221	0.5980861	36.5228882	21.8556461	0.8208139	-0.8575531
49.4800873	-0.5887301	0.0162310	35.9507790	21.2737923	0.8338761	-0.7889849
50.2654839	-0.0177491	-0.6131703	36.5217590	20.6443901	0.8208393	-0.8574187
51.0508804	0.6036100	0.0173351	37.1431198	21.2748966	0.8071076	-0.9306855
51.8362808	-0.0151622	0.5623448	36.5243454	21.8199062	0.8207812	-0.8577261
52.6216774	-0.5540695	0.0148274	35.9854393	21.2723885	0.8330729	-0.7931700
53.4070778	-0.0161776	-0.5766323	36.5233307	20.6809292	0.8208039	-0.8576057
54.1924744	0.5681658	0.0158252	37.1076736	21.2733860	0.8078786	-0.9265392
54.9778748	-0.0139050	0.5306469	36.5256042	21.7882080	0.8207529	-0.8578759
55.7632713	-0.5232750	0.0136123	36.0162354	21.2711735	0.8323606	-0.7968851
56.5486679	-0.0148283	-0.5441912	36.5246811	20.7133694	0.8207736	-0.8577662
57.3340683	0.5366415	0.0145221	37.0761490	21.2720833	0.8085655	-0.9228480
58.1194649	-0.0128094	0.5023422	36.5266991	21.7599030	0.8207283	-0.8580059
58.9048653	-0.4957332	0.0125548	36.0437775	21.2701149	0.8317245	-0.8002049
59.6902618	-0.0136525	-0.5151959	36.5258560	20.7423649	0.8207472	-0.8579059
60.4756584	0.5084220	0.0133842	37.0479317	21.2709446	0.8091813	-0.9195414
61.2610588	-0.0118541	0.4769124	36.5276566	21.7344723	0.8207068	-0.8581198
62.0464554	-0.4709536	0.0116311	36.0685539	21.2691917	0.8311532	-0.8031892

TABLE A4-1 (cont.)

$2\pi a/\lambda$	$R - R_\infty$ (OHMS)	$X - X_\infty$ (OHMS)	R (OHMS)	X (OHMS)	d($\pi/2$) (NUMERIC)	D($\pi/2$) (dBi)
98.1747742	0.3115083	0.0064025	36.8510170	21.2639637	0.8135052	-0.8963966
98.9601746	-0.0058392	0.2968726	36.5336685	21.5544338	0.8205717	-0.8588346
99.7455673	-0.2945590	0.0057676	36.2449493	21.2633286	0.8271082	-0.8243767
100.5309677	-0.0061717	-0.3041370	36.5333366	20.9534245	0.8205792	-0.8587951
101.3163681	0.3017562	0.0061040	36.8412666	21.2636642	0.8137205	-0.8952473
102.1017609	-0.0055717	0.2878293	36.5339355	21.5453892	0.8205657	-0.8588665
102.8871613	-0.2856542	0.0055081	36.2538528	21.2630692	0.8269051	-0.8254435
103.6725616	-0.0058889	-0.2948331	36.5336189	20.9627285	0.8205728	-0.8588289
104.4579620	0.2925949	0.0058280	36.8321037	21.2633896	0.8139229	-0.8941671
105.2433548	-0.0053280	0.2793218	36.5341797	21.5368824	0.8205602	-0.8588954
106.0287552	-0.2772731	0.0052675	36.2622375	21.2628288	0.8267139	-0.8264478
106.8141556	-0.0056270	-0.2860803	36.5338821	20.9714813	0.8205669	-0.8588602
107.5995483	0.2839723	0.0055678	36.8234825	21.2631283	0.8141135	-0.8931505
108.3849487	-0.0051016	0.2713037	36.5344086	21.5288639	0.8205551	-0.8589226
109.1703491	-0.2693710	0.0050439	36.2701378	21.2626038	0.8265338	-0.8273939
109.9557495	-0.0053838	-0.2778310	36.5341263	20.9797306	0.8205614	-0.8588892
110.7411423	0.2758422	0.0053304	36.8153496	21.2628918	0.8142934	-0.8921910
111.5265427	-0.0048909	0.2637343	36.5346184	21.5212955	0.8205504	-0.8589475
112.3119431	-0.2619078	0.0048357	36.2775993	21.2623959	0.8263638	-0.8282872
113.0973358	-0.0051617	-0.2700431	36.5343475	20.9875183	0.8205565	-0.8589154
113.8827362	0.2681635	0.0051093	36.8076706	21.2626705	0.8144633	-0.8912850
114.6681366	-0.0046944	0.2565766	36.5348129	21.5141373	0.8205460	-0.8589709
115.4535370	-0.2548479	0.0046414	36.2846603	21.2622013	0.8262030	-0.8291324
116.2389297	-0.0049507	-0.2626789	36.5345573	20.9948826	0.8205518	-0.8589402
117.0243301	0.2608999	0.0049032	36.8004074	21.2624645	0.8146240	-0.8904280
117.8097305	-0.0045108	0.2497980	36.5349998	21.5073586	0.8205418	-0.8589929
118.5951233	-0.2481594	0.0044634	36.2913513	21.2620239	0.8260506	-0.8299333
119.3805237	-0.0047535	-0.2557049	36.5347557	21.0018559	0.8205473	-0.8589639
120.1659241	0.2540186	0.0047106	36.7935257	21.2622719	0.8147764	-0.8896159
120.9513168	-0.0043353	0.2433694	36.5351753	21.5009308	0.8205379	-0.8590138
121.7367172	-0.2418137	0.0042932	36.2976952	21.2618542	0.8259063	-0.8306924
122.5221176	-0.0045691	-0.2490909	36.5349388	21.0084705	0.8205432	-0.8589857
123.3075180	0.2474902	0.0045303	36.7869987	21.2620907	0.8149209	-0.8888455
124.0929108	-0.0041742	0.2372638	36.5353355	21.4948254	0.8205343	-0.8590326
124.8783112	-0.2357852	0.0041335	36.3037224	21.2616940	0.8257692	-0.8314133
125.6637115	-0.0043961	-0.2428097	36.5351143	21.0147514	0.8205392	-0.8590065
126.4491043	0.2412883	0.0043575	36.7807961	21.2619190	0.8150584	-0.8881130
127.2345047	-0.0040230	0.2314578	36.5354843	21.4890194	0.8205310	-0.8590504
128.0198975	-0.2300505	0.0039869	36.3094597	21.2615471	0.8256387	-0.8320996
128.8052979	-0.0042375	-0.2368367	36.5352707	21.0207233	0.8205357	-0.8590251
129.5906982	0.2353889	0.0041988	36.7748985	21.2617588	0.8151891	-0.8874166
130.3760986	-0.0038806	0.2259296	36.5356293	21.4834900	0.8205277	-0.8590677
131.1614990	-0.2245889	0.0038422	36.3149185	21.2614021	0.8255146	-0.8327526
131.9468994	-0.0040811	-0.2311501	36.5354271	21.0264111	0.8205322	-0.8590438
132.7322998	0.2297704	0.0040531	36.7692795	21.2616138	0.8153136	-0.8867530

TABLE A4-1 (cont.)

$2\pi a/\lambda$	$R - R_\infty$ (OHMS)	$X - X_\infty$ (OHMS)	R (OHMS)	X (OHMS)	d(π/2) (NUMERIC)	D(π/2) (dBi)
62.8318558	-0.0126236	-0.4891256	36.5268860	20.7684345	0.8207241	-0.8580283
63.6172523	0.4830137	0.0123905	37.0225220	21.2699509	0.8097367	-0.9165620
64.4026489	-0.0110089	0.4539403	36.5284996	21.7115002	0.8206878	-0.8582201
65.1880493	-0.4485402	0.0108128	36.0909691	21.2683735	0.8306370	-0.8058872
65.9734497	-0.0117172	-0.4655595	36.5277901	20.7920017	0.8207037	-0.8581358
66.7588425	0.4600173	0.0115103	36.9995270	21.2690716	0.8102400	-0.9138634
67.5442429	-0.0102629	0.4330855	36.5292473	21.6906471	0.8206710	-0.8583090
68.3296432	-0.4281690	0.0100862	36.1113396	21.2676468	0.8301684	-0.8083379
69.1150436	-0.0109139	-0.4441538	36.5285950	20.8134079	0.8206857	-0.8582314
69.9004364	0.4391052	0.0107323	36.9786148	21.2682934	0.8106982	-0.9114082
70.6858368	-0.0095979	0.4140679	36.5299110	21.6716290	0.8206561	-0.8583879
71.4712372	-0.4095726	0.0094376	36.1299362	21.2669983	0.8297411	-0.8105739
72.2566299	-0.0102046	-0.4246245	36.5293045	20.8329353	0.8206697	-0.8583160
73.0420303	0.4200068	0.0100381	36.9595146	21.2675991	0.8111171	-0.9091644
73.8274307	-0.0090020	0.3966545	36.5305061	21.6542149	0.8206428	-0.8584585
74.6128311	-0.3925286	0.0088557	36.1469803	21.2664165	0.8293499	-0.8126222
75.3982239	-0.0095630	-0.4067361	36.5299454	20.8508244	0.8206553	-0.8583920
76.1836243	0.4024961	0.0094156	36.9420052	21.2669754	0.8115016	-0.9071064
76.9690247	-0.0084656	0.3806503	36.5310440	21.6382103	0.8206307	-0.8585226
77.7544174	-0.3768499	0.0083369	36.1626587	21.2658978	0.8289903	-0.8145053
78.5398178	-0.0089857	-0.3902901	36.5305214	20.8672714	0.8206424	-0.8584604
79.3252182	0.3863834	0.0088547	36.9258919	21.2664146	0.8118557	-0.9052117
80.1106186	-0.0079807	0.3658908	36.5315285	21.6234512	0.8206198	-0.8585804
80.8960114	-0.3623789	0.0078620	36.1771317	21.2654228	0.8286587	-0.8162431
81.6814117	-0.0084639	-0.3751190	36.5310440	20.8824425	0.8206307	-0.8585226
82.4668121	0.3715079	0.0083475	36.9110184	21.2659073	0.8121828	-0.9034622
83.2522049	-0.0075354	0.3522363	36.5319748	21.6097965	0.8206097	-0.8586333
84.0376053	-0.3489809	0.0074307	36.1905289	21.2649918	0.8283519	-0.8178512
84.8230057	-0.0079905	-0.3610804	36.5315170	20.8964806	0.8206201	-0.8585787
85.6083984	0.3577328	0.0078812	36.8972435	21.2654419	0.8124861	-0.9018409
86.3937988	-0.0071347	0.3395666	36.5323753	21.5971279	0.8206007	-0.8586810
87.1791992	-0.3365408	0.0070373	36.2029686	21.2645988	0.8280673	-0.8193436
87.9645996	-0.0075594	-0.3480521	36.5319481	20.9095078	0.8206103	-0.8586302
88.7499924	0.3449399	0.0074616	36.8844490	21.2650223	0.8127679	-0.9003348
89.5353928	-0.0067684	0.3277787	36.5327415	21.5853386	0.8205925	-0.8587244
90.3207932	-0.3249591	0.0066775	36.2145500	21.2642384	0.8278025	-0.8207328
91.1061859	-0.0071708	-0.3359287	36.5323372	20.9216328	0.8206016	-0.8586765
91.8915863	0.3330283	0.0070780	36.8725357	21.2646389	0.8130305	-0.8989318
92.6769867	-0.0064327	0.3167838	36.5330772	21.5743446	0.8205850	-0.8587642
93.4623871	-0.3141500	0.0063474	36.2253571	21.2639084	0.8275555	-0.8220285
94.2477798	-0.0068096	-0.3246197	36.5326996	20.9329414	0.8205935	-0.8587194
95.0331802	0.3219100	0.0067261	36.8614197	21.2642860	0.8132756	-0.8976224
95.8185806	-0.0061239	0.3065044	36.5333862	21.5640659	0.8205780	-0.8588012
96.6039734	-0.3040383	0.0060481	36.2354698	21.2636089	0.8273246	-0.8232408
97.3893738	-0.0064776	-0.3140456	36.5330315	20.9435158	0.8205860	-0.8587589

TABLE A4-1 (cont.)

$2\pi a/\lambda$	$R - R_\infty$ (OHMS)	$X - X_\infty$ (OHMS)	R (OHMS)	X (OHMS)	d($\pi/2$) (NUMERIC)	D($\pi/2$) (dBi)
133.5176849	-0.0037433	0.2206601	36.5357666	21.4782200	0.8205246	-0.8590841
134.3030853	-0.2193809	0.0037124	36.3201294	21.2612724	0.8253961	-0.8333758
135.0884857	-0.0039409	-0.2257294	36.5355682	21.0318317	0.8205290	-0.8590605
135.8738861	0.2244135	0.0039089	36.7639236	21.2614689	0.8154324	-0.8861202
136.6592865	-0.0036202	0.2156310	36.5358887	21.4731922	0.8205218	-0.8590986
137.4446869	-0.2144097	0.0035835	36.3250999	21.2611446	0.8252832	-0.8339702
138.2300873	-0.0038018	-0.2205569	36.5357056	21.0370045	0.8205259	-0.8590769
139.0154724	0.2193003	0.0037729	36.7588081	21.2613335	0.8155459	-0.8855162
139.8008728	-0.0034976	0.2108267	36.5360107	21.4683876	0.8205191	-0.8591131
140.5862732	-0.2096590	0.0034680	36.3298492	21.2610283	0.8251753	-0.8345379
141.3716736	-0.0036771	-0.2156154	36.5358315	21.0419445	0.8205231	-0.8590918
142.1570740	0.2144142	0.0036510	36.7539215	21.2612114	0.8156543	-0.8849385
142.9424744	-0.0033878	0.2062321	36.5361214	21.4637928	0.8205166	-0.8591264
143.7278748	-0.2051150	0.0033526	36.3343925	21.2609138	0.8250721	-0.8350810
144.5132599	-0.0035593	-0.2108901	36.5359497	21.0466709	0.8205205	-0.8591059
145.2986603	0.2097409	0.0035293	36.7492485	21.2610893	0.8157580	-0.8843864
314.1591797	-0.0011151	-0.0965002	36.5383949	21.1610603	0.8204656	-0.8593965
314.9445801	0.0962590	0.0010780	36.6357689	21.2586384	0.8182849	-0.8709548
471.2387695	-0.0006121	-0.0642021	36.5388985	21.1933594	0.8204542	-0.8594564
472.0241699	0.0640954	0.0005776	36.6036034	21.2581387	0.8190039	-0.8671401
628.3183594	-0.0004024	-0.0480928	36.5391083	21.2094669	0.8204495	-0.8594813
629.1037598	0.0480331	0.0003687	36.5875435	21.2579288	0.8193634	-0.8652343
785.3981934	-0.0002733	-0.0384425	36.5392342	21.2191181	0.8204467	-0.8594962
786.1835938	0.0384038	0.0002777	36.5779114	21.2578392	0.8195792	-0.8640909
942.4777832	-0.0002102	-0.0320156	36.5392990	21.2255459	0.8204452	-0.8595040
943.2631836	0.0319889	0.0002084	36.5714989	21.2577686	0.8197229	-0.8633294
10**99	0.0000000	0.0000000	36.5395090	21.2575610	0.8204405	-0.85952902
INFINITY	0.0000000	0.0000000	36.5395090	21.2575610	3.2818500	5.16118730

A.5 METHOD OF MOMENTS COMBINED WITH GEOMETRICAL THEORY OF DIFFRACTION

TABLE A5-1; $2\pi a/\lambda = 20.0$, $h/\lambda = 0.25$, $b/\lambda = 10^{-6}$

TABLE A5-1

AWADALLA RESULT
EPSLN= 20.00000000 (NORMALIZED GROUNDPLANE RADIUS)
MONOPLE LENGTH= 0.25000000 (WAVELENGTH)
MONOPLE RADIUS= 0.00000100 (WAVELENGTH)
INPUT IMPEDANCE= 0.393019028E+02 +j 0.201706104E+02 (OHMS)

ELEVATION ANGLE (DEG)	DIRECTIVITY (NUMERIC)	DIRECTIVITY (dBi)	RELATIVE POWER (dB)	ELEVATION ANGLE (DEG)	DIRECTIVITY (NUMERIC)	DIRECTIVITY (dBi)	RELATIVE POWER (dB)
0	0.00000	-INFINITE	-INFINITE				
1	0.06796	-11.67747	-17.72616	45	0.47273	-3.25390	-9.30259
2	0.24612	-6.08851	-12.13720	46	0.41968	-3.77077	-9.81946
3	0.46721	-3.30488	-9.35356	47	0.39627	-4.02008	-10.06877
4	0.64815	-1.88322	-7.93190	48	0.39948	-3.98501	-10.03370
5	0.72163	-1.41685	-7.46554	49	0.42857	-3.67975	-9.72844
6	0.66119	-1.79673	-7.84542	50	0.48542	-3.13881	-9.18750
7	0.49049	-3.09366	-9.14235	51	0.57395	-2.41129	-8.45998
8	0.27358	-5.62909	-11.67778	52	0.69883	-1.55627	-7.60496
9	0.09050	-10.43371	-16.48240	53	0.86399	-0.63489	-6.68358
10	0.00793	-21.00809	-27.05678	54	1.07116	0.29855	-5.75014
11	0.05607	-12.51250	-18.56119	55	1.31888	1.20206	-4.84663
12	0.21974	-6.58088	-12.62956	56	1.60211	2.04691	-4.00177
13	0.44608	-3.50587	-9.55456	57	1.91237	2.81572	-3.23297
14	0.66481	-1.77302	-7.82170	58	2.23848	3.49953	-2.54916
15	0.81272	-0.90058	-6.94927	59	2.56756	4.09521	-1.95348
16	0.85329	-0.68901	-6.73770	60	2.88624	4.60332	-1.44536
17	0.78509	-1.05078	-7.09947	61	3.18178	5.02670	-1.02198
18	0.63734	-1.95629	-8.00498	62	3.44309	5.36949	-0.67920
19	0.45585	-3.41174	-9.46043	63	3.66141	5.63649	-0.41220
20	0.28575	-5.44013	-11.48882	64	3.83072	5.83280	-0.21588
21	0.15747	-8.02805	-14.07674	65	3.94784	5.96359	-0.08509
22	0.08068	-10.93260	-16.98129	66	4.01229	6.03393	-0.01476
23	0.04696	-13.28296	-19.33165	67	4.02595	6.04869	0.00000
24	0.03888	-14.10309	-20.15178	68	3.99261	6.01257	-0.03612
25	0.04090	-13.88312	-19.93181	69	3.91742	5.93000	-0.11869
26	0.04763	-13.22128	-19.26997	70	3.80641	5.80516	-0.24353
27	0.06646	-11.77436	-17.82305	71	3.66605	5.64198	-0.40671
28	0.11410	-9.42717	-15.47585	72	3.50277	5.44411	-0.60458
29	0.20892	-6.80028	-12.84897	73	3.32274	5.21496	-0.83373
30	0.36230	-4.40930	-10.45799	74	3.13158	4.95763	-1.09105
31	0.57232	-2.42364	-8.47233	75	2.93426	4.67499	-1.37370
32	0.82178	-0.85246	-6.90115	76	2.73502	4.36961	-1.67908
33	1.08125	0.33926	-5.70943	77	2.53733	4.04378	-2.00491
34	1.31575	1.19172	-4.85697	78	2.34396	3.69949	-2.34919
35	1.49288	1.74024	-4.30845	79	2.15698	3.33846	-2.71023
36	1.59002	2.01404	-4.03465	80	1.97791	2.96206	-3.08663
37	1.59865	2.03752	-4.01116	81	1.80774	2.57137	-3.47732
38	1.52488	1.83236	-4.21632	82	1.64707	2.16712	-3.88157
39	1.38678	1.42007	-4.62862	83	1.49615	1.74974	-4.29895
40	1.20929	0.82529	-5.22339	84	1.35497	1.31929	-4.72940
41	1.01860	0.08005	-5.96864	85	1.22335	0.87551	-5.17318
42	0.83732	-0.77111	-6.81980	86	1.10099	0.41782	-5.63087
43	0.68137	-1.66615	-7.71484	87	0.98749	-0.05466	-6.10335
44	0.55920	-2.52435	-8.57304	88	0.88246	-0.54303	-6.59172

Table A5-1 (cont.)

AWADALLA RESULT
EPSLN= 20.00000000 (NORMALIZED GROUNDPLANE RADIUS)
MONOPLE LENGTH= 0.25000000 (WAVELENGTH)
MONOPLE RADIUS= 0.00000100 (WAVELENGTH)
INPUT IMPEDANCE= 0.393019028E+02 +j 0.201706104E+02 (OHMS)

ELEVATION ANGLE (DEG)	DIRECTIVITY (NUMERIC)	DIRECTIVITY (dBi)	RELATIVE POWER (dB)
89	0.78549	-1.04857	-7.09726
90	0.69622	-1.57253	-7.62122
91	0.61434	-2.11591	-8.16460
92	0.53962	-2.67910	-8.72779
93	0.47193	-3.26121	-9.30990
94	0.41123	-3.85919	-9.90788
95	0.35756	-4.46647	-10.51516
96	0.31109	-5.07113	-11.11982
97	0.27203	-5.65377	-11.70246
98	0.24068	-6.18564	-12.23433
99	0.21733	-6.62873	-12.67742
100	0.20230	-6.94002	-12.98871
101	0.19581	-7.08158	-13.13027
102	0.19798	-7.03373	-13.08242
103	0.20873	-6.80419	-12.85288
104	0.22770	-6.42630	-12.47499
105	0.25423	-5.94781	-11.99650
106	0.28720	-5.41809	-11.46678
107	0.32510	-4.87985	-10.92854
108	0.36589	-4.36650	-10.41519
109	0.40710	-3.90301	-9.95170
110	0.44585	-3.50811	-9.55680
111	0.47901	-3.19654	-9.24523
112	0.50337	-2.98112	-9.02980
113	0.51591	-2.87424	-8.92293
114	0.51411	-2.88940	-8.93809
115	0.49631	-3.04248	-9.09117
116	0.46200	-3.35354	-9.40223
117	0.41219	-3.84904	-9.89773
118	0.34949	-4.56565	-10.61434
119	0.27820	-5.55648	-11.60517
120	0.20404	-6.90292	-12.95161
121	0.13372	-8.73790	-14.78659
122	0.07426	-11.29272	-17.34140
123	0.03201	-14.94763	-20.99632
124	0.01174	-19.30150	-25.35019
125	0.01569	-18.04321	-24.09189
126	0.04282	-13.68314	-19.73183
127	0.08860	-10.52583	-16.57452
128	0.14525	-8.37898	-14.42767
129	0.20273	-6.93074	-12.97943
130	0.25029	-6.01552	-12.06420
131	0.27836	-5.55393	-11.60261
132	0.28057	-5.51954	-11.56823
133	0.25539	-5.92788	-11.97657
134	0.20692	-6.84207	-12.89075
135	0.14450	-8.40143	-14.45012

ELEVATION ANGLE (DEG)	DIRECTIVITY (NUMERIC)	DIRECTIVITY (dBi)	RELATIVE POWER (dB)
134	0.20692	-6.84207	-12.89075
135	0.14450	-8.40143	-14.45012
136	0.08112	-10.90860	-16.95728
137	0.03067	-15.13315	-21.18184
138	0.00458	-23.39319	-29.44189
139	0.00877	-20.57190	-26.62059
140	0.04160	-13.80860	-19.85729
141	0.09375	-10.28045	-16.32914
142	0.15012	-8.23576	-14.28444
143	0.19379	-7.12669	-13.17537
144	0.21090	-6.75930	-12.80799
145	0.19511	-7.09717	-13.14586
146	0.15030	-8.23055	-14.27923
147	0.09012	-10.45171	-16.50040
148	0.03442	-14.63222	-20.68091
149	0.00300	-25.23113	-31.27982
150	0.00881	-20.54938	-26.59806
151	0.05272	-12.77989	-18.82858
152	0.12206	-9.13432	-15.18301
153	0.19391	-7.12389	-13.17258
154	0.24268	-6.14973	-12.19842
155	0.24939	-6.03114	-12.07983
156	0.20970	-6.78402	-12.83271
157	0.13696	-8.63419	-14.68288
158	0.05879	-12.30668	-18.35537
159	0.00757	-21.20760	-27.25629
160	0.00782	-21.06523	-27.11391
161	0.06544	-11.84184	-17.89052
162	0.16323	-7.87204	-13.92073
163	0.26562	-5.75747	-11.80616
164	0.33154	-4.79458	-10.84327
165	0.33138	-4.79670	-10.84538
166	0.26114	-5.83118	-11.87987
167	0.14774	-8.30502	-14.35371
168	0.04190	-13.77805	-19.82674
169	0.00020	-37.02296	-43.07164
170	0.06237	-12.05005	-18.09874
171	0.23273	-6.33144	-12.38013
172	0.47380	-3.24409	-9.29278
173	0.71599	-1.45095	-7.49964
174	0.88100	-0.55025	-6.59894
175	0.91069	-0.40631	-6.45500
176	0.79043	-1.02138	-7.07007
177	0.55740	-2.53837	-8.58706
178	0.28966	-5.38113	-11.42982
179	0.07940	-11.00158	-17.05026
180	0.00000	-81.87164	-87.92033

Appendix B
Listings of Computer Programs

B.1 BARDEEN

```
C
C                           PROGRAM "BARDEEN"
C**************************************************************
C                                                             *
C      THE PROGRAM "BARDEEN", WRITTEN IN FORTRAN, COMPUTES    *
C      THE RADIATION RESISTANCE AND THE DIRECTIVE             *
C      GAIN FOR A MONOPOLE ELEMENT OF LENGTH, h , MOUNTED     *
C      AT THE CENTER OF AN ELECTRICALLY-SMALL CIRCULAR        *
C      GROUNDPLANE OF RADIUS, a , UTILIZING A BARDEEN-MITRE   *
C      INTEGRAL EQUATION METHOD.                              *
C                                                             *
C**************************************************************
C      THE INPUT VARIABLES OF THIS PROGRAM ARE:               *
C                                                             *
C          L = 2πh/λ   (ELEMENT LENGTH IN WAVENUMBERS)        *
C                                                             *
C  ε = EPSLN = 2πa/λ   (GROUNDPLANE RADIUS IN WAVENUMBERS)    *
C                                                             *
C      WHERE                                                  *
C          λ = EXCITATION WAVELENGTH                          *
C                                                             *
C**************************************************************
C      THE OUTPUT VARIABLES OF THIS PROGRAM ARE:              *
C                                                             *
C              THETAD = θ , ELEVATION ANGLE  (DEGREES)        *
C                                                             *
C              RADRES = R , RADIATION RESISTANCE (OHMS)       *
C                                                             *
C               DIREC = d(θ), DIRECTIVITY (NUMERIC)           *
C                                                             *
C              DIRECD = D(θ), DIRECTIVITY (DBI)               *
C**************************************************************
      REAL*8 Z,Z1,Z2,G,G1,G2,Y,Y1,Y2,F1(250),F2(250),T
      REAL*8 I,Q,EPSLN,DCADRE,A,B,GSUM1,GSUM2,H,HEIGHT,L
      REAL*8 TSUM1,TSUM2,TYSUM1,TYSUM2,BRKT1,YSUM1,YSUM2
      REAL*8 BRKT2,BRKT(89),W(12),THTA(12),BRKT3,BRKT4,BRKT5
      REAL*8 THETA(89),PI,CNVRT,MBRKT(89),DENOM
      REAL*8 RADRES,DIREC(89),THETAD(89)
      REAL*8 DIRECR(89),DIRECD(89),DIRMAX
      PI=3.1415926536D0
      L=PI/2.0D0
      HEIGHT=L/(2*PI)
C**************CONVERT DEGREE TO RADIAN*******************************
      DO 1 II=1,89
      K=II*2
      THETAD(II)=DFLOAT(K)
      THETA(II)=PI*THETAD(II)/180.0D0
    1 CONTINUE
C*********************************************************************
      EPSLN=2.25D0
      DO 26 II=1,5
C*********************COMPUTE WITH RESPECT TO EVERY TWO DEGREE*******
      DO 2 INDEX=1,89
      K=1
      T=0.000001D0
      IF (EPSLN.EQ.0.0D0) I=0.0D0
      IF (EPSLN.EQ.0.0D0) Q=0.0D0
```

```
      IF (EPSLN.EQ.0.0D0) GO TO 44
 33   Z1=DSINH(T)*DSINH(EPSLN*DCOS(THETA(INDEX)))
      Z2=DSINH(EPSLN)*DSINH(T*DCOS(THETA(INDEX)))
      Z=(Z1-Z2)/DSINH(EPSLN)
      Y1=(H*DSINH(T)*DSIN(L)+T*DCOS(L)*DCOSH(T))/(L*L+T*T)
      Y2=DCOS(L)*DCOSH(T)/T
      Y=Y1-Y2
      G1=(DSINH(T)*DSIN(L)-T*DCOSH(T)*DSIN(L))/(T*T)
      G2=(T*DCOSH(T)*DSIN(L)-H*DSINH(T)*DCOS(L))/(L*L+T*T)
      G=G1+G2
      F1(K)=Z*Y
      F2(K)=Z*G
      T=T+0.01D0
      K=K+1
      IF(T.LE.EPSLN)GO TO 33
      GSUM1=0.0D0
      YSUM1=0.0D0
      M=K-1
      DO 10 J=1,M
      YSUM1=YSUM1+F1(J)
      GSUM1=GSUM1+F2(J)
 10   CONTINUE
      I=YSUM1*0.01
      Q=GSUM1*0.01
C******COMPUTE RADIATION RESISTANCE AND DIRECTIVE GAIN***************
 44   BRKT3=DSIN(L*DCOS(THETA(INDEX)))-DCOS(THETA(INDEX))*DSIN(L)-
     *2.0D0*I/PI
      BRKT4=0.0D0
      BRKT5=0.0D0
      IF (EPSLN.EQ.0.0D0) GO TO 23
      BRKT4=DSIN(L)*DCOS(THETA(INDEX))
      BRKT5=DSIN(L)*DSINH(EPSLN*DCOS(THETA(INDEX)))/DSINH(EPSLN)
 23   BRKT1=BRKT3+BRKT4-BRKT5
      BRKT2=DCOS(L*DCOS(THETA(INDEX)))-2.0D0*Q/PI-DCOS(L)
      BRKT(INDEX)=BRKT1*BRKT1+BRKT2*BRKT2
      MBRKT(INDEX)=BRKT(INDEX)/DSIN(THETA(INDEX))
    2 CONTINUE
      DENOM=0.0D0
      DO 3 K=1,89
      DENOM=DENOM+MBRKT(K)
    3 CONTINUE
      DENOM=DENOM*THETA(1)
      RADRES=14.98962514D0*DENOM/(DSIN(L)*DSIN(L))
      DIRMAX=0.0D0
      DO 4 K=1,89
      DIREC(K)=2.0D0*BRKT(K)/((DSIN(THETA(K)))**2.0D0*DENOM)
      DIRECD(K)=10.D0*DLOG10(DIREC(K))
      IF(DIRMAX.LT.DIRECD(K)) DIRMAX=DIRECD(K)
    4 CONTINUE
      DO 15 J=1,89
      DIRECR(J)=DIRECD(J)-DIRMAX
   15 CONTINUE
C***************WRITE STATMENTS******************************
      WRITE(6,80)
   80 FORMAT(1H1,'INTEGRAL EQUATION METHOD')
      WRITE(6,81) EPSLN
   81 FORMAT(1X,'EPSILON= ',F6.4,'(NORMALIZED GROUNDPLANE RADIUS)')
      WRITE(6,82) HEIGHT
   82 FORMAT(1X,'MONOPOLE LENGTH= ',F8.6,'(WAVELENGTH)')
      WRITE(6,83)
   83 FORMAT(1X,'INFINITELY THIN MONOPOLE RADIUS')
```

```
      WRITE(6,84) RADRES
   84 FORMAT(1X,'RADIATION RESISTANCE = ',F10.6,' (OHMS)')
      WRITE(6,85)
   85 FORMAT(//,1X,'ELEVATION',53X,'ELEVATION')
      WRITE(6,86)
   86 FORMAT(1X,2('ANGLE',6X,'DIRECTIVITY',5X,'DIRECTIVITY',5X
     *,'RELATIVE POWER',5X))
      WRITE(6,87)
   87 FORMAT(1X,2('(DEG)',6X,'(NUMERIC)',7X,'(DBI)',11X
     *,'(DB)',15X))
      WRITE(6,88)
   88 FORMAT(//,2X,'   0. ',8X,'0.00000',5X,'-INFINITY',6X,'-INFINITY')
      DO 16 L=1,44
      WRITE(6,89) THETAD(L),DIREC(L),DIRECD(L),DIRECR(L)
     *,THETAD(L+45),DIREC(L+45),DIRECD(L+45),DIRECR(L+45)
   89 FORMAT(1X,2(1X,F5.1,6X,F9.5,5X,F9.5,6X,F9.5,11X))
   16 CONTINUE
      WRITE(6,90) THETAD(45),DIREC(45),DIRECD(45),DIRECR(45)
   90 FORMAT(1X,1X,F5.1,6X,F9.5,5X,F9.5,6X,F9.5,12X,'180.  '
     *,8X,'0.00000',5X,'-INFINITY',6X,'-INFINITY')
      EPSLN=EPSLN+0.25D0
   26 CONTINUE
      STOP
      END
```

B.2 RICHMD1 AND RICHMD2

```
C       PROGRAM "RICHMD1"
C***************************************************************************
C       THIS COMPUTER PROGRAM, IN FORTRAN LANGUAGE, WAS WRITTEN BY         *
C  PROF. RICHMOND OF OHIO STATE UNIVERSITY.  IT USES A SINUSOIDAL-         *
C  GALERKIN METHOD OF MOMENTS TO COMPUTE THE INPUT IMPEDANCE AND           *
C  THE CURRENT DISTRIBUTIONS OF A MONOPOLE ELEMENT OF LENGTH               *
C  h AND RADIUS b AT THE CENTER OF A CIRCULAR GROUNDPLANE OF RADIUS a.     *
C  A DETAILED DERIVATION IS PUBLISHED IN REFERENCE 1.   IN                 *
C  ORDER TO ADAPT THE PROGRAM INTO OUR COMPUTER SYSTEM                     *
C  (VAX 11/780), THE PROGRAM HAS BEEN CHANGED FROM SINGLE PRECISION TO     *
C  DOUBLE PRECISION.                                                       *
C                                                                          *
C       UTILIZING THE CURRENT DISTRIBUTIONS DETERMINED FROM THIS           *
C  PROGRAM, A RADIATION PATTERN CAN BE OBTAINED FROM THE MITRE             *
C  COMPUTER PROGRAM "RICHMD2".  AN INPUT FILE, "INPUT1.DAT", TO THE        *
C  COMPUTER PROGRAM "RICHMD2" IS AUTOMATICALLY GENERATED FROM THIS         *
C  PROGRAM.                                                                *
C                                                                          *
C     REFERENCE                                                            *
C                                                                          *
C     1)    J.H.RICHMOND,"MONOPOLE ANTENNA ON CIRCULAR DISK"               *
C                OHIO STATE UNIVERSITY,ELECTROSCIENCE LABORATORY,          *
C                TECHNICAL REPORT 711639-1, JULY 1979.                     *
C                                                                          *
C***************************************************************************
C       THIS COMPUTER PROGRAM REQUIRES SEVEN INPUTS WHICH ARE ENTERED      *
C  FROM AN INPUT FILE NAMED "MON.DAT" WITH FORMAT 17.                      *
C                                                                          *
C                b = AL = MONOPOLE RADIUS IN WAVELENGTHS                   *
C                h = HL = MONOPOLE LENGTH IN WAVELENGTHS                   *
C     ϵ = 2πa/λ = EPSLN = NORMALIZED GROUNDPLANE RADIUS IN WAVENUMBERS     *
C            b1/b = BAR = RATIO OF OUTER TO INNER CONDUCTOR RADII          *
C                          OF THE COAXIAL LINE FEED                        *
C                N = NSW = NUMBER OF MONOPOLE SEGMENTS                     *
C                M = NSD = NUMBER OF GROUNDPLANE CONCENTRIC ANNULAR ZONES  *
C                X = IX = NUMBER OF SUBDIVIDED ANNULAR ZONES FOR EACH NSD  *
C                                                                          *
C       NOTE: THE MAXIMUN NUMBER OF SEGMENTS (NSD+NSW) CAN BE USED IN      *
C             THIS PROGRAM IS 31.  IF NUMBER OF SEGMENTS USED IS OVER      *
C             31, THEN ALL THE ARRAYS OF DIMENSION OF 30 AND 30x30         *
C             SHOULD BE CHANGED TO ARRAYS OF DIMENSION OF (NSD+NSW)        *
C             AND (NSD+NSW)x(NSD+NSW).                                     *
C                                                                          *
C***************************************************************************
      IMPLICIT REAL*8 (A-H),(P-Z)
      COMPLEX*16 CJ(30),ZJ(30),VIJ(30,30),ZIJ(30,30),J11
      COMPLEX*16 Y11,EC,D11,D12,D21,D22,DZ1J,DV1,W11
      COMPLEX*16 P11,P12,P21,P22,ZDD,ZDM,ZMD,ZMM,Z11,ZD,Z22,Z12,Z21
      DIMENSION FB(500)
      OPEN(UNIT=4,FILE='INPUT1',STATUS='NEW')
      OPEN(UNIT=9,FILE='RICH',STATUS='NEW')
      OPEN(UNIT=6,FILE='MON.DAT',STATUS='OLD',READONLY)
      DATA EO,UO/8.85418533677E-12,1.25663706144E-6/
      DATA ETA,PI,TP/376.730366239,3.14159265359,6.28318530718/
      DATA ICC,IFB/30,500/
  1   FORMAT(1X,2I5,2E15.8)
  2   FORMAT(/16X,2F15.4)
  3   FORMAT(1X,2I5,4F10.4)
  5   FORMAT(1HO)
C*********************************************************************
C       INPUTS                                                       *
```

```
C*********************************************************************
      READ(6,17)AL,HL,EPSLN,BAR,NSD,NSW,IX
 17   FORMAT(3F12.9,F6.3,3I2)
 101  BL=EPSLN/(2.*PI)
C*********************************************************************
C        INPUT WRITE STATMENTS                                       *
C*********************************************************************
 103  WRITE (9,102)
 102  FORMAT (1H1,14X,'DOUBLE PRECISION')
      WRITE (9,80) AL
 80   FORMAT (15X,'MONOPOLE RADIUS(WAVELENGTH), AL=   ',F15.8)
      WRITE (9,82) HL
 82   FORMAT (15X,'MONOPOLE LENGTH(WAVELENGTH), HL=   ',F5.4)
      WRITE (9,91)EPSLN
 91   FORMAT (15X,'DISC RADIUS(WAVELENGTH), BL= ',G,'/(2*PI)')
      WRITE (9,84)NSW
 84   FORMAT (15X,'NUMBER OF MONOPOLE SEGMENTS, NSW= ',I2)
      WRITE (9,85)NSD
 85   FORMAT (15X,'NUMBER OF DISC SEGMENTS, NSD=',I2)
      WRITE (9,88)BAR
 88   FORMAT (15X,'COAXIAL RADIUS RATIO, BAR= ',F4.2)
      NLI=NSD+NSW
      WRITE(4,666)AL,HL,EPSLN,NSD,NSW,NLI,IX
 666  FORMAT(E12.4,2X,F8.6,2X,F6.4,2X,4(I2,2X))
 81   CONTINUE
      DO 400 IBL=10,10
      TL=AL/100.
      IWCJ=1
      IZIJ=0
      IWZ=0
      NPH=6
      NEQ=NSD+NSW-1
      AK=TP*AL
      BK=TP*BL
      HK=TP*HL
      TK=TP*TL
      DKD=(BK-AK)/NSD
      DKW=HK/NSW
      RH2=AK+DKD
      IF(RH2.LT.BAR*AK)GO TO 400
      TDKD=2.*DKD
      CDKD=DCOS(DKD)
      SDKD=DSIN(DKD)
      CDK=DCOS(DKW)
      SDK=DSIN(DKW)
      MAX=NSW-1
      NA=NSD+1
      CALL QMM(AK,DKD,DKW,CDKD,SDKD,CDK,SDK,TK,IWZ,NPH,Z11)
      ZIJ(1,1)=Z11
      IF(NSD.LE.1)GO TO 100
      S1=AK
      DO 60 J=2,NSD
      S2=S1+DKD
      S3=S1+TDKD
      T1=AK
      DO 50 I=2,J
      T2=T1+DKD
      T3=T1+TDKD
      CALL QDD(CDKD,SDKD,S1,S3,T1,T3,TK,IWZ,NPH,Z22)
      ZIJ(I,J)=Z22
  50  T1=T1+DKD
```

```
      CALL QDM(AK, DKD, DKW, CDKD, SDKD, SDK, S1, S3, TK, IWZ, NPH, Z12)
      ZIJ(1, J)=Z12
   60 S1=S1+DKD
  100 IF(NSW.LE.1)GO TO 200
      CALL SPART(AK, DKD, DKW, MAX, IWZ, ZJ, CJ)
      L=0
      DO 160 I=NA, NEQ
      DO 150 J=I, NEQ
      K=J-I+1
  150 ZIJ(I, J)=ZJ(K)
      L=L+1
      ZIJ(1, I)=CJ(L)
  160 CONTINUE
  178 IF(NSD.LE.1)GO TO 200
      Z2=.ODO
      DO 190 J=NA, NEQ
      Z2=Z2+DKW
      S1=Z2-DKW
      S3=Z2+DKW
      RH2=AK
      DO 180 I=2, NSD
      RH2=RH2+DKD
      T1=RH2-DKD
      T3=RH2+DKD
      CALL SKEWT(AK, S1, S3, T1, T3, CDK, SDK, CDKD, SDKD, IWZ, Z12)
  180 ZIJ(I, J)=Z12
  190 CONTINUE
  200 CALL GRILL(AK, BAR, DKD, DKW, NEQ, NSD, NSW, TK, CJ)
      DO 210 I=1, NEQ
      DO 210 J=I, NEQ
  210 VIJ(I, J)=ZIJ(I, J)
C**********************************************************************
C        MATRIX INVERSION FOR A NSW+NSD-1 BY NSW+NSD-1 SQUARE MATRIX *
C**********************************************************************
      CALL CROUT(ZIJ, CJ, ICC, O, IWCJ, 1, NEQ)
      Y11=CJ(1)
      Z11=1./Y11
  320 CONTINUE
      J11=Z11
C**********************************************************************
C        OUTPUT WRITE STATEMENTS                                      *
C**********************************************************************
  400 CONTINUE
      WRITE(9, 92)
      WRITE(9, 93)
   92 FORMAT (//15X, 'INPUT IMPEDANCE Z11 (OHMS)')
   93 FORMAT (24X, 'RE(Z11)', 8X, 'IM(Z11)')
      WRITE(9, 2)J11
      CLOSE(UNIT=9, STATUS='KEEP')
      CLOSE(UNIT=6, STATUS='KEEP')
  500 CALL EXIT
      END
```

```
C*****************************************************************
C       SUBROUTINE CISI                                          *
C*****************************************************************
      SUBROUTINE CISI(CI,CIN,SI,X)
      IMPLICIT REAL*8 (A-H),(P-Z)
      DATA GAM,P2/.57721566,1.57079632/
      A=DABS(X)
      IF(A.GT.4.)GO TO 10
      IF(A.GT..1)GO TO 3
      IF(A.GT.0.)GO TO 2
      CI=.0D0
      CIN=.0D0
      SI=.0D0
      RETURN
  2   X2=A*A
      SI=X*((.03*X2-1.)*X2/18.+1.)
      CIN=.25*X2*((X2/45.-1.)*X2/24.+1.)
      GO TO 8
  3   Y=(4.-A)*(4.+A)
      SI=X*(((((1.753141E-9*Y+1.568988E-7)*Y+1.374168E-5)*Y+6.939889E-4)
     C*Y+1.964882E-2)*Y+4.395509E-1)
      CIN=       A*A*(((((1.386985E-10*Y+1.584996E-8)*Y
     C+1.725752E-6)*Y+1.185999E-4)*Y+4.990920E-3)*Y+1.315308E-1)
  8   CI=GAM+DLOG(A)-CIN
      RETURN
 10   SI=DSIN(A)
      Y=DCOS(A)
      Z=4./A
      U=((((((((4.048069E-3*Z-2.279143E-2)*Z+5.515070E-2)*Z-7.261642E-2)
     C*Z+4.987716E-2)*Z-3.332519E-3)*Z-2.314617E-2)*Z-1.134958E-5)*Z
     C+6.250011E-2)*Z+2.583989E-10
      V=((((((((-5.108699E-3*Z+2.819179E-2)*Z-6.537283E-2)*Z
     C+7.902034E-2)*Z-4.400416E-2)*Z-7.945556E-3)*Z+2.601293E-2)*Z
     C-3.764000E-4)*Z-3.122418E-2)*Z-6.646441E-7)*Z+2.500000E-1
      CI=Z*(SI*V-Y*U)
      SI=-Z*(SI*U+Y*V)+P2
      IF(X.LT..0D0)SI=-SI
      CIN=GAM+DLOG(A)-CI
      RETURN
      END
```

```
C***************************************************************
C       SUBROUTINE CROUT                                       *
C***************************************************************
      SUBROUTINE CROUT(C,S,ICC,ISYM,IWR,I12,N)
      IMPLICIT REAL*8 (A-H),(P-Z)
      COMPLEX*16 C(ICC,ICC),S(1)
      COMPLEX*16 F,P,SS,T
  1   FORMAT(15X,'LIST OF THE LOOP CURRENTS CJ(I)')
  2   FORMAT(5X,'I',3X,'MAGNITUDE',3X,'PHASE',5X,'REAL'
     2,5X,'IMAGINARY')
  3   FORMAT(1X,I5,F10.4,F10.1,2E12.4)
  5   FORMAT(1HO)
      IF(I12.NE.1)GO TO 22
      IF(N.EQ.1)S(1)=S(1)/C(1,1)
      IF(N.EQ.1)GO TO 39
      IF(ISYM.NE.ODO)GO TO 8
      DO 6 I=1,N
      DO 6 J=I,N
  6   C(J,I)=C(I,J)
  8   F=C(1,1)
      DO 10 L=2,N
  10  C(1,L)=C(1,L)/F
      DO 20 L=2,N
      LLL=L-1
      DO 20 I=L,N
      F=C(I,L)
      DO 11 K=1,LLL
  11  F=F-C(I,K)*C(K,L)
      C(I,L)=F
      IF(L.EQ.I)GO TO 20
      P=C(L,L)
      IF(ISYM.EQ.ODO)GO TO 15
      F=C(L,I)
      DO 12 K=1,LLL
  12  F=F-C(L,K)*C(K,I)
      C(L,I)=F/P
      GO TO 20
  15  F=C(I,L)
      C(L,I)=F/P
  20  CONTINUE
  22  DO 30 L=1,N
      P=C(L,L)
      T=S(L)
      IF(L.EQ.1)GO TO 30
      LLL=L-1
      DO 25 K=1,LLL
  25  T=T-C(L,K)*S(K)
  30  S(L)=T/P
      DO 38 L=2,N
      I=N-L+1
      II=I+1
      T=S(I)
      DO 35 K=II,N
  35  T=T-C(I,K)*S(K)
  38  S(I)=T
  39  IF(IWR.LE.ODO) GO TO 100
      WRITE(9,5)
      WRITE(9,1)
      WRITE(9,2)
      CNOR=.ODO
      DO 40 I=1,N
```

```
      SA=CDABS(S(I))
 40   IF(SA.GT.CNOR)CNOR=SA
      IF(CNOR.LE.0.)CNOR=1.
      DO 44 I=1,N
      SS=S(I)
      SA=CDABS(SS)
      SNOR=SA/CNOR
      PH=.ODO
      IF(SA.GT.0.)PH=57.29578*DATAN2(DIMAG(SS),DREAL(SS))
      WRITE(9,3)I,SNOR,PH,SS
 44   WRITE(4,111)SS
111   FORMAT(2E12.4)
      WRITE(9,5)
 100  RETURN
      END
```

```
C***********************************************************************
C         SUBROUTINE EXPJ                                              *
C***********************************************************************
      SUBROUTINE EXPJ(V1,V2,W12)
      IMPLICIT REAL*8 (A-H),(P-Z)
      COMPLEX*16 EC,E15,S,T,UC,VC,V1,V2,W12,Z
      DIMENSION V(21),W(21),D(16),E(16)
      DATA V/   0.22284667E 00,
     20.11889321E 01,0.29927363E 01,0.57751436E 01,0.98374674E 01,
     20.15982874E 02,0.93307812E-01,0.49269174E 00,0.12155954E 01,
     20.22699495E 01,0.36676227E 01,0.54253366E 01,0.75659162E 01,
     20.10120228E 02,0.13130282E 02,0.16654408E 02,0.20776479E 02,
     20.25623894E 02,0.31407519E 02,0.38530683E 02,0.48026086E 02/
      DATA W/   0.45896460E 00,
     20.41700083E 00,0.11337338E 00,0.10399197E-01,0.26101720E-03,
     20.89854791E-06,0.21823487E 00,0.34221017E 00,0.26302758E 00,
     20.12642582E 00,0.40206865E-01,0.85638778E-02,0.12124361E-02,
     20.11167440E-03,0.64599267E-05,0.22263169E-06,0.42274304E-08,
     20.39218973E-10,0.14565152E-12,0.14830270E-15,0.16005949E-19/
      DATA D/   0.22495842E 02,
     2 0.74411568E 02,-0.41431576E 03,-0.78754339E 02, 0.11254744E 02,
     2 0.16021761E 03,-0.23862195E 03,-0.50094687E 03,-0.68487854E 02,
     2 0.12254778E 02,-0.10161976E 02,-0.47219591E 01, 0.79729681E 01,
     2-0.21069574E 02, 0.22046490E 01, 0.89728244E 01/
      DATA E/   0.21103107E 02,
     2-0.37959787E 03,-0.97489220E 02, 0.12900672E 03, 0.17949226E 02,
     2-0.12910931E 03,-0.55705574E 03, 0.13524801E 02, 0.14696721E 03,
     2 0.17949528E 02,-0.32981014E 00, 0.31028836E 02, 0.81657657E 01,
     2 0.22236961E 02, 0.39124892E 02, 0.81636799E 01/
      Z=V1
      DO 100 JIM=1,2
      X=DREAL(Z)
      Y=DIMAG(Z)
      E15=(.ODO,.ODO)
      AB=CDABS(Z)
      IF(AB.EQ.0.)GO TO 90
      IF(X.GE.0.  .AND.  AB.GT.10.)GO TO 80
      YA=DABS(Y)
      IF(X.LE.0.  .AND.  YA.GT.10.)GO TO 80
      IF(YA-X.GE.17.5.OR.YA.GE.6.5.OR.X+YA.GE.5.5.OR.X.GE.3.)GO TO 20
      IF(X.LE.-9.)GO TO 40
      IF(YA-X.GE.2.5)GO TO 50
      IF(X+YA.GE.1.5)GO TO 30
   10 N=6.+3.*AB
      E15=1./(N-1.)-Z/N**2
   15 N=N-1
      E15=1./(N-1.)-Z*E15/N
      IF(N.GE.3)GO TO 15
      E15=Z*E15-DCMPLX(.577216+DLOG(AB),DATAN2(Y,X))
      GO TO 90
   20 J1=1
      J2=6
      GO TO 31
   30 J1=7
      J2=21
   31 S=(.ODO,.ODO)
      YS=Y*Y
      DO 32 I=J1,J2
      XI=V(I)+X
      CF=W(I)/(XI*XI+YS)
   32 S=S+DCMPLX(XI*CF,-YA*CF)
```

```
      GO TO 54
40    T3=X*X-Y*Y
      T4=2.*X*YA
      T5=X*T3-YA*T4
      T6=X*T4+YA*T3
      UC=DCMPLX(D(11)+D(12)*X+D(13)*T3+T5-E(12)*YA-E(13)*T4,
     2 E(11)+E(12)*X+E(13)*T3+T6+D(12)*YA+D(13)*T4)
      VC=DCMPLX(D(14)+D(15)*X+D(16)*T3+T5-E(15)*YA-E(16)*T4,
     2 E(14)+E(15)*X+E(16)*T3+T6+D(15)*YA+D(16)*T4)
      GO TO 52
50    T3=X*X-Y*Y
      T4=2.*X*YA
      T5=X*T3-YA*T4
      T6=X*T4+YA*T3
      T7=X*T5-YA*T6
      T8=X*T6+YA*T5
      T9=X*T7-YA*T8
      T10=X*T8+YA*T7
      UC=DCMPLX(D(1)+D(2)*X+D(3)*T3+D(4)*T5+D(5)*T7+T9-(E(2)*YA+E(3)*T4
     2+E(4)*T6+E(5)*T8),E(1)+E(2)*X+E(3)*T3+E(4)*T5+E(5)*T7+T10+
     3(D(2)*YA+D(3)*T4+D(4)*T6+D(5)*T8))
      VC=DCMPLX(D(6)+D(7)*X+D(8)*T3+D(9)*T5+D(10)*T7+T9-(E(7)*YA+E(8)*T4
     2+E(9)*T6+E(10)*T8),E(6)+E(7)*X+E(8)*T3+E(9)*T5+E(10)*T7+T10+
     3(D(7)*YA+D(8)*T4+D(9)*T6+D(10)*T8))
52    EC=UC/VC
      S=EC/DCMPLX(X,YA)
54    EX=DEXP(-X)
      T=EX*DCMPLX(DCOS(YA),-DSIN(YA))
      E15=S*T
56    IF(Y.LT.0.)E15=DCONJG(E15)
      GO TO 90
80    E15=.409319/(Z+.193044)+.421831/(Z+1.02666)+.147126/(Z+2.56788)+
     2.206335E-1/(Z+4.90035)+.107401E-2/(Z+8.18215)+.158654E-4/(Z+
     312.7342)+.317031E-7/(Z+19.3957)
      E15=E15*CDEXP(-Z)
90    IF(JIM.EQ.1)W12=E15
100   Z=V2
      Z=V2/V1
      TH=DATAN2(DIMAG(Z),DREAL(Z))-DATAN2(DIMAG(V2),DREAL(V2))
     2+DATAN2(DIMAG(V1),DREAL(V1))
      AB=DABS(TH)
      IF(AB.LT.1.)TH=.0D0
      IF(TH.GT.1.)TH=6.2831853
      IF(TH.LT.-1.)TH=-6.2831853
      W12=W12-E15+DCMPLX(.0D0,TH)
      RETURN
      END
```

```
C*******************************************************************
C       SUBROUTINE GRILL                                            *
C*******************************************************************
      SUBROUTINE GRILL(AK,BAR,DKD,DKW,NEQ,NSD,NSW,TK,VJ)
      IMPLICIT REAL*8 (A-H),(P-Z)
      COMPLEX*16 EGZ,GM,GP,GI(20),VJ(1),GII,QST,WST,VJ1
      DOUBLE PRECISION DZS,RS1,RS2
      DATA PI,TP/3.14159265359,6.28318530718/
      IDM=20
      DO 20 I=1,NEQ
   20 VJ(I)=(.ODO,.ODO)
      VJ(1)=(1.,.ODO)
      IF(BAR.LE.1.)RETURN
      VJ(1)=(.ODO,.ODO)
      DK=DKW
      SDK=DSIN(DK)
      CDK=DCOS(DK)
      BAL=DLOG(BAR)
      QST=DCMPLX(.ODO,1./(4.*BAL*SDK))
      BK=AK*BAR
      AKS=AK*AK
      BKS=BK*BK
      LIM=NSW+1
      IF(LIM.GT.IDM)LIM=IDM
      NPH=6
      NPH=2*(NPH/2)
      NPP=NPH+1
      PHA=.0174533*20.
      DPH=PHA/NPH
      PH=.ODO
      DO 90 LPH=1,2
      WST=DPH*QST/(3.*PI)
      SGN=-1.
      DO 80 IPH=1,NPP
      WF=3.+SGN
      IF(IPH.EQ.1)WF=1.
      IF(IPH.EQ.NPP)WF=1.
      CPH=DCOS(PH)
      IF(IPH.GT.1)GO TO 40
      IF(LPH.GT.1)GO TO 40
      CPH=DCOS(DPH/10.)
   40 RS1=2.*AKS*(1.-CPH)
      RS2=AKS+BKS-2.*AK*BK*CPH
      RH1=DSQRT(RS1)
      RH2=DSQRT(RS2)
      CALL CISI(CA,CIN,SA,RH1)
      CALL CISI(CB,CIN,SB,RH2)
      GI(1)=2.*DCMPLX(CB-CA,SA-SB)
      DO 50 I=2,LIM
      DZ=DK*(I-1)
      DZS=DZ*DZ
      RA=DSQRT(RS1+DZS)
      RB=DSQRT(RS2+DZS)
      CALL CISI(C1,CIN,S1,RA+DZ)
      CALL CISI(C2,CIN,S2,RB+DZ)
      GP=DCMPLX(C2-C1,S1-S2)
      RAM=RS1/(RA+DZ)
      RBM=RS2/(RB+DZ)
      CALL CISI(C1,CIN,S1,RAM)
      CALL CISI(C2,CIN,S2,RBM)
      GM=DCMPLX(C2-C1,S1-S2)
```

```
      EGZ=DCMPLX(DCOS(DZ),DSIN(DZ))
50    GI(I)=GP*EGZ+GM/EGZ
      VJ(1)=VJ(1)+WF*WST*(GI(2)-CDK*GI(1))
      IF(NSW.LE.1)GO TO 78
      K1=0
      IA=NSD+1
      DO 60 I=IA,NEQ
      K1=K1+1
      K2=K1+1
      K3=K2+1
      IF(K3.GT.IDM)GO TO 60
      GP=GI(K1)-2.*CDK*GI(K2)+GI(K3)
      VJ(I)=VJ(I)+WF*WST*GP
60    CONTINUE
78    SGN=-SGN
80    PH=PH+DPH
      DPH=(PI-PHA)/NPH
90    PH=PHA
      CALL CISI(CA,CIN,SA,AK)
      CALL CISI(CB,CIN,SB,BK)
      R2=AK+DKD
      SR2=DSIN(R2)
      CR2=DCOS(R2)
      SDKD=DSIN(DKD)
      V11=(SR2*(CB-CA)-CR2*(SB-SA))/(2.*BAL*SDKD)
      VJ(1)=V11+VJ(1)
      IF(NSD.LE.1)RETURN
      V22=(DSIN(AK)*(CB-CA)-DCOS(AK)*(SB-SA))/(2.*BAL*SDKD)
      VJ(2)=DCMPLX(V22,.ODO)
      RETURN
      END
```

```
C*****************************************************************
C       SUBROUTINE QDD                                           *
C*****************************************************************
      SUBROUTINE QDD(CDKD,SDKD,S1,S3,T1,T3,TK,IWZ,NPH,Z22)
      IMPLICIT REAL*8 (A-H),(P-Z)
      COMPLEX*16 Z12,Z22,ZD
      DATA PI,P3/3.14159265359,9.42477796077/
    2 FORMAT(1X,8F10.2)
    5 FORMAT(1H0)
    6 FORMAT(5X,'PRINTOUT FROM QDD')
    7 FORMAT(5X,'DISK DIPOLE TO DISK DIPOLE')
      IF(IWZ.LE.0)GO TO 10
      WRITE(8,6)
      WRITE(8,7)
      WRITE(8,5)
   10 PHA=.0174533*2.
      PHB=.0174533*20.
      NPH=2*(NPH/2)
      DPH=PHA/NPH
      PH=.0D0
      NPHP=NPH+1
      Z22=(.0D0,.0D0)
      DO 80 IPH=1,3
      ZD=(.0D0,.0D0)
      SGI=-1.
      DO 70 I=1,NPHP
      WF=3.+SGI
      IF(I.EQ.1)WF=1.
      IF(I.EQ.NPHP)WF=1.
      CPH=DCOS(PH)
      IF(I.EQ.1 .AND. IPH.GT.1)GO TO 60
C  NEXT:  DISK DIPOLE TO DISK DIPOLE.
      CALL SKEWS(S1,S3,T1,T3,TK,CDKD,SDKD,CDKD,SDKD,CPH,Z12)
   60 PHD=57.29578*PH
      IF(IWZ.GT.0D0)WRITE(8,2)PHD,Z12
      SGI=-SGI
      PH=PH+DPH
   70 ZD=ZD+WF*Z12
      Z22=Z22+DPH*ZD/P3
      PH=PHA
      DPH=(PHB-PHA)/NPH
      IF(IPH.EQ.1)GO TO 80
      PH=PHB
      DPH=(PI-PHB)/NPH
   80 CONTINUE
      IF(IWZ.GT.0D0)WRITE(8,5)
      RETURN
      END
```

```
C**********************************************************************
C       SUBROUTINE QDM                                                *
C**********************************************************************
      SUBROUTINE QDM(AK, DKD, DKW, CDKD, SDKD, SDK, S1, S3, TK, IWZ, NPH, Z12)
      IMPLICIT REAL*8 (A-H), (P-Z)
      COMPLEX*16 ZD, Z12, Z21, ZDM, ZM, P12, ZDD, PDM
      DATA PI, P3/3. 14159265359, 9. 42477796077/
  2   FORMAT(1X, 8F10. 2)
  5   FORMAT(1HO)
  6   FORMAT(5X, 'PRINTOUT FROM QDM')
  7   FORMAT(5X, 'DISK DIPOLE TO DISK MONOPOLE')
  8   FORMAT(5X, 'DISK DIPOLE TO WIRE MONOPOLE')
      IF(IWZ. LE. 0)GO TO 10
      WRITE(8, 6)
      WRITE(8, 7)
      WRITE(8, 5)
  10  PHA=. 0174533*2.
      PHB=. 0174533*20.
      NPH=2*(NPH/2)
      NPHP=NPH+1
      DPH=PHA/NPH
      IDM=1
      IF(S1. GT. 10. *AK)IDM=0
      PH=. 0D0
      ZDD=(. 0D0, . 0D0)
      T2=AK+DKD
      DO 40 IPH=1, 3
      ZD=(. 0D0, . 0D0)
      SGI=-1.
      DO 30 I=1, NPHP
      WF=3. +SGI
      IF(I. EQ. 1)WF=1.
      IF(I. EQ. NPHP)WF=1.
      CPH=DCOS(PH)
      IF(I. EQ. 1 . AND.  IPH. GT. 1)GO TO 20
C  NEXT:   DISK-DIPOLE TO DISK-MONOPOLE.
      CALL ZSDM(S1, S3, AK, T2, TK, CDKD, SDKD, SDKD, CPH, P12)
  20  PHD=57. 29578*PH
      IF(IWZ. GT. 0)WRITE(8, 2)PHD, P12
      SGI=-SGI
      PH=PH+DPH
  30  ZD=ZD+WF*P12
      ZDD=ZDD+DPH*ZD/P3
      PH=PHA
      DPH=(PHB-PHA)/NPH
      IF(IPH. EQ. 1)GO TO 40
      PH=PHB
      DPH=(PI-PHB)/NPH
  40  CONTINUE
      IF(IWZ. GT. 0)WRITE(8, 5)
      W2=TK+DKW
      IF(IDM. EQ. 0)GO TO 100
      IF(IWZ. GT. 0)WRITE(8, 8)
      LPH=6
      LPH=2*(LPH/2)
      LPP=LPH+1
      PHA=. 0174533*20.
      DPH=PHA/LPH
      PH=. 0D0
      ZDM=(. 0D0, . 0D0)
      RMN=AK/100.
```

```
      DO 90 IPH=1,2
      PDM=(.ODO,.ODO)
      SGI=-1.
      DO 80 I=1,LPP
      WF=3.+SGI
      IF(I.EQ.1)WF=1.
      IF(I.EQ.LPP)WF=1.
      CPH=DCOS(PH)
      RH=AK*DSIN(PH)
      IF(I.EQ.1)RH=RMN
      AC=AK*CPH
      V1=S1-AC
      V3=S3-AC
      IF(I.EQ.1 .AND. IPH.GT.1)GO TO 70
C  NEXT:   DISK-DIPOLE TO WIRE-MONOPOLE.
      CALL ZSDM(V1,V3,TK,W2,RH,CDKD,SDKD,SDK,.ODO,P12)
   70 PHD=57.29578*PH
      IF(IWZ.GT.ODO)WRITE(8,2)PHD,P12
      SGI=-SGI
      PH=PH+DPH
   80 PDM=PDM+WF*P12
      ZDM=ZDM+DPH*PDM/P3
      DPH=(PI-PHA)/LPH
   90 PH=PHA
      Z12=ZDM-ZDD
      IF(IWZ.GT.ODO)WRITE(8,5)
      RETURN
C  NEXT:   DISK-DIPOLE TO WIRE-MONOPOLE.
  100 CALL ZSDM(S1,S3,TK,W2,AK,CDKD,SDKD,SDK,.ODO,ZDM)
      Z12=ZDM-ZDD
      RETURN
      END
```

```
C**********************************************************************
C        SUBROUTINE QMM                                               *
C**********************************************************************
      SUBROUTINE QMM(AK,DKD,DKW,CDKD,SDKD,CDK,SDK,TK,IWZ,NPH,Z11)
      IMPLICIT REAL*8 (A-H),(P-Z)
      COMPLEX*16 FDM,FMD,FMM,PDM,PMD,PMM
      COMPLEX*16 ZD,ZDD,ZDM,ZMD,ZMM,Z11,P11
      DOUBLE PRECISION DRG
      DATA PI,P3/3.14159265359,9.42477796077/
  1   FORMAT(1X,8F8.0)
  2   FORMAT(1X,8F10.2)
  5   FORMAT(1H0)
  6   FORMAT(5X,'PRINTOUT FROM QMM')
  7   FORMAT(5X,'DISK MONOPOLE TO DISK MONOPOLE')
      IF(IWZ.LE.0)GO TO 10
      WRITE(8,6)
      WRITE(8,5)
  10  AKS=AK*AK
      DKMP=TK+DKW
      CDKMP=DCOS(DKMP)
      SDKMP=DSIN(DKMP)
      ZMM=(.0D0,.0D0)
      ZMD=(.0D0,.0D0)
      ZDM=(.0D0,.0D0)
      LPH=6
      LPH=2*(LPH/2)
      LPP=LPH+1
      PHA=.0174533*20.
      DPH=PHA/LPH
      RMN=AK*DPH/10.
      PH=.0D0
      DO 44 IPH=1,2
      FDM=(.0D0,.0D0)
      FMD=(.0D0,.0D0)
      FMM=(.0D0,.0D0)
      SGI=-1.
      DO 40 I=1,LPP
      WF=3.+SGI
      IF(I.EQ.1)WF=1.
      IF(I.EQ.LPP)WF=1.
      IF(I.EQ.1 .AND. IPH.GT.1)GO TO 38
      CPH=DCOS(PH)
      SPH=DSIN(PH)
      DRG=2.*(1.-CPH)
      R=AK*DSQRT(DRG)
      IF(I.EQ.1)R=R+RMN
C  NEXT:  WIRE MONOPOLE TO WIRE MONOPOLE.
      CALL ZSMM(-TK,DKW,.0D0,DKW,R,CDKMP,SDKMP,SDK,1.,PMM)
      R=AK*SPH
      IF(I.EQ.1)R=R+RMN
      T1=AK*(1.-CPH)
      T2=T1+DKD
C  NEXT:  WIRE MONOPOLE TO DISK MONOPOLE.
      CALL ZSMM(-TK,DKW,T1,T2,R,CDKMP,SDKMP,SDKD,.0D0,PMD)
      S1=AK*(1.-CPH)
      S2=S1+DKD
C  NEXT:  DISK MONOPOLE TO WIRE MONOPOLE.
      CALL ZSMM(S1,S2,TK,TK+DKW,R,CDKD,SDKD,SDK,.0D0,PDM)
  38  FMM=FMM+WF*PMM
      FMD=FMD+WF*PMD
      FDM=FDM+WF*PDM
```

```
      PHD=57.29578*PH
      IF(IWZ.GT.O)WRITE(8,1)PHD,PMM,PMD,PDM
      SGI=-SGI
   40 PH=PH+DPH
      ZMM=ZMM+DPH*FMM/P3
      ZMD=ZMD+DPH*FMD/P3
      ZDM=ZDM+DPH*FDM/P3
      RMN=.ODO
      DPH=(PI-PHA)/LPH
   44 PH=PHA
      IF(IWZ.GT.O)WRITE(8,5)
      IF(IWZ.GT.O)WRITE(8,7)
      PHA=.0174533*2.
      PHB=.0174533*20.
      NPH=2*(NPH/2)
      NPHP=NPH+1
      DPH=PHA/NPH
      PH=.ODO
      ZDD=(.ODO,.ODO)
      S2=AK+DKD
      DO 60 IPH=1,3
      ZD=(.ODO,.ODO)
      SGI=-1.
      DO 50 I=1,NPHP
      WF=3.+SGI
      IF(I.EQ.1)WF=1.
      IF(I.EQ.NPHP)WF=1.
      CPH=DCOS(PH)
      IF(I.EQ.1 .AND. IPH.GT.1)GO TO 48
C  NEXT:  DISK MONOPOLE TO DISK MONOPOLE.
      CALL ZSMM(AK,S2,AK,S2,TK,CDKD,SDKD,SDKD,CPH,P11)
   48 PHD=57.29578*PH
      IF(IWZ.GT.O)WRITE(8,2)PHD,P11
      SGI=-SGI
      PH=PH+DPH
   50 ZD=ZD+WF*P11
      ZDD=ZDD+DPH*ZD/P3
      PH=PHA
      DPH=(PHB-PHA)/NPH
      IF(IPH.EQ.1)GO TO 60
      PH=PHB
      DPH=(PI-PHB)/NPH
   60 CONTINUE
      Z11=ZDD-ZDM-ZMD+ZMM
      IF(IWZ.GT.O)WRITE(8,5)
      RETURN
      END
```

```
C*********************************************************************
C        SUBROUTINE SKEWS                                            *
C*********************************************************************
      SUBROUTINE SKEWS(S1,S3,T1,T3,RHK,CDK,SDK,CDKT,SDKT,CPSI,Z12)
      IMPLICIT REAL*8 (A-H),(P-Z)
      COMPLEX*16 Z12,EIN,EGDZ,CQX,EJXX,EM,EP
      DIMENSION S(3),T(3)
      DOUBLE PRECISION R,SI,SIS,TJ,TJS,XXD,S,T,CI,CJ,FK,FL,DZ
      DOUBLE PRECISION RHS,DZS,DDZ,RIJ,DPSI,CPSS,SPSI,RH2
      DATA ETA,PI,TP/376.730366239,3.14159265359,6.28318530718/
      S(1)=S1
      S(2)=(S1+S3)/2.
      S(3)=S3
      T(1)=T1
      T(2)=(T1+T3)/2.
      T(3)=T3
      Z12=(.0D0,.0D0)
      DPSI=CPSI
      IF(DABS(CPSI).LT..999999)GO TO 10
      RHS=RHK*RHK
      CPSS=DPSI*DPSI
      IF(CPSS.GT.1.D0)CPSS=1.D0
      SPSI=DSQRT(1.D0-CPSS)
      RH2=SPSI*(T1+T3)/2.
      RHS=RHS+RH2*RH2
      SGN=1.
      IF(CPSI.GT..0D0)GO TO 60
      SGN=-1.
      T(1)=-T3
      T(2)=-T(2)
      T(3)=-T1
      GO TO 60
   10 DO 50 I=1,3
      SI=S(I)
      SIS=SI*SI
      CI=1.
      IF(I.EQ.2)CI=-2.*CDK
      DO 50 J=1,3
      TJ=T(J)
      TJS=TJ*TJ
      R=DSQRT(SIS+TJS-2.*SI*TJ*DPSI)
      CJ=1.
      IF(J.EQ.2)CJ=-2.*CDKT
      CQX=(.0D0,.0D0)
      DO 40 K=1,2
      FK=(-1.)**K
      DO 40 L=1,2
      FL=(-1.)**L
      XXD=FK*SI+FL*TJ
      XX=XXD
      EJXX=DCMPLX(DCOS(XX),DSIN(XX))
      XXX=R+XXD
      X=DABS(XXX)
      CALL CISI(COSI,CIN,SINI,X)
      IF(XXX.LT..0D0)SINI=-SINI
      CQX=CQX+DCMPLX(COSI,-SINI)*EJXX*FK*FL
   40 CONTINUE
      Z12=Z12+CQX*CI*CJ
   50 CONTINUE
      Z12=-ETA*Z12/(8.*PI*SDK*SDKT)
      RETURN
```

```
60    DO 80 I=1,3
      SI=S(I)
      CI=1.
      IF(I.EQ.2)CI=-2.*CDK
      CQX=(.0D0,.0D0)
      DO 70 J=1,3
      TJ=T(J)
      CJ=1.
      IF(J.EQ.2)CJ=-2.*CDKT
      DZ=TJ-SI
      R=DSQRT(RHS+DZ*DZ)
      X=R+DZ
      IF(DZ.LT..0D0)X=RHS/(R-DZ)
      CALL CISI(COSI,CIN,SINI,X)
      EP=DCMPLX(COSI,-SINI)
      X=R-DZ
      IF(DZ.GT..0D0)X=RHS/(R+DZ)
      CALL CISI(COSI,CIN,SINI,X)
      EM=DCMPLX(COSI,-SINI)
      EGDZ=DCMPLX(DCOS(DZ),DSIN(DZ))
70    CQX=CQX+CJ*(EP*EGDZ+EM/EGDZ)
80    Z12=Z12+CI*CQX
      Z12=SGN*ETA*Z12/(8.*PI*SDK*SDKT)
      RETURN
      END
```

```
C*******************************************************************
C        SUBROUTINE SKEWT                                          *
C*******************************************************************
      SUBROUTINE SKEWT(AK, S1, S3, T1, T3, CDK, SDK, CDKD, SDKD, IWZ, Z12)
      IMPLICIT REAL*8 (A-H), (P-Z)
      COMPLEX*16 P12, Z12, Q12
      DATA PI, P3/3. 14159265359, 9. 42477796077/
  2   FORMAT(1X, 8F10. 2)
  5   FORMAT(1H0)
  6   FORMAT(5X, 'PRINTOUT FROM SKEWT')
  7   FORMAT(5X, 'WIRE DIPOLE TO DISK DIPOLE')
      IF(IWZ. LE. 0)GO TO 20
      WRITE(8, 6)
      WRITE(8, 7)
      WRITE(8, 5)
 20   RMN=AK/100.
      NPH=6
      NPH=2*(NPH/2)
      NPP=NPH+1
      PHA=. 0174533*20.
      DPH=PHA/NPH
      Z12=(. 0D0, . 0D0)
      PH=. 0D0
C  WIRE DIPOLE TO DISK DIPOLE.
      DO 80 IPH=1, 2
      Q12=(. 0D0, . 0D0)
      SGI=-1.
      DO 60 I=1, NPP
      WF=3. +SGI
      IF(I. EQ. 1)WF=1.
      IF(I. EQ. NPP)WF=1.
      RH=AK*DSIN(PH)
      IF(I. EQ. 1)RH=RH+RMN
      AC=AK*DCOS(PH)
      V1=T1-AC
      V3=T3-AC
      IF(I. EQ. 1 . AND. IPH. EQ. 2)GO TO 50
      CALL SKEW (S1, S3, V1, V3, RH, CDK, SDK, CDKD, SDKD, . 0D0, P12)
 50   Q12=Q12+WF*P12
      PHD=57. 29578*PH
      IF(IWZ. GT. 0D0)WRITE(8, 2)PHD, P12
      SGI=-SGI
 60   PH=PH+DPH
      Z12=Z12+DPH*Q12/P3
      RMN=. 0D0
      DPH=(PI-PHA)/NPH
 80   PH=PHA
      IF(IWZ. GT. 0D0)WRITE(8, 5)
      RETURN
      END
```

```
C***********************************************************************
C        SUBROUTINE SPART                                               *
C***********************************************************************
      SUBROUTINE SPART(AK, DKD, DKW, MAX, IWZ, Z, Z1)
      IMPLICIT REAL*8 (A-H), (P-Z)
      COMPLEX*16 EID(20), EM(20), EP(20), Z(1), Z1(1)
      COMPLEX*16 CEM, CEP, EMD, EPD, EMD2, EPD2, Z11, Z22, G11, Q11
      DIMENSION CID(20), SID(20), CM(20), CP(20), SM(20), SP(20)
      DOUBLE PRECISION DRG
      DATA GAM, P2/. 577215664, 1. 57079632/
      DATA ETA, PI/376. 727, 3. 14159/
      IDM=20
 1    FORMAT(3X, 'MUST INCREASE DIMENSIONS IN SUBROUTINE SPART')
 2    FORMAT(3X, 'ACTUAL DIMENSION IDM = ', I5, 6X,
     2'REQUIRED DIMENSION MAX2 = ', I5)
 3    FORMAT(1X, 8F10. 2)
 5    FORMAT(1HO)
 6    FORMAT(5X, 'PRINTOUT FROM SPART')
 7    FORMAT(5X, 'FIRST:   WIRE DIPOLE TO MODE ONE')
 8    FORMAT(5X, 'THEN:   WIRE DIPOLE TO WIRE DIPOLE')
      IF(MAX. LE. O)RETURN
      MAX2=MAX+2
      DO 14 I=1, MAX
      Z1(I)=(. ODO, . ODO)
 14   Z(I)=(. ODO, . ODO)
      IF(MAX2. LE. IDM)GO TO 16
      WRITE(9, 1)
      WRITE(9, 2)IDM, MAX2
      RETURN
 16   DK=DKW
      IF(IWZ. LE. O)GO TO 18
      WRITE(8, 6)
      WRITE(8, 7)
      WRITE(8, 8)
      WRITE(8, 5)
 18   TDK=2. *DK
      SDKD=DSIN(DKD)
      S11=. ODO
      S13=TDK
      S21=DK
      S23=3. *DK
      DO 20 N=1, MAX2
      I=N-1
      DZ=I*DK
      CID(N)=DCOS(DZ)
      SID(N)=DSIN(DZ)
 20   EID(N)=DCMPLX(CID(N), SID(N))
      CDK=DCOS(DK)
      SDK=DSIN(DK)
      EPD=DCMPLX(CDK, SDK)
      EMD=DCMPLX(CDK, -SDK)
      EPD2=EPD*EPD
      EMD2=EMD*EMD
      CEM=2. *CDK+EMD
      CEP=2. *CDK+EPD
      AK2=AK*AK
      CSS=ETA/(8. *PI*SDK*SDK)
      NPH=6
      NPH=2*(NPH/2)
      NPP=NPH+1
      PHA=. 0174533*20.
```

```
      DPH=PHA/NPH
      PH=.ODO
      DO 100 JPH=1,2
      CST=DPH*ETA/(24.*PI*PI*SDK*SDK)
      C22=DPH/(3.*PI)
      SGN=-1.
      DO 80 IPH=1,NPP
      CPH=DCOS(PH)
      SPH=DSIN(PH)
      IF(IPH.GT.1)GO TO 30
      IF(JPH.GT.1)GO TO 30
      PHO=DPH/10.
      CPH=DCOS(PHO)
      SPH=DSIN(PHO)
30    T1=AK*(1.-CPH)
      T2=T1+DKD
      RH=AK*SPH
      DRG=2.*AK2*(1.-CPH)
      RK=DSQRT(DRG)
      RS=DRG
      WF=3.+SGN
      IF(IPH.EQ.1)WF=1.
      IF(IPH.EQ.NPP)WF=1.
      WST=WF*CST
      W22=WF*C22
      DO 40 N=1,MAX2
      I=N-1
      DZ=I*DK
      DZS=DZ*DZ
      R=DSQRT(RS+DZS)
      ARG=R+DZ
      IF(N.EQ.1)ARG=RK
      CALL CISI(CP(N),CIN,SP(N),ARG)
      EP(N)=DCMPLX(CP(N),-SP(N))
      IF(N.GT.1)GO TO 38
      CM(1)=CP(1)
      SM(1)=SP(1)
      EM(1)=EP(1)
      GO TO 40
38    ARG=RS/ARG
      CALL CISI(CM(N),CIN,SM(N),ARG)
      EM(N)=DCMPLX(CM(N),-SM(N))
40    CONTINUE
      R=4.*(-CM(2)+2.*CP(1)-CP(2))
     A+2.*CID(3)*(+CM(3)-2.*CM(2)+2.*CP(1)-2.*CP(2)+CP(3))
     B+2.*SID(3)*(-SM(3)+2.*SM(2)-2.*SP(2)+SP(3))
      X=4.*(SM(2)-2.*SP(1)+SP(2))
     C+2.*CID(3)*(-SM(3)+2.*SM(2)-2.*SP(1)+2.*SP(2)-SP(3))
     D+2.*SID(3)*(-CM(3)+2.*CM(2)-2.*CP(2)+CP(3))
      Q11=CSS*DCMPLX(R,X)
      Z(1)=Z(1)+WST*DCMPLX(R,X)
      Z11=-6.*CDK*EP(1)+2.*(EPD+CDK)*EP(2)-EPD*EP(3)+
     22.*(EMD+CDK)*EM(2)-EMD*EM(3)
      CALL ZSDM(S11,S13,T1,T2,RH,CDK,SDK,SDKD,.ODO,Z22)
      G11=CSS*Z11-Z22
      Z1(1)=Z1(1)+WST*Z11-W22*Z22
      PHD=57.29578*PH
      IF(IWZ.GT.O)WRITE(8,3)PHD,G11,Q11
      IF(MAX.EQ.1)GO TO 70
      R=2.*CID(2)*(-CM(3)+3.*CM(2)-4.*CP(1)+3.*CP(2)-CP(3))
     E+2.*SID(2)*(+SM(3)-2.*SM(2)+2.*SP(2)-SP(3))
```

```
    F+CID(4)*(+CM(4)-2.*CM(3)+CM(2)+CP(2)-2.*CP(3)+CP(4))
    G+SID(4)*(-SM(4)+2.*SM(3)-SM(2)+SP(2)-2.*SP(3)+SP(4))
     X=2.*CID(2)*(SM(3)-3.*SM(2)+4.*SP(1)-3.*SP(2)+SP(3))
    H +2.*SID(2)*(CM(3)-2.*CM(2)+2.*CP(2)-CP(3))
    I+CID(4)*(-SM(4)+2.*SM(3)-SM(2)-SP(2)+2.*SP(3)-SP(4))
    J+SID(4)*(-CM(4)+2.*CM(3)-CM(2)+CP(2)-2.*CP(3)+CP(4))
     Z(2)=Z(2)+WST*DCMPLX(R,X)
     Z11=2.*EP(1)+CEM*(EMD*EM(3)-EPD*EP(2))+
    2CEP*(EPD*EP(3)-EMD*EM(2))-EPD2*EP(4)-EMD2*EM(4)
     CALL ZSDM(S21,S23,T1,T2,RH,CDK,SDK,SDKD,.ODO,Z22)
     Z1(2)=Z1(2)+WST*Z11-W22*Z22
     IF(MAX.EQ.2)GO TO 70
     S1=DK
     DO 60 N=3,MAX
     M1=N-1
     M2=N-2
     N1=N+1
     N2=N+2
     Z11=EP(M1)*EID(M1)+EM(M1)/EID(M1)+CEP*(EP(N1)*EID(N)-EM(N)/EID(N))
    2-CEM*(EP(N)*EID(N)-EM(N1)/EID(N))-EP(N2)*EID(N1)-EM(N2)/EID(N1)
     S1=S1+DK
     S3=S1+TDK
     CALL ZSDM(S1,S3,T1,T2,RH,CDK,SDK,SDKD,.ODO,Z22)
     Z1(N)=Z1(N)+WST*Z11-W22*Z22
     CPA=CP(M2)-2.*CP(M1)+CP(N)
     CPB=2.*CP(N)-CP(M1)-CP(N1)
     CPC=CP(N2)-2.*CP(N1)+CP(N)
     CMA=CM(M2)-2.*CM(M1)+CM(N)
     CMB=2.*CM(N)-CM(N1)-CM(M1)
     CMC=CM(N2)-2.*CM(N1)+CM(N)
     SPA=SP(M2)-2.*SP(M1)+SP(N)
     SPB=2.*SP(N)-SP(M1)-SP(N1)
     SPC=SP(N2)-2.*SP(N1)+SP(N)
     SMA=SM(M2)-2.*SM(M1)+SM(N)
     SMB=2.*SM(N)-SM(N1)-SM(M1)
     SMC=SM(N2)-2.*SM(N1)+SM(N)
     R=CID(M2)*(CPA+CMA)+2.*CID(N)*(CPB+CMB)+2.*SID(N)*(SPB-SMB)
    K +CID(N2)*(CPC+CMC)+SID(N2)*(SPC-SMC)
     IF(N.GT.3)R=R+SID(M2)*(SPA-SMA)
     X=-CID(M2)*(SPA+SMA)-2.*CID(N)*(SPB+SMB)+2.*SID(N)*(CPB-CMB)
    L -CID(N2)*(SPC+SMC)+SID(N2)*(CPC-CMC)
     IF(N.GT.3)X=X+SID(M2)*(CPA-CMA)
60   Z(N)=Z(N)+WST*DCMPLX(R,X)
70   PH=PH+DPH
80   SGN=-SGN
     DPH=(PI-PHA)/NPH
100  PH=PHA
     IF(IWZ.GT.0)WRITE(8,5)
     RETURN
     END
```

```
C***********************************************************************
C       SUBROUTINE ZSDM                                                *
C***********************************************************************
      SUBROUTINE ZSDM(S1,S3,T1,T2,RHK,CDK,SDK,SDKT,CPSI,Z12)
      IMPLICIT REAL*8 (A-H),(P-Z)
      COMPLEX*16 E(2,2),F(2,2),ES2,ET2,EXPA,EXPB,EGZI,ES1
      COMPLEX*16 CQX,EJXX,Z12,EP1,EM1,P11,P21,EGDZ,EM,EP
      DIMENSION S(3)
      DOUBLE PRECISION TD1,TD2,SDI,ST1,ST2,R1,R2,SK,ZD,TL1,TL2,SA,SB
      DOUBLE PRECISION R,SI,SIS,TJ,TJS,XXD,S,T,CI,CJ,FK,FL
      DOUBLE PRECISION RHS,DZS,DDZ,RIJ,DPSI,CPSS,SPSI,RH2
      DATA ETA,PI,TP/376.730366239,3.14159265359,6.28318530718/
      S(1)=S1
      S2=(S1+S3)/2.
      S(2)=S2
      S(3)=S3
      Z12=(.0D0,.0D0)
      DPSI=CPSI
      CPSS=DPSI*DPSI
      IF(CPSS.GT.1.D0)CPSS=1.D0
      SPSI=DSQRT(1.D0-CPSS)
      IF(DABS(CPSI).LT..999999)GO TO 10
      RHS=RHK*RHK
      RH2=SPSI*(T1+T2)/2.
      RHS=RHS+RH2*RH2
      SGN=1.
      IF(CPSI.GT..0D0)GO TO 80
      S(1)=-S3
      S(2)=-S2
      S(3)=-S1
      SGN=-1.
      GO TO 80
   10 D=RHK
      DSQ=D*D
      CD=D/SPSI
      BD=CD*DPSI
      EB=DEXP(-BD)
      EC=DEXP(-CD)
      TD1=T1
      TD2=T2
      TS1=TD1*TD1
      TS2=TD2*TD2
      CST=-ETA/(16.*PI*SDK*SDKT)
      SA=S1
      SB=S2
      ET2=DCMPLX(DCOS(T2),DSIN(T2))
      DO 60 ISS=1,2
      IF(ISS.EQ.1)ES1=DCMPLX(DCOS(SA),DSIN(SA))
      IF(ISS.EQ.2)ES2=DCMPLX(DCOS(SB),DSIN(SB))
      DO 20 K=1,2
      DO 20 L=1,2
   20 E(K,L)=(.0D0,.0D0)
      SI=SA
      DO 50 I=1,2
      FI=(-1)**I
      SDI=SI
      SIS=SDI*SDI
      ST1=2.*SDI*TD1*DPSI
      ST2=2.*SDI*TD2*DPSI
      R1=DSQRT(DSQ+SIS+TS1-ST1)
      R2=DSQRT(DSQ+SIS+TS2-ST2)
```

```
      EK=EB
      DO 40 K=1,2
      FK=(-1)**K
      SK=FK*SDI
      EL=EC
      DO 30 L=1,2
      FL=(-1)**L
      EKL=EK*EL
      XX=FK*BD+FL*CD
      TL1=FL*TD1
      TL2=FL*TD2
      RR1=R1+SK+TL1
      RR2=R2+SK+TL2
      CALL EXPJ(DCMPLX(XX,RR1),DCMPLX(XX,RR2),EXPA)
      CALL EXPJ(DCMPLX(-XX,RR1),DCMPLX(-XX,RR2),EXPB)
      E(K,L)=E(K,L)+FI*(EXPA*EKL+EXPB/EKL)
30    EL=1./EC
40    EK=1./EB
      IF(I.EQ.ISS)GO TO 50
      ZD=SDI*DPSI
      ZC=ZD
      EGZI=DCMPLX(DCOS(ZC),DSIN(ZC))
      RR1=R1+ZD-TD1
      RR2=R2+ZD-TD2
      CALL EXPJ(DCMPLX(.ODO,RR1),DCMPLX(.ODO,RR2),EXPB)
      RR1=R1-ZD+TD1
      RR2=R2-ZD+TD2
      CALL EXPJ(DCMPLX(.ODO,RR1),DCMPLX(.ODO,RR2),EXPA)
      F(I,1)=(.ODO,2.)*SDK*EXPA/EGZI
      F(I,2)=(.ODO,2.)*SDK*EXPB*EGZI
50    SI=SB
      IF(ISS.EQ.1)
     AP21=CST*((-F(2,1)-E(2,2)*ES1+E(1,2)/ES1)*ET2
     B+( F(2,2)+E(2,1)*ES1-E(1,1)/ES1)/ET2)
      IF(ISS.EQ.2)
     CP11=CST*(( F(1,1)+E(2,2)*ES2-E(1,2)/ES2)*ET2
     D+(-F(1,2)-E(2,1)*ES2+E(1,1)/ES2)/ET2)
      SA=S2
60    SB=S3
      Z12=P11+P21
      RETURN
80    DO 100 I=1,3
      CI=1.
      IF(I.EQ.2)CI=-2.*CDK
      SI=S(I)
      TJ=T1
      DO 90 J=1,2
      DZ=TJ-SI
      R=DSQRT(RHS+DZ*DZ)
      X=R+DZ
      IF(DZ.LT..ODO)X=RHS/(R-DZ)
      CALL CISI(COSI,CIN,SINI,X)
      EP=DCMPLX(COSI,-SINI)
      X=R-DZ
      IF(DZ.GT..ODO)X=RHS/(R+DZ)
      CALL CISI(COSI,CIN,SINI,X)
      EM=DCMPLX(COSI,-SINI)
      IF(J.EQ.2)GO TO 90
      EP1=EP
      EM1=EM
90    TJ=T2
```

```
      X=T2-SI
      EGDZ=DCMPLX(DCOS(X),DSIN(X))
      Z12=Z12+CI*((EP-EP1)*EGDZ+(EM-EM1)/EGDZ)
  100 CONTINUE
      Z12= SGN*ETA*Z12/(8.*PI*SDK*SDKT)
      RETURN
      END
```

```
C**********************************************************************
C       SUBROUTINE ZSMM                                               *
C**********************************************************************
      SUBROUTINE ZSMM(S1,S2,T1,T2,D,CDS,SDS,SDT,CPSI,P11)
      IMPLICIT REAL*8 (A-H),(P-Z)
      DOUBLE PRECISION BD,CD,DD,DO,DZ,DPQ,DSQ,DPSI,CPSS
      DOUBLE PRECISION R,R1,R2,SDI,SIS,SK,SPSI,ST1,ST2
      DOUBLE PRECISION TJ,TD1,TD2,TL1,TL2,TS1,TS2,ZD
      COMPLEX*16 E(2,2),F(2,2),GAM,P11,P12,P21,P22
      COMPLEX*16 EGZI,ES1,ES2,ET1,ET2,EXPA,EXPB
      COMPLEX*16 EGDZ,EM,EP,EM1,EP1
      DATA ETA,GAM,PI/376.730366239,(.ODO,1.),3.14159265359/
      DD=D
      DPQ=DD*DD
      DPSI=CPSI
      CPSS=DPSI*DPSI
      IF(CPSS.GT.1.DO)CPSS=1.DO
      SPSI=DSQRT(1.DO-CPSS)
      SGDS=SDS
      IF(S2.LT.S1)SGDS=-SDS
      SGDT=SDT
      IF(T2.LT.T1)SGDT=-SDT
      IF(DABS(CPSI).LT..999999)GO TO 6
      DO=SPSI*(T1+T2)/2.
      DSQ=DPQ+DO*DO
      GO TO 110
 6    ES1=DCMPLX(DCOS(S1),DSIN(S1))
      ES2=DCMPLX(DCOS(S2),DSIN(S2))
      ET1=DCMPLX(DCOS(T1),DSIN(T1))
      ET2=DCMPLX(DCOS(T2),DSIN(T2))
      ID=1
      IF(D.EQ..ODO)ID=0
      TD1=T1
      TD2=T2
      CD=DD/SPSI
      C=CD
      BD=CD*DPSI
      B=BD
      EB=.ODO
      EC=.ODO
      IF(ID.EQ.0)GO TO 8
      EB=DEXP(-B)
      EC=DEXP(-C)
 8    DO 10 K=1,2
      DO 10 L=1,2
 10   E(K,L)=(.ODO,.ODO)
      TS1=TD1*TD1
      TS2=TD2*TD2
      SI=S1
      DO 100 I=1,2
      FI=(-1)**I
      SDI=SI
      SIS=SDI*SDI
      ST1=2.*SDI*TD1*DPSI
      ST2=2.*SDI*TD2*DPSI
      R1=DSQRT(DPQ+SIS+TS1-ST1)
      R2=DSQRT(DPQ+SIS+TS2-ST2)
      EK=EB
      DO 50 K=1,2
      FK=(-1)**K
      SK=FK*SDI
```

```
      EL=EC
      DO 40 L=1,2
      FL=(-1)**L
      EKL=EK*EL
      XX=FK*BD+FL*CD
      TL1=FL*TD1
      TL2=FL*TD2
      RR1=R1+SK+TL1
      RR2=R2+SK+TL2
      AXX=DABS(XX)
      IF(AXX.GT.DABS(RR1)/100.)GO TO 28
      IF(AXX.GT.DABS(RR2)/100.)GO TO 28
      IF(AXX.GT..001)GO TO 28
      IF(RR1/RR2.LT..ODO)GO TO 28
      CALL CISI(COS1,CIN,SIN1,RR1)
      CALL CISI(COS2,CIN,SIN2,RR2)
      EXPA=DCMPLX(COS2-COS1,SIN1-SIN2)
      E(K,L)=E(K,L)+FI*EXPA*(EKL+1./EKL)
      GO TO 40
28    CALL EXPJ(DCMPLX(XX,RR1),DCMPLX(XX,RR2),EXPA)
      CALL EXPJ(DCMPLX(-XX,RR1),DCMPLX(-XX,RR2),EXPB)
      E(K,L)=E(K,L)+FI*(EXPA*EKL+EXPB/EKL)
40    EL=1./EC
50    EK=1./EB
      IF(I.EQ.2)GO TO 100
      ZD=SDI*DPSI
      ZC=ZD
      EGZI=DCMPLX(DCOS(ZC),DSIN(ZC))
      RR1=R1+ZD-TD1
      RR2=R2+ZD-TD2
      CALL CISI(COS1,CIN,SIN1,RR1)
      CALL CISI(COS2,CIN,SIN2,RR2)
      EXPB=DCMPLX(COS2-COS1,SIN1-SIN2)
      RR1=R1-ZD+TD1
      RR2=R2-ZD+TD2
      CALL CISI(COS1,CIN,SIN1,RR1)
      CALL CISI(COS2,CIN,SIN2,RR2)
      EXPA=DCMPLX(COS2-COS1,SIN1-SIN2)
      F(I,1)=2.*SGDS*(.ODO,1.)*EXPA/EGZI
      F(I,2)=2.*SGDS*(.ODO,1.)*EXPB*EGZI
100   SI=S2
      CST=-ETA/(16.*PI*SGDS*SGDT)
      P11=CST*(( F(1,1)+E(2,2)*ES2-E(1,2)/ES2)*ET2
     A+(-F(1,2)-E(2,1)*ES2+E(1,1)/ES2)/ET2)
      RETURN
110   IF(CPSI.LT.0.)GO TO 120
      TA=T1
      TB=T2
      GO TO 130
120   TA=-T1
      TB=-T2
      SGDT=-SGDT
130   SI=S1
      CI=-CDS
      P11=(.ODO,.ODO)
      DO 150 I=1,2
      TJ=TA
      DO 140 J=1,2
      DZ=TJ-SI
      R=DSQRT(DSQ+DZ*DZ)
      X=R+DZ
```

```
      IF(DZ.LT..ODO)X=DSQ/(R-DZ)
      CALL CISI(COSI,CIN,SINI,X)
      EP=DCMPLX(COSI,-SINI)
      X=R-DZ
      IF(DZ.GT..ODO)X=DSQ/(R+DZ)
      CALL CISI(COSI,CIN,SINI,X)
      EM=DCMPLX(COSI,-SINI)
      IF(J.EQ.2)GO TO 140
      EP1=EP
      EM1=EM
  140 TJ=TB
      X=TB-SI
      EGDZ=DCMPLX(DCOS(X),DSIN(X))
      P11=P11+CI*((EP-EP1)*EGDZ+(EM-EM1)/EGDZ)
      CI=1.
  150 SI=S2
      P11=ETA*P11/(8.*PI*SGDS*SGDT)
      RETURN
      END
```

```
C***********************************************************************
C       SUBROUTINE SKEW                                                *
C***********************************************************************
      SUBROUTINE SKEW (S1,S3,T1,T3,RHK,CDK,SDK,CDKT,SDKT,CPSI,Z12)
      IMPLICIT REAL*8 (A-H),(P-Z)
      COMPLEX*16 Z12,EIN,EGDZ,CQX,EJXX,EM,EP
      COMPLEX*16 E(2,2),F(2,2),ES1,ES2,ET1,ET2,EXPA,EXPB,EGZI
      COMPLEX*16 P11,P12,P21,P22
      DIMENSION S(3),T(3)
      DOUBLE PRECISION R,SI,SIS,TJ,TJS,XXD,S,T,CI,CJ,FK,FL,DZ
      DOUBLE PRECISION RHS,DZS,DDZ,RIJ,DPSI,CPSS,SPSI,RH2
      DOUBLE PRECISION TD1,TD2,SDI,ST1,ST2,R1,R2,SK,ZD,TL1,TL2,SA,SB
      DATA ETA,PI,TP/376.730366239,3.14159265359,6.28318530718/
      S(1)=S1
      S2=(S1+S3)/2.
      S(2)=S2
      S(3)=S3
      T(1)=T1
      T2=(T1+T3)/2.
      T(2)=T2
      T(3)=T3
      Z12=(.0D0,.0D0)
      DPSI=CPSI
      CPSS=DPSI*DPSI
      IF(CPSS.GT.1.D0)CPSS=1.D0
      SPSI=DSQRT(1.D0-CPSS)
      IF(DABS(CPSI).LT..999999)GO TO 10
      RHS=RHK*RHK
      RH2=SPSI*(T1+T3)/2.
      RHS=RHS+RH2*RH2
      SGN=1.
      IF(CPSI.GT..0D0)GO TO 80
      SGN=-1.
      T(1)=-T3
      T(2)=-T2
      T(3)=-T1
      GO TO 80
 10   D=RHK
      DSQ=D*D
      CD=D/SPSI
      BD=CD*DPSI
      EB=DEXP(-BD)
      EC=DEXP(-CD)
      CST=-ETA/(16.*PI*SDK*SDKT)
      TA=T1
      TB=T2
      DO 70 ITT=1,2
      IF(ITT.EQ.1)ET1=DCMPLX(DCOS(TA),DSIN(TA))
      IF(ITT.EQ.2)ET2=DCMPLX(DCOS(TB),DSIN(TB))
      TD1=TA
      TD2=TB
      TS1=TD1*TD1
      TS2=TD2*TD2
      SA=S1
      SB=S2
      DO 60 ISS=1,2
      IF(ISS.EQ.1)ES1=DCMPLX(DCOS(SA),DSIN(SA))
      IF(ISS.EQ.2)ES2=DCMPLX(DCOS(SB),DSIN(SB))
      DO 20 K=1,2
      DO 20 L=1,2
 20   E(K,L)=(.0D0,.0D0)
```

```
      SI=SA
      DO 50 I=1,2
      FI=(-1)**I
      SDI=SI
      SIS=SDI*SDI
      ST1=2.*SDI*TD1*DPSI
      ST2=2.*SDI*TD2*DPSI
      R1=DSQRT(DSQ+SIS+TS1-ST1)
      R2=DSQRT(DSQ+SIS+TS2-ST2)
      EK=EB
      DO 40 K=1,2
      FK=(-1)**K
      SK=FK*SDI
      EL=EC
      DO 30 L=1,2
      FL=(-1)**L
      EKL=EK*EL
      XX=FK*BD+FL*CD
      TL1=FL*TD1
      TL2=FL*TD2
      RR1=R1+SK+TL1
      RR2=R2+SK+TL2
      CALL EXPJ(DCMPLX(XX,RR1),DCMPLX(XX,RR2),EXPA)
      CALL EXPJ(DCMPLX(-XX,RR1),DCMPLX(-XX,RR2),EXPB)
      E(K,L)=E(K,L)+FI*(EXPA*EKL+EXPB/EKL)
30    EL=1./EC
40    EK=1./EB
      IF(I.EQ.ISS)GO TO 50
      ZD=SDI*DPSI
      ZC=ZD
      EGZI=DCMPLX(DCOS(ZC),DSIN(ZC))
      RR1=R1+ZD-TD1
      RR2=R2+ZD-TD2
      CALL EXPJ(DCMPLX(.0D0,RR1),DCMPLX(.0D0,RR2),EXPB)
      RR1=R1-ZD+TD1
      RR2=R2-ZD+TD2
      CALL EXPJ(DCMPLX(.0D0,RR1),DCMPLX(.0D0,RR2),EXPA)
      F(I,1)=(.0D0,2.)*SDK*EXPA/EGZI
      F(I,2)=(.0D0,2.)*SDK*EXPB*EGZI
50    SI=SB
      IF(ITT.EQ.2)GO TO 54
      IF(ISS.EQ.1)P22=CST*((F(2,1)+E(2,2)*ES1-E(1,2)/ES1)*ET1
     A+(-F(2,2)-E(2,1)*ES1+E(1,1)/ES1)/ET1)
      IF(ISS.EQ.2)P12=CST*((-F(1,1)-E(2,2)*ES2+E(1,2)/ES2)*ET1
     B+( F(1,2)+E(2,1)*ES2-E(1,1)/ES2)/ET1)
      GO TO 58
54    IF(ISS.EQ.1)P21=CST*((-F(2,1)-E(2,2)*ES1+E(1,2)/ES1)*ET2
     C+( F(2,2)+E(2,1)*ES1-E(1,1)/ES1)/ET2)
      IF(ISS.EQ.2)P11=CST*((F(1,1)+E(2,2)*ES2-E(1,2)/ES2)*ET2
     D+(-F(1,2)-E(2,1)*ES2+E(1,1)/ES2)/ET2)
58    SA=S2
60    SB=S3
      TA=T2
70    TB=T3
      Z12=P11+P12+P21+P22
      RETURN
80    DO 100 I=1,3
      SI=S(I)
      CI=1.
      IF(I.EQ.2)CI=-2.*CDK
      CQX=(.0D0,.0D0)
```

```
      DO 90 J=1,3
      TJ=T(J)
      CJ=1.
      IF(J.EQ.2)CJ=-2.*CDKT
      DZ=TJ-SI
      R=DSQRT(RHS+DZ*DZ)
      X=R+DZ
      IF(DZ.LT..0D0)X=RHS/(R-DZ)
      CALL CISI(COSI,CIN,SINI,X)
      EP=DCMPLX(COSI,-SINI)
      X=R-DZ
      IF(DZ.GT..0D0)X=RHS/(R+DZ)
      CALL CISI(COSI,CIN,SINI,X)
      EM=DCMPLX(COSI,-SINI)
      EGDZ=DCMPLX(DCOS(DZ),DSIN(DZ))
90    CQX=CQX+CJ*(EP*EGDZ+EM/EGDZ)
100   Z12=Z12+CI*CQX
      Z12= SGN*ETA*Z12/(8.*PI*SDK*SDKT)
      RETURN
      END
```

```
C
C       PROGRAM "RICHMD2"
C       ***************************************************************
C       *       THIS COMPUTER PROGRAM, IN FORTRAN LANGUAGE, UTILIZES  *
C       *  THE CURRENT DISTRIBUTIONS DETERMINED FROM THE "RICHMD1"    *
C       *  PROGRAM TO OBTAIN THE RADIATION PATTERN OF A MONOPOLE      *
C       *  ELEMENT OF LENGTH h AND RADIUS b AT THE CENTER OF A        *
C       *  CIRCULAR GROUNDPLANE OF RADIUS a.  AN INPUT FILE,          *
C       *  INPUT1.DAT, IS AUTOMATICALLY GENERATED FROM THE            *
C       *  "RICHMD1" PROGRAM.                                         *
C       *                                                             *
C       *       THIS PROGRAM REQUIRES THE USE OF THE MMBSJO           *
C       *  SUBROUTINE FUNCTION OF THE IMSL SOFTWARE LIBRARY.  MMBSJ1  *
C       *  IS THE MODIFIED BESSEL FUNCTION OF THE FIRST KIND AND      *
C       *  ORDER ZERO.                                                *
C       ***************************************************************
        IMPLICIT REAL*8 (A-H),(P-Z)
        INTEGER*4 X,IER
        REAL*8 THETAD(89),THETA(89),PLOT(89),L
        REAL*8 BRKT(89),MBRKT(89),DIREC(89),DIRECD(89),DIRECR(89)
        REAL MMBSJO
        COMPLEX*16 C(50),E(50),ACUR1,ACUR2
        COMPLEX*16 ACUR(300),UX,UY,U,USUM
        COMPLEX*16 SX,SY,S,SSUM
        COMPLEX*16 GP,GSUM,TSUM
        OPEN(UNIT=5,STATUS='OLD',FILE='INPUT1.DAT',READONLY)
        OPEN(UNIT=6,STATUS='NEW',FILE='TEMP')
C***********************************************************************
C       READ AND WRITE STATEMENTS                                      *
C***********************************************************************
        READ(5,8)AL,H,EPSLN,NSD,NSW,N,X
    8   FORMAT(E12.4,2X,F8.6,2X,F7.4,2X,4(I2,2X))
        FLAG=1.0
        GO TO 70
   71   WRITE(6,74)
   74   FORMAT(/,1X,'INPUT GROUNDPLANE CURRENTS,I')
        WRITE(6,75)
   75   FORMAT(1X,' RE(I)',7X,'IM(I)')
        DO 1 I=1,N-1
        READ(5,2) C(I)
    2   FORMAT(2E12.4)
        IF (I.EQ.1)THEN
        E(I)=-C(I)
        ELSE
        E(I)=C(I)
    1   CONTINUE
        DO 77 I=1,NSD
        WRITE(6,76)E(I),I
   76   FORMAT(2E12.4,2X,'AT RHO ',I2)
   77   CONTINUE
        WRITE (6,78)
   78   FORMAT(/,1X,'INPUT MONOPOLE CURRENTS, I')
        WRITE (6,75)
        DO 555 I=N,NSD+1,-1
        C(I)=C(I-1)
  555   CONTINUE
        C(NSD+1)=C(1)
        DO 41 I=NSD+1,N
        MM=I-NSD
        WRITE(6,42)C(I),MM
   42   FORMAT(2E12.4,2X,'AT Z ',I2)
```

```
   41    CONTINUE
         PI=3.1415926536D0
         D=H/NSW
         DD=(EPSLN/(2*PI)-AL)/NSD
         E(NSD+1)=DCMPLX(0.0,0.0)
C***********************************************************************
C        COMPUTE THE AVERAGE CURRENT IN EACH ANNULAR ZONE              *
C***********************************************************************
         DELTA=DD/X
         K=1
         J=1
         P1=AL
         P2=AL+DELTA
   10    ACUR1=E(K)*(DCOS(2*PI*(K*DD+AL-P2))-DCOS(2*PI*(K*DD+AL-P1)))
         ACUR2=E(K+1)*(DCOS(2*PI*(P2-(K-1)*DD-AL))-DCOS(2*PI*(P1-(K-1)*DD-AL)))
         ACUR(J)=(ACUR1-ACUR2)/(2*PI*DELTA*DSIN(2*PI*DD))
         J=J+1
         P1=P1+DELTA
         P2=P2+DELTA
         G=(J-1)/(K*X)
         IF(G.EQ.1) K=K+1
         IF(K.LE.NSD) GO TO 10
C***********************************************************************
C        CONVERT DEGREE INTO RADIAN                                    *
C***********************************************************************
         DO 12 II=1,89
         K=II*2
         THETAD(II)=DFLOAT(K)
         THETA(II)=PI*THETAD(II)/180.0D0
   12    CONTINUE
C***********************************************************************
C        COMPUTE WITH RESPECT TO EVERY TWO DEGREE                      *
C***********************************************************************
         DO 30 INDEX=1,89
C***********************************************************************
C        FIELD DUE TO THE MONOPOLE                                     *
C***********************************************************************
         USUM=DCMPLX(0.0,0.0)
         NN=1
   40    UXR=DSIN(NN*2*PI*D*DCOS(THETA(INDEX)))
         UXIM=DCOS(NN*2*PI*D*DCOS(THETA(INDEX)))
         UX=DCMPLX(-UXR,UXIM)
         UYR1=1-DCOS(2*PI*D)*DCOS(2*PI*D*DCOS(THETA(INDEX)))
         UYR2=DCOS(THETA(INDEX))*DSIN(2*PI*D)*DSIN(2*PI*D*DCOS(THETA(INDEX)))
         UYR=UYR1-UYR2
         UYIM1=DCOS(2*PI*D)*DSIN(2*PI*D*DCOS(THETA(INDEX)))
         UYIM2=DCOS(THETA(INDEX))*DSIN(2*PI*D)*DCOS(2*PI*D*DCOS(THETA(INDEX)))
         UYIM=UYIM1-UYIM2
         UY=DCMPLX(UYR,UYIM)
         U=UX*UY*C(NN+NSD)
         USUM=USUM+U
         NN=NN+1
         IF(NN.LE.NSW) GO TO 40
         USUM=USUM/DSIN(2*PI*D)
C***********************************************************************
         SSUM=DCMPLX(0.0,0.0)
         MM=2
   50    SXR=DSIN(2*PI*D*(MM-2)*DCOS(THETA(INDEX)))
         SXIM=DCOS(2*PI*D*(MM-2)*DCOS(THETA(INDEX)))
         SX=DCMPLX(SXR,-SXIM)
         SYR1=DCOS(2*PI*D)*DCOS(2*PI*D*DCOS(THETA(INDEX)))
```

```
      SYR2=DCOS(THETA(INDEX))*DSIN(2*PI*D)*DSIN(2*PI*D*DCOS(THETA(INDEX)))-1
      SYR=SYR1+SYR2
      SYIM1=DCOS(2*PI*D)*DSIN(2*PI*D*DCOS(THETA(INDEX)))
      SYIM2=DCOS(THETA(INDEX))*DSIN(2*PI*D)*DCOS(2*PI*D*DCOS(THETA(INDEX)))
      SYIM=SYIM1-SYIM2
      SY=DCMPLX(SYR,SYIM)
      S=C(NSD+MM)*SY*SX
      SSUM=SSUM+S
      MM=MM+1
      IF(MM.LE.NSW) GO TO 50
      SSUM=SSUM/DSIN(2*PI*D)
C***********************************************************************
C     FIELD DUE TO THE GROUNDPLANE                                     *
C***********************************************************************
      GSUM=DCMPLX(0.0,0.0)
      M=1
      P1=AL
      P2=AL+DELTA
   60 X1=2*PI*P1*DSIN(THETA(INDEX))
      B1=MMBSJ0(X1,IER)
      X2=2*PI*P2*DSIN(THETA(INDEX))
      B2=MMBSJ0(X2,IER)
      GP=DCOS(THETA(INDEX))*ACUR(M)*(B1-B2)
      GSUM=GSUM+GP
      P1=P1+DELTA
      P2=P2+DELTA
      M=M+1
      IF(M.LE.(NSD*X)) GO TO 60
C***********************************************************************
C     COMPUTE THE RADIATION RESISTANCE AND DIRECTIVE GAIN              *
C***********************************************************************
      TSUM=GSUM+USUM+SSUM
      RP=DREAL(TSUM)
      P=DIMAG(TSUM)
      BRKT(INDEX)=RP*RP+P*P
      MBRKT(INDEX)=BRKT(INDEX)/DSIN(THETA(INDEX))
   30 CONTINUE
      DENOM=0.0D0
      DO 3 K=1,89
      DENOM=DENOM+MBRKT(K)
    3 CONTINUE
      DENOM=DENOM*THETA(1)
      DIRMAX=0.0D0
      DO 4 K=1,89
      DIREC(K)=2.0D0*BRKT(K)/((DSIN(THETA(K)))**2.0D0*DENOM)
      DIRECD(K)=10.0D0*DLOG10(DIREC(K))
      IF(DIRMAX.LT.DIRECD(K))DIRMAX=DIRECD(K)
    4 CONTINUE
      DO 15 J=1,89
      DIRECR(J)=DIRECD(J)-DIRMAX
   15 CONTINUE
      RADRES=15.0*DENOM/(CDABS(C(NSD+1)))**2
C***********************************************************************
C     WRITE STATMENTS                                                  *
C***********************************************************************
   70 WRITE(6,80) EPSLN
   80 FORMAT(1H1,'EPSLN= ',F7.4,'(NORMALIZED GROUNDPLANE RADIUS)')
      WRITE(6,81) H
   81 FORMAT(1X,'MONOPOLE LENGTH= ',F8.6,'(WAVELENGTH)')
      WRITE(6,82) AL
   82 FORMAT(1X,'MONOPOLE RADIUS= ',E12.4,'(WAVELENGTH)')
```

```
      WRITE(6,83) NSW
83    FORMAT(1X,'NUMBER OF MONOPOLE SEGMENTS= ',I2)
      WRITE(6,84) NSD
84    FORMAT(1X,'NUMBER OF DISK SEGMENTS= ',I2)
      FLAG=FLAG+1.0
      IF (FLAG.LE.2)GO TO 71
      WRITE(6,73)X
73    FORMAT(1X,'NUMBER OF ANNULAR ZONES PER DISK SEGEMENT= ',I2)
      WRITE(6,96) RADRES
96    FORMAT(1X,'RADIATION RESISTANCE DETERMINED BY PATTERN= ',
   1  E12.6,' OHMS')
      WRITE(6,47)
47    FORMAT(//,1X,'ELEVATION',53X,'ELEVATION')
      WRITE(6,85)
85    FORMAT(1X,2('ANGLE',6X,'DIRECTIVITY',5X,'DIRECTIVITY',5X
   1   ,'RELATIVE POWER',5X))
      WRITE(6,86)
86    FORMAT(1X,2('(DEG)',6X,'(NUMERIC)',7X,'(dBi)',11X
   1   ,'(dB)',15X))
      WRITE(6,87)
87    FORMAT(//,2X,'   0. ',9X,'0.00000',5X,'-INFINITY',6X,'-INFINITY')
      DO 16 L=1,44
      WRITE(6,88) THETAD(L),DIREC(L),DIRECD(L),DIRECR(L)
   1   ,THETAD(L+45),DIREC(L+45),DIRECD(L+45),DIRECR(L+45)
88    FORMAT(1X,2(1X,F5.1,6X,F9.5,5X,F9.5,6X,F9.5,11X))
16    CONTINUE
      WRITE(6,89) THETAD(45),DIREC(45),DIRECD(45),DIRECR(45)
89    FORMAT(1X,1X,F5.1,6X,F9.5,5X,F9.5,6X,F9.5,12X,'180.'
   1  ,9X,'0.00000',5X,'-INFINITY',6X,'-INFINITY')
23    CONTINUE
      CLOSE(UNIT=5,STATUS='KEEP')
      CLOSE(UNIT=6,STATUS='KEEP')
      END
```

B.3 MONOPL

```
C***********************************************************************
C                                                                      *
C                            PROGRAM  "MONOPL"                         *
C                                                                      *
C***********************************************************************
C                                                                      *
C    THIS PROGRAM IS WRITTEN IN FORTRAN FOR A DEC PDP 11/70.           *
C    IT CALCULATES, BY THE METHOD OF OBLATE SPHEROIDAL WAVE FUNCTIONS, *
C    THE DIRECTIVITY (D,DDBI) AND THE RADIATION RESISTANCE             *
C    (RADRES) IN OHMS FOR A QUARTERWAVE MONOPOLE ANTENNA MOUNTED       *
C    AT THE CENTER OF CIRCULAR GROUNDPLANE OF RADIUS a IN FREE         *
C    SPACE, WHERE a IS OF THE ORDER OF A WAVELENGTH.                   *
C                                                                      *
C***********************************************************************
C                                                                      *
C    THE INPUT VARIABLE IS:  ε  (EPSLN) GIVEN BY                       *
C      ε = 2πa/λ (NORMALIZED GROUNDPLANE RADIUS)                       *
C    WHERE λ IS THE EXCITATION WAVELENGTH. THIS IS ENTERED IN THE      *
C    MAIN PROGRAM ON LINE 12.                                          *
C                                                                      *
C***********************************************************************
C                                                                      *
C    THE OUTPUT VARIABLES ARE:                                         *
C                                                                      *
C      RADRES -- THE ANTENNA'S RADIATION RESISTANCE (OHMS)             *
C      D      -- THE ANTENNA'S DIRECTIVITY (NUMERIC)                   *
C      DDBI   -- THE ANTENNA'S DIRECTIVITY (DB)                        *
C      THETA  -- ANGLE FROM THE ZENITH (RADIANS)                       *
C      THETAD -- ANGLE FROM THE ZENITH (DEGREES)                       *
C                                                                      *
C***********************************************************************
C                                                                      *
C    REFERENCES:                                                       *
C                                                                      *
C    1)    A. LEITNER AND R.D. SPENCE,"THE OBLATE SPHEROIDAL WAVE      *
C              FUNCTIONS", JOURNAL OF THE FRANKLIN INSTITUTE,          *
C              VOL. 249, NUMBER 4, APRIL 1950, PAGES 299-321.          *
C                                                                      *
C    2)    A. LEITNER AND R.D. SPENCE, "EFFECT OF A CIRCULAR           *
C              GROUNDPLANE ON ANTENNA RADIATION", JOURNAL OF APPLIED   *
C              PHYSICS, VOL. 21, OCTOBER 1950, PAGES 1001-1006.        *
C                                                                      *
C    3)    M.M. WEINER, S.P. CRUZE, C.C. LI, W.J. WILSON,"MONOPOLE     *
C              ELEMENT AT THE CENTER OF A CIRCULAR GROUNDPLANE OF      *
C              ARBITRARY RADIUS", MITRE TECHNICAL REPORT MTR-9622,     *
C              SECTION 3.6, MAY 1985.                                  *
C                                                                      *
C***********************************************************************
C                                                                      *
C    THIS PROGRAM CONSISTS OF THE MAIN ROUTINE AND THE FOLLOWING       *
C    SUBROUTINES: GAMMA, AALPHA, CEE, ANGLAR, FACTOR, ODDEVN,          *
C     NORMAL, QUE, FACTR2, BAETA, QU, RAD30, DBLYU2, BEE,              *
C     DIRECT, AEY, BBEE, GGEE, RAD1, RAD2, RADRES, IMPROV,             *
C     ODD, EVEN.                                                       *
C                                                                      *
C***********************************************************************
```

```
C
C
C                       **  MAIN ROUTINE  **
C
C
      COMPLEX*8 Q1(20,1),B(20),V30(20,1)
      REAL*8 U(20,1),NUE,C(40,20,1),V1(20,1),PSI,BETA(20,1)
      REAL*8 GAM(20,1),F(8,20,1),ALPHA(20,1),EPSLN,N(20,1)
      REAL*8 Q2(20,1),V2(20,1),W(20,1),D,A(40,20,1)
      REAL*8 BB(40,20,1),B0(20,1),G(20,1),R,V22(20,1),GAMIMP(20,1)
      REAL*8 INCR,THETA,DDBI,THETAD
      INDEX=10
      JNDEX=24
      KNDEX=12
      LNDEX=18
      MNDEX=18
C
C
C  INPUT:
C
      EPSLN=3.0D0
C
C
      WRITE(4,4) EPSLN
      WRITE(4,5)
    5 FORMAT(1X,'ANGLE FROM ZENITH(DEGREES)',3X,'ANGLE FROM ZENITH
     *(RADIANS)',3X,'DIRECTIVITY(DBI)',3X,'DIRECTIVITY(NUMERIC)'/)
      INCR=1.0D0
    2 CONTINUE
      THETA=INCR*3.1415926536D0/90.0D0
      THETAD=THETA*180.0D0/3.1415926536D0
      NUE=DCOS(THETA)
C
C  CALL SUBROUTINES
C
      CALL GAMMA(F,GAM,EPSLN,INDEX)
      CALL IMPROV(GAM,GAMIMP,EPSLN,INDEX)
      CALL AALPHA(ALPHA,GAMIMP,INDEX)
      CALL CEE(C,GAMIMP,EPSLN,INDEX,JNDEX)
      CALL ANGLAR(C,U,NUE,INDEX,KNDEX)
      CALL NORMAL(C,N,INDEX,KNDEX)
      CALL QUE(Q1,C,EPSLN,INDEX,KNDEX)
      CALL BAETA(C,BETA,INDEX,KNDEX)
      CALL QU(Q2,BETA,Q1,EPSLN,INDEX)
      CALL AEY(A,ALPHA,EPSLN,INDEX,LNDEX)
      CALL RAD1(EPSLN,A,V1,INDEX,LNDEX)
      CALL BBEE(BB,EPSLN,ALPHA,A,BETA,B0,INDEX,LNDEX,MNDEX)
      CALL GGEE(EPSLN,BB,G,B0,INDEX,MNDEX)
      CALL RAD2(V2,EPSLN,G,V1,Q2,INDEX)
      CALL RAD30(BETA,Q2,V30,INDEX)
      CALL DBLYU2(W,EPSLN,V1,V2,Q2,BETA,INDEX)
      CALL BEE(B,W,ALPHA,N,Q1,V30,INDEX)
      CALL DIRECT(B,U,EPSLN,D,N,INDEX,DDBI)
      CALL RADRES(B,N,EPSLN,R,INDEX)
C
C  OUTPUT:
C
      WRITE(4,1) THETAD,THETA,DDBI,D
C
      IF(INCR.EQ.89.0D0) GOTO 3
      INCR=INCR+1.0D0
      GOTO 2
    6 FORMAT(///1X,'RADIATION RESISTANCE = ',F10.5,1X,'OHMS')
    4 FORMAT(1X,'EPSILON = ',F5.2///)
    1 FORMAT(10X,F5.1,24X,F7.5,16X,F9.5,11X,F9.5)
    3 CONTINUE
      WRITE(4,6) R
      STOP
      END
```

```
C
C                  ** SUBROUTINE GAMMA **
C
C THIS SUBROUTINE CALCULATES THE RAW EIGENVALUES FROM EQUATION 21,
C REFERENCE 1. THE F'S CORRESPOND TO THE F'S FROM EQUATIONS 22,23,24,AND
C 25. THE H CORRESPONDS TO THE H FROM EQUATION 26.
C THIS SUBROUTINE IS CALLED FROM MAIN AND DOES NOT CALL ANY OTHER
C SUBROUTINES.
C INPUTS: EPSLN,INDEX
C OUTPUT: GAM
C
      SUBROUTINE GAMMA(F,GAM,EPSLN,INDEX)
      REAL*8 F(8,20,1),GAM(20,1),H,A,B,EPSLN,AJ
      REAL*8 Z1,Z2,Z3,Z4,Z5,Z6,Z7,Z8
      H(A)=(A*A-1.0D0)/(4.0D0*A*A-1.0D0)
      DO 4 N=1,INDEX
      GAM(N,1)=0.0D0
    4 CONTINUE
      B=1.0D0
      DO 1 L=1,INDEX
      F(2,L,1)=2.0D0*(B*(B+1.0D0))/((2.0D0*B-1.0D0)*(2.0D0*B+3.0D0))
      F(4,L,1)=H(B-1.0D0)*H(B)/(2.0D0*(2.0D0*B-1.0D0))-
     *H(B+1.0D0)*H(B+2.0D0)/(2.0D0*(2.0D0*B+3.0D0))
      F(6,L,1)=(3.0D0/((2.0D0*B-1.0D0)*(2.0D0*B+3.0D0)))*(H(B-1.0D0)
     **H(B)/(((2.0D0*B-1.0D0)**2.0D0)*(2.0D0*B-5.0D0))-H(B+1.0D0)
     **H(B+2.0D0)/(((2.0D0*B+3.0D0)**2.0D0)*(2.0D0*B+7.0D0)))
      Z1=H(B-1.0D0)*H(B)/(2.0D0*(2.0D0*B-1.0D0))
      Z2=36.0D0/((2.0D0*B-5.0D0)**2.0D0*(2.0D0*B-1.0D0)**4.0D0
     **(2.0D0*B+3.0D0)**2.0D0)
      Z3=F(4,L,1)/(2.0D0*(2.0D0*B-1.0D0))
      Z4=H(B-3.0D0)*H(B-2.0D0)/(8.0D0*(2.0D0*B-1.0D0)*
     *(2.0D0*B-3.0D0))
      Z5=H(B+1.0D0)*H(B+2.0D0)/(2.0D0*(2.0D0*B+3.0D0))
      Z6=36.0D0/((2.0D0*B-1.0D0)**2.0D0*(2.0D0*B+3.0D0)**4.0D0
     **(2.0D0*B+7.0D0)**2.0D0)
      Z7=F(4,L,1)/(2.0D0*(2.0D0*B+3.0D0))
      Z8=H(B+3.0D0)*H(B+4.0D0)/(8.0D0*(2.0D0*B+3.0D0)*
     *(2.0D0*B+5.0D0))
      F(8,L,1)=Z1*(Z2-Z3+Z4)-Z5*(Z6+Z7+Z8)
      B=B+1.0D0
    1 CONTINUE
      DO 2 L=1,INDEX
      AJ=2.0D0
      DO 3 J=2,8,2
      GAM(L,1)=GAM(L,1)+F(J,L,1)*EPSLN**AJ
      AJ=AJ+2.0D0
    3 CONTINUE
    2 CONTINUE
      RETURN
      END
```

```
C
C               ** SUBROUTINE AALPHA **
C
C THIS SUBROUTINE CALCULATES ALPHA FROM EQUATION 15, REFERENCE 1.
C GAM IS THE REFINED EIGENVALUE WHICH IS INPUT TO THE SUBROUTINE.
C THIS SUBROUTINE IS CALLED FROM MAIN.
C INPUTS: GAM,INDEX
C OUTPUT: ALPHA
C
      SUBROUTINE AALPHA(ALPHA,GAM,INDEX)
      REAL*8 ALPHA(20,1),GAM(20,1),AL
      AL=1.0D0
      DO 1 L=1,INDEX
      ALPHA(L,1)=AL*(AL+1.0D0)+GAM(L,1)
      AL=AL+1.0D0
    1 CONTINUE
      RETURN
      END
C
C                ** SUBROUTINE CEE **
C
C THIS SUBROUTINE CALCULATES THE C'S IN EQUATIONS 16 AND 17 IN
C REFERENCE 1.
C THIS SUBROUTINE IS CALLED FROM MAIN AND DOES NOT CALL ANY OTHER
C SUBROUTINES.
C INPUTS: GAM,EPSLN,INDEX,JNDEX
C OUTPUT: C
C
      SUBROUTINE CEE(C,GAM,EPSLN,INDEX,JNDEX)
      REAL*8 C(40,20,1),GAM(20,1),AL,AJ,EPSLN
      AL=1.0D0
      DO 1 L=1,(INDEX-1),2
      C(2,L,1)=((1.0D0-AL)*(2.0D0+AL)-GAM(L,1))/8.0D0
      C(4,L,1)=(((3.0D0-AL)*(4.0D0+AL)-GAM(L,1))*C(2,L,1)+
     *EPSLN**2.0D0)/24.0D0
      AJ=6.0D0
      DO 2 J=6,JNDEX,2
      C(J,L,1)=(((AJ-1.0D0-AL)*(AJ+AL)-GAM(L,1))*C(J-2,L,1)+
     *C(J-4,L,1)*EPSLN**2.0D0)/(AJ*(AJ+2.0D0))
      AJ=AJ+2.0D0
    2 CONTINUE
      AL=AL+2.0D0
    1 CONTINUE
      AL=2.0D0
      DO 3 L=2,INDEX,2
      C(2,L,1)=((2.0D0-AL)*(3.0D0+AL)-GAM(L,1))/8.0D0
      C(4,L,1)=(((4.0D0-AL)*(5.0D0+AL)-GAM(L,1))*C(2,L,1)+
     *EPSLN**2.0D0)/24.0D0
      AJ=6.0D0
      DO 4 J=6,JNDEX,2
      C(J,L,1)=(((AJ-AL)*(AJ+1.0D0+AL)-GAM(L,1))*C(J-2,L,1)
     *+C(J-4,L,1)*EPSLN**2.0D0)/(AJ*(AJ+2.0D0))
      AJ=AJ+2.0D0
    4 CONTINUE
      AL=AL+2.0D0
    3 CONTINUE
      RETURN
      END
```

```
C
C                         ** SUBROUTINE ANGLAR **
C
C THIS SUBROUTINE CALCULATES THE ANGULAR SPHEROIDAL WAVE FUNCTIONS
C AS PER EQUATION 13 AND 14, REFERENCE 1. THE ANGULAR SPHEROIDAL
C WAVE FUNCTIONS ARE REPRESENTED BY U AND THE COEFFICIENTS BY C.
C THIS SUBROUTINE IS CALLED FROM MAIN AND DOES NOT CALL ANY SUBROUTINES.
C INPUTS: C,NUE,INDEX,KNDEX
C OUTPUT: U
C
      SUBROUTINE ANGLAR(C,U,NUE,INDEX,KNDEX)
      REAL*8 C(40,20,1),U(20,1),NUE,AJ
      DO 3 N=1,INDEX
      U(N,1)=0.0D0
    3 CONTINUE
      DO 1 L=1,(INDEX-1),2
      AJ=1.0D0
      DO 2 J=1,KNDEX
      U(L,1)=U(L,1)+C(2*J,L,1)*(1.0D0-NUE**2.0D0)**AJ
      AJ=AJ+1.0D0
    2 CONTINUE
      U(L,1)=DSQRT(1.0D0-NUE**2.0D0)*(1.0D0+U(L,1))
    1 CONTINUE
      DO 4 L=2,INDEX,2
      AJ=1.0D0
      DO 5 J=1,KNDEX
      U(L,1)=U(L,1)+C(2*J,L,1)*(1.0D0-NUE**2.0D0)**AJ
      AJ=AJ+1.0D0
    5 CONTINUE
      U(L,1)=DSQRT(1.0D0-NUE**2.0D0)*NUE*(1.0D0+U(L,1))
    4 CONTINUE
      RETURN
      END
C
C                         ** SUBROUTINE FACTOR **
C
C THIS SUBROUTINE CALCULATES THE FRACTIONAL PART OF EQUATIONS 28 AND
C 29 IN REFERENCE 1.
C THIS SUBROUTINE IS CALLED FROM NORMAL AND DOES NOT CALL ANY
C SUBROUTINES.
C INPUTS: I,J,ITERM,AI,AJ
C OUTPUTS: AF,DENOM,B
C
      SUBROUTINE FACTOR(ITERM,AI,AJ,DENOM,AF,I,J,B)
      REAL*8 AI,AJ,DENOM,AF,AN,B,AM
      AF=1.0D0
      DENOM=1.0D0
      AN=1.0D0
      DO 1 N=1,(1+I+J)
      AF=AF*AN
      AN=AN+1.0D0
    1 CONTINUE
      AM=1.0D0
      DO 2 M=1,(2*(I+J+1)+ITERM),2
      DENOM=DENOM*AM
      AM=AM+2.0D0
    2 CONTINUE
      B=2.0D0**(2.0D0+AI+AJ)
      RETURN
      END
```

```
C
C                          ** SUBROUTINE ODDEVN **
C
C THIS SUBROUTINE FIGURES OUT WHETHER A ONE OR A THREE BELONGS IN THE
C DENOMINATOR OF EQUATIONS 28 AND 29 OF REFERENCE 1. THE PROGRAM INPUTS
C L, DETERMINES WHETHER L-1 IS ODD OR EVEN AND OUTPUTS THE CORRESPONDING
C 3 OR 1.
C THIS SUBROUTINE IS CALLED FROM NORMAL AND CALLS NO OTHER SUBROUTINES.
C INPUTS: L
C OUTPUTS: ITERM
C
      SUBROUTINE ODDEVN(L,ITERM)
      ITERM=1
      B=L
      I=L/2
      C=B/2.0+0.6
      J=IINT(C)
      IF(J.EQ.I) ITERM=3
      RETURN
      END
C
C                       ** SUBROUTINE NORMAL **
C
C THIS SUBROUTINE CALCULATES THE NORMALIZATION FACTOR,N, FOR THE
C ANGULAR FUNCTIONS,U. THIS SUBROUTINE CALCULATES EQUATIONS 28
C AND 29 OF REFERENCE 1.
C THIS SUBROUTINE CALLS SUBROUTINES FACTOR AND ODDEVN, AND IS CALLED
C FROM MAIN.
C INPUTS: C,INDEX,KNDEX
C OUTPUTS: N
C
      SUBROUTINE NORMAL(C,N,INDEX,KNDEX)
      REAL*8 N1,AJ,AI,AF,DENOM,B,N2,N(20,1),C(40,20,1)
      DO 4 L=1,INDEX
      CALL ODDEVN(L,ITERM)
      I=0
      AJ=1.0D0
      AI=0.0D0
      N1=0.0D0
      DO 1 J=1,KNDEX
      CALL FACTOR(ITERM,AI,AJ,DENOM,AF,I,J,B)
      N1=N1+C(2*J,L,1)*B*AF/DENOM
      AJ=AJ+1.0D0
    1 CONTINUE
      N1=2.0D0*N1
      AI=1.0D0
      N2=0.0D0
      DO 2 I=1,KNDEX
      AJ=1.0D0
      DO 3 J=1,KNDEX
      CALL FACTOR(ITERM,AI,AJ,DENOM,AF,I,J,B)
      N2=N2+C(2*J,L,1)*C(2*I,L,1)*B*AF/DENOM
      AJ=AJ+1.0D0
    3 CONTINUE
      AI=AI+1.0D0
    2 CONTINUE
      CALL FACTOR(ITERM,0.0D0,0.0D0,DENOM,AF,0,0,B)
      N(L,1)=AF*B/DENOM+N1+N2
    4 CONTINUE
      RETURN
      END
```

```
C
C                          ** SUBROUTINE QUE **
C
C THIS SUBROUTINE CALCULATES THE JOINING FACTOR Q, USING EQUATIONS
C 69 AND 70 IN REFERENCE 1.
C THIS SUBROUTINE CALLS SUBROUTINE FACTR2 AND IS CALLED FROM MAIN.
C INPUTS: C,EPSLN,INDEX,KNDEX
C OUTPUTS: Q1
C
      SUBROUTINE QUE(Q1,C,EPSLN,INDEX,KNDEX)
      REAL IMPRT
      COMPLEX Q1(20,1)
      REAL*8 AF,DENOM,B,AK,TERM,C(40,20,1)
      DO 1 L=1,(INDEX-1),2
      AK=1.0D0
      TERM=0.0D0
      DO 2 K=1,KNDEX
      CALL FACTR2(AK,K,DENOM,AF,B,1)
      TERM=TERM+C(2*K,L,1)*AF*B/DENOM
      AK=AK+1.0D0
    2 CONTINUE
      CALL FACTR2(0.0D0,0,DENOM,AF,B,1)
      TERM=TERM+B*AF/DENOM
      RLPRT=1.0/TERM
      Q1(L,1)=CMPLX(RLPRT,0.0)
    1 CONTINUE
      DO 3 L=2,INDEX,2
      AK=1.0D0
      TERM=0.0D0
      DO 4 K=1,KNDEX
      CALL FACTR2(AK,K,DENOM,AF,B,3)
      TERM=TERM+C(2*K,L,1)*AF*B/DENOM
      AK=AK+1.0D0
    4 CONTINUE
      CALL FACTR2(0.0D0,0,DENOM,AF,B,3)
      TERM=TERM+B*AF/DENOM
      IMPRT=-1.0/(EPSLN*TERM)
      Q1(L,1)=CMPLX(0.0,IMPRT)
    3 CONTINUE
      RETURN
      END
C
C                     ** SUBROUTINE FACTR2 **
C
C THIS SUBROUTINE CALCULATES THE FRACTIONAL PART OF EQUATIONS 69
C AND 70 IN REFERENCE 1.
C THIS SUBROUTINE IS CALLED FROM SUBROUTINE QUE, AND CALLS NO
C OTHER SUBROUTINES.
C INPUTS: AK,K,ITERM
C OUTPUTS: DENOM,AF,B
C
      SUBROUTINE FACTR2(AK,K,DENOM,AF,B,ITERM)
      REAL*8 AK,DENOM,AF,B,AN
      AF=1.0D0
      DENOM=1.0D0
      AN=1.0D0
      DO 1 N=1,(1+K)
      AF=AF*AN
      AN=AN+1.0D0
    1 CONTINUE
      AM=1.0D0
      DO 2 M=1,(2*(K+1)+ITERM),2
      DENOM=DENOM*AM
      AM=AM+2.0D0
    2 CONTINUE
      B=2.0D0**(1.0D0+AK)
      RETURN
      END
```

```
C
C                    ** SUBROUTINE BAETA **
C
C THIS SUBROUTINE CALCULATES AN INTERMEDIATE RESULT BASED ON EQUATIONS
C 44 AND 45 FROM REFERENCE 1. BETA CAN ALSO BE DERIVED FROM EQUATIONS
C 84,87, AND 88 IN REFERENCE 1.
C THIS SUBROUTINE IS CALLED FROM MAIN, AND CALLS NO OTHER SUBROUTINES.
C INPUTS: C,INDEX,KNDEX
C OUTPUT: BETA
C
      SUBROUTINE BAETA(C,BETA,INDEX,KNDEX)
      REAL*8 C(40,20,1),BETA(20,1),TERM
      DO 2 L=1,(INDEX-1),2
      TERM=1.0D0
      DO 1 I=1,KNDEX
      TERM=TERM+C(2*I,L,1)
    1 CONTINUE
      TERM=TERM*TERM
      BETA(L,1)=(0.5D0-2.0D0*C(2,L,1))*TERM
    2 CONTINUE
      DO 4 L=2,INDEX,2
      TERM=1.0D0
      DO 3 I=1,KNDEX
      TERM=TERM+C(2*I,L,1)
    3 CONTINUE
      TERM=TERM*TERM
      BETA(L,1)=(1.5D0-2.0D0*C(2,L,1))*TERM
    4 CONTINUE
      RETURN
      END
C
C                ** SUBROUTINE QU **
C
C THIS SUBROUTINE EVALUATES EQUATIONS 87 AND 88 IN REFERENCE 1.
C THIS SUBROUTINE IS CALLED FROM MAIN AND CALLS NO SUBROUTINES.
C INPUTS: BETA,Q1,EPSLN,INDEX
C OUTPUT: Q2
C
      SUBROUTINE QU(Q2,BETA,Q1,EPSLN,INDEX)
      REAL*8 Q2(20,1),BETA(20,1),EPSLN
      COMPLEX Q1(20,1)
      DO 1 L=1,INDEX
      Q2(L,1)=4.0D0*BETA(L,1)*(CABS(Q1(L,1))**2.0)/(EPSLN**3.0D0)

    1 CONTINUE
      RETURN
      END
C
C                    ** SUBROUTINE RAD30 **
C
C THIS SUBROUTINE CALCULATES THE DERIVATIVE OF THE RADIAL FUNCTION
C OF THE THIRD KIND, EVALUATED AT ZERO. SEE EQUATION 57 REFERENCE 1.
C THIS SUBROUTINE IS CALLED FROM MAIN AND CALLS NO OTHER SUBROUTINES.
C INPUTS: Q2,BETA,INDEX
C OUTPUTS: V30
C
      SUBROUTINE RAD30(BETA,Q2,V30,INDEX)
      REAL*8 BETA(20,1),Q2(20,1)
      COMPLEX V30(20,1)
      DO 1 L=1,(INDEX-1),2
      A=Q2(L,1)/BETA(L,1)
      V30(L,1)=CMPLX(0.0,A)
    1 CONTINUE
      DO 2 L=2,INDEX,2
      B=-Q2(L,1)*1.570796327D0
      V30(L,1)=CMPLX(1.0,B)
    2 CONTINUE
      RETURN
      END
```

```
C
C                   ** SUBROUTINE DBLYU2 **
C
C THIS SUBROUTINE EVALUATES EQUATION 26 OF REFERENCE 2, USING THE
C PORTION OF EQUATION 17, REFERENCE 2, WHICH IS IN BRACKETS.
C THIS SUBROUTINE IS CALLED FROM MAIN AND CALLS NO OTHER SUBROUTINE
C INPUTS: EPSLN,V1,V2,Q2,BETA,INDEX
C OUTPUT: W
C
      SUBROUTINE DBLYU2(W,EPSLN,V1,V2,Q2,BETA,INDEX)
      REAL*8 W(20,1),PSI,V1(20,1),V2(20,1)
      REAL*8 Q2(20,1),BETA(20,1),EPSLN,PI2
      PI2=1.570796327D0
      PSI=PI2/EPSLN
      DO 1 L=1,(INDEX-1),2
      W(L,1)=DSQRT(1.0D0+PSI*PSI)*(Q2(L,1)*V1(L,1)/
     *BETA(L,1))
    1 CONTINUE
      DO 2 L=2,INDEX,2
      W(L,1)=DSQRT(1.0D0+PSI*PSI)*(-Q2(L,1)*1.570796327D0
     **V1(L,1)-V2(L,1))
    2 CONTINUE
      RETURN
      END
C
C                   ** SUBROUTINE BEE **
C
C THIS SUBROUTINE CALCULATES THE B'S IN EQUATION 35 OF REFERENCE 2.
C THIS SUBROUTINE IS CALLED FROM MAIN AND CALLS NO SUBROUTINES.
C INPUTS: W,ALPHA,N,Q1,V30,INDEX
C OUTPUT: B
C
      SUBROUTINE BEE(B,W,ALPHA,N,Q1,V30,INDEX)
      COMPLEX Q1(20,1),V30(20,1),B(20)
      REAL*8 W(20,1),ALPHA(20,1),N(20,1)
      DO 1 L=1,(INDEX-1),2
      B(L)=W(L,1)/(ALPHA(L,1)*N(L,1)*Q1(L,1)*V30(L,1))
    1 CONTINUE
      DO 2 L=2,INDEX,2
      B(L)=-W(L,1)/(ALPHA(L,1)*N(L,1)*Q1(L,1)*V30(L,1))
    2 CONTINUE
      RETURN
      END
C
C                   ** SUBROUTINE DIRECT **
C
C THIS PROGRAM CALCULATES THE DIRECTIVITY,D, AND THE DIRECTIVITY,
C DDBI, IN DBI, USING EQUATIONS FOR DIRECTIVITY DEVELOPED IN THE
C WORKING PAPER. THESE EQUATIONS, IN TURN, COME FROM EQUATIONS 43 AND
C 45 OF REFERENCE 2.
C THIS SUBROUTINE IS CALLED FROM MAIN AND CALLS NO SUBROUTINES.
C INPUTS: U,EPSLN,N,INDEX,B
C OUTPUTS: D,DDBI
C
      SUBROUTINE DIRECT(B,U,EPSLN,D,N,INDEX,DDBI)
      COMPLEX B(20),SI
      REAL*8 U(20,1),EPSLN,D,S,P,N(20,1),DDBI
      SI=CMPLX(0.0,0.0)
      DO 1 L=1,INDEX
      SI=SI+B(L)*U(L,1)
    1 CONTINUE
      S=(CABS(SI))**2.0D0
      P=0.0D0
      DO 2 J=1,INDEX
      P=P+N(J,1)*(CABS(B(J))**2.0)
    2 CONTINUE
      D=S/P*2.0D0
      DDBI=10.0D0*DLOG10(D)
      RETURN
      END
```

```
C
C                  ** SUBROUTINE AEY **
C
C THIS SUBROUTINE EVALUATES THE A'S IN EQUATION 35 REFERENCE 1.
C THIS SUBROUTINE IS CALLED FROM MAIN AND CALLS NO OTHER SUBROUTINES.
C INPUTS: ALPHA,EPSLN,INDEX,LNDEX
C OUTPUT: A
C
      SUBROUTINE AEY(A,ALPHA,EPSLN,INDEX,LNDEX)
      REAL*8 A(40,20,1),ALPHA(20,1),EPSLN,AN
      DO 1 L=1,(INDEX-1),2
      A(2,L,1)=(ALPHA(L,1)-EPSLN*EPSLN-2.0D0)/2.0D0
      A(4,L,1)=((ALPHA(L,1)-EPSLN*EPSLN-12.0D0)*A(2,L,1)
     *-EPSLN*EPSLN)/12.0D0
      AN=6.0D0
      DO 2 N=6,LNDEX,2
      A(N,L,1)=((ALPHA(L,1)-EPSLN*EPSLN-2.0D0-(AN-2.0D0)
     **(AN+1.0D0))*A(N-2,L,1)-EPSLN*EPSLN*A(N-4,L,1))/
     *(AN*(AN-1.0D0))
      AN=AN+2.0D0
    2 CONTINUE
    1 CONTINUE
      DO 3 L=2,INDEX,2
      A(1,L,1)=1.0D0
      A(3,L,1)=(ALPHA(L,1)-EPSLN*EPSLN-6.0D0)/6.0D0
      AN=5.0D0
      DO 4 N=5,(LNDEX-1),2
      A(N,L,1)=((ALPHA(L,1)-EPSLN*EPSLN-2.0D0-(AN-2.0D0)
     **(AN+1.0D0))*A(N-2,L,1)-EPSLN*EPSLN*A(N-4,L,1))/
     *(AN*(AN-1.0D0))
      AN=AN+2.0D0
    4 CONTINUE
    3 CONTINUE
      RETURN
      END
```

```
C
C                        ** SUBROUTINE BBEE **
C
C THIS SUBROUTINE EVALUATES THE B'S, HERE CALLED BB, IN EQUATION 50,
C REFERENCE 1.
C THIS SUBROUTINE IS CALLED FROM MAIN AND DOES NOT CALL ANY SUBROUTINES
C ITSELF.
C INPUTS: EPSLN,ALPHA,A,BETA,INDEX,LNDEX,MNDEX
C OUTPUTS: BB,BO
C
      SUBROUTINE BBEE(BB,EPSLN,ALPHA,A,BETA,BO,INDEX,LNDEX,MNDEX)
      REAL*8 BB(40,20,1),EPSLN,AN,ALPHA(20,1),A(40,20,1)
      REAL*8 BETA(20,1),BO(20,1),DENOM
      DENOM=6.0D0
      DO 1 L=1,(INDEX-1),2
      BB(1,L,1)=1.0D0/BETA(L,1)-1.0D0
      BB(3,L,1)=(-4.0D0*(A(2,L,1)+1.0D0)+BB(1,L,1)*
     *(ALPHA(L,1)-EPSLN*EPSLN-4.0D0))/DENOM
      AN=5.0D0
      DO 2 N=5,(MNDEX-1),2
      BB(N,L,1)=(BB(N-2,L,1)*(ALPHA(L,1)-EPSLN*EPSLN-
     *(AN-2.0D0)*(AN+1.0D0))-2.0D0*(AN-1.0D0)*
     *(A(N-1,L,1)+A(N-3,L,1))-EPSLN*EPSLN*BB(N-4,L,1))
     */(AN*(AN-1.0D0))
      AN=AN+2.0D0
    2 CONTINUE
    1 CONTINUE
      DO 3 L=2,INDEX,2
      BO(L,1)=-1.0D0/BETA(L,1)
      BB(2,L,1)=(-2.0D0+(ALPHA(L,1)-EPSLN*EPSLN)*BO(L,1))/2.0D0
      BB(4,L,1)=(-6.0D0*(A(3,L,1)+1.0D0)-EPSLN*EPSLN*BO(L,1)+
     *BB(2,L,1)*(ALPHA(L,1)-10.0D0-EPSLN*EPSLN))/12.0D0
      AN=6.0D0
      DO 4 N=6,MNDEX,2
      BB(N,L,1)=(BB(N-2,L,1)*(ALPHA(L,1)-EPSLN*EPSLN-
     *(AN-2.0D0)*(AN+1.0D0))-2.0D0*(AN-1.0D0)*
     *(A(N-1,L,1)+A(N-3,L,1))-EPSLN*EPSLN*BB(N-4,L,1))
     */(AN*(AN-1.0D0))
      AN=AN+2.0D0
    4 CONTINUE
    3 CONTINUE
      RETURN
      END
```

```
C
C                   ** SUBROUTINE GGEE **
C
C THIS SUBROUTINE CALCULATES THE G'S IN EQUATION 48 REFERENCE 1.
C THIS SUBROUTINE IS CALLED BY MAIN AND DOES NOT CALL ANY SUBROUTINES
C ITSELF.
C INPUTS: EPSLN,BB,BO,INDEX,MNDEX
C OUTPUT: G
C
      SUBROUTINE GGEE(EPSLN,BB,G,BO,INDEX,MNDEX)
      REAL*8 EPSLN,BB(40,20,1),G(20,1),AN,TERM,BO(20,1),PSI
      PSI=3.1415926536D0/(2.0D0*EPSLN)
      DO 1 L=2,INDEX,2
      TERM=BO(L,1)
      AN=2.0D0
      DO 2 N=2,MNDEX,2
      TERM=TERM+BB(N,L,1)*(PSI**AN)
      AN=AN+2.0D0
    2 CONTINUE
      G(L,1)=TERM/DSQRT(1.0D0+PSI**2.0D0)
    1 CONTINUE
      DO 3 L=1,(INDEX-1),2
      TERM=0.0D0
      AN=1.0D0
      DO 4 N=1,(MNDEX-1),2
      TERM=TERM+BB(N,L,1)*(PSI**AN)
      AN=AN+2.0D0
    4 CONTINUE
      G(L,1)=TERM/DSQRT(1.0D0+PSI**2.0D0)
    3 CONTINUE
      RETURN
      END
C
C                  ** SUBROUTINE RAD1 **
C
C THIS SUBROUTINE CALCULATES THE RADIAL FUNCTION OF THE FIRST KIND,
C V1, USING EQUATION 34 IN REFERENCE 1.
C THIS SUBROUTINE IS CALLED FROM MAIN BUT DOES NOT CALL ANY SUBROUTINES
C ITSELF.
C INPUTS: EPSLN,A,INDEX,LNDEX
C OUTPUT: V1
C
      SUBROUTINE RAD1(EPSLN,A,V1,INDEX,LNDEX)
      REAL*8 EPSLN,A(40,20,1),V1(20,1),AN,TERM,PSI
      PSI=3.1415926536D0/(2.0D0*EPSLN)
      DO 1 L=2,INDEX,2
      TERM=0.0D0
      AN=1.0D0
      DO 2 N=1,(LNDEX-1),2
      TERM=TERM+A(N,L,1)*(PSI**AN)
      AN=AN+2.0D0
    2 CONTINUE
      V1(L,1)=TERM*DSQRT(1.0D0+PSI**2.0D0)
    1 CONTINUE
      DO 3 L=1,(INDEX-1),2
      TERM=1.0D0
      AN=2.0D0
      DO 4 N=2,LNDEX,2
      TERM=TERM+A(N,L,1)*(PSI**AN)
      AN=AN+2.0D0
    4 CONTINUE
      V1(L,1)=TERM*DSQRT(1.0D0+PSI**2.0D0)
    3 CONTINUE
      RETURN
      END
```

```
C
C                  ** SUBROUTINE RAD2 **
C
C THIS SUBROUTINE CALCULATES THE RADIAL FUNCTION OF THE SECOND KIND,
C V2, USING EQUATION 46, REFERENCE 1.
C THIS SUBROUTINE IS CALLED BY MAIN BUT DOES NOT CALL ANYTHING ITSELF.
C INPUTS: EPSLN,G,V1,Q2,INDEX
C OUTPUT: V2
C
      SUBROUTINE RAD2(V2,EPSLN,G,V1,Q2,INDEX)
      REAL*8 V2(20,1),G(20,1),V1(20,1),Q2(20,1),EPSLN,C,PSI
      PSI=3.1415926536D0/(2.0D0*EPSLN)
      C=-3.1415926536D0/2.0D0
      DO 1 L=1,INDEX
      V2(L,1)=Q2(L,1)*(V1(L,1)*(DATAN(PSI)+C)+G(L,1))
    1 CONTINUE
      RETURN
      END
C
C                 ** SUBROUTINE RADRES **
C
C THIS SUBROUTINE CALCULATES THE RADIATION RESISTANCE,R,USING
C EQUATION 45, REFERENCE 2.
C THIS SUBROUTINE IS CALLED BY MAIN AND DOES NOT CALL ANY
C SUBROUTINES ITSELF.
C INPUTS: N,EPSLN,B,INDEX
C OUTPUT: R
C
      SUBROUTINE RADRES(B,N,EPSLN,R,INDEX)
      REAL*8 R,N(20,1),EPSLN,TERM
      COMPLEX*8 B(20)
      TERM=0.0D0
      DO 1 L=1,INDEX
      TERM=TERM+N(L,1)*(CABS(B(L))**2.0)
    1 CONTINUE
      R=(EPSLN**4.0D0)*TERM*60.0D0
      R=R*0.99930833D0
      RETURN
      END
```

```
C
C              ** SUBROUTINE IMPROV **
C
C THIS SUBROUTINE CALCULATES THE EXACT EIGENVALUES USING THE
C RECURSIVE HALF INTERVAL METHOD. FIRST IT COMPUTES AN ERROR
C BETWEEN A PREVIOUS EIGENVALUE(GAM) AND AN EIGENVALUE RETURNED
C FROM THE CONTINUED FRACTION(ZE OR ZO). ZE OR ZO IS OBTAINED BY
C PLUGGING  GAM INTO THE CONTINUED FRACTION. POSITIVE OR NEGATIVE
C ERROR TELLS THE PROGRAM TO SEARCH DOWN OR UP. ONCE A POSITIVE
C ERROR AND NEGATIVE ERROR HAVE BEEN FOUND, CORRESPONDING TO TWO INPUTS
C TO THE CONT. FRAC., THE PROGRAM DECLARES THAT IT HAS FOUND THE
C INTERVAL WHERE THE ROOT(PRESUMABLY THE EIGENVALUE) OF THE CONTINUED
C FRACTION LIES.
C FROM HERE THE HALF INTERVAL METHOD TAKES OVER. THE HALF INTERVAL
C METHOD OPERATES BY CUTTING THE INTERVAL IN HALF, FIGURING  OUT IN
C WHICH INTERVAL THE ROOT LIES AND CUTTING IT IN HALF AGAIN. THIS
C REPEATS UNTIL THE ERROR IS LESS THAN SOME TOLERANCE(TOLER).
C THIS SUBROUTINE IS CALLED BY MAIN, AND CALLS SUBROUTINE EVEN AND
C SUBROUTINE ODD.
C INPUTS: EPSLN,INDEX,GAM
C OUTPUT: GAMIMP
C
      SUBROUTINE IMPROV(GAM,GAMIMP,EPSLN,INDEX)
      REAL*8 GAM(20,1),ZO(20,1),ZE(20,1),AL,EPSLN,XL(20,1)
      REAL*8 GAMIMP(20,1),ERROR(20,1),TOLER,XM(20,1),XU(20,1)
      TOLER=0.00000001D0
C
C  L ODD,(L-1) EVEN
C
      AL=1.0D0
      DO 1 L=1,(INDEX-1),2
      I=1
      J=1
      CALL EVEN(ZE(L,1),AL,L,GAM(L,1),EPSLN)
      ERROR(L,1)=GAM(L,1)-ZE(L,1)
      IF(ERROR(L,1).LE.TOLER.AND.ERROR(L,1).GE.-TOLER) GOTO 2
      IF(ERROR(L,1).LT.-TOLER) GOTO 3
C
C  ERROR IS POSITIVE
C
    5 GAM(L,1)=GAM(L,1)-0.5D0
      I=I+1
      IF(I.GT.1000) GOTO 4
      CALL EVEN(ZE(L,1),AL,L,GAM(L,1),EPSLN)
      ERROR(L,1)=GAM(L,1)-ZE(L,1)
      IF(ERROR(L,1).LE.TOLER.AND.ERROR(L,1).GE.-TOLER) GOTO 2
      IF(ERROR(L,1).GT.TOLER) GOTO 5
C
C  FOUND THE INTERVAL
C
C     WRITE(4,7) I,L
    7 FORMAT(1X,'FOUND THE INTERVAL',2X,2I5)
      GOTO 6
```

```
C
C  ERROR IS NEGATIVE
C
    3 GAM(L,1)=GAM(L,1)+0.5D0
      I=I+1
      IF(I.GT.1000) GOTO 4
      CALL EVEN(ZE(L,1),AL,L,GAM(L,1),EPSLN)
      ERROR(L,1)=GAM(L,1)-ZE(L,1)
      IF(ERROR(L,1).LE.TOLER.AND.ERROR(L,1).GE.-TOLER) GOTO 2
      IF(ERROR(L,1).LT.-TOLER) GOTO 3
C
C  FOUND THE INTERVAL
C
C     WRITE(4,7) I,L
C
C  HALF INTERVAL METHOD
C
      XU(L,1)=GAM(L,1)
      XL(L,1)=GAM(L,1)-0.5D0
      GOTO 8
    6 XU(L,1)=GAM(L,1)+0.5D0
      XL(L,1)=GAM(L,1)
    8 XM(L,1)=(XU(L,1)+XL(L,1))/2.0D0
      J=J+1
      IF(J.GT.100) GOTO 14
      CALL EVEN(ZE(L,1),AL,L,XM(L,1),EPSLN)
      ERROR(L,1)=XM(L,1)-ZE(L,1)
      IF(ERROR(L,1).LE.TOLER.AND.ERROR(L,1).GE.-TOLER) GOTO 2
      IF(ERROR(L,1).LT.-TOLER) GOTO 9
      XU(L,1)=XM(L,1)
      GOTO 8
    9 XL(L,1)=XM(L,1)
      GOTO 8
C
C  CONVERGENCE
C
    2 CONTINUE
   11 FORMAT(1X,'CONVERGENCE',2X,E12.5,1X,F20.15,1X,I4)
      GAMIMP(L,1)=XM(L,1)
      AL=AL+2.0D0
      GOTO 1
C
C  DID NOT FIND THE INTERVAL
C
    4 WRITE(4,12)
   12 FORMAT(1X,'DID NOT FIND THE INTERVAL')
      AL=AL+2.0D0
      GOTO 1
C
C  NONCONVERGENCE
C
   14 WRITE(4,13)
   13 FORMAT(1X,'HALF INTERVAL METHOD DID NOT CONVERGE')
      AL=AL+2.0D0
    1 CONTINUE
C
C
C  L EVEN,(L-1) ODD
C
C
      AL=2.0D0
      DO 21 L=2,INDEX,2
      I=1
      J=1
      CALL ODD(ZO(L,1),AL,L,GAM(L,1),EPSLN)
      ERROR(L,1)=GAM(L,1)-ZO(L,1)
      IF(ERROR(L,1).LE.TOLER.AND.ERROR(L,1).GE.-TOLER) GOTO 22
      IF(ERROR(L,1).LT.-TOLER) GOTO 23
```

```
C
C  ERROR IS POSITIVE
C
   25 GAM(L,1)=GAM(L,1)-0.5D0
      I=I+1
      IF(I.GT.1000) GOTO 24
      CALL ODD(ZO(L,1),AL,L,GAM(L,1),EPSLN)
      ERROR(L,1)=GAM(L,1)-ZO(L,1)
      IF(ERROR(L,1).LE.TOLER.AND.ERROR(L,1).GE.-TOLER) GOTO 22
      IF(ERROR(L,1).GT.TOLER) GOTO 25
C
C  FOUND THE INTERVAL
C
C     WRITE(4,27) I,L
   27 FORMAT(1X,'FOUND THE INTERVAL',2X,2I5)
      GOTO 26
C
C  ERROR IS NEGATIVE
C
   23 GAM(L,1)=GAM(L,1)+0.5D0
      I=I+1
      IF(I.GT.1000) GOTO 24
      CALL ODD(ZO(L,1),AL,L,GAM(L,1),EPSLN)
      ERROR(L,1)=GAM(L,1)-ZO(L,1)
      IF(ERROR(L,1).LE.TOLER.AND.ERROR(L,1).GE.-TOLER) GOTO 22
      IF(ERROR(L,1).LT.-TOLER) GOTO 23
C
C  FOUND THE INTERVAL
C
C     WRITE(4,27) I,L
C
C  HALF INTERVAL METHOD
C
      XU(L,1)=GAM(L,1)
      XL(L,1)=GAM(L,1)-0.5D0
      GOTO 28
   26 XU(L,1)=GAM(L,1)+0.5D0
      XL(L,1)=GAM(L,1)
   28 XM(L,1)=(XU(L,1)+XL(L,1))/2.0D0
      J=J+1
      IF(J.GT.100) GOTO 214
      CALL ODD(ZO(L,1),AL,L,XM(L,1),EPSLN)
      ERROR(L,1)=XM(L,1)-ZO(L,1)
      IF(ERROR(L,1).LE.TOLER.AND.ERROR(L,1).GE.-TOLER) GOTO 22
      IF(ERROR(L,1).LT.-TOLER) GOTO 29
      XU(L,1)=XM(L,1)
      GOTO 28
   29 XL(L,1)=XM(L,1)
      GOTO 28
C
C  CONVERGENCE
C
   22 CONTINUE
  211 FORMAT(1X,'CONVERGENCE',2X,E12.5,1X,F20.15,1X,I4)
      GAMIMP(L,1)=XM(L,1)
      AL=AL+2.0D0
      GOTO 21
C
C  DID NOT FIND THE INTERVAL
C
   24 WRITE(4,212)
  212 FORMAT(1X,'DID NOT FIND THE INTERVAL')
      AL=AL+2.0D0
      GOTO 21
C
C  NONCONVERGENCE
C
  214 WRITE(4,213)
  213 FORMAT(1X,'HALF INTERVAL METHOD DID NOT CONVERGE')
      AL=AL+2.0D0
   21 CONTINUE
      RETURN
      END
```

```
C
C               ** SUBROUTINE ODD **
C
C THIS SUBROUTINE CALCULATES THE CONTINUED FRACTION AS PER
C EQUATION 20,REFERENCE 1. TERM1 AND TERM2 ARE THE FRACTION TERMS
C IN THE CONTINUED FRACTION. EVALUATION IS STARTED WITH THE
C LAST FRACTION TERM AND IS RECURSIVELY WORKED FORWARD UNTIL
C THE ENTIRE CONT. FRAC. IS EVALUATED. THE TWO CONTINUED FRACTIONS
C ARE THEN ADDED TOGETHER TO GIVE THE FINAL RESULT.
C THIS SUBROUTINE IS CALLED BY IMPROV AND DOES NOT CALL ANY
C SUBROUTINES.
C INPUTS: AL,L,GAM,EPSLN
C OUTPUT: ZO
C
      SUBROUTINE ODD(ZO,AL,L,GAM,EPSLN)
      REAL*8 A1,A2,A3,A4,ZO,TERM1,TERM2,AL,GAM,EP,EPSLN,ALM
      REAL*8 M
      M=1.0D0
      ALM=AL*2.0D0
      A1=201.0D0
      A2=200.0D0
      A3=199.0D0
      A4=198.0D0
      EP=EPSLN*EPSLN
      TERM1=0.0D0
      TERM2=0.0D0
      DO 1 I=1,100
      TERM1=(AL+A3+M)*(AL+A4)*EP/(A2*(ALM+A1)-GAM+TERM1)
      TERM2=(AL-A3+M)*(AL-A2)*EP/(A2*(ALM-A3)+GAM+TERM2)
      A1=A1-2.0D0
      A2=A2-2.0D0
      A3=A3-2.0D0
      A4=A4-2.0D0
    1 CONTINUE
      ZO=TERM1-TERM2
      RETURN
      END
```

```
C
C                    ** SUBROUTINE EVEN **
C
C THIS SUBROUTINE CALCULATES THE CONTINUED FRACTION AS PER
C EQUATION 19, REFERENCE 1. TERM1 AND TERM2 ARE THE FRACTION TERMS
C IN THE CONTINUED FRACTION. EVALUATION IS STARTED WITH THE LAST
C FRACTION TERM AND IS RECURSIVELY WORKED FORWARD UNTIL THE ENTIRE
C CONTINUED FRACTION IS EVALUATED. THE TWO CONTINUED FRACTIONS
C ARE THEN ADDED TOGETHER TO GIVE THE FINAL RESULT.
C THIS SUBROUTINE IS CALLED BY IMPROV AND DOES NOT CALL ANY
C OTHER SUBROUTINES.
C INPUTS: AL,L,GAM,EPSLN
C OUTPUT: ZE
C
      SUBROUTINE EVEN(ZE,AL,L,GAM,EPSLN)
      REAL*8 A1,A2,A3,A4,ZE,TERM3,TERM4,AL,GAM,EP,EPSLN,ALM
      REAL*8 M
      M=1.0D0
      ALM=AL*2.0D0
      EP=EPSLN*EPSLN
      A1=201.0D0
      A2=200.0D0
      A3=199.0D0
      A4=198.0D0
      TERM3=0.0D0
      TERM4=0.0D0
      DO 1 I=1,100
      TERM3=(AL+A2+M)*(AL+A3)*EP/(A2*(ALM+A1)-GAM+TERM3)
      TERM4=(AL-A4+M)*(AL-A3)*EP/(A2*(ALM-A3)+GAM+TERM4)
      A1=A1-2.0D0
      A2=A2-2.0D0
      A3=A3-2.0D0
      A4=A4-2.0D0
    1 CONTINUE
      ZE=TERM3-TERM4
      RETURN
      END
>
```

B.4 MONOSTOR

```
C       THIS PROGRAM, IN FORTRAN LANGUAGE, CALCULATES THE INPUT
C       IMPEDANCE Z OF A 1/4 WAVELENGTH MONOPOLE ANTENNA MOUNTED
C       ON A CIRCULAR GROUNDPLANE OF RADIUS A FOR E=2*PI*A/LAMBDA
C       GREATER THAN APPROXIMATELY 30, WHERE LAMBDA IS THE FREE
C       SPACE WAVELENGTH.
C
C       REFERENCE; J.A.STORER,"THE IMPEDANCE OF AN ANTENNA OVER A
C       LARGE CIRCULAR SCREEN",J.APPL.PHYS.,VOL 22,NO.8,PP.1058-1066,
C       AUGUST 1951,EQUATION PRECEDING EQUATION 20.
C
C       LET     FS=FREE SPACE WAVE IMPEDANCE
C               R=RE(Z)=RADIATION RESISTANCE
C               X=IM(Z)=REACTANCE
C               ZINF=IMPEDANCE FOR E=INFINITY
C               RINF=RADIATION RESISTANCE FOR E=INFINITY
C               XINF=REACTANCE FOR E=INFINITY
C               K=D(PI/2)NUMERIC=DIRECTIVE GAIN (NUMERIC) FOR
C                 ELEVATION ANGLE OF PI/2 RADIANS
C               K1=D(PI/2)DBI=DIRECTIVE GAIN (DBI) FOR
C                 ELEVATION ANGLE OF PI/2 RADIANS
C               RO=RADIATION RESISTANCE OF MONOPOLE FOR E=O
C               DO=DIRECTIVE GAIN AT AN ELEVATION ANGLE PI/2
C                 RADIANS OF MONOPOLE FOR E=O
C*********************************************************************
      REAL E(200)
      REAL K(200),K1(200),G(200)
      COMPLEX C1,C2,C3,C4,C5,ZINF,ZF(200),Z(200),C6
      FS=376.73037
      PI=3.141592654
      DO=1.54259
      RO=19.43387
      WRITE (6,15)
   15 FORMAT ('1',//////)
      DO 10 M=1,150
      READ (5,20)E(M)
   20 FORMAT(F15.7)
   10 CONTINUE
      M=1
   40 C1=CMPLX (O.,-FS/(4.*PI*E(M)))
      C3=CMPLX (COS(2.*E(M)+3.*PI/4.),SIN(2.*E(M)+3.*PI/4.))
      C2=CMPLX (COS(2.*E(M)),SIN(2.*E(M)))
      C4=C3/SQRT(4*PI*E(M))
      C5=1+C4
      C6=1/C5
      ZF(M)=C1*C2*C6
      ZINF=CMPLX (36.539509,O.)
      Z(M)=ZF(M)+ZINF
      G(M)=REAL(ZD(M))
      K(M)=DO*RO/G(M)
      K1(M)=10.*ALOG10(K(M))
      WRITE (6,25)E(M),ZF(M),Z(M),K(M),K1(M)
   25 FORMAT (1X,7F15.7)
      IF (M.GT.149) GO TO 50
      M=M+1
      GO TO 40
   50 STOP
      END
```

B.5 AWADALLA

```
C       PROGRAM AWADALLA
C**********************************************************************
C       THIS COMPUTER PROGRAM, IN FORTRAN LANGUAGE,                   *
C  WAS WRITTEN BY PROF. AWADALLA OF MANOUFIA UNIVERSITY, EGYPT.       *
C  IT USES THE METHOD OF MOMENTS AND THE GEOMETRIC THEORY OF          *
C  DIFFRACTION TO COMPUTE THE INPUT IMPEDANCE AND THE DIRECTIVITY     *
C  PATTERN OF A MONOPOLE ELEMENT OF LENGTH h AND RADIUS b AT THE      *
C  CENTER OF A CIRCULAR GROUNDPLANE OF RADIUS a. DETAILED             *
C  DERIVATIONS ARE PUBLISHED    IN REFERENCES 1 AND 2.                *
C                                                                     *
C       THIS PROGRAM REQUIRES THREE INPUTS WHICH ARE ENTERED FROM     *
C  AN INPUT FILE NAMED "AIN.DAT" WITH FORMAT 172.                     *
C                                                                     *
C           b = AL = MONOPOLE RADIUS IN WAVELENGTHS                   *
C           h = AH = MONOPOLE LENGTH IN WAVELENGTHS                   *
C    2πa/λ = ε = EPSLN = NORMALIZED GROUNDPLANE RADIUS IN WAVENUMBERS *
C                                                                     *
C       THIS PROGRAM ALSO REQUIRES THE USE OF THE MMBSJ1 SUBROUTINE   *
C  FUNCTION OF THE IMSL SOFTWARE LIBRARY.  MMBSJ1 IS THE MODIFIED     *
C  BESSEL FUNCTION OF THE FIRST KIND AND ORDER ONE.                   *
C                                                                     *
C      REFERENCE:                                                     *
C                                                                     *
C      1. K.H. AWADALLA AND T.S.M. MACLEAN, "INPUT IMPEDANCE OF       *
C      A MONOPOLE ANTENNA AT THE CENTER OF A FINITE GROUNDPLANE",     *
C      IEEE TRANS. ON ANTENNAS AND PROPAGATION, VOL. AP-26, NO. 2,    *
C      MARCH 1978, PP. 244-248                                        *
C                                                                     *
C      2. K.H. AWADALLA AND T.S.M. MACLEAN, "MONOPOLE ANTENNA         *
C      AT THE CENTER OF CIRCULAR GROUNDPLANE: INPUT IMPEDANCE         *
C      AND RADIATION PATTERN", IEEE TRANS. ON ANTENNAS AND            *
C      PROPAGATION, VOL. AP-27, NO. 2, MARCH 1979, PP. 151-153        *
C                                                                     *
C       NOTE: THE NUMBER OF MONOPOLE SEGMENTS USED IN THE PROGRAM     *
C             IS 30. IF THE NUMBER OF MONOPOLE SEGMENTS IS CHANGED    *
C             TO N, THEN THE DIMENSION OF ALL THE ARRAYS OF 30, AND   *
C             30x30 SHOULD BE CHANGED INTO N, AND NxN.                *
C**********************************************************************
C
        INTEGER IER
        REAL MMBSJ1
        COMPLEX Z(30,30),ZAZ(30,30),B(30),ZC(30),FF,JHF(100)
     1  ,EP,C1,C2,EPPM,EPMP,ZT,PSI,ZMG,YIN,ZIN,CDH,DH,EDP,SMP,ZT2
     2  ,SM2,CMEV,ZT1,JLD(100),YAY
        DIMENSION IH(30),AUTDB(182),Y2(182),AUT(182),Y1(182)
        DIMENSION AUT2(182),DIRECT(182),DIRDB(182),Y(182)
        COMMON DL,RHO1,PI,AK
        OPEN(UNIT=6,FILE='AOUT',STATUS='NEW')
        OPEN(UNIT=5,FILE='AIN.DAT',STATUS='OLD',READONLY)
        DO 555 LI=1,35
        N=30
        FMHZ=1000.0
        FREQ=FMHZ*1.0E6
        ALAMDA=3.0E8/FREQ
C*****************************************************************
C       INPUT READ STATEMENT                                     *
C*****************************************************************
        READ(5,172)AL,AH,EPSLN
 172    FORMAT(3F12.8)
        WRITE(6,990)
 990    FORMAT(1H1,'AWADALLA RESULT')
```

```
      WRITE(6,171)EPSLN,AH,AL
171   FORMAT(1X,'EPSLN= ',1F12.8,' (NORMALIZED GROUNDPLANE RADIUS)',
     1 /1X,'MONOPLE LENGTH= ',1F12.8,' (WAVELENGTH)'
     2 ,/1X,'MONOPLE RADIUS= ',1F12.8,' (WAVELENGTH)')
      R=AL*ALAMDA
      PI=4.0*ATAN(1.0)
      OM=2.0*PI*FREQ
      AK=2.0*PI/ALAMDA
      AMU=4.0*PI*1.0E-7
      EPSI=1.0/(36.0*PI*1.0E9)
      DL=AH*ALAMDA/(FLOAT(N)+0.5)
      A=DL/2.0
      C1=CMPLX(0.0,OM*AMU*DL*DL)
      C2=CMPLX(0.0,-1.0/(OM*EPSI))
      RHO1=EPSLN*ALAMDA/(2.*PI)
      DO 333 I=1,N
      B(I)=CMPLX(0.,0.)
333   IH(I)=0
      EP=PSI(0.0,R,A,AK,PI)
      EPPM=PSI(DL,R,A,AK,PI)
      ZT=C1*EP+2.0*C2*(EP-EPPM)
      Z(1,1)=ZT
      RMNPM=DL
      NN=N+N
      DO 12 IN=2,NN
      EPMP=EP
      EP=EPPM
      RMNPM=RMNPM+DL
      EPPM=PSI(RMNPM,R,A,AK,PI)
      ZT=C2*(EP+EP-EPMP-EPPM)
      IF(IN.GT.N)GO TO 13
      Z(1,IN)=C1*EP+ZT
      GO TO 12
13    IC=IN-N
      ZC(IC)=C1*EP+ZT
12    CONTINUE
      DO 16 J=1,N
      DO 11 I=J,N
      IA=I-J+1
      Z(J,I)=Z(1,IA)
11    Z(I,J)=Z(J,I)
16    CONTINUE
      DO 22 K=2,N
      DO 20 I=1,N
      IB=I+K
      IF(IB.GT.N)GO TO 21
      Z(K,I)=Z(K,I)+Z(1,IB)
      GO TO 20
21    IC=IB-N
      Z(K,I)=Z(K,I)+ZC(IC)
20    CONTINUE
22    CONTINUE
      NF=N-1
      DO 200 I=1,NF
      IZ=I+1
200   Z(1,I)=Z(1,I)+Z(1,IZ)
      Z(1,N)=Z(1,N)+ZC(1)
      DO 121 I=1,N
      DO 122 J=1,N
      ZAZ(I,J)=Z(I,J)
122   CONTINUE
```

```
 121      CONTINUE
C*********************************************************************
C         DIFFRACTION EFFECT OF THE EDGE OF GROUNDPLANE        *
C*********************************************************************
          DO 30 I=1,N
          ZS=DL*(FLOAT(I)-0.5)
          DO 32 J=I,N
          ZO=DL*(FLOAT(J)-0.5)
          Z(I,J)=Z(I,J)+ZMG(ZS,ZO)
  32      Z(J,I)=Z(I,J)
  30      CONTINUE
C*********************************************************************
C         CALCULATION OF THE ADMITTANCE                        *
C*********************************************************************
 150      CALL MTXINV(Z,N,B,IH)
 151      ZIN=1.0/Z(1,1)
          WRITE(6,23)ZIN
  23      FORMAT(1X,'INPUT IMPEDANCE= ',E15.9,' +j ',E15.9,2X,'(OHMS)'//)
          WRITE(6,999)
 999      FORMAT(/1X,'ELEVATION',53X,'ELEVATION',/1X,2(
     1    'ANGLE',6X,'DIRECTIVITY',5X,'DIRECTIVITY',5X,
     2    'RELATIVE POWER',5X),/1X,2('(DEG)',6X,'(NUMERIC)',
     3    7X,'(dBi)',11X,'(dB)',15X))
C*********************************************************************
C         COMPUTING THE E FIELD RADIATION PATTERN              *
C*********************************************************************
          CDH=-CMPLX(1.,1.)/SQRT(2.*PI)
          AKA=AK*RHO1
          PID2=PI/2.
          PI4=4.*PI
          PID4=PI/4.
          SAKP=SQRT(AK/(PI*8.))
          DTH=PI/180.
          TH=0.
          RHO2=RHO1*RHO1
        ZT=CMPLX(0.,0.)
        NG=RHO1/(0.05*ALAMDA)
        DR=RHO1/NG
        RG=DR/2.0
        CJHF=-2.*DL/PI4
        DO 40 IK=1,NG
          ZT1=CMPLX(0.,0.)
          ZT=CMPLX(0.,0.)
          H=A
        DO 41 J=1,N
          RG2=RG*RG
          H2=H*H
          RA2=RG2+H2
          RA=SQRT(RA2)
          SNA=RG/RA
          AKRA=-AK*RA
          CSAKRA=COS(AKRA)
          SNAKRA=SIN(AKRA)
          ZT=ZT+Z(1,J)*(SNA/RA)*CMPLX(CSAKRA,SNAKRA)*CMPLX(1.0/RA,AK)
          RI2=RHO2+H2
          RI=SQRT(RI2)
          AKRI=-AK*RI
          AK1=-AK*(RHO1-RG)
          ZT1=ZT1+Z(1,J)*RHO1*DL*CMPLX(COS(AKRI),SIN(AKRI))*CMPLX(1./RI,AK)
     1    *CMPLX(COS(AK1),SIN(AK1))/(PI4*RI)
          H=H+DL
```

```
   41   CONTINUE
         JHF(IK)=ZT*CJHF+ZT1*2.0
         JLD(IK)=ZT1*2.0
         RG=RG+DR
   40   CONTINUE
         FCR=RHO1*SQRT(2.*AK)/PI*DL
        DO 33 I=1,181
         SNTH=SIN(TH)
         CSTH=COS(TH)
         H=A
         EDP=CMPLX(0.,0.)
        DO 34 J=1,N
         S=SQRT(H*H+RHO2)
         PHI=ATAN(RHO1/H)
         SB=PI-PHI
         RB=PID2-PHI
         AI=1.0
  100    ZKC=AK*H*CSTH
         EDP=EDP+Z(1,J)*SNTH*DL*AI*CMPLX(COS(ZKC),SIN(ZKC))
   34    H=H+DL
         THA=TH*180.0/PI
         ZT=CMPLX(0.,0.)
         RG=DR/2.
         CONST=-2.*PI*DR*CSTH
         AKSN=AK*SNTH
        DO 42 K=1,NG
         X=AKSN*RG
         IF(X.LT.0.1E-6)GO TO 44
         SABF=MMBSJ1(X,IER)
         GO TO 45
  44     SABF=0.0
  45     ZT=ZT+JHF(K)*SABF*CMPLX(0.,CONST)*RG
  47     RG=RG+DR
  42     CONTINUE
  43     EDP=EDP+ZT
         AEDP=CABS(EDP)
         AUT(I)=AEDP
         AUT2(I)=AEDP*AEDP
         IF(AEDP.EQ.0.0) GO TO 33
         PHASE=180.0*ATAN2(AIMAG(EDP),REAL(EDP))/PI
  33     TH=TH+DTH
         X=0.0
         DO 70 K=1,181
         CHO=AUT(K)
  70     IF (CHO.GT.X) X=CHO
         DO 71 I=1,181
         Y(I)=AUT(I)/X
         Y1(I)=Y(I)*Y(I)
         IF(Y1(I).EQ.0.)THEN
         Y2(I)=999
         AUTDB(I)=999
         ELSE
         AUTDB(I)=10*ALOG10(AUT(I))
         Y2(I)=10*ALOG10(Y1(I))
         END IF
  71     CONTINUE
C*****************************************************
C        COMPUTE THE DIRECTIVITY PATTERN             *
C*****************************************************
         PR=0.0
         DO 995 I=2,181
```

```
      DEL=(I-1)*DTH
      PR=PR+AUT2(I)*SIN(DEL)*DTH
995   CONTINUE
      DO 994 I=1,181
      DIRECT(I)=2*AUT2(I)/PR
      IF(DIRECT(I).EQ.0.)THEN
      DIRDB(I)=999
      ELSE
      DIRDB(I)=10*ALOG10(DIRECT(I))
      END IF
994   CONTINUE
      DO 998 I=1,45
      II=I-1
      II2=I+43
      II3=II2+1
      IF(I.EQ.1)THEN
      WRITE(6,997)II,AUT(I)
997   FORMAT(2X,I3,8X,F9.5,5X,'-INFINITE',6X,'-INFINITE')
      ELSE
      WRITE(6,996)II,DIRECT(I),DIRDB(I),Y2(I),II2,DIRECT(II3)
     1 ,DIRDB(II3),Y2(II3)
996   FORMAT(1X,2(1X,I3,8X,F9.5,5X,F9.5,6X,F9.5,11X))
      END IF
998   CONTINUE
      WRITE(6,990)
      WRITE(6,171)EPSLN,AH,AL
      WRITE(6,23)ZIN
      WRITE(6,999)
      DO 992 I=90,136
      II=I-1
      II2=I+44
      II3=II2+1
      WRITE(6,996)II,DIRECT(I),DIRDB(I),Y2(I),II2,DIRECT(II3)
     1 ,DIRDB(II3),Y2(II3)
992   CONTINUE
555   CONTINUE
      CLOSE(UNIT=6,STATUS='KEEP')
      CLOSE(UNIT=5,STATUS='KEEP')
      STOP
      END
```

```
C*************************************************************
C       SUBROUTINE ZMG                                       *
C*************************************************************
        COMPLEX FUNCTION ZMG(ZS, ZO)
        COMMON DL, RHO1, PI, AK
        ZO2=ZO*ZO
        ZS2=ZS*ZS
        RH2=RHO1*RHO1
        RS2=RH2+ZS2
        RO2=RH2+ZO2
        RS=SQRT(RS2)
        RO=SQRT(RO2)
        AKA=AK*(RS+RO)
        PS=ATAN(ZS/RHO1)
        PO=ATAN(ZO/RHO1)
        ZMG=15.*DL*DL*RHO1*RH2/(AK*RS2*RO2)*CMPLX(0.,-1.)*
     1 CEXP(CMPLX(0.,-AKA))*(1./COS((PO-PS)/2.)+1./COS((PO+PS)/2.))
     2 *CMPLX(1./RS,AK)*CMPLX(1./RO,AK)
 35     RETURN
        END
```

```
C*************************************************************
C       SUBROUTINE PSI                                       *
C                                                            *
C       THIS FUNCTION IS TO CALCULATE THE SELF AND MUTUAL    *
C       IMPEDANCES FOR A STRAIGHT WIRE ANTENNA               *
C*************************************************************
        COMPLEX FUNCTION PSI(H, R, A, AK, PI)
        IMPLICIT REAL*8 (A-H), (P-Z)
        COMPLEX*16 P1, P2, P3, P4, PII
        R2=R*R
        D=DSQRT(H*H+R2)
        S1=DSQRT(R2+(H+A)**2)
        S2=DSQRT(R2+(H-A)**2)
        AI1=DLOG((H+A+S1)/(H-A+S2))
        AI2=2.*A
        AI3=(A+H)/2.*S1+(A-H)/2.*S2+R2/2.*AI1
        AI4=2.*A*R2+(2.*A*A*A+6.*A*H*H)/3.
        AKA=-AK*D
        PIA=8.*PI*A
        P4=DCMPLX(0.0D0,AKA)
        P1=CDEXP(P4)/PIA
        P2=AI1-(AI2-D*AI1)*DCMPLX(0.0D0,AK)
        P=(AK*AK/2.)*(AI3-2.*D*AI2+AI1*D*D)
        P3=DCMPLX(0.0D0,AK*AK*AK/6.)
        PP=AI4-3.*D*AI3+3.*D*D*AI2
        PP1=D*D*D*AI1
        P5=PP-PP1
        PII=P1*(P2-P+P3*P5)
        PSI=CMPLX(PII)
        RETURN
        END
```

```
C*****************************************************************
C        SUBROUTINE MTXINV                                      *
C                                                               *
C        THIS SUBROUTINE IS FOR THE INVERSION OF A MATRIX A(N,N) *
C        THE MATRIX SHOULD BE A SQUARE ONE OF ORDER "N"          *
C*****************************************************************
         SUBROUTINE MTXINV(A,N,B,IH)
         COMPLEX A(N,N),B(N),Q
         DIMENSION IH(N)
         DO 24 I=1,N
         B(I)=CMPLX(0.0,0.0)
 24      IH(I)=0
         DO 1 K=1,N
         IK=0
 12      Z=CABS(A(K,K))
         IF(Z.EQ.0.0) GO TO 6
         Q=1.0/A(K,K)
         GO TO 7
  6      IF(K.EQ.N)GO TO 13
         IH(K)=K
         DO 8 I=1,N
  8      B(I)=A(K,I)
         NA=N-1
         DO 9 I=K,NA
         II=I+1
         DO 10 J=1,N
 10      A(I,J)=A(II,J)
  9      CONTINUE
         DO 11 J=1,N
 11      A(N,J)=B(J)
         IK=IK+1
         IF(IK.EQ.1) GO TO 12
         WRITE (6,22)K,K
 22      FORMAT(1X,'MTXINV FAIL BECAUSE ELEMENT Z(',I2,',',I2,
     1   ') AND THE NEXT IN THE SAME COLUMN ARE ZEROS'/)
         GO TO 23
  7      DO 2 I=1,N
         IF(I.EQ.K) GO TO 2
         DO 3 J=1,N
         IF(J.EQ.K)GO TO 3
         A(I,J)=A(I,J)-A(I,K)*A(K,J)*Q
  3      CONTINUE
  2      CONTINUE
         DO 4 I=1,N
         IF(I.EQ.K) GO TO 4
         A(I,K)=A(I,K)*Q
  4      CONTINUE
         DO 5 I=1,N
         IF(I.EQ.K) GO TO 5
         A(K,I)=-A(K,I)*Q
  5      CONTINUE
         A(K,K)=Q
  1      CONTINUE
         N1=N+1
         DO 15 I=1,N
         NN=N1-I
         IF(IH(NN).EQ.0) GO TO 15
         DO 16 J=1,N
 16      B(J)=A(N,J)
         K=IH(NN)
         KK=N-K
```

```
      DO 17 J=1,KK
      N2=N1-J
      N3=N2-1
      DO 18 L=1,N
18    A(N2,L)=A(N3,L)
17    CONTINUE
      DO 19 L=1,N
19    A(K,L)=B(L)
15    CONTINUE
      GO TO 20
13    WRITE (6,14)
14    FORMAT(1X,'A(N,N)=0: PROGRAM FAILURE.')
20    RETURN
23    STOP
      END
```

Index